T0203892

Missing and Modified Data in Nonparametric Estimation

With R Examples

MONOGRAPHS ON STATISTICS AND APPLIED PROBABILITY

General Editors

F. Bunea, V. Isham, N. Keiding, T. Louis, R. L. Smith, and H. Tong

1. Stochastic Population Models in Ecology and Epidemiology *M.S. Barlett* (1960)
2. Queues *D.R. Cox and W.L. Smith* (1961)
3. Monte Carlo Methods *J.M. Hammersley and D.C. Handscomb* (1964)
4. The Statistical Analysis of Series of Events *D.R. Cox and P.A.W. Lewis* (1966)
5. Population Genetics *W.J. Ewens* (1969)
6. Probability, Statistics and Time *M.S. Barlett* (1975)
7. Statistical Inference *S.D. Silvey* (1975)
8. The Analysis of Contingency Tables *B.S. Everitt* (1977)
9. Multivariate Analysis in Behavioural Research *A.E. Maxwell* (1977)
10. Stochastic Abundance Models *S. Engen* (1978)
11. Some Basic Theory for Statistical Inference *E.J.G. Pitman* (1979)
12. Point Processes *D.R. Cox and V. Isham* (1980)
13. Identification of Outliers *D.M. Hawkins* (1980)
14. Optimal Design *S.D. Silvey* (1980)
15. Finite Mixture Distributions *B.S. Everitt and D.J. Hand* (1981)
16. Classification *A.D. Gordon* (1981)
17. Distribution-Free Statistical Methods, 2nd edition *J.S. Maritz* (1995)
18. Residuals and Influence in Regression *R.D. Cook and S. Weisberg* (1982)
19. Applications of Queueing Theory, 2nd edition *G.F. Newell* (1982)
20. Risk Theory, 3rd edition *R.E. Beard, T. Pentikäinen and E. Pesonen* (1984)
21. Analysis of Survival Data *D.R. Cox and D. Oakes* (1984)
22. An Introduction to Latent Variable Models *B.S. Everitt* (1984)
23. Bandit Problems *D.A. Berry and B. Fristedt* (1985)
24. Stochastic Modelling and Control *M.H.A. Davis and R. Vinter* (1985)
25. The Statistical Analysis of Composition Data *J. Aitchison* (1986)
26. Density Estimation for Statistics and Data Analysis *B.W. Silverman* (1986)
27. Regression Analysis with Applications *G.B. Wetherill* (1986)
28. Sequential Methods in Statistics, 3rd edition *G.B. Wetherill and K.D. Glazebrook* (1986)
29. Tensor Methods in Statistics *P. McCullagh* (1987)
30. Transformation and Weighting in Regression *R.J. Carroll and D. Ruppert* (1988)
31. Asymptotic Techniques for Use in Statistics *O.E. Bandorff-Nielsen and D.R. Cox* (1989)
32. Analysis of Binary Data, 2nd edition *D.R. Cox and E.J. Snell* (1989)
33. Analysis of Infectious Disease Data *N.G. Becker* (1989)
34. Design and Analysis of Cross-Over Trials *B. Jones and M.G. Kenward* (1989)
35. Empirical Bayes Methods, 2nd edition *J.S. Maritz and T. Lwin* (1989)
36. Symmetric Multivariate and Related Distributions *K.T. Fang, S. Kotz and K.W. Ng* (1990)
37. Generalized Linear Models, 2nd edition *P. McCullagh and J.A. Nelder* (1989)
38. Cyclic and Computer Generated Designs, 2nd edition *J.A. John and E.R. Williams* (1995)
39. Analog Estimation Methods in Econometrics *C.F. Manski* (1988)
40. Subset Selection in Regression *A.J. Miller* (1990)
41. Analysis of Repeated Measures *M.J. Crowder and D.J. Hand* (1990)
42. Statistical Reasoning with Imprecise Probabilities *P. Walley* (1991)
43. Generalized Additive Models *T.J. Hastie and R.J. Tibshirani* (1990)
44. Inspection Errors for Attributes in Quality Control *N.L. Johnson, S. Kotz and X. Wu* (1991)
45. The Analysis of Contingency Tables, 2nd edition *B.S. Everitt* (1992)
46. The Analysis of Quantal Response Data *B.J.T. Morgan* (1992)
47. Longitudinal Data with Serial Correlation—A State-Space Approach *R.H. Jones* (1993)

48. Differential Geometry and Statistics *M.K. Murray and J.W. Rice* (1993)

49. Markov Models and Optimization *M.H.A. Davis* (1993)

50. Networks and Chaos—Statistical and Probabilistic Aspects
    O.E. Barndorff-Nielsen, J.L. Jensen and W.S. Kendall (1993)

51. Number-Theoretic Methods in Statistics *K.-T. Fang and Y. Wang* (1994)

52. Inference and Asymptotics *O.E. Barndorff-Nielsen and D.R. Cox* (1994)

53. Practical Risk Theory for Actuaries *C.D. Daykin, T. Pentikäinen and M. Pesonen* (1994)

54. Biplots *J.C. Gower and D.J. Hand* (1996)

55. Predictive Inference—An Introduction *S. Geisser* (1993)

56. Model-Free Curve Estimation *M.E. Tarter and M.D. Lock* (1993)

57. An Introduction to the Bootstrap *B. Efron and R.J. Tibshirani* (1993)

58. Nonparametric Regression and Generalized Linear Models *P.J. Green and B.W. Silverman* (1994)

59. Multidimensional Scaling *T.F. Cox and M.A.A. Cox* (1994)

60. Kernel Smoothing *M.P. Wand and M.C. Jones* (1995)

61. Statistics for Long Memory Processes *J. Beran* (1995)

62. Nonlinear Models for Repeated Measurement Data *M. Davidian and D.M. Giltinan* (1995)

63. Measurement Error in Nonlinear Models *R.J. Carroll, D. Rupert and L.A. Stefanski* (1995)

64. Analyzing and Modeling Rank Data *J.J. Marden* (1995)

65. Time Series Models—In Econometrics, Finance and Other Fields
    D.R. Cox, D.V. Hinkley and O.E. Barndorff-Nielsen (1996)

66. Local Polynomial Modeling and its Applications *J. Fan and I. Gijbels* (1996)

67. Multivariate Dependencies—Models, Analysis and Interpretation *D.R. Cox and N. Wermuth* (1996)

68. Statistical Inference—Based on the Likelihood *A. Azzalini* (1996)

69. Bayes and Empirical Bayes Methods for Data Analysis *B.P. Carlin and T.A Louis* (1996)

70. Hidden Markov and Other Models for Discrete-Valued Time Series *I.L. MacDonald and W. Zucchini* (1997)

71. Statistical Evidence—A Likelihood Paradigm *R. Royall* (1997)

72. Analysis of Incomplete Multivariate Data *J.L. Schafer* (1997)

73. Multivariate Models and Dependence Concepts *H. Joe* (1997)

74. Theory of Sample Surveys *M.E. Thompson* (1997)

75. Retrial Queues *G. Falin and J.G.C. Templeton* (1997)

76. Theory of Dispersion Models *B. Jørgensen* (1997)

77. Mixed Poisson Processes *J. Grandell* (1997)

78. Variance Components Estimation—Mixed Models, Methodologies and Applications *P.S.R.S. Rao* (1997)

79. Bayesian Methods for Finite Population Sampling *G. Meeden and M. Ghosh* (1997)

80. Stochastic Geometry—Likelihood and computation
    O.E. Barndorff-Nielsen, W.S. Kendall and M.N.M. van Lieshout (1998)

81. Computer-Assisted Analysis of Mixtures and Applications—Meta-Analysis, Disease Mapping and Others
    D. Böhning (1999)

82. Classification, 2nd edition *A.D. Gordon* (1999)

83. Semimartingales and their Statistical Inference *B.L.S. Prakasa Rao* (1999)

84. Statistical Aspects of BSE and vCJD—Models for Epidemics *C.A. Donnelly and N.M. Ferguson* (1999)

85. Set-Indexed Martingales *G. Ivanoff and E. Merzbach* (2000)

86. The Theory of the Design of Experiments *D.R. Cox and N. Reid* (2000)

87. Complex Stochastic Systems *O.E. Barndorff-Nielsen, D.R. Cox and C. Klüppelberg* (2001)

88. Multidimensional Scaling, 2nd edition *T.F. Cox and M.A.A. Cox* (2001)

89. Algebraic Statistics—Computational Commutative Algebra in Statistics
    G. Pistone, E. Riccomagno and H.P. Wynn (2001)

90. Analysis of Time Series Structure—SSA and Related Techniques
    N. Golyandina, V. Nekrutkin and A.A. Zhigljavsky (2001)

91. Subjective Probability Models for Lifetimes *Fabio Spizzichino* (2001)

92. Empirical Likelihood *Art B. Owen* (2001)

93. Statistics in the 21st Century *Adrian E. Raftery, Martin A. Tanner, and Martin T. Wells* (2001)

94. Accelerated Life Models: Modeling and Statistical Analysis
    Vilijandas Bagdonavicius and Mikhail Nikulin (2001)

95. Subset Selection in Regression, Second Edition *Alan Miller* (2002)
96. Topics in Modelling of Clustered Data *Marc Aerts, Helena Geys, Geert Molenberghs, and Louise M. Ryan* (2002)
97. Components of Variance *D.R. Cox and P.J. Solomon* (2002)
98. Design and Analysis of Cross-Over Trials, 2nd Edition *Byron Jones and Michael G. Kenward* (2003)
99. Extreme Values in Finance, Telecommunications, and the Environment
    Bärbel Finkenstädt and Holger Rootzén (2003)
100. Statistical Inference and Simulation for Spatial Point Processes
    Jesper Møller and Rasmus Plenge Waagepetersen (2004)
101. Hierarchical Modeling and Analysis for Spatial Data
    Sudipto Banerjee, Bradley P. Carlin, and Alan E. Gelfand (2004)
102. Diagnostic Checks in Time Series *Wai Keung Li* (2004)
103. Stereology for Statisticians *Adrian Baddeley and Eva B. Vedel Jensen* (2004)
104. Gaussian Markov Random Fields: Theory and Applications *Håvard Rue and Leonhard Held* (2005)
105. Measurement Error in Nonlinear Models: A Modern Perspective, Second Edition
    Raymond J. Carroll, David Ruppert, Leonard A. Stefanski, and Ciprian M. Crainiceanu (2006)
106. Generalized Linear Models with Random Effects: Unified Analysis via H-likelihood
    Youngjo Lee, John A. Nelder, and Yudi Pawitan (2006)
107. Statistical Methods for Spatio-Temporal Systems
    Bärbel Finkenstädt, Leonhard Held, and Valerie Isham (2007)
108. Nonlinear Time Series: Semiparametric and Nonparametric Methods *Jiti Gao* (2007)
109. Missing Data in Longitudinal Studies: Strategies for Bayesian Modeling and Sensitivity Analysis
    Michael J. Daniels and Joseph W. Hogan (2008)
110. Hidden Markov Models for Time Series: An Introduction Using R
    Walter Zucchini and Iain L. MacDonald (2009)
111. ROC Curves for Continuous Data *Wojtek J. Krzanowski and David J. Hand* (2009)
112. Antedependence Models for Longitudinal Data *Dale L. Zimmerman and Vicente A. Núñez-Antón* (2009)
113. Mixed Effects Models for Complex Data *Lang Wu* (2010)
114. Intoduction to Time Series Modeling *Genshiro Kitagawa* (2010)
115. Expansions and Asymptotics for Statistics *Christopher G. Small* (2010)
116. Statistical Inference: An Integrated Bayesian/Likelihood Approach *Murray Aitkin* (2010)
117. Circular and Linear Regression: Fitting Circles and Lines by Least Squares *Nikolai Chernov* (2010)
118. Simultaneous Inference in Regression *Wei Liu* (2010)
119. Robust Nonparametric Statistical Methods, Second Edition
    Thomas P. Hettmansperger and Joseph W. McKean (2011)
120. Statistical Inference: The Minimum Distance Approach
    Ayanendranath Basu, Hiroyuki Shioya, and Chanseok Park (2011)
121. Smoothing Splines: Methods and Applications *Yuedong Wang* (2011)
122. Extreme Value Methods with Applications to Finance *Serguei Y. Novak* (2012)
123. Dynamic Prediction in Clinical Survival Analysis *Hans C. van Houwelingen and Hein Putter* (2012)
124. Statistical Methods for Stochastic Differential Equations
    Mathieu Kessler, Alexander Lindner, and Michael Sørensen (2012)
125. Maximum Likelihood Estimation for Sample Surveys
    R. L. Chambers, D. G. Steel, Suojin Wang, and A. H. Welsh (2012)
126. Mean Field Simulation for Monte Carlo Integration *Pierre Del Moral* (2013)
127. Analysis of Variance for Functional Data *Jin-Ting Zhang* (2013)
128. Statistical Analysis of Spatial and Spatio-Temporal Point Patterns, Third Edition *Peter J. Diggle* (2013)
129. Constrained Principal Component Analysis and Related Techniques *Yoshio Takane* (2014)
130. Randomised Response-Adaptive Designs in Clinical Trials *Anthony C. Atkinson and Atanu Biswas* (2014)
131. Theory of Factorial Design: Single- and Multi-Stratum Experiments *Ching-Shui Cheng* (2014)
132. Quasi-Least Squares Regression *Justine Shults and Joseph M. Hilbe* (2014)
133. Data Analysis and Approximate Models: Model Choice, Location-Scale, Analysis of Variance, Nonparametric
    Regression and Image Analysis *Laurie Davies* (2014)
134. Dependence Modeling with Copulas *Harry Joe* (2014)
135. Hierarchical Modeling and Analysis for Spatial Data, Second Edition *Sudipto Banerjee, Bradley P. Carlin,
    and Alan E. Gelfand* (2014)

136. Sequential Analysis: Hypothesis Testing and Changepoint Detection *Alexander Tartakovsky, Igor Nikiforov, and Michèle Basseville* (2015)

137. Robust Cluster Analysis and Variable Selection *Gunter Ritter* (2015)

138. Design and Analysis of Cross-Over Trials, Third Edition *Byron Jones and Michael G. Kenward* (2015)

139. Introduction to High-Dimensional Statistics *Christophe Giraud* (2015)

140. Pareto Distributions: Second Edition *Barry C. Arnold* (2015)

141. Bayesian Inference for Partially Identified Models: Exploring the Limits of Limited Data *Paul Gustafson* (2015)

142. Models for Dependent Time Series *Granville Tunnicliffe Wilson, Marco Reale, John Haywood* (2015)

143. Statistical Learning with Sparsity: The Lasso and Generalizations *Trevor Hastie, Robert Tibshirani, and Martin Wainwright* (2015)

144. Measuring Statistical Evidence Using Relative Belief *Michael Evans* (2015)

145. Stochastic Analysis for Gaussian Random Processes and Fields: With Applications *Vidyadhar S. Mandrekar and Leszek Gawarecki* (2015)

146. Semialgebraic Statistics and Latent Tree Models *Piotr Zwiernik* (2015)

147. Inferential Models: Reasoning with Uncertainty *Ryan Martin and Chuanhai Liu* (2016)

148. Perfect Simulation *Mark L. Huber* (2016)

149. State-Space Methods for Time Series Analysis: Theory, Applications and Software
Jose Casals, Alfredo Garcia-Hiernaux, Miguel Jerez, Sonia Sotoca, and A. Alexandre Trindade (2016)

150. Hidden Markov Models for Time Series: An Introduction Using R, Second Edition
Walter Zucchini, Iain L. MacDonald, and Roland Langrock (2016)

151. Joint Modeling of Longitudinal and Time-to-Event Data
Robert M. Elashoff, Gang Li, and Ning Li (2016)

152. Multi-State Survival Models for Interval-Censored Data
Ardo van den Hout (2016)

153. Generalized Linear Models with Random Effects: Unified Analysis via H-likelihood, Second Edition
Youngjo Lee, John A. Nelder, and Yudi Pawitan (2017)

154. Absolute Risk: Methods and Applications in Clinical Management and Public Health
Ruth M. Pfeiffer and Mitchell H. Gail (2017)

155. Asymptotic Analysis of Mixed Effects Models: Theory, Applications, and Open Problems
Jiming Jiang (2017)

156. Missing and Modified Data in Nonparametric Estimation: With R Examples
Sam Efromovich (2017)

Monographs on Statistics and Applied Probability 156

Missing and Modified Data in Nonparametric Estimation

With R Examples

Sam Efromovich

The University of Texas at Dallas

CRC Press
Taylor & Francis Group
Boca Raton London New York

CRC Press is an imprint of the
Taylor & Francis Group, an **informa** business

A CHAPMAN & HALL BOOK

CRC Press
Taylor & Francis Group
6000 Broken Sound Parkway NW, Suite 300
Boca Raton, FL 33487-2742

First issued in paperback 2020

ISBN-13: 978-0-367-57198-6 (pbk)
ISBN-13: 978-1-138-05488-2 (hbk)

Visit the Taylor & Francis Web site at
http://www.taylorandfrancis.com

and the CRC Press Web site at
http://www.crcpress.com

Contents

Preface   **xiii**

1   Introduction   **1**
     1.1   Density Estimation for Missing and Modified Data   2
     1.2   Nonparametric Regression with Missing Data   7
     1.3   Notions and Notations   14
     1.4   Software   21
     1.5   Inside the Book   23
     1.6   Exercises   26
     1.7   Notes   28

2   Estimation for Directly Observed Data   **31**
     2.1   Series Approximation   31
     2.2   Density Estimation for Complete Data   38
     2.3   Nonparametric Regression   47
     2.4   Bernoulli Regression   51
     2.5   Multivariate Series Estimation   53
     2.6   Confidence Bands   56
     2.7   Exercises   59
     2.8   Notes   65

3   Estimation for Basic Models of Modified Data   **67**
     3.1   Density Estimation for Biased Data   67
     3.2   Regression with Biased Responses   72
     3.3   Regression with Biased Predictors and Responses   74
     3.4   Ordered Grouped Responses   80
     3.5   Mixture   83
     3.6   Nuisance Functions   85
     3.7   Bernoulli Regression with Unavailable Failures   87
     3.8   Exercises   93
     3.9   Notes   97

4   Nondestructive Missing   **99**
     4.1   Density Estimation with MCAR Data   101
     4.2   Nonparametric Regression with MAR Responses   107
     4.3   Nonparametric Regression with MAR Predictors   112
     4.4   Conditional Density Estimation   117
     4.5   Poisson Regression with MAR Data   124
     4.6   Estimation of the Scale Function with MAR Responses   127
     4.7   Bivariate Regression with MAR Responses   129
     4.8   Additive Regression with MAR Responses   133
     4.9   Exercises   137

      4.10 Notes   142

5 Destructive Missing   **145**
      5.1   Density Estimation When the Availability Likelihood is Known   146
      5.2   Density Estimation with an Extra Sample   151
      5.3   Density Estimation with Auxiliary Variable   156
      5.4   Regression with MNAR Responses   160
      5.5   Regression with MNAR Predictors   169
      5.6   Missing Cases in Regression   172
      5.7   Exercises   175
      5.8   Notes   179

6 Survival Analysis   **181**
      6.1   Hazard Rate Estimation for Direct Observations   183
      6.2   Censored Data and Hazard Rate Estimation   188
      6.3   Truncated Data and Hazard Rate Estimation   192
      6.4   LTRC Data and Hazard Rate Estimation   197
      6.5   Estimation of Distributions for RC Data   203
      6.6   Estimation of Distributions for LT Data   208
      6.7   Estimation of Distributions for LTRC Data   219
      6.8   Nonparametric Regression with RC Responses   225
      6.9   Nonparametric Regression with RC Predictors   229
      6.10 Exercises   232
      6.11 Notes   240

7 Missing Data in Survival Analysis   **243**
      7.1   MAR Indicator of Censoring in Estimation of Distribution   243
      7.2   MNAR Indicator of Censoring in Estimation of Distribution   249
      7.3   MAR Censored Responses   253
      7.4   Censored Responses and MAR Predictors   258
      7.5   Censored Predictors and MAR Responses   262
      7.6   Truncated Predictors and MAR Responses   264
      7.7   LTRC Predictors and MAR Responses   269
      7.8   Exercises   274
      7.9   Notes   280

8 Time Series   **281**
      8.1   Discrete-Time Series   281
      8.2   Spectral Density and Its Estimation   285
      8.3   Bernoulli Missing   291
      8.4   Amplitude-Modulated Missing   297
      8.5   Censored Time Series   301
      8.6   Probability Density Estimation   304
      8.7   Nonparametric Autoregression   310
      8.8   Exercises   313
      8.9   Notes   320

9 Dependent Observations   **323**
      9.1   Nonparametric Regression   323
      9.2   Stochastic Process   327
      9.3   Nonstationary Time Series With Missing Data   333
      9.4   Decomposition of Amplitude-Modulated Time Series   343

9.5   Nonstationary Amplitude-Modulation   348
9.6   Nonstationary Autocovariance and Spectral Density   351
9.7   The Simpson Paradox   355
9.8   Sequential Design   359
9.9   Exercises   363
9.10 Notes   369

10 Ill-Posed Modifications   **371**
10.1 Measurement Errors in Density Estimation   372
10.2 Density Deconvolution with Missing Data   379
10.3 Density Deconvolution for Censored Data   386
10.4 Current Status Censoring   388
10.5 Regression with Measurement Errors in Predictors   399
10.6 MEP Regression with Missing Responses   404
10.7 Estimation of Derivative   409
10.8 Exercises   414
10.9 Notes   421

References   **423**

Author Index   **439**

Subject Index   **445**

Preface

Missing and modified data are familiar complications in statistical analysis, while statistical literature and college classes are primarily devoted to the case of direct observations when the data are elements of a matrix. Missing makes some elements of the matrix not available, while a modification replaces elements by modified values. Classical examples of missing are missing completely at random when an element of the direct data is missed by a pure chance regardless of values of the underlying direct data, or missing not at random when the likelihood of missing depends on values of missed data. Classical examples of data modifications are biased data, truncated and censored data in survival analysis, amplitude-modulation of time series and measurement errors. Of course, missing may be considered as a particular case of data modification, while a number of classical data modifications, including biased and truncated data, are created by hidden missing mechanisms. This explains why it is prudent to consider missing and modification together. Further, these statistical complications share one important feature which is the must know. A missing or modification of data may be destructive and imply impossibility of consistent (feasible) estimation. If the latter is the case, this should be recognized and dealt with appropriately. On the other hand, it is a tradition in statistical literature to separate missing from modification. Further, because missing may affect already modified data and vice versa, it is convenient to know how to deal with both complications in a unified way.

This book is devoted solely to nonparametric curve estimation with main examples being estimation of the probability density, regression, scale (volatility) function, conditional and joint densities, hazard rate function, survival function, and spectral density of a time series. Nonparametric curve estimation means that no assumption about shape of a curve (like in linear regression or in estimation of a normal density) is made. The unique feature of the nonparametric estimation component of the book is that for all considered statistical models the same nonparametric series estimator is used whose statistical methodology is based on estimation of Fourier coefficients by a corresponding sample mean estimator. While the approach is simple and straightforward, the asymptotic theory shows that no other estimator can produce a better nonparametric estimation under standard statistical criteria. Further, using the same nonparametric estimator will allow the reader to concentrate on different models of missing and data modification rather than on the theory and methods of nonparametric statistics. The used approach, however, has a couple of downsides. There is a number of popular methods of dealing with missing data like maximum likelihood, expectation-maximization, partial deletion, imputation and multiple imputation, or the product-limit (Kaplan–Meier) methodology of dealing with censored data. Similarly, there exist equally interesting and potent methods of nonparametric curve estimation like kernel, spline, wavelet and local polynomials. These topics are not covered in the book and only corresponding references for further reading are provided.

The complementary R package is an important part of the book and it makes the learning instructional and brings the discussed topics in the realm of reproductive research. Virtually every claim and development mentioned in the book are illustrated with graphs that are available for the reader to reproduce and modify. This makes the material fully transparent and allows one to study it interactively. It is important to stress that no knowledge of R is needed for using the R package and the Introduction explains how to install and use it.

The book is self-contained, has a brief review of necessary facts from probability, parametric and nonparametric statistics, and is appropriate for a one-semester course for diverse classes with students from statistics and other sciences including engineering, business, social, medical, and biological among others (in this case a traditional intermediate calculus course plus an introductory course in probability, on the level of the book by Ross (2015), are the prerequisites). It also may be used for teaching graduate students in statistics (in this case an intermediate course in statistical inference, on the level of the book by Casella and Berger (2002), is the prerequisite). A large number of exercises, with different levels of difficulty, will guide the reader and help an instructor to test understanding of the material. Some exercises are based on using the R package.

There are 10 chapters in the book. Chapter 1 presents examples of basic models, contains a brief review of necessary facts from the probability and statistics, and explains how to install and use the R software. Chapter 2 presents the recommended nonparametric estimation methodology for the case of direct observations. Its Section 2.2 is a must read because it explains construction of the nonparametric estimator which is used throughout the book. Further, from a statistical point of view, the only fact that the reader should believe in is that a population mean may be reliably estimated by a sample mean. Chapter 3 is primarily devoted to the case of biased data that serve as a bridge between direct data and missed, truncated and censored data. Chapters 4 and 5 consider nonparametric curve estimation for missing data. While Chapter 4 is devoted to the cases when a consistent estimation is possible based solely on missing data, Chapter 5 explores more complicated models of missing when an extra information is necessary for consistent estimation. Chapters 6 and 7 are devoted to classical truncated and censored data modifications traditionally studied in survival analysis, and in Chapter 7 the modified data may be also missed. Modified and missing time series are discussed in Chapters 8 and 9. Models of missing and modified data, considered in Chapters 3-9, do not slow down the rate of traditional statistical criteria. But some modifications, like measurement errors or current status censoring, do affect the rate. The corresponding problems, called ill-posed, are discussed in Chapter 10.

Let us also note that the main emphasis of the book is placed on the study of small samples and data-driven estimation. References on asymptotic results and further reading may be found in the "Notes" sections.

For the reader who would like to use this book for self-study and who is venturing for the first time into this area, the advice is as follows. Begin with Section 1.3 and review basic probability facts. Sections 2.1 and 2.2 will explain the used nonparametric estimator. Please keep in mind that the same estimator is used for all problems. Then read Section 3.1 that explains the notion of biased data, how to recover the distribution from biased data, and that a consistent estimation, based solely on biased data, is impossible. Also, following the explanation of Section 1.4, install the book's R package and test it using figures in the above-mentioned sections. Now you are ready to read the material of your interest in any chapter. During your first reading of the material of interest, you may skip probability formulas and go directly to simulations and data analysis based on figures and R package; this will make the learning process easier. Please keep in mind that a majority of figures are based on simulations, and hence you know the underlying model and can appreciate the available data and nonparametric estimates.

All updates to this book will be posted on the web site http://www.utdallas.edu/~efrom and the author may be contacted by electronic mail at efrom@utdallas.edu.

Acknowledgments

I thank everyone who, in various ways, has had an influence on this book. My biggest thanks go to my family for the constant support and understanding. My students, colleagues and three reviewers graciously read and gave comments on a draft of the book.

John Kimmel provided invaluable assistance through the publishing process. The support by NSF Grants DMS-0906790 and DMS-1513461, NSA Grant H982301310212, and actuarial research Grants from the CAS, TAF and CKER are greatly appreciated.

Sam Efromovich
Dallas, Texas, USA, 2017

Chapter 1

Introduction

Nonparametric curve estimation allows one to analyze data without assuming the shape of an estimated curve. Methods of nonparametric estimation are well developed for the case of directly observed data, much less is known for the case of missing and modified data. The chapter presents several examples that shed light upon the nature of missing and modified data and raises important questions that will be answered in the following chapters. Section 1.1 explores a number of examples with missing, truncated and censored data when the problem is to estimate an underlying probability density. The examples help us to understand that missing and/or modified data not only complicate the estimation problem but also may prevent us from the possibility to consistently estimate an underlying density, and this makes a correct treatment and understanding specifics of the data paramount. It is also explained why statistical simulations should be a part of the learning process. The latter is the primary role of the book's R package which allows the reader to repeat and modify simulations presented in the book. Section 1.2 explains the problem of nonparametric regression with missing data via real and simulated examples. It is stressed that different estimation procedures are needed for cases where either responses or predictors are missed. Further, the problem of estimation of nuisance functions is also highlighted. Main notions and notations, used in the book, are collected in Section 1.3. It also contains a brief primer in elementary probability and statistics. Here the interested reader may also find references on good introductory texts, but overall material of the book is self-explanatory. Installation of the book's R package and a short tutorial on its use can be found in Section 1.4. Section 1.5 briefly reviews what is inside other chapters. Exercises in Section 1.6 will allow the reader to review basics of probability and statistics.

Before proceeding to the above-outlined sections, let us make two general remarks. The first remark is about the used nonparametric approach for missing and modified data versus a parametric approach. For the case of direct data, it is often possible to argue that an underlying function of interest is known up to a parameter (which may be a vector) and then estimate the parameter. Missing and/or modified data require estimation of a number of nuisance functions characterizing missing and/or modification, and then a parametric approach would require assuming parametric models for those nuisance functions. The latter is a challenging problem on its own. Nonparametric estimation is based solely on data and does not require parametric models neither for the function of interest nor for nuisance functions. This is what makes this book's discussion of the nonparametric approach so valuable for missing and modified data. Further, even if a parametric approach is recommended for estimation of a function of interest, it is prudent to use nonparametric estimation for more elusive nuisance functions describing data modification. The second remark is about the used terminology of missing and modified data. Missing data are, of course, modified data. Further, one may argue that truncated or censored data, our main examples of modified data, are special cases of missing data. Nonetheless, it is worthwhile to separate the classical notion of missing data, when some observations are not available (missed), from data modified by truncation, censoring, biasing, measurement errors, etc. Further, it is important to

stress that historically different branches of statistical science study missing and modified
data. In the book a unified nonparametric estimation methodology is used for statistical
analysis of missing and modified data, but at the same time it is stressed that each type of
data has its own specifics that must be taken into account.

1.1 Density Estimation for Missing and Modified Data

Suppose that a researcher would like to know the distribution of the ratio of alcohol in
the blood of liquor-intoxicated drivers traveling along a particular highway. The police is
helping the researcher and over a fixed period of time T every driver is stopped and then
the ratio of alcohol in the blood is measured. Because we are interested only in the ratio
for liquor-intoxicated drivers, only positive ratios are of interest and hence collected. As a
result, the researcher gets a sample $X_1^*, X_2^*, \ldots, X_n^*$ of the ratios of alcohol in the blood.
This is a *direct* sample of size n from the random variable of interest X^* of the ratio of
alcohol, and it is known that the ratio is scaled in such a way that $X^* \in (0, 1]$.

The top diagram in Figure 1.1 presents a simulated sample of size $n = 400$ shown by the
(frequency) histogram. A histogram is a simple and attractive statistical tool for presenting
a sample where the height of a histogram at a point x is proportional to the number of
observations in the same bin as x, and then the heights are rescaled in such a way that the
integral of the histogram is equal to one. The latter implies that the heights of a histogram
show us a probability density (because as a function in x the height is nonnegative and
integrated to one). Histogram is one of the most widely used probability density estimators,
and it is also the only density estimator that is studied in all undergraduate classes. To
construct a histogram, we have to choose an origin (the first left bin) and a bin width. As a
result, histogram is a *nonparametric* density estimator because no assumption about shape
of an underlying density is made. On the contrary, if for instance one assumes that an underlying density is normal, then estimation of the density is called *parametric* because a normal
density is completely specified by its mean and variance, and hence the density's estimation
is converted into estimation of the underlying mean and variance (two parameters).

Now let us look again at the histogram in the top diagram in Figure 1.1. What does it
tell us about an underlying density? Is it unimodal or bimodal or none of the above? Is it
symmetric or skewed? These are basic questions that a density estimate should answer, and
probably the questions are confusing a bit given the wiggly pattern of the histogram. You
may try to smooth the histogram or make it sharper, but still it is difficult to understand
if your action is correct. Furthermore, when only data are available, it is very difficult (if
almost impossible) to learn about and be trained in density estimation because we do not
know an underlying density. Instead, it is simpler and faster to become well trained in
understanding and dealing with statistical problems via learning from simulated data. In
particular, let us spell a "secret" that the sample in the top diagram is generated from a
known density. The second from the top diagram shows us the same histogram, only now
it is overlaid by an underlying density (the solid line). Have you noticed the pronounced
difference made by the underlying density? Now you can visualize what the histogram
exhibits, how it deviates from the underlying density, and what action should be done to
smooth the histogram. Further, note that the sample size is $n = 400$, it is "huge" by all
measures, and nonetheless the sample (and respectively the histogram) does not perfectly
follow the density. We may conclude that the second diagram, where we see both data and
the underlying density, sheds light on the density estimation problems, helps us to gain
experience in dealing with data and in understanding what we can and cannot estimate for
this or that sample sizes. In other words, while in real life data analysts are dealing with
samples from unknown distributions that should be estimated, during a learning period it
is prudent to use simulated examples for testing our understanding of data and different
methods of statistical analysis and inference. This pedagogical tool of simulated examples

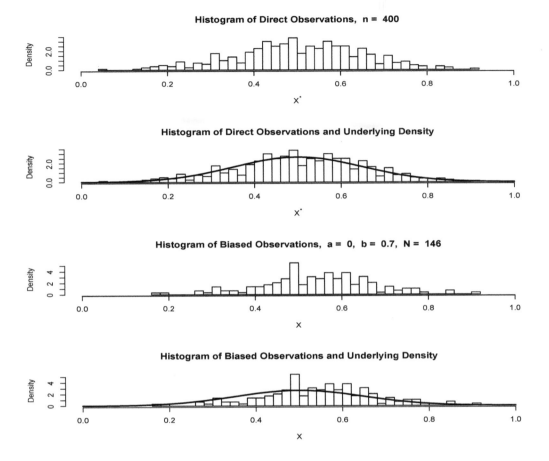

Figure 1.1 *Directly observed and biased data exhibited by two top and two bottom histograms, correspondingly. Sample size of direct observations is n while N is the number of biased observations. Direct observations of X^* are generated by the Normal density defined in Section 2.1, the biasing function is $\max(0, \min(a + bx, 1))$ with a = 0 and b = 0.7. {Here and in all other captions information in curly brackets is devoted to explanation of arguments of the figure that can be changed. For Figure 1.1 the simulation can be repeated and results visualized by using the book's R package (see how to install and use it in Section 1.4). When the R package is installed, start R and simply enter (after the R prompt >) ch1(fig=1). Note that using ch1(fi=1) or ch1(f=1) yields the same outcome. Default values of arguments, used by the package for Figure 1.1 are shown in the square brackets. All indicated default values may be changed. For instance, after entering ch1(fig=1,n=100,a=0.1,b=0.5) a sample of size n = 100 with the biasing function $\max(0, \min(0.1 + 0.5x, 1))$ will be generated and exhibited.} [n = 400, a = 0, b = 0.7].*

is used throughout the book and the reader will be able to repeat simulations and change parameters of estimators using the R package discussed in Section 1.4. Further, each figure, in its caption, contains information about an underlying simulation and its parameters that may be changed.

Now let us return to the discussion of the study of the ratio of alcohol in the blood of liquor-intoxicated drivers. The above-described scenario of collecting data is questionable because the police must stop all drivers during the time period of collecting data. A more realistic scenario is that the researcher gets data available from routine police reports on arrested drivers charged with driving under the influence of alcohol (a routine report means

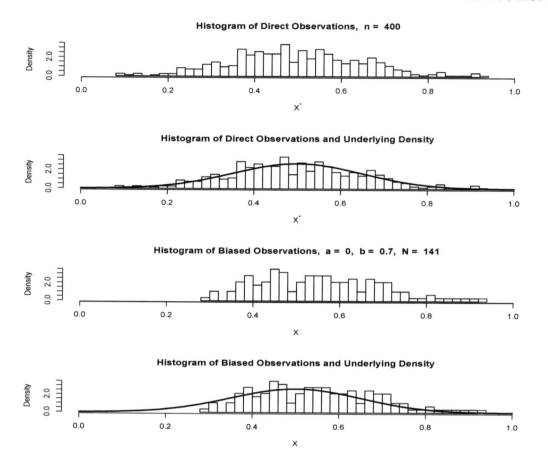

Figure 1.2 *Repeated simulation of Figure 1.1.* { *This figure is created by the call* > **ch1(f=1).** }

that there are no special police operations to reveal all intoxicated drivers). Because a drunker driver has a larger chance of attracting the attention of the police, it is clear that the data will be biased toward higher ratios of alcohol in the blood. To understand this phenomenon, let us simulate a corresponding sample using the direct data shown in the top diagram in Figure 1.1. Assume that if $X^* = x$ is the ratio of alcohol, then with the probability $\max(0, \min(a + bx, 1))$ the driver will be stopped by the police. The above-presented formula looks complicated, but it simply reflects the fact that the probability of any event is nonnegative and less or equal to one. The third from the top diagram in Figure 1.1 shows us results of the simulation which created the biased observations, here the particular values $a = 0$ and $b = 0.7$ are used. As it could be expected, the histogram shows that available observations are skewed to the right with respect to the direct data because a "drunker" driver has a larger probability to be stopped. Further, the sample size of the biased sample $X_1, \ldots, X_N$ is just $N = 146$, in other words, among $n = 400$ drivers using the highway during the period of time T of the study, only $N = 146$ were stopped. Note that the value of N, as well as values of parameters a and b, are shown in the title. An important conclusion from the histogram is that it is impossible to restore the underlying density of X^* based solely on the biased data. This message becomes even more clear from the bottom diagram where the same histogram of biased observations is overlaid by the underlying density. Further, we again see how the underlying density sheds light on the data and the estimation problem, and note that this learning tool is available only in a simulation.

What we have observed in Figure 1.1 is the example of modified by biasing data and also a specific example of missing data because now only $N = 146$ observations, from the hidden direct sample of size $n = 400$, are available. It is clear that a reliable (in statistics we may say consistent) estimator must take this modification of direct data into account. How to do this will be explained in Chapters 3-5.

The book's R package (see Section 1.4 on how to install and use it) allows us to repeat the simulation of Figure 1.1. To illustrate this possibility, diagrams in Figure 1.2 present another simulation produced by R function *ch1(fig=1)* of the package. Arguments n, a and b, used to create Figure 1.2, have the same default values of $n = 400$, $a = 0$ and $b = 0.7$ as in Figure 1.1. Note that values of these parameters are shown in Figure 1.2. As a result, diagrams in the two figures are different only due to stochastic nature of underlying simulations. Let us look at diagrams in Figure 1.2. The direct sample creates a relatively flat middle part of the histogram, and it clearly does not look symmetric about 0.5. Further, the biased data is even more pronouncedly skewed to the right. Just imagine how difficult it would be for an estimator to restore the symmetric and bell-shaped underlying density shown by the solid line. The reader is advised to repeat this (and other figures) multiple times to learn possible patterns of random samples.

Now let us consider another example of data modification. Suppose that we are interested in the distribution of an insured loss X^* due to hail. We can get information about payments for the losses from insurance companies, but the issue is that insurance companies do not know all insured losses due to the presence of a deductible D in each hail insurance policy. A deductible is the amount a policyholder agrees to pay out-of-pocket after making a claim covered by the insurance policy. For example, if the deductible is \$5,000 and the accident damage costs \$25,000, the policyholder will pay \$5,000 and the insurance company will pay the remaining \$20,000. The presence of deductible in an insurance policy implies that if the loss X^* is not larger than the deductible D then no payments occurs, while otherwise the compensation (payment) for the loss is equal to $X^* - D$. As a result losses that do not exceed policy deductibles are typically missed (there is no sense to make a claim for a loss less than the deductible because not only the policyholder gets no compensation but also the next insurance premium more likely will go up). Further, for a policy with a payment we can calculate the loss as the payment plus the policy deductible. We conclude that, based on policies with payments, we observe the loss $X = X^*$ only given $X^* > D^*$. The latter is an important actuarial example of biased and missing data.

In statistical literature, the above-described effect of a deductible on availability of a hidden loss is called a *left truncation* or simply a truncation. (The only pure notational difference is that traditionally, under the left truncation, the loss is observed if it is larger or equal to the deductible. In what follows we are going to use this approach.) In short, truncation is a special missing mechanism which is defined by the variable of interest and another truncating variable (here the deductible). Further, deductibles in insurance policies may imply a destructive modification of data when no consistent estimation of the density of losses is possible. Indeed, if the deductible in all policies is larger than a positive constant d_0, then the density $f^{X^*}(x)$ of the loss cannot be recovered for $x < d_0$ because we always miss the small losses. The latter is an important property of data truncation that should be taken care of.

Deductible is not the only complication that one may deal with while working with insurance data. An insurance policy may also include a limit L on the payment which implies that a compensation for insurable loss cannot exceed the limit L. If a policy with limit L made a payment on loss X^*, then the available information is the pair (V, Δ) where $V := \min(X^*, L)$ and $\Delta := I(X^* \leq L)$ (here and in what follows, $I(B)$ is the indicator of event B which is equal to one if the event B occurs and it is equal to zero otherwise).

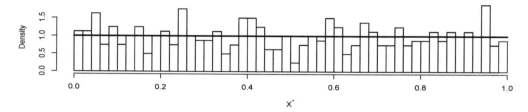

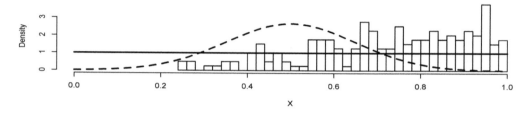

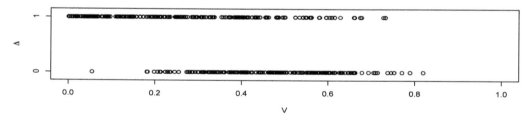

Figure 1.3 *Hidden, truncated and censored insured losses due to hail. The top diagram exhibits the histogram of hidden losses $X_1^*, \ldots, X_n^*$ distributed according to the density shown by the solid line. The middle diagram shows the histogram of these losses truncated by deductibles $D_1, \ldots, D_n$ that are generated from the density shown by the dashed line. In other words, this is the histogram of $X_1, \ldots, X_N$ which is a subsample of X_l satisfying $X_l^* \geq D_l$, $l = 1, \ldots, n$ and the number of available truncated losses is $N := \sum_{l=1}^n I(X_l \geq D_l) = 198$ as shown in the title. The bottom diagram shows right censored losses. It is based on the underlying losses shown in the top diagram and a new sample $L_1, \ldots, L_n$ of policy limits generated from the density shown by the dashed line in the middle diagram. The observations are n pairs $(V_1, \Delta_1), \ldots, (V_n, \Delta_n)$ shown by circles in the bottom diagram. Here $V_l := \min(X_l, L_l)$, $\Delta_l := I(X_l \leq L_l)$. The title also shows the number of uncensored losses $M := \sum_{l=1}^n \Delta_l = 206$. {In all figures that may be repeated using the book's R package, at the end of their caption the reader can find in square brackets arguments that may be changed. For instance, this Figure 1.3 allows one to repeat simulations with different sample sizes n. All other parameters of the above-outlined simulation remain the same.} [n = 400].*

In statistics this type of data modification is called the *right censoring*, and note that the censoring does not decrease the total number of observations but it does decrease the number of observed values of X^*. Further, if $\mathbb{P}(L > l_0) = 0$, then the distribution of X^* beyond l_0 cannot be restored.

The above-presented actuarial example, as well as notions of truncation and censoring, may be confusing a bit. Let us complement the explanation by a simulated example shown in Figure 1.3. The top diagram in Figure 1.3 exhibits the histogram of simulated losses

$X_1^*, \ldots, X_n^*$ generated according to the density shown by the solid line. The middle diagram in Figure 1.3 shows us the histogram of left truncated losses that are generated as follows. Deductibles $D_1, \ldots, D_n$ are simulated according to the density shown by the dashed line, and then for each $l = 1, \ldots, n$ if $X_l^* < D_l$ then X_l^* is skipped and otherwise X_l^* is observed and added to the sample of truncated losses. As a result, we get the truncated sample $X_1, \ldots, X_N$ of size $N := \sum_{l=1}^{n} I(X_l \geq D_l)$. As it could be expected, the middle diagram indicates that the left truncated losses are skewed to the right and the histogram does not resemble the underlying density. We will discuss in Chapters 6 and 7 how to restore the underlying density of hidden losses based on truncated data.

The bottom diagram in Figure 1.3 exhibits the example of right censored losses with the limit L distributed according to the density shown by the dashed line in the middle diagram and the hidden losses shown in the top diagram. The observed pairs (V_l, Δ_l), $l = 1, \ldots, n$ are shown by circles, and recall that $V_l := \min(X_l^*, L_l)$ and $\Delta_l := I(X_l \leq L_l)$. The top row of circles shows us underlying (uncensored) losses corresponding to cases $X_l \leq L_l$, while the bottom row of circles shows us limits on payments corresponding to cases $X_l > L_l$. Note that the available losses are clearly biased, they are skewed to the left with respect to the hidden losses, and their number $M := \sum_{l=1}^{n} \Delta_l = 206$ is dramatically smaller than the number $n = 400$ of the hidden losses.

Clearly, special statistical methods should be used to recover the distribution of a random variable whose observations are modified by truncation and/or censoring, and corresponding statistical methods are collectively known as *survival analysis*. While survival data and missing data have a lot in common, in the book we are following the tradition to consider them separately; at the same time let us stress that the proposed method of nonparametric analysis is the same for both types of data. Furthermore, we will consider missing survival data when these two mechanisms of data modification perform together.

Let us make a final remark. The above-presented examples are simulated. The author is a strong believer in learning statistics via simulated examples because in those examples you know the underlying model and then can appreciate complexity of the problem and quality of its solution. This is why the approach of simulated examples is used in the book. Further, the book's R software will allow the reader to test theoretical assertions, change parameters of underlying statistical models and choose parameters of recommended estimators. This will help the reader to gain the necessary experience in dealing with missing and modified data.

1.2   Nonparametric Regression with Missing Data

Consider a pair of random variables (X, Y). Suppose that we are interested in prediction of Y given $X = x$ based on a sample from the pair. Ideally, we would like to propose a function m such that given a particular value of $X = x$ the function $m(x)$ indicates the correct value of Y. Unfortunately, in a majority of statistical problems there is no functional relation between the predictor (independent, explanatory variable) X and the response (dependent variable) Y. As a result, a number of different approaches have been suggested. One of them is to propose a function (optimal predictor) $m(x)$ which minimizes the conditional mean squared error $\mathbb{E}\{(Y - \mu(x))^2 | X = x\}$ among all possible predictors $\mu(x)$. A function $m(x)$, which minimizes this risk, is the conditional expectation

$$m(x) := \mathbb{E}\{Y | X = x\}. \tag{1.2.1}$$

If no assumption about the shape of $m(x)$ is made, then (1.2.1) is called the nonparametric regression of response Y on predictor X. Another popular interpretation of (1.2.1) is the equation $Y = m(X) + \varepsilon$ where ε is the regression error satisfying $\mathbb{E}\{\varepsilon | X\} = 0$. The familiar alternative to the nonparametric regression is a parametric linear regression when it is

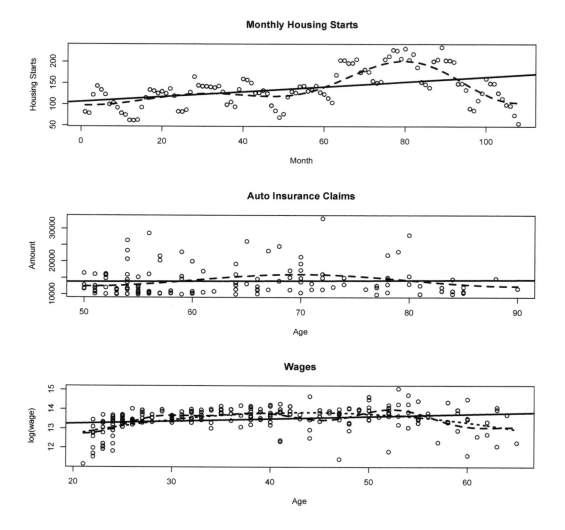

Figure 1.4 *Parametric and nonparametric regressions for three real datasets. Observations are shown by circles, linear regression is shown by the solid line, nonparametric regression is shown by the dashed line, and in the bottom diagram quadratic regression is shown by the dotted line. Sample sizes in the top, middle and bottom diagrams are 108, 124, and 205, respectively.*

assumed that $m(x) := \beta_0 + \beta_1 x$ and then parameters β_0 and β_1 are estimated. While parametric regressions are well known, the nonparametric regression is less familiar and not studied in undergraduate and often even in graduate statistics classes. As a result, let us begin with considering several examples that shed light on nonparametric regression.

Figure 1.4 presents three classical datasets. For now let us ignore curves and concentrate on observations (pairs (X_l, Y_l), $l = 1, 2, \ldots, n$) shown by circles. A plot of pairs (X_l, Y_l) in the xy-plane, called scattergram or scatter plot, is a useful tool to get a first impression about studied data. Let us begin with analysis of the top diagram which presents monthly housing starts in the USA from January 1966 to December 1974. Note that the predictor is deterministic and its realizations are equidistant meaning that $X_{l+1} = X_l + 1$. It is also reasonable to assume that monthly housing starts are depended variables, and then we observe a classical time series of housing starts. The simplest classical decomposition model of a time series is

$$Y_l = m(X_l) + S(X_l) + \varepsilon_l, \tag{1.2.2}$$

where $m(x)$ is a slowly changing function known as a trend component, $S(x)$ is a periodic function with period T (that is, $S(x+T) = S(x)$) known as a seasonal (cyclical) component (it is also customarily assumed that the sum of its values over the period is zero), and ε_l are random and possibly dependent components with zero mean; more about time series and its statistical analysis can be found in Chapters 8 and 9. While a traditional time series problem is to analyze the random components, here we are interested in estimation of the trend. Note that $\mathbb{E}(Y_l|X_l = x) = m(x) + S(x)$, by its definition the trend is a "slowly" changing (with respect to the seasonal component) function, and therefore the problem of interest is the regression problem with the so-called fixed design when we are interested in a low-frequency component of the regression. Now, please use your imagination and try to draw the trend $m(x)$ via the scattergram. Note that the period of seasonal component is 12 months and this may simplify the task. Now look at the solid line which exhibits the classical least-squares linear regression. Linear regression is a traditional tool used in time series analysis, and here it may look like a reasonable curve to predict the housing market. Unfortunately, back in the seventies, too many believed that this is the curve which describes the housing market. The dashed line shows the estimated nonparametric trend whose construction will be explained in the following chapters. The nonparametric trend exhibits the famous boom and the tragic collapse of the housing market in the seventies, and it nicely fits the scattergram by showing two modes in the trend. Further, note how well the nonparametric trend allows us to visualize the seasonal component with the period of 12 months.

The middle diagram presents a regression dataset of independent pairs of observations. Its specifics is that the predictors are distributed not uniformly and the volatility of the response depends on the predictor. Such a regression is called heteroscedastic. The scattergram exhibits automobile insurance claims data. The dependent variable Y is the amount paid on a closed claim, in (US) dollars, and the predictor X is the age of the driver. Only claims larger than \$10,000 are analyzed. Because the predictor is the random variable, the regression (1.2.1) may be referred to as a random design regression. An appealing nature of the regression problem is that one can easily appreciate its difficulty. To do this, try to draw a curve $m(x)$ through the middle of the cloud of circles in the scattergram that, according to your own understanding of the data, gives a good fit (describes a relationship between X and Y) according to the model (1.2.1). Clearly such a curve depends on your imagination as well as your understanding of the data at hand. Now we are ready to compare our imagination with performance of estimators. The solid line shows us the linear regression. It indicates that there is no statistically significant relationship between the age and the amount paid on a closed claim (the estimated slope of 14.22 is insignificant with p-value equal to 0.7). Using linear regression for this data looks like a reasonable approach, but let us stress that this is up to the data analyst to justify that relationship between the amount paid on a claim and the age of the operator is linear and not of any other shape. Now let us look at the dashed line which exhibits a nonparametric estimate whose shape is defined by the data. How this estimate is constructed will be explained shortly in Chapter 2. The nonparametric regression exhibits a pronounced shape which implies an interesting conclusion: the amount paid on closed claims is largest for drivers around 68 years old and then it steadily decreases for both younger and older drivers. (Of course, it is possible that drivers of this age buy higher limits of insurance, or there are other lurking variables that we do not know. If these variables are available, then a multivariate regression should be used.) Now, when we have an opinion of the nonparametric estimator, please look one more time at the data and you may notice that the estimator's conclusion has merit.

The bottom diagram in Figure 1.4 presents another classical regression dataset of 1971 wages as a function of age. The linear regression (the solid line) indicates an increase in wages with age, which corresponds to a reasonable opinion that wages reflect the worker's experience. In economics, the traditional age wage equation is modeled as a quadratic in age, and the corresponding parametric quadratic regression $m_2(x) := \beta_0 + \beta_1 x + \beta_2 x^2$ is shown by the dotted line. The quadratic regression indicates that the wages do initially increase with age (experience) but then they eventually go down to the level of the youngest workers. Note that the two classical parametric models present different pictures of the age wage patterns. The dashed line presents the nonparametric regression. Note that overall it follows the pattern of the quadratic regression but adds some important nuances about modes in the age wage relationship.

Figure 1.4 allows us to understand performance of statistical estimators. At the same time, it is important to stress that analysis of real datasets does not allow us to appreciate accuracy of statistical estimates because an underlying regression function is unknown, and hence we can only speculate about the quality of this or that estimate.

Similarly to our previous conclusion in Section 1.1 for density estimation, analysis of real data does not allow us to fully appreciate how well a particular regression estimator performs. To overcome this drawback, we are going to use numerical simulations with a known underlying regression function. Let us consider several simulated examples.

We begin with the study of the likelihood (probability) of an insurable event, which may be a claim, payment, accident, early prepayment on mortgage, default on a payment, reinsurance event, early retirement, theft, loss of income, etc. Let Y be the indicator of an insurable event (claim), that is $Y = 1$ if the event occurs and $Y = 0$ otherwise, and X be a covariate which may affect the probability of claim; for instance, X may be general economic inflation, or deductible, or age of roof, or credit score, etc. We are interested in estimation of the conditional probability $\mathbb{P}(Y = 1|X = x) =: m(x)$. On first glance, this problem has nothing to do with regression, but as soon as we realize that Y is a Bernoulli random variable, then we get the regression $\mathbb{E}\{Y|X = x\} = \mathbb{P}(Y = 1|X = x) = m(x)$. The top diagram in Figure 1.5 illustrates this problem. Here X is uniformly distributed on $[0, 1]$ and then $n = 200$ pairs of independent observations are generated according to Bernoulli distribution with $\mathbb{P}(Y = 1|X) = m(X)$ and $m(x)$ is shown by the dotted line. Because the regression function is known, we may learn how to recognize it in the scattergram. Linear regression (the solid line), as it could be expected, gives us no hint about the underlying regression while the nonparametric estimate (the dashed line), which will be defined in Section 2.4, nicely exhibits the unimodal shape of the regression.

Now let us consider our second example where Y is the number of claims (events) of interest, or it may be the number of: noncovered losses, payments on an insurance contract, payments by the reinsurer, defaults on mortgage, early retirees, etc. For a given $X = x$, the number of claims is modeled by Poisson distribution with parameter $\lambda(x)$; that is, $\mathbb{P}(Y = k|X = x) = e^{-\lambda(x)}[\lambda(x)]^k/k!$, $k = 0, 1, 2, \ldots$ The aim is to estimate $\lambda(x)$, and because $\mathbb{E}\{Y|X = x\} = \lambda(x)$ this problem again can be considered as a regression problem. A corresponding simulation with $n = 200$ observations is shown in the bottom diagram of Figure 1.5. The underlying regression function (the dotted line) is bimodal. Try to use your imagination and draw a regression curve through the scattergram, or even simpler, using the dotted line try to understand the scattergram. The nonparametric estimate (the dashed line) exhibits two modes. The estimate is not perfect but it does fit the scattergram. The linear regression (the solid line) sheds no light on the data.

In summary, nonparametric regression can be used for solving a large and diverse number of statistical problems. We will continue our discussion of the nonparametric regression in Chapter 2, and now let us turn our attention to regression estimation with missing data when values of responses or predictors may be not available (missed).

Likelihood of Claim

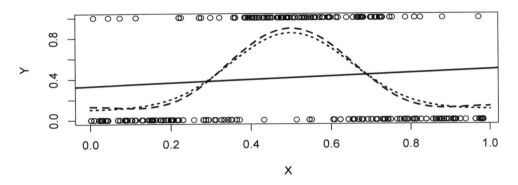

Number of Claims

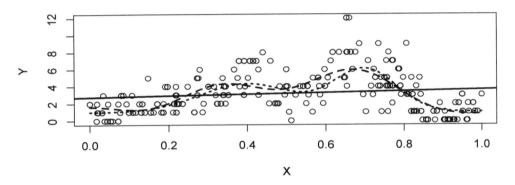

Figure 1.5 *Linear and nonparametric regressions for two simulated regression datasets, $n = 200$. In each diagram, observations are shown by circles overlaid by the underlying regression function (the dotted line), linear regression (the solid line) and nonparametric regression (the dashed line).*

Let us recall our example with intoxicated drivers discussed in the previous section when the police stops with some likelihood drivers and then the ratio of alcohol is measured. A Bernoulli random variable A describes the stopping mechanism and if $A = 1$, then a car is stopped and if $A = 0$, then the car passes by. In other words, if $A = 1$ then the ratio of alcohol is available and otherwise it is missed. Further, in general the availability A and the ratio of alcohol X are dependent random variables. Then, as it is explained in Section 1, based solely on available observations the density of the ratio of alcohol cannot be consistently estimated. One possibility to overcome this obstacle is to obtain extra information. Let us conjecture that the probability of a car to be stopped is a function of the car speed S, and during the experiment the police measures speeds of all passing cars. Further, let us additionally assume that the speed is a function of the ratio of alcohol. To be specific in our simulated example, set $S := c + dX + v(U - 0.5)$ where U is a standard uniform random variable supported on $[0, 1]$ and X is the ratio of alcohol level. Then we can write for the availability likelihood,

$$\mathbb{P}(A = 1 | X = x, S = s) = \mathbb{P}(A = 1 | S = s) =: w(s). \qquad (1.2.3)$$

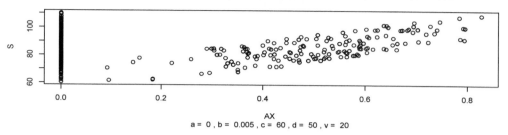

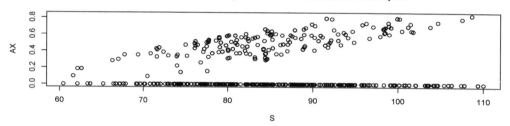

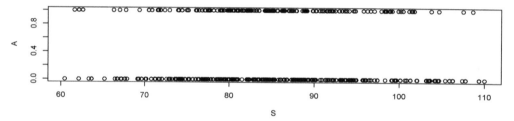

Figure 1.6 *Missing data in intoxicated drivers example when the availability A is a function of car speed S. The ratio of alcohol X is generated as in Figure 1.1, the car speed $S = c + dX + v(U − 0.5)$ where U is the Uniform random variable on $[0, 1]$ and independent of X, and given $S = s$ the availability random variable A is equal to one with the probability $\max(0, \min(a + bs, 1))$. The top diagram shows $n = 400$ pairs $(A_1X_1, S_1), \ldots, (A_nX_n, S_n)$ where only $N = 165$ pairs are complete (the ratio of alcohol is available). The middle scattergram shows pairs $(S_1, A_1X_1), \ldots, (S_n, A_nX_n)$. The bottom diagram is a traditional scattergram of n complete pairs $(S_1, A_1), \ldots, (S_n, A_n)$. {Parameters n, a, b, c, d and v can be changed. Used (default) values of the parameters are shown in the subtitle of the top diagram.} [n = 400, a = 0, b = 0.005, c = 60, d = 50, v = 20].*

Under this scenario, available observations are $(S_1, A_1X_1), (S_2, A_2X_2), \ldots, (S_n, A_nX_n)$ from the underlying pair of random variables (S, AX). Note that the speed S and the availability A are known for all n drivers, while the ratio of alcohol X is known only for drivers who were stopped. As we shall see in Chapter 5, in this case consistent estimation of the density of X is possible without knowing the availability likelihood $w(s)$ because it can be estimated using nonparametric regression of A on S.

Figure 1.6 presents a particular simulation of this example, and its caption describes both the simulation and used parameters. The top diagram allows us to look at $n = 400$ realizations of the pair (AX, S) where now the availability A depends on the speed according to (1.2.3) with the specific availability likelihood $w(s) = 0.005s$. Let us stress that the availability A and the ratio of alcohol X are dependent (because X and the speed S are dependent), but conditionally, given the speed S, the availability A and the ratio of alcohol X are independent. Using the top diagram, we can consider several classical statistical problems like estimation of the regression of the speed S on the ratio of alcohol X or estimation of the conditional density of the speed given the ratio of alcohol. Note that these are not trivial problems. First of all, a majority of underlying ratios of alcohol are not available, and instead we have only a large number (here $n - N = 235$) of speeds corresponding to not available ratios of alcohol shown by the vertical column of circles with $A_l X_l = 0$. Of course, we have $N = 165$ complete pairs (X, S) of observations, but are they biased or unbiased? Can they be used for consistent estimation, or in other words, can a complete case estimator be consistent? It will be explained in Chapter 4 that the answer is "no," the complete pairs are biased, and traditional statistical procedures developed for direct observations are not applicable here. Nonetheless, a special regression estimator, defined in Chapter 4 and based on all n pairs of observations (including incomplete ones) does produce a consistent estimation.

The middle diagram in Figure 1.6 exhibits the same missing data only now we can visualize AX versus S. Note the special layout of incomplete pairs and that we can better visualize speeds of cars that were not stopped by police. The diagram raises a natural statistical question about the possibility to estimate the regression of the ratio of alcohol X on the speed S and estimation of the conditional probability of X given the speed S. As it will be shown in Chapter 4, in this case a consistent estimation is possible based solely on complete cases. Furthermore, the asymptotic theory shows that incorporating incomplete cases does not improve the estimation.

These two contradictory conclusions, made for the two top diagrams, sound confusing and require some explanation. First of all, let us stress that in general the regression of S on X and the regression of X on S are different problems and knowing one regression may not help us in defining another. Just to give a simple explanation of this phenomenon, consider the case of independent X and S. Then the regression of S on X is a constant equal to the mean of S while the regression of X on S is again a constant but now the constant is equal to the mean of X. Clearly from knowing one regression we cannot infer about another. Second, the difference in the regressions for the top and middle diagrams is that in the top diagram the predictor X is missed while the response S is always observed, while in the middle diagram the predictor S is always observed while the response X may be missed. This is what differentiates these two settings and implies different types of solution.

We may conclude from this discussion of the two top diagrams that missing data create interesting statistical problems where the same data may imply different solutions for related statistical problems. This is the type of problems that will be explained in detail in the book.

Now examine the bottom diagram. This is a classical scattergram for direct (complete) pairs of observations, and it allows us to understand the underlying missing mechanism and estimate the availability likelihood $w(s) = \mathbb{P}(A = 1|S = s) = \mathbb{E}\{A|S = s\}$ which is a classical nonparametric regression problem. As a result, as it will be explained in Chapter 5, for this sample we will be able to consistently estimate the probability density of X and solve the above-formulated regression problems.

Finally, let us make the following remark. In statistics, a nuisance parameter or function is any parameter or function which is not of immediate interest but which must be accounted for in the analysis of the estimand. For instance, in a nonparametric regression the design density of predictor is typically not of immediate interest but it affects nonparametric estimation. Similarly, for regression with missing data the availability likelihood function is not of immediate interest but it must be accounted for. Further, nuisance functions may be of interest on their own, and in what follows it will be explained how to estimate and analyze them.

1.3 Notions and Notations

Nonparametric curve estimation is founded on the theory of probability and statistics. In the book traditional definitions, notions and notations are used, and the aim of this section is to briefly review them. While basic probability results are typically reviewed and reminded of in each chapter, it is also convenient to have them all together in one section. The reader interested in a more detailed and comprehensive account of the probability and statistics may refer to books by Ross (2015) and Casella and Berger (2002).

The theory of probability assigns a *probability* (likelihood) to events. If B is an event, then we denote by $\mathbb{P}(B)$ the probability of this event. Two events are called independent if the probability that they both occur is equal to the product of their individual probabilities. In other words, if B_1 and B_2 are two events and we introduce a new event $B_1 \cap B_2$ which occurs only if both these events occur, then the two events are independent if $\mathbb{P}(B_1 \cap B_2) = \mathbb{P}(B_1)\mathbb{P}(B_2)$. Otherwise the events are called *dependent*.

In the book we denote by uppercase letters X, Y, Z, etc. random variables, and by lowercase letters x, y, z, etc. their values. A random variable X is completely defined by its *cumulative distribution function*

$$F^X(x) := \mathbb{P}(X \le x) \tag{1.3.1}$$

or the *survival function*

$$G^X(x) := \mathbb{P}(X > x) = 1 - F^X(x). \tag{1.3.2}$$

Note that in (1.3.2) the sign ":=" means *equal by definition* and "=" denotes a traditional equality, and we can also use the sign "=:" to define the right side of a relation, for instance, we may write $\mathbb{P}(X \le x) =: F^X(x)$.

A random variable that can take at most a countable number of possible values is said to be a *discrete* random variable. A discrete random variable X is completely characterized by the *probability mass function* $p^X(x) := \mathbb{P}(X = x)$. Consider a function $g(x)$, for instance $5x^2$ or $\sin(\pi x)$, and we may use notation $g^{-1}(x) := 1/g(x)$ for its reciprocal. Then $g(X)$ is also a discrete random variable, and its *expectation* is defined as

$$\mathbb{E}\{g(X)\} := \sum_{\{x:\ p^X(x)>0\}} g(x)p^X(x). \tag{1.3.3}$$

The expectation is also called the first moment, and in general the *kth moment* of $g(X)$ is defined as $\mathbb{E}\{[g(X)]^k\}$. Assuming that the second moment of $g(X)$ exists (finite), the *variance* of $g(X)$ is defined as

$$\mathbb{V}(g(X)) := \mathbb{E}\{[g(X) - \mathbb{E}\{g(X)\}]^2\} = \mathbb{E}\{[g(X)]^2\} - [\mathbb{E}\{g(X)\}]^2. \tag{1.3.4}$$

The equality on the right side of (1.3.4) is a famous relation between the variance of $g(X)$ and its two first moments.

If $\mathcal{B}$ is a set of numbers, then we may use notation $I(X \in \mathcal{B})$ for the *indicator function* (or simply *indicator*) of the event $X \in \mathcal{B}$ (which reads as "X belongs to set $\mathcal{B}$"), that is by definition $I(X \in \mathcal{B}) = 1$ if the event $X \in \mathcal{B}$ occurs and $I(X \in \mathcal{B}) = 0$ if the event $X \in \mathcal{B}$ fails to occur. Then the following relation between the expectation and the probability holds,

$$\mathbb{P}(X \in \mathcal{B}) = \mathbb{E}\{I(X \in \mathcal{B})\}. \tag{1.3.5}$$

Several classical discrete random variables are used in the book. The *Bernoulli* random variable takes on only two values that are traditionally classified as either a "success" or a "failure." If A is a Bernoulli random variable, then a success is traditionally written as $A = 1$ and a failure is traditionally written as $A = 0$. We may say that the set $\{0, 1\}$ of the two numbers is the *support* of a Bernoulli probability mass function because it is positive only on these two numbers. A Bernoulli random variable is completely described by the probability of success $w := \mathbb{P}(A = 1)$; note that the probability of failure is $1 - w$. Further, we have $\mathbb{E}\{A\} = \mathbb{P}(A = 1) = w$ and $\mathbb{V}(A) = w(1 - w)$.

If we have k mutually independent and identically distributed (iid) Bernoulli random variables $A_1, \ldots, A_k$, then their sum $B := \sum_{l=1}^{k} A_l$ is called a *Binomial* random variable, whose distribution is completely defined by parameters k and w, its mean is kw and the variance is $kw(1 - w)$. The support of the Binomial random variable is the set $\{0, 1, \ldots, k\}$ and on this set the probability mass function is $p^B(x) = [k!/(x!(k - x)!]w^x(1 - w)^{k-x}$.

Another discrete random variable of interest, which takes on nonnegative integer numbers, is called *Poisson*. It is characterized by a single parameter λ which is both the mean and the variance of this random variable. If X is Poisson with parameter λ, we can simply write Poisson(λ). The corresponding probability mass function is $p^X(x) = e^{-\lambda}\lambda^x/x!$, $x \in \{0, 1, 2, \ldots\}$.

A random variable X is called *continuous* if there exists a nonnegative function $f^X(x)$ such that $\int_{-\infty}^{\infty} f^X(x)dx = 1$ and the cumulative distribution function $F^X(x)$ may be written as

$$F^X(x) = \int_{-\infty}^{x} f^X(y)dy. \tag{1.3.6}$$

The function $f^X(x)$ is called the *probability density function* or simply *density*. The *support* of a continuous random variable X is the set of points x where the density $f^X(x)$ is positive. In other words, X may take values only from its support. In what follows the lower and upper bounds of the support of X are denoted as α_X and β_X, respectively. From the definition of a continuous random variable X we get that $\mathbb{P}(X = x) = 0$ for any number x, and this represents the main distinction between continuous and discrete random variables. Both the cumulative distribution function and the density give a complete description of the corresponding continuous random variable. The mean of a function $g(X)$ is defined as

$$\mathbb{E}\{g(X)\} := \int_{-\infty}^{\infty} g(x)f^X(x)dx = \int_{-\infty}^{\infty} g(x)dF^X(x), \tag{1.3.7}$$

and its variance is defined by (1.3.4). Let us also recall that a square root of the variance is called a *standard deviation*, and a standard deviation has the same units as the corresponding random variable.

Let us define several specific continuous random variables. A random variable X is said to be *uniformly* distributed over the interval $[a, a+b]$ if $f^X(x) = b^{-1}I(x \in [a, a+b])$; we may write that X is Uniform($a, a + b$). Note that $[a, a + b]$ is the support of X, $\mathbb{E}\{X\} = a + b/2$ and $\mathbb{V}(X) = b^2/12$. A random variable X is a *normal random variable* with mean μ and variance σ^2 (in short, it is Normal(μ, σ^2)) if its density is

$$f^X(x) = (2\pi\sigma^2)^{1/2}e^{-(x-\mu)^2/(2\sigma^2)} =: d_{\mu,\sigma}(x), \quad -\infty < x < \infty. \tag{1.3.8}$$

The density is unimodal, the support is a real line, its mean, median and mode coincide and equal to μ, the variance is equal to σ^2, the graph of the density is symmetric about μ and bell-shaped, and the density practically vanishes (becomes very small) whenever $|x - \mu| > 3\sigma$, which is the so-called rule of three sigma. A useful property to know is that the sum of two independent normal random variables with parameters (μ_1, σ_1^2) and (μ_2, σ_2^2) is again a normal random variable with parameters $(\mu_1 + \mu_2, \sigma_1^2 + \sigma_2^2)$. Further, a normal random variable is called *standard normal* if its mean is zero and variance is equal to 1. A random variable X is *Laplace* (double exponential) with parameter b if its density is $f^X(x) = (1/2b)e^{-|x|/b}$, $-\infty < x < \infty$. Its mean is zero and variance is $2b^2$.

If we would like to consider two random variables X and Y, then their joint distribution is completely defined by the *joint cumulative distribution function*

$$F^{X,Y}(x,y) := \mathbb{P}((X \le x) \cap (Y \le y)) =: \mathbb{P}(X \le x, Y \le y). \qquad (1.3.9)$$

Note that the right side of (1.3.9) simply introduces a convenient and traditional notation for the probability of the intersection of two events. The joint cumulative distribution function allows us to introduce *marginal* cumulative distribution functions for X and Y,

$$F^X(x) := F^{X,Y}(x,\infty), \quad F^Y(y) := F^{X,Y}(\infty,y). \qquad (1.3.10)$$

Random variables X and Y are *independent* if $F^{X,Y}(x,y) = F^X(x)F^Y(y)$ for all x and y, and otherwise they are *dependent*. All these notions are straightforwardly extended to the case of k variables $X_1, \ldots, X_k$, for instance

$$F^{X_1,\ldots,X_k}(x_1,\ldots,x_k) := \mathbb{P}((X_1 \le x_1) \cap \ldots \cap (X_k \le x_k))$$

$$=: \mathbb{P}(X_1 \le x_1, \ldots, X_k \le x_k). \qquad (1.3.11)$$

Random variables X and Y are *jointly continuous* if there exists a bivariate nonnegative function $f^{X,Y}(x,y)$ on a plane $(-\infty,\infty) \times (-\infty,\infty)$ (*two-dimensional* or *bivariate probability density*) such that

$$F^{X,Y}(x,y) =: \int_{-\infty}^{x} \int_{-\infty}^{y} f^{X,Y}(u,v) dv du. \qquad (1.3.12)$$

Similarly to the marginal cumulative distribution functions (1.3.10), we may define marginal densities of X and Y,

$$f^X(x) := \int_{-\infty}^{\infty} f^{X,Y}(x,u) du \quad \text{and} \quad f^Y(y) := \int_{-\infty}^{\infty} f^{X,Y}(u,y) du. \qquad (1.3.13)$$

Random variables X and Y are independent if $f^{X,Y}(x,y) = f^X(x)f^Y(y)$ for all x and y and otherwise they are dependent. The *conditional probability density* $f^{X|Y}(x|y)$ of X given $Y = y$ is defined from the relation

$$f^{X,Y}(x,y) =: f^Y(y)f^{X|Y}(x|y). \qquad (1.3.14)$$

(If $f^Y(y_0) = 0$, then $f^{X|Y}(x|y_0)$ may be formally defined as a function in x which is nonnegative and integrated to 1.) Given $Y = y$, the conditional density $f^{X|Y}(x|y)$ becomes a regular density and has all its properties. In particular, the *conditional expectation* of $g(X,Y)$ given $Y = y$ is calculated by the formula

$$\mathbb{E}\{g(X,Y)|Y = y\} := \int_{-\infty}^{\infty} g(x,y)f^{X|Y}(x|y) dx. \qquad (1.3.15)$$

Using (1.3.15) we can also introduce a new random variable $\mathbb{E}\{g(X,Y)|Y\}$ which is a function of Y.

Conditional expectation often helps us to calculate probabilities and expectations for bivariate events and functions, namely for a set $\mathbf{B} \in (-\infty,\infty) \times (-\infty,\infty)$ we can write

$$\mathbb{P}((X,Y) \in \mathbf{B}) = \mathbb{E}\{I((X,Y) \in \mathbf{B})\} = \int_{\mathbf{B}} f^Y(y)f^{X|Y}(x|y) dx dy$$

$$= \int_{-\infty}^{\infty} \left[\int_{-\infty}^{\infty} I((x,y) \in \mathbf{B})f^{X|Y}(x|y) dx \right] f^Y(y) dy = \mathbb{E}\{\mathbb{E}\{I((X,Y) \in \mathbf{B})|Y\}\}, \qquad (1.3.16)$$

and similarly for a bivariate function $g(x, y)$,

$$\mathbb{E}\{g(X,Y)\} = \mathbb{E}\{\mathbb{E}\{g(X,Y)|Y\}\}. \tag{1.3.17}$$

In some statistical problems we are dealing with a vector of random variables where some components are continuous and others are discrete. For instance, consider a pair (X, Z) where X is continuous and Z is discrete and takes on nonnegative integer numbers. Then we can introduce a joint *mixed* density $f^{X,Z}(x, z)$ defined on $(-\infty, \infty) \times \{0, 1, \ldots\}$ and such that

$$F^{X,Z}(x, z) =: \sum_{k=0}^{z} \int_{-\infty}^{x} f^{X,Z}(u, k) du. \tag{1.3.18}$$

Then for a bivariate function $g(x, z)$ we have

$$E\{g(X, Z)\} = \sum_{k=0}^{\infty} \int_{-\infty}^{\infty} g(x, k) f^{X,Z}(x, k) dx. \tag{1.3.19}$$

Furthermore, the marginal density and the marginal probability mass function are defined as

$$f^X(x) := \sum_{k=0}^{\infty} f^{X,Z}(x, k) \quad \text{and} \quad p^Z(z) := \int_{-\infty}^{\infty} f^{X,Z}(x, z) dx, \tag{1.3.20}$$

and the conditional density and the conditional probability mass function are

$$f^{X,Z}(x, z) =: p^Z(z) f^{X|Z}(x|z) \quad \text{and} \quad f^{X,Z}(x, z) =: f^X(x) p^{Z|X}(z|x). \tag{1.3.21}$$

To finish the probability part, let us stress that much of the previous material carries over to the case of a k-dimensional *random vector* $(X_1, \ldots, X_n)$.

Now let us recall basic concepts of *parametric statistics* that deal with a *sample* of n independent and identically distributed random variables $X_1, \ldots, X_n$. If X has the same distribution as these variables, then we may say that X_l is the lth realization of X or that it is generated according to the distribution of X. If the realizations (sample values) are placed in ascending order from the smallest to the largest, then they are called *ordered statistics* and denoted by $X_{(1)}, \ldots, X_{(n)}$, where $X_{(1)} \leq X_{(2)} \leq \ldots \leq X_{(n)}$. The main assumption of parametric statistics is that the cumulative distribution function $F_\theta^X(x)$ of X is known up to the parameter $\theta \in \Theta$ where the set Θ is known. Mean of a normal distribution with known variance is a classical example of unknown parameter.

The main aim of parametric statistics is to estimate an unknown parameter θ of the distribution of X based on a sample from X. The parameter may be referred to as the *estimand* because this is the quantity that we would like to estimate. The *estimator* of a parameter θ is a function of a sample and, given values of observations in the sample, the value of the estimator is called the *estimate* (in the literature these two notions are often used interchangeably). Different diacritics above θ such as $\bar{\theta}$, $\hat{\theta}$, $\tilde{\theta}$, etc. are used to denote estimators or *statistics* based on a sample. Similarly, if any functional or function is estimated, then the estimators are denoted by a diacritic above the estimand, say if the density $f^X(x)$ is the estimand then $\hat{f}^X(x)$ denotes its estimator.

The *mean squared error* $MSE(\hat{\theta}, \theta) := \mathbb{E}_\theta\{(\hat{\theta} - \theta)^2\}$ is traditionally used to measure the goodness of estimating θ by an estimator $\hat{\theta}$, and it is one of possible risks. Here $\mathbb{E}_\theta\{\cdot\}$ denotes the expectation according to the cumulative distribution function F_θ^X, and we use the subscript to stress that the underlying distribution is defined by the parameter θ. At the same time, when skipping the subscript does not cause a confusion, it may be skipped.

One of the attractive methods of constructing a parametric estimator is the *sample mean* method. It works as follows. Suppose that there exists a function $g(x)$ such that an estimand may be written as the expectation of $g(X)$, namely

$$\theta = \mathbb{E}_\theta\{g(X)\}, \quad \theta \in \Theta. \tag{1.3.22}$$

Then the *sample mean estimator*, based on a sample of size n from X, is defined as

$$\hat{\theta} := n^{-1} \sum_{l=1}^{n} g(X_l). \tag{1.3.23}$$

If the function $g(x)$ is unknown but may be estimated by some statistic $\tilde{g}(x)$, then we may use a *plug-in* sample mean estimator

$$\tilde{\theta} := n^{-1} \sum_{l=1}^{n} \tilde{g}(X_l). \tag{1.3.24}$$

An estimator $\tilde{\theta}$ is called *unbiased* if $\mathbb{E}_\theta\{\tilde{\theta}\} = \theta$ for all $\theta \in \Theta$, that is on average an unbiased estimator is equal to the estimand. An estimator $\tilde{\theta}_n$ is called *consistent* if for any positive number t

$$\mathbb{P}_\theta(|\tilde{\theta}_n - \theta| > t) = o_n(1), \quad \theta \in \Theta, \tag{1.3.25}$$

where $o_n(1)$ denotes a generic sequence in n such that $\lim_{n\to\infty} o_n(1) = 0$. Consistency is the theoretical property, and it tells us that asymptotically, as sample size increases, a consistent estimator approaches its estimand. Consistency is the minimal property of a feasible estimator that statisticians check.

There are several inequalities in the theory of probability that are useful in the analysis of estimators and in particular sample mean estimators.

Markov's inequality. Let X be a nonnegative random variable and $\mathbb{E}\{X\}$ exists. Then for any positive t,

$$\mathbb{P}(X \geq t) \leq \frac{\mathbb{E}\{X\}}{t}. \tag{1.3.26}$$

Chebyshev's inequality. Let the variance $\mathbb{V}(X)$ exist. Then for any positive t,

$$\mathbb{P}(|X - \mathbb{E}\{X\}| \geq t) \leq \frac{\mathbb{V}(X)}{t^2}. \tag{1.3.27}$$

A classical application of the Chebyshev inequality is to establish consistency of a sample mean estimator (1.3.23). Indeed, if the variance of $g(X)$ is finite, then $\mathbb{V}(\hat{\theta}) = n^{-1}\mathbb{V}(g(X))$, and the Chebyshev inequality yields consistency of the sample mean estimator.

Generalized Chebyshev's inequality. Consider an increasing and positive function $g(x)$, $x \in [0, \infty)$. Suppose that $\mathbb{E}\{g(|X|)\}$ exists. Then for any positive t,

$$\mathbb{P}(|X| \geq t) \leq \frac{\mathbb{E}\{g(|X|)\}}{g(t)}. \tag{1.3.28}$$

Hoeffding's inequality. Consider a sample $X_1, \ldots, X_n$ from X such that there exist two finite numbers a and b such that $\mathbb{P}(a \leq X \leq b) = 1$. Denote by $\bar{X} := n^{-1}\sum_{l=1}^{n} X_l$ the sample mean estimator of $\theta := \mathbb{E}\{X\}$. Then, for any positive t,

$$\mathbb{P}(\bar{X} - \theta \geq t) \leq e^{-2nt^2/(b-a)^2}. \tag{1.3.29}$$

Hoeffding's inequality for Bernoulli(w). Consider a sample $X_1, \ldots, X_n$ from X which is Bernoulli(w). Then for any positive t,

$$\mathbb{P}(\bar{X} - w \geq t) \leq e^{-2nt^2}. \tag{1.3.30}$$

Bernstein's inequality. Let $X_1, \ldots, X_n$ be independent zero-mean random variables. Suppose that $\mathbb{P}(|X_l| > M) = 0$ for $l = 1, 2, \ldots, n$. Then for any positive t,

$$\mathbb{P}(\sum_{l=1}^{n} X_l > t) \leq \exp\left(-\frac{t^2/2}{\sum_{l=1}^{n} \mathbb{E}\{X_l^2\} + Mt/3}\right). \qquad (1.3.31)$$

Two more inequalities are often used in statistical analysis.

Cauchy–Schwarz inequality. Consider a pair (X, Y) of random variables with finite second moments. Then

$$\mathbb{E}\{|XY|\} \leq [\mathbb{E}\{X^2\}]^{1/2}[\mathbb{E}\{Y^2\}]^{1/2}. \qquad (1.3.32)$$

The following inequality for two square-integrable functions $g_1(x)$ and $g_2(x)$,

$$\int_{-\infty}^{\infty} |g_1(x)g_2(x)|dx \leq [\int_{-\infty}^{\infty} g_1^2(x)dx]^{1/2}[\int_{-\infty}^{\infty} g_2^2(x)dx]^{1/2}, \qquad (1.3.33)$$

is another useful version of Cauchy-Schwarz inequality.

Jensen's inequality. If a function g is convex, then

$$\mathbb{E}\{g(X)\} \geq g(\mathbb{E}\{X\}). \qquad (1.3.34)$$

If g is concave, then

$$\mathbb{E}\{g(X)\} \leq g(\mathbb{E}\{X\}). \qquad (1.3.35)$$

In nonparametric statistics, specifically in nonparametric curve estimation, the estimand (recall that the estimand is what we want to estimate) is not a parameter but a function (curve). For instance, the probability density $f^X(x)$ may be the estimand. Similarly, let $Y := m(X) + \varepsilon$ where X and ε are independent with known distributions. Then $m(x)$ may be the estimand. Our main nonparametric estimation approach is based on an orthonormal series approximation. As an example, suppose that X is supported on $[0, 1]$. Introduce a classical *cosine orthonormal basis* $\varphi_0(x) := 1$, $\varphi_j(x) := 2^{1/2} \cos(\pi j x)$, $j = 1, 2, \ldots$ on $[0, 1]$. Note that $\int_0^1 \varphi_j(x)\varphi_i(x)dx = I(j = i)$. Then any square integrable on $[0, 1]$ function $m(x)$, that is a function satisfying $\int_0^1 [m(x)]^2 dx < \infty$, can be written as a *Fourier series*

$$m(x) = \sum_{j=0}^{\infty} \theta_j \varphi_j(x). \qquad (1.3.36)$$

In (1.3.36) parameters

$$\theta_j := \int_0^1 m(x)\varphi_j(x)dx \qquad (1.3.37)$$

are called *Fourier coefficients* of function $m(x)$. Note that knowing $m(x)$ is equivalent to knowing the infinite number of Fourier coefficients, and this fact explains the notion of nonparametric curve estimation. Traditional risk, used for evaluation quality of estimating $m(x)$, $x \in [0, 1]$ by an estimator $\tilde{m}(x)$, is the MISE *(mean integrated squared error)* which is defined as

$$\text{MISE}(\tilde{m}, m) := \mathbb{E}_m\{\int_0^1 [\tilde{m}(x) - m(x)]^2 dx\}. \qquad (1.3.38)$$

Similarly to the parametric statistics, an estimator $\tilde{m}(x)$, $x \in [0, 1]$ is called consistent if for any positive t, $x \in [0, 1]$ and underlying function $m(x)$ the relation $\mathbb{P}_m(|\tilde{m}(x) - m(x)| > t) = o_n(1)$ holds. In what follows, whenever no confusion occurs, we may skip the subscript m and write $\mathbb{P}(\cdot)$ and $\mathbb{E}\{\cdot\}$ in place of $\mathbb{P}_m(\cdot)$ and $\mathbb{E}_m\{\cdot\}$, respectively.

A convenient tool for the analysis of the integral of a squared function (1.3.36) or the integral of the product of two square-integrable functions is the *Parseval identity*. Consider the Fourier series $g(x) = \sum_{j=0}^{\infty} \kappa_j \varphi_j(x)$ of a square-integrable on $[0,1]$ function $g(x)$. The Parseval identity is the following relation between the integral of the product $m(x)g(x)$ of two functions and the sum of products of their Fourier coefficients,

$$\int_0^1 m(x)g(x)dx = \sum_{j=0}^{\infty} \theta_j \kappa_j. \tag{1.3.39}$$

There are two important corollaries from (1.3.39). The former is that the integrated squared function can be expressed via the sum of its squared Fourier coefficients, namely the Parseval identity implies that

$$\int_0^1 [m(x)]^2 dx = \sum_{j=0}^{\infty} \theta_j^2, \tag{1.3.40}$$

and note that (1.3.40) is also traditionally referred to as the Parseval identity. The latter is that the Parseval identity implies the following relation for the MISE,

$$\mathbb{E}\Big\{ \int_0^1 [\tilde{m}(x) - m(x)]^2 dx \Big\} = \sum_{j=0}^{\infty} \mathbb{E}\{(\tilde{\theta}_j - \theta_j)^2\}, \tag{1.3.41}$$

where $\tilde{\theta}_j := \int_0^1 \tilde{m}(x)\varphi_j(x)dx$ are Fourier coefficients of the estimator $\tilde{m}(x)$. Relation (1.3.41) is the foundation of the theory of *series* estimation when a function is estimated via its Fourier coefficients.

Finally, let us explain notions and terminology used in dealing with randomly missing data. Consider an example of regression when a hidden underlying sample (H-sample) $(X_1, Y_1), \ldots, (X_n, Y_n)$ from (X, Y) cannot be observed, and instead a sample with missing data (M-sample) $(X_1, A_1 Y_1, A_1), \ldots, (X_n, A_n Y_n, A_n)$ from (X, AY, A) is observed. Here X is the predictor, Y is the response, A is the *availability* which is a Bernoulli random variable such that

$$\mathbb{P}(A = 1 | X = x, Y = y) =: w(x, y), \tag{1.3.42}$$

and the function $w(x, y)$ is called the *availability likelihood*. If $A_l = 1$, then we observe (X_l, Y_l) and the case (realization, observation) is called *complete*, and otherwise if $A_l = 0$, then we observe $(X_l, 0)$ (or in R language this will be written as (X_l, NA) where the R string "NA" stands for *Not Available*) and hence the case (realization, observation) is called *incomplete*.

Depending on the availability likelihood, three basic types of random missing mechanisms are defined. (i) *Missing completely at random* (MCAR) when missing a variable occurs by chance that does not depend on underlying variables. In our particular case (1.3.42), this means that $w(x, y) = w$, that is the availability likelihood is constant. (ii) *Missing at random* (MAR) when missing a variable occurs by chance depending only on other always observed variables. In our example this means that $w(x, y) = w(x)$. Note that MCAR is a particular case of MAR. (iii) *Missing not at random* (MNAR) when missing a variable occurs by chance which depends on its value and may also depend on other variables. In our example this means that the availability likelihood $w(x, y)$ depends on y and may also depend on x. Let us make several comments about MNAR because the notion may be confusing. First, MNAR simply serves as a complement to MAR, that is if a missing is not MAR, then it is called MNAR. Second, the term MNAR means that the probability of missing may be defined by factors that are simply unknown. Further, in no way MNAR means that an underlying missing mechanism is not stochastic, and in the book we are considering only random missing mechanisms.

We will also use several inequalities for sums. The first one is the analog of (1.3.33) and it is also called the Cauchy–Schwarz inequality,

$$|\sum_{l=1}^{n} u_l v_l|^2 \le \sum_{l=1}^{n} u_l^2 \sum_{l=1}^{n} v_l^2. \tag{1.3.43}$$

Here $u_1,\ldots,u_n$ and $v_1,\ldots,v_n$ are numbers. The Cauchy inequality

$$(u+v)^2 \le (1+\gamma)u^2 + (1+\gamma^{-1})v^2, \quad \gamma > 0 \tag{1.3.44}$$

is another useful inequality, and here $\gamma^{-1} := 1/\gamma$. The Minkowski inequality for numbers is

$$\Big[\sum_{k=1}^{n} |u_k + v_k|^r\Big]^{1/r} \le \Big[\sum_{k=1}^{n} |u_k|^r\Big]^{1/r} + \Big[\sum_{k=1}^{n} |v_k|^r\Big]^{1/r}, \quad r \ge 1. \tag{1.3.45}$$

There is also a useful Minkowski inequality for the sum of two random variables,

$$\Big[\mathbb{E}\{|X+Y|^r\}\Big]^{1/r} \le \Big[\mathbb{E}\{|X|^r\}\Big]^{1/r} + \Big[\mathbb{E}\{|Y|^r\}\Big]^{1/r}, \quad r \ge 1. \tag{1.3.46}$$

The inequality for the sum of numbers

$$|\sum_{k=1}^{n} u_k|^r \le n^{r-1} \sum_{k=1}^{n} |u_k|^r, \quad r \ge 1 \tag{1.3.47}$$

implies the following inequality for not necessarily independent random variables $X_1,\ldots,X_n$,

$$\mathbb{E}\{|\sum_{k=1}^{n} X_k|^r\} \le n^{r-1} \sum_{k=1}^{n} \mathbb{E}\{|X_k|^r\}, \quad r \ge 1. \tag{1.3.48}$$

If additionally the random variables $X_1,\ldots,X_n$ are independent and zero mean, that is $\mathbb{E}\{X_k\} = 0$ for $k = 1,\ldots,n$, then

$$\mathbb{E}\{|\sum_{k=1}^{n} X_k|^r\} \le C(r)\Big[\sum_{k=1}^{n} \mathbb{E}\{|X_k|^r\} + \Big(\sum_{k=1}^{n} \mathbb{E}\{X_k^2\}\Big)^{r/2}\Big], \quad r \ge 2, \tag{1.3.49}$$

and

$$\mathbb{E}\{|\sum_{k=1}^{n} X_k|^r\} \le C(r)n^{\frac{r}{2}-1} \sum_{k=1}^{n} \mathbb{E}\{|X_k|^r\}, \quad r \ge 2, \tag{1.3.50}$$

where $C(r)$ is a positive and finite function in r.

There are four corner (test) functions that are used throughout the book. They are the *Uniform*, the *Normal*, the *Bimodal* and the *Strata*. These functions are defined in Section 2.1 and Figure 2.1 exhibits them.

1.4 Software

R is a commonly used free statistics software. R allows you to carry out statistical analyses in an interactive mode, as well as allowing simple programming. You can find detailed installation instructions in the R Installation and Administration manual on CRAN (www.r-project.org, http://cran.r-project.org). After installing R (or if it was already installed), you need to choose your working directory for the book's software package. For instance, for

MAC you can use "/Users/Me/Book2". Download to this directory file book2.r from the author's web site www.utdallas.edu/~efrom. The web site also contains relevant information about the book. By downloading the package, the user agrees to consider it as a "black-box" and employ it for educational purposes only.

Now you need to start the R on your computer. This should bring up a new window, which is the R console. In the R console you will see:

>

This is the R prompt. Now, only once while you are working with the book, you need to install several standard R packages. Type the following command in the R console to install required packages

> install.packages("MASS")
> install.packages("mvtnorm")
> install.packages("survival")
> install.packages("scatterplot3d")

These packages are installed just once, you do not need to repeat this step when you start your next R session.

Next you need to source (install) the book's package, and you do this with R operating in the chosen working directory. To do this, first type

> getwd()

This R command will allow you to see a working directory in which R is currently operating. If it is not your chosen working directory "/Users/Me/Book2", type

> setwd("/Users/Me/Book2")

and then R will operate in the chosen directory for the book. Then type

> source("book2.r")

and the book's R software will be installed in the chosen directory and you are ready to use it. This sourcing should be done every time when you start a new R session.

For the novice, it is important to stress that no knowledge of R is needed to use the package and repeat/modify figures. Nonetheless, a bit of information may be useful to understand the semantics of calling a figure. What you are typing is the name of an R function and then in its parentheses you assign values to its arguments. For instance,

> ch1(fig=1, n=300)

is the call to R function ch1 that will be run with arguments fig=1 and n=300, and all other arguments of the function will be equal to their default values (here a=0 and b=0.7) indicated in the caption of the corresponding Figure 1.1. Note that ch1 indicates that the figure is from Chapter 1 while its argument fig=1 indicates that it is the first figure in Chapter 1. In other words, ch1(fig=1,n=300) runs Figure 1.1 with the sample size $n = 300$. In R, arguments of a function may be either scalars (for instance, a=5 implies that a function will use value 5 for argument a), or vectors (for instance, vec = c(2,4) implies that a function will use vector vec with the first element equal to 2 and the second element equal to 4, and note that c() is a special R function called "combine" that creates vectors), or a string (say den= "normal" implies that a function will use name *normal* for its argument den). R functions are smart in terms of using shorter names for its arguments, for instance both fig=5, fi=5 and f=5 will be correctly recognized and imply the same value 5 for the argument fig unless there is a confusion with other arguments.

In general, to repeat Figure k.j, which is jth figure in Chapter k, and to use the default values of its arguments outlined in square brackets of the caption of Figure k.j, type

> chk(f=j)

If you want to use new values for two arguments, say n=130 and a=5, type

> chk(f=j, n=130, a=5)

Note that values of other arguments will be equal to default ones. Further, recall that typing f=j and fig=j implies the same outcome.

To finish the R session, type

> q()

Finally a short comment about useful captions and figures. Figure 2.1 exhibits four corner (test) functions: the Uniform, the Normal, the Bimodal and the Strata. Figure 2.2 shows first eight elements of the cosine basis. The caption to Figure 2.3 explains how to create a custom-made corner function. Figure 2.4 shows the sequence of curves and their colors on the monitor. The caption to Figure 2.5 explains arguments used by the E-estimator.

1.5 Inside the Book

There are always two parts in nonparametric estimation based on missing and modified data. The first one is how to construct a feasible nonparametric estimator for the case of direct data, and the second is how to address the issue of missing and modified data. The book takes the following approach used for all considered problems. The same orthogonal series estimator, called an E-estimator and defined in Section 2.2 for the case of direct data, is also used without any "nonparametric" modification for indirectly observed data. The attractiveness of this approach is that the book is less about how to estimate a nonpara- metric function (curve) and more about how the E-estimator may be used in a variety of settings with missing and modified data. This methodology is supported by the asymp- totic theory which states that for considered settings no other estimator can outperform the E-estimation methodology. As a result, other known methods for dealing with missing and modified data are not considered, but a relevant literature can be found at the end of each chapter in the section Notes. Further, the interested reader can find a book-length treatment of nonparametric series estimators for direct data as well as a discussion of other popular nonparametric estimators in the author's book Efromovich (1999a), *Nonparametric Curve Estimation: Methods, Theory and Applications*.

Let us explain why and how a single (and we may say universal) E-estimator may be used for the totality of considered problems. Suppose that our aim is to estimate a bounded function $m(x)$ over an interval $[0,1]$. Then, according to (1.3.36) and (1.3.37) it can be written as a Fourier series

$$m(x) = \sum_{j=0}^{\infty} \theta_j \varphi_j(x), \tag{1.5.1}$$

where $\varphi_j(x)$ are elements of the cosine basis and

$$\theta_j := \int_0^1 m(x)\varphi_j(x)dx \tag{1.5.2}$$

are Fourier coefficients of $m(x)$.

The main step of E-estimation is to propose a sample mean estimator $\hat{\theta}_j$ of Fourier coefficients. If this step is done, then there is a straightforward and universal (it depends on neither the estimand, nor the setting, nor the data) procedure of plugging $\hat{\theta}_j$ in (1.5.1). This procedure, called the E-estimator, is defined in Section 2.2. As a result, a more compli- cated nonparametric problem is converted into a simpler parametric problem of proposing a sample mean estimator for Fourier coefficients.

Let us present several examples that shed light on construction of sample mean esti- mators. Consider the problem of estimation of the probability density $f^X(x)$ of a random variable X supported on $[0,1]$. Suppose that a sample $X_1,\ldots,X_n$ from X is available. Then the sample mean estimator of Fourier coefficient $\theta_j := \int_0^1 f^X(x)\varphi_j(x)dx = \mathbb{E}\{\varphi_j(X)\}$ is

$$\hat{\theta}_j := n^{-1}\sum_{l=1}^{n} \varphi_j(X_l). \tag{1.5.3}$$

If some observations of X are missed with a known positive availability likelihood $w(x)$ and we observe a sample $(A_1X_1, A_1), \ldots, (A_nX_n, A_n)$ from (AX, A) where the availability A is Bernoulli and $\mathbb{P}(A = 1|X = x) = w(x)$, then the sample mean Fourier estimator is

$$\hat{\theta}_j := n^{-1} \sum_{l=1}^{n} \frac{A_l \varphi_j(A_l X_l)}{w(A_l X_l)}. \tag{1.5.4}$$

Finally, if the availability likelihood is unknown but may be estimated, then its estimator is plugged in (1.5.4). The same approach is used for all other problems considered in the book, and then the main statistical issue to explore is how to construct a sample mean estimator for Fourier coefficients.

Now let us briefly review the context of chapters and begin with a general remark. Each chapter contains exercises (more difficult ones highlighted by the asterisk) and it ends with section Notes that contain short historical reviews, useful references for further reading, and discussion of relevant results.

Chapter 2 presents a brief review of basic nonparametric estimation problems with direct data. The reader familiar with the topic, and specifically with the book Efromovich (1999a), may skip this chapter with the exception of Sections 2.2 and 2.6. Section 2.2 defines the E-estimator and its parameters that are used in all figures. Let us stress that proposed E-estimator may be improved a bit for specific problems, and Efromovich (1999a) explains how to do that. On the other hand, the E-estimator is simple, reliable and it performs well for all studied settings. Further, Section 2.6 describes confidence bands used in the book. A novice to nonparametric curve estimation is advised to become familiar with Sections 2.1, 2.3 and 2.4, get used to the book's software and do the exercises because this is a necessary preparation for next chapters.

Chapter 3 serves as a bridge between settings with direct data and missing and modified data. Sections 3.1, 3.2 and 3.3 are a must read because they discuss pivotal problems of density and regression estimation based on biased data. Another section to study is Section 3.7 which explores the possibility of estimating Bernoulli regression with unavailable failures. As we will see shortly, this problem is the key for understanding a proposed solution for a number of applied regression problems with missing data. All other sections explore special topics that often occur in applications.

Chapters 4 and 5 are devoted solely to nonparametric estimation with missing data. Chapter 4 explores cases of nondestructive missing when consistent estimation of an estimand is possible solely on missing data. Let us stress that while all proposed procedures are explained for the case of small sample sizes and illustrated via simulated examples, they are supported by asymptotic theory which indicates optimality of proposed solutions among all possible statistical estimators. Section 4.1 is devoted to the classical topic of density estimation. Regression problems with missing responses and predictors are considered in separate Sections 4.2 and 4.3 because proposed solutions are different. For the case of missing responses, the proposed estimator is based on complete cases and it ignores incomplete cases. This is an interesting proposition because there is a rich literature on how to use incomplete cases for "better" estimation. For the case of missing predictors, a special estimation procedure is proposed that takes into account incomplete cases. At the same time, it is also explained what may be done if incomplete cases are not available. All remaining sections are devoted to special topics that expand a class of possible practical applications.

Chapter 5 is devoted to cases of destructive missing. As it follows from the definition, here we explore problems when knowing missing data alone is not enough for consistent estimation. Hence, the main emphasis is on a minimal additional information sufficient for consistent estimation. As a result, the first three sections are devoted to density estimation with different types of extra information. Here all issues, related to destructive missing, are explained in great details. Other sections are devoted to regression problems with destructive missing.

Chapter 6 is devoted to nonparametric curve estimation for truncated and censored data. The reader familiar with survival analysis knows that product-limit estimators, with Kaplan–Meier estimator being the most popular one, are the main tool for solution practically all nonparametric problems related to truncation and/or censoring. While being a powerful and comprehensive methodology, it is complicated for understanding and making an inference about. To overcome this complexity and to replace the product-limit methodology by E-estimation, we need to begin with a new (but very familiar in survival analysis) problem of estimating the hazard rate. The hazard rate, similarly to the probability density and cumulative distribution function, presents a complete description of a random variable. Further, it is of a great practical interest on its own even for the case of direct observations. The reason why we begin with the hazard rate is that E-estimation is a perfect fit for this problem and because it is plain to propose a reliable sample mean estimator of its Fourier coefficients for the case of direct observations. Further, the E-estimator naturally extends to the case of truncated and censored data. As a result of this approach, estimation of the hazard rate is considered in the first 4 sections. Then Sections 6.5-6.9 are devoted to estimation of distributions and regressions.

What will be if some of truncated and/or censored observations are missed? This is the topic of Chapter 7 where a large number of possible models are considered. The overall conclusion is that the E-estimator again may be used to solve these complicated problems.

So far all considered problems were devoted to classical samples with independent observations. What will be if observations are dependent? This is the topic of the science called the theory of time series. Chapter 8 explores a number of stationary time series problems with missing data, both rich in history and new ones. Section 8.1 contains a brief review of the theory of classical time series. One of the classical tools for the analysis of stationary time series is spectral analysis. It is discussed in Section 8.2. Several models of missing data in time series are discussed in Sections 8.3 and 8.4. Time series with censored observations are discussed in Section 8.5. More special topics are discussed in other sections.

Chapter 9 continues the discussion of time series, only here the primary interest is in analysis of not stationary (changing in time) time series and missing mechanisms. Also, Section 9.2 introduces the reader into the world of continuous in time stochastic processes. Special topics include the Simpson paradox and sequential design.

The last Chapter 10 considers more complicated cases of modified data that slow down the rate of the MISE convergence. This phenomenon is called ill-posedness. As we will see, measurement errors in data typically imply ill-posedness, and this explains practical importance of the topic. Further, some types of censoring, for instance the current status censoring, may make the problem (data) ill-posed. In some cases data do not correctly fit an estimand, and this also causes ill-posedness. For instance, in a regression setting we have a data that fits the problem of estimating the regression function. At the same time, the regression data do not fit the problem of estimating the derivative of the regression function, and the latter is another example of ill-posed problem (data). Another example is a sample from a continuous random variable X. This sample perfectly fits the problem of estimating the cumulative distribution function $F^X(x)$, which can be estimated with the parametric rate n^{-1} of the mean squared error (MSE) convergence. At the same time, this rate slows down if we want to estimate the probability density $f^X(x)$. The rate slows even more if we want to estimate derivative of the density. As a result, we get a ladder of ill-posed problems, and this and other related issues are considered in the chapter. The main conclusion of Chapter 10 is that while ill-posedness is a serious issue, for small samples the proposed E-estimators may be reasonable for relatively simple shapes of underlying curves. On the other hand, there is no chance to visualize curves with more nuanced shapes like ones with closely located and similar in magnitude modes. This is where the R package will help the reader to gain practical experience in dealing with these complicated statistical problems and understanding what can and cannot be done for ill-posed problems.

1.6 Exercises

Here and in all other chapters, the asterisk denotes a more difficult exercise.

1.1.1 What is the difference between direct and biased observations in Figure 1.1?

1.1.2* How are direct and biased samples in Figure 1.1 generated? Suggest a probability model for observations generated by a continuous random variable.

1.1.3 In Figure 1.1, is the biased sample skewed to the right or to the left with respect to the direct sample? Why?

1.1.4 Repeat Figure 1.1 with arguments that will imply a biased sample skewed to the left.

1.1.5 Repeat Figure 1.1 with sample sizes $n \in \{25, 50, 100, 200\}$. For what sample size does the histogram better exhibit shape of the underlying Normal density?

1.1.6 What is a histogram? Is it a nonparametric or parametric estimator?

1.1.7 Do you think that histograms in Figure 1.1 are over-smoothed or under-smoothed? If you could change the number of bins, would you increase or decrease the number?

1.1.8 Can the generated by Figure 1.1 biased sample be considered as a hidden sample of direct observations with missing values?

1.1.9 Explain how the biased sample in Figure 1.1 can be generated using the availability random variable A. Is the availability a Bernoulli random variable?

1.1.10 What is the definition of the availability likelihood function?

1.1.11* Is it possible to estimate the density of a hidden variable based on its biased observations? If the answer is negative, propose a feasible solution based on additional information.

1.1.12 Consider the data exhibited in Figure 1.2. Can the left tail of the underlying density $f^{X^*}(x)$ be estimated?

1.1.13 Define a deductible in an insurance policy. Does it increase or decrease payments for insurable losses?

1.1.14 Can we say that a deductible causes left truncation of the payment for insurable losses?

1.1.15* If X is the truncated by deductible D loss X^*, then what is the formula for $F^X(x)$? What is the expected payment? Hint: Assume that the distributions of D and X^* are known.

1.1.16 Suppose that you got data about payments for occurred losses from an insurance company. Can you realize from the data that they are truncated?

1.1.17 What is the definition of a limit in insurance policy? Does the limit increase or decrease a payment for insurable losses?

1.1.18 Give an example of right censored data.

1.1.19* Suppose that $(V_1, \Delta_1), \ldots, (V_n, \Delta_n)$ is a sample of censored variables. How are the joint distribution of V and Δ related to the joint distribution of an underlying (hidden) variable of interest X and a censoring variable L? Find the expectation of V. Hint: Assume that the variables are continuous and the joint density is known.

1.1.20 What is the probability that a variable X will be censored by a variable L?

1.1.21 Explain how data in Figure 1.3 are simulated.

1.1.22 In Figure 1.3 the truncated losses are skewed to the right. Why? Is it always the case for left truncated data?

1.1.23 What is shown in the bottom diagram in Figure 1.3?

1.1.24 Look at the bottom diagram in Figure 1.3. Why are there no losses with values greater than 0.85? Is it a typical or atypical outcome? Explain, and then repeat Figure 1.3 and check your conclusion.

1.1.25* The numbers N and M of available hidden losses after truncation or censoring, shown in Figure 1.3, are only about 50% of the size $n = 400$ of hidden losses. Why are the numbers so small? How can one change the simulation to increase these numbers?

1.2.1* Verify the assertion that $m(x)$, defined in (1.2.1), minimizes the conditional mean squared error $\mathbb{E}\{(Y - \mu(x))^2 | X = x\}$ among all possible predictors $\mu(x)$. Hint: Begin with the proof that $\mathbb{E}\{(X - \mathbb{E}\{X\})^2\} \leq \mathbb{E}\{(X - c)^2\}$ for any constant c.

1.2.2 What is the difference, if any, between parametric and nonparametric regressions?

1.2.3 What is a linear regression? How many parameters do you need to specify for a linear regression?

1.2.4 Are nonparametric regressions, shown in Figure 1.4, justified by data?

1.2.5* Explain components of the time series decomposition (1.2.2). Then propose a method for their estimation. Hint: See Chapters 8 and 9.

1.2.6 Do you think that responses in the top diagram in Figure 1.4 are dependent or independent?

1.2.7 Let Y be the indicator of an insurable event. Suppose that Y is a Bernoulli random variable with $\mathbb{P}(Y = 1) = w$. Verify that $\mathbb{E}\{Y\} = \mathbb{P}(Y = 1)$.

1.2.8 In Figure 1.5 the number of insurance claims is generated by a Poussin random variable. Do you think that this is a reasonable distribution for a number of claims?

1.2.9 Explain the underlying idea behind assumption (1.2.3).

1.2.10 Compare the assumption (1.2.3) with the assumption when the left side of (1.2.3) is equal to $w(x)$. What is the difference between these missing mechanisms? Which one creates more complications?

1.2.11 Explain simulations that created Figure 1.6.

1.2.12* What changes may be expected in Figure 1.6 if one chooses negative argument d ?

1.2.13 What can the bottom diagram in Figure 1.6 be used for?

1.2.14 Repeat Figure 1.6 with smaller sample sizes. Does this affect visualization of underlying regression functions?

1.3.1 Give definitions of the cumulative distribution function and the survival function. Which one is decreasing and which one increasing?

1.3.2 Verify the equality in (1.3.4). Do you need any assumption for its validity?

1.3.3 Prove that $\mathbb{V}(X) \leq \mathbb{E}\{X^2\}$. When do we have the equality?

1.3.4 Is $\mathbb{E}\{X^2\}$ smaller than $[E\{X\}]^2$, or vice versa, or impossible to say?

1.3.5 Verify (1.3.5).

1.3.6 Consider two independent Bernoulli random variables with probabilities of success being w_1 and w_2. What are the mean and variance of their difference?

1.3.7* Give the definition of a Poisson(λ) random variable. What are its mean, second moment, variance, and third moment?

1.3.8 What is the definition of a continuous random variable? Is it defined by the probability density?

1.3.9 Suppose that function $f(x)$ is a bona fide probability density. What are its properties?

1.3.10 Verify the equality in (1.3.7).

1.3.11 What is the relationship between variance and standard deviation?

1.3.12 Consider a pair (X, Y) of continuous random variables with the joint cumulative distribution function $F^{X,Y}(x, y)$. What is the marginal cumulative distribution function of Y? What is the probability that $X = Y$? What is the probability that $X < Y$?

1.3.13 It is given that $F^{X,Y}(1, 2) = 0.06$, $F^X(1) = 0.3$, $F^Y(1) = 0.1$ and $F^Y(2) = 0.2$. Are the random variables X and Y independent, dependent or not enough information to make a conclusion?

1.3.14 Suppose that $f^Y(y)$ is positive. Define the conditional probability density $f^{X|Y}(x|y)$ via the joint density of (X, Y) and the marginal density of Y.

1.3.15 Verify (1.3.16).

1.3.16 Verify (1.3.17).

1.3.17 Explain the notion of mixed probability density.

1.3.18 Assume that Z is Bernoulli(w), and given $Z = z$ the random variable X has normal distribution with mean z and unit variance. What is the formula for the joint cumulative distribution function of (X, Z)?

1.3.19 What is the definition of a sample of size n from a random variable X?

1.3.20 What is the difference (if any) between an estimand and an estimator?

1.3.21 What does the abbreviation MSE stand for?

1.3.22 What is the definition of an unbiased estimator?

1.3.23 Consider a sample mean estimator $\hat{\theta} := n^{-1} \sum_{l=1}^{n} g(X_l)$ proposed for a sample from X. It is known that this estimator is unbiased. Suppose that X is distributed according to the density $f_\theta^X(x)$. What is the estimand for $\hat{\theta}$?

1.3.24* Consider a sample mean estimator (1.3.23). What is its mean and variance? Then solve the same problem for the plug-in estimator (1.3.24). Hint: Make convenient assumptions.

1.3.25 Present a condition under which a sample mean estimator is consistent.

1.3.26 Prove the generalized Chebyshev inequality. Hint: Begin with the case of a continuous random variable, write

$$\mathbb{P}(|X| \geq t) = \int_{|x| \geq t} f^X(x)dx = \int_{-\infty}^{\infty} I(g(|x|) \geq g(t))f^X(x)dx,$$

and then continue with the aim to establish (1.3.28).

1.3.27 Prove (1.3.30) using (1.3.29).

1.3.28 Prove (1.3.32) using (1.3.33).

1.3.29 What does the abbreviation MISE stand for?

1.3.30 Formulate the Parseval identity.

1.3.31 Let $\varphi_j(x)$ be elements of the cosine basis defined in the paragraph above line (1.3.36). Show that $\int_0^1 \varphi_i(x)\varphi_j(x)dx = I(i = j)$.

1.3.32* Prove the Cauchy-Schwarz inequality (1.3.43).

1.7   Notes

This book is a natural continuation of the book Efromovich (1999a) where the case of direct data is considered and a number of possible series estimators, as well as other nonparametric estimators like spline, kernel, nearest neighbor, etc., are discussed. A relatively simple, brief and introductory level discussion of nonparametric curve estimation can be found in Wasserman (2006). Mathematically more rigorous statistical theory of series estimation can be found in Tsybakov (2009) and Johnstone (2017).

There is a large choice of good books devoted to missing data where the interested reader may find the theory and practical recommendations. Let us mention Allison (2002), Little and Rubin (2002), Longford (2005), Tsiatis (2006), Molenberghs and Kenward (2007), Tan, Tian and Ng (2009), Enders (2010), Graham (2012), van Buuren (2012), Bouza-Herrera (2013), Berglund and Heeringa (2014), Molenberghs et al. (2014), O'Kelly and Ratitch (2014), Zhou, Zhou and Ding (2014), and Raghunathan (2016). These books discuss a large number of statistical methods and methodologies of dealing with missing data with emphasis on parametric and semiparametric models. Imputation and multiple imputation are the hottest topics in the literature, and some procedures are proved to be optimal. On the other hand, for nonparametric models it is theoretically established that no estimator can outperform the E-estimation methodology. In other words, any other estimator may at best be on par with an E-estimator.

There are many good sources to read about survival analysis. The literature is primarily devoted to the limit-product methodology and practical aspects of survival analysis in different sciences. Let us mention books by Miller (1981), Cox and Oakes (1984), Kalbfleisch and Prentice (2002), Klein and Moeschberger (2003), Gill (2006), Martinussen and Scheike (2006), Aalen, Borgan and Gjessing (2008), Hosmer, Lemeshow and May (2008), Allison (2010, 2014), Guo (2010), Mills (2011), Royston and Lambert (2011), van Houwelingen and Putter (2011), Chen, Sun and Peace (2012), Crowder (2012), Kleinbaum and Klein (2012), Liu (2012), Klein et al. (2014), Lee and Wang (2013), Li and Ma (2013), Collett (2014), Zhou (2015), and Moore (2016).

A nice collection of results for the sum of random variables can be found in Petrov (1975).

Finally, let us mention the classical mathematical book Tikhonov (1998) and more recent Kabanikhin (2011) on ill-posed problems. Nonparametric statistical analysis of ill-posed problems may be found in Meister (2009) and Groeneboom and Jongbloed (2014).

1.2 In Figure 1.4, the "Monthly Housing Starts" dataset is discussed in Efromovich (1999a), the "Auto Insurance Claims" dataset is discussed in Efromovich (2016c), and the "Wages" dataset is discussed in the R **np** package.

Chapter 2

Estimation for Directly Observed Data

This chapter presents pivotal results for the classical case of directly observed data. It is explicitly assumed that observations are neither modified nor missing. All presented results will serve as a foundation for more complicated cases considered in other chapters. The chapter is intended to: (i) overview basics of orthonormal series approximation; (ii) present a universal method of orthonormal series estimation of nonparametric curves which is used throughout the book; and (iii) explain adaptive estimation of the probability density and regression function for the case of complete data. Section 2.1 considers a cosine series approximation which is used throughout the book. It also reminds the reader how to use the book's R package and how to repeat and modify graphics. Section 2.2 explains the problem of nonparametric density estimation. Here the nonparametric E-estimator, which will be used for all considered in the book problems, is introduced and explained. Section 2.3 is devoted to the classical problem of nonparametric regression estimation. It is explained how the E-estimator, proposed for the density model, can be used for the regression model. As a result, even if the reader is familiar with regression problems, it is worthwhile to read this section and understand the underlying idea of E-estimation methodology.

In many applied settings with modified and/or missing data, a special type of a non-parametric regression, called a Bernoulli (binary) regression, plays a key role. This is why Section 2.4 is devoted to this important topic. E-estimator, used for estimation of multivariate functions, is defined and discussed in Section 2.5. Nonparametric estimation of functions, similarly to the classical parametric inference, may be complemented by confidence bands. This topic is discussed in Section 2.6.

2.1 Series Approximation

In this section the cosine orthonormal basis on the unit interval $[0, 1]$, introduced in Section 1.3 and which will be our main mathematical tool, is discussed via visualization of its approximations and via its theoretical properties. For the performance assessment, we choose a set of *corner* (*test*) functions. Corner functions should represent different functions of interest that are expected to occur in practice. In this book four specific corner functions with some pronounced characteristics are used. The set is shown in Figure 2.1, and additionally, as explained in the caption of Figure 2.3, it is possible to consider any custom-made corner function.

To make the discussion of all topics simpler, the corner functions are some specific probability densities supported on $[0, 1]$. Let us define and discuss each corner function in turn.

1. *Uniform.* This is a uniform density on $[0, 1]$, that is, $f_1(x) := I(0 \leq x \leq 1)$. Note that $f_1(x)$ is a very simple and smooth on its support density, and we will see that despite its triviality, statistical estimation of the Uniform is challenging. Moreover, this function plays a central role in asymptotic theory, and it is an excellent tool for debugging different types of errors.

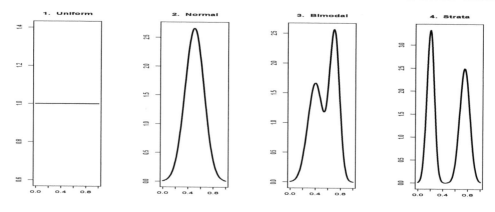

Figure 2.1 *The corner functions. {This set may be seen on the monitor by calling (after the R prompt)* > **ch2(f=1)**. *A corner function may be substituted by a custom-made one, see explanation in the caption of Figure 2.3.}*

2. *Normal.* This is a normal density with mean 0.5 and standard deviation 0.15, that is, $f_2(x) := d_{0.5,0.15}(x)$ defined in (1.3.8). The normal (bell-shaped) curve is the most widely recognized curve. Recall the rule of three standard deviations, which states that a normal density $d_{\mu,\sigma}(x)$ practically vanishes whenever $|x - \mu| > 3\sigma$. This rule helps us to understand the curve. It also explains why we do not divide f_2 by its integral over the unit interval, because this integral is very close to 1.

3. *Bimodal.* This is a mixture of two normal densities, $f_3(x) := 0.5d_{0.4,0.12}(x) + 0.5d_{0.7,0.08}(x)$. The curve has two closely located modes. As we will see throughout the book, this is one of the more challenging corner functions.

4. *Strata.* This is a function supported over two separated subintervals. In the case of a density, this corresponds to two distinct strata in the population. This is what differentiates the Strata from the Bimodal. The curve is obtained by a mixture of two normal densities, namely, $f_4(x) := 0.5d_{0.2,0.06}(x) + 0.5d_{0.7,0.08}(x)$. (Note how the rule of three standard deviations was used to choose the parameters of the normal densities in the mixture.)

Now, let us recall that a *function* $f(x)$ defined on an interval (the *domain*) is a rule that assigns to each point x from the domain exactly one element from the *range* of the function. Three traditional methods to define a function are a table, a formula, and a graph. For instance, we used both formulae and graphs to define the corner functions.

The fourth (unconventional) method of describing a function $f(x)$ is via a series expansion. Here and in what follows we always assume that the domain is $[0, 1]$ and a function is square integrable on the unit interval, that is $\int_0^1 [f(x)]^2 dx < \infty$. The latter is a mild restriction because in statistical applications we are primarily dealing with bounded functions. Then

$$f(x) = \sum_{j=0}^{\infty} \theta_j \varphi_j(x), \ x \in [0, 1] \ \text{ where } \ \theta_j := \int_0^1 f(x)\varphi_j(x)dx \,. \qquad (2.1.1)$$

Here the functions $\varphi_j(x)$ are known, fixed, and referred to as the *orthonormal functions* or *elements* of the *orthonormal basis* (or simply *basis*) $\{\varphi_0, \varphi_1, \ldots\}$, and the θ_j are called the *Fourier* coefficients of $f(x)$, $x \in [0, 1]$. A system of functions is called *orthonormal* if the integral $\int_0^1 \varphi_s(x)\varphi_j(x)dx = 0$ for $s \neq j$ and $\int_0^1 (\varphi_j(x))^2 dx = 1$ for all j.

Note that to describe a function via an infinite orthogonal series expansion (2.1.1) one needs to know the infinite number of Fourier coefficients. No one can store or deal with an infinite number of coefficients. Instead, a *truncated (finite) orthonormal series* (or so-called

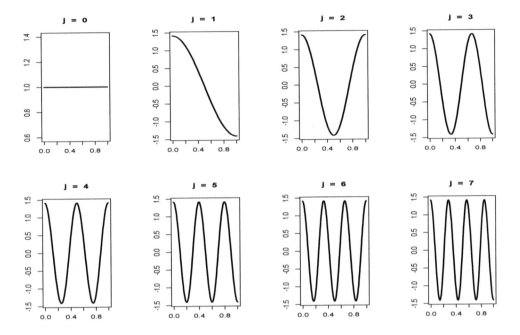

Figure 2.2 *The first eight elements of the cosine system.*

partial sum)

$$f_J(x) := \sum_{j=0}^{J} \theta_j \varphi_j(x) \tag{2.1.2}$$

is used to approximate f, namely the integrated squared error (ISE) $\int_0^1 [f_J(x) - f(x)]^2 dx \to$ 0 as $J \to \infty$. The integer parameter J is called the *cutoff*. Also, here and in what follows the mathematical symbol := means "equal by definition"; in other words (2.1.2) is the definition of the partial sum $f_J(x)$.

The advantage of this approach is the possibility to compress the data and describe a function using just several Fourier coefficients. In statistical applications this also leads to the estimation of a relatively small number of Fourier coefficients. Roughly speaking, the main statistical issue will be how to choose a cutoff J and estimate Fourier coefficients θ_j. Correspondingly, the rest of this section is devoted to the issue of how a choice of J affects visualization of series approximations and what are the known mathematical results which shed light on the choice. This will give us a necessary understanding and experience in choosing reasonable cutoffs.

In what follows, the cosine orthonormal basis on $[0, 1]$

$$\varphi_0(x) := 1 \quad \text{and} \quad \varphi_j(x) := \sqrt{2} \, \cos(\pi j x) \quad \text{for} \quad j = 1, 2, \ldots . \tag{2.1.3}$$

is used. The first eight elements are shown in Figure 2.2. It is not easy to believe that such elements may be good building blocks for approximating different functions, and this is where our corner functions become so handy.

Series approximations (2.1.2) for $J = 0, 1, 2, 3, 4, 5, 6, 7, 8$ are shown in Figure 2.3. In this and all other figures an underlying corner function is always shown by the solid line. As a result, some other curves may be "hidden" behind a solid line whenever they coincide. Further, it is always better to visualize figures using the R software because of colored curves.

We do not show approximation of the Uniform function because it is the element $\varphi_0(x)$ of the basis, so it is perfectly approximated by any partial sum. Indeed, the Uniform is clearly described by the single Fourier coefficient $\theta_0 = 1$, all other θ_j being equal to zero because $\int_0^1 \varphi_j(x)dx = 0$ whenever $j > 0$ (recall that the antiderivative of $\cos(\pi jx)$ is $(1/\pi j)\sin(\pi jx)$; thus $\int_0^1 \sqrt{2}\cos(\pi jx)dx = \sqrt{2}(\pi j)^{-1}[\sin(\pi j1) - \sin(\pi j0)] = 0$ for any positive integer j). Thus, there is no surprise that the Uniform is perfectly fitted by the cosine system; this is why we can skip the study of its approximations. At the same time, surprisingly enough, as we shall see shortly in the next chapters, this function is a difficult one for reliable statistical estimation. The reason for this is that if any estimated Fourier coefficient θ_j, $j > 0$ is not zero, it is easy to realize that the corresponding estimate is wrong because it is not a flat curve.

In Figure 2.3 the Uniform is replaced by the custom-made function, and the caption explains how any custom-made function can be created and then analyzed using the software. The function has a pronounced shape (look at the solid line), it is aperiodic, not differentiable, and its right tail is flat. Beginning with $J = 3$ we get a clear understanding of the underlying shape, and $J = 4$ gives us a very satisfactory visualization. This corner function also allows us to discuss the approximation of a function near the boundary points. As we see in the left bottom diagram, the partial sums are flattened out near the edges. This is because derivatives of any partial sum (2.1.2) are zeros at the boundary points (derivatives of $\cos(\pi jx)$ are equal to $-\pi j \sin(\pi jx)$ and therefore they are zeros for $x = 0$ and $x = 1$). In other words, the visualization of a cosine partial sum always reveals small flat plateaus near the edges (you may notice them in all approximations). Increasing the cutoff helps to decrease the length of the plateaus and improve the visualization. This is the *boundary* effect which exists, in this or that form, for all bases. A number of methods have been proposed on how to improve visualization of approximations near boundaries, see Chapters 2 and 3 in Efromovich (1999a); at the same time if we are aware about boundary effects and know how to recognize them, it is better to simply ignore them. Overall, here and in what follows we are going to use often the famous Voltaire's aphorism "...better is the enemy of the good..." as a guide in finding reasonable solutions.

Returning to Figure 2.3, approximation of the Normal is a great success story for the cosine system. Even the approximation based on the cutoff $J = 3$, where only 4 Fourier coefficients are used, gives us a fair visualization of the underlying function, and the cutoff $J = 5$ gives us an almost perfect fit. Just think about a possible compression of the data in a familiar table for a normal density into only several Fourier coefficients.

Now let us consider the approximations of the Bimodal and the Strata. Note that here partial sums with small cutoffs "hide" the modes. This is especially true for the Bimodal, whose modes are less pronounced and separated. In other words, approximations with small cutoffs oversmooth an underlying curve. Overall, about ten Fourier coefficients are necessary to get a fair approximation. On the other hand, even the cutoff $J = 5$ gives us a correct impression about a possibility of two modes for the Bimodal and clearly indicates two strata for the Strata. Note that you can clearly see the dynamic in approximations as J increases. Further, if we know that an underlying function is nonnegative, then by truncating negative values of approximations (just imagine that you replace negative values by zero) the approximations are dramatically improved. We will always use this and other opportunities to improve estimates.

Note that while cosine approximations are not perfect for some corner functions, understanding how these partial sums perform may help us to "read" messages of these approximations and guess about underlying functions. Overall, for the given set of corner functions, the cosine system does an impressive job in both representing the functions and the data compression.

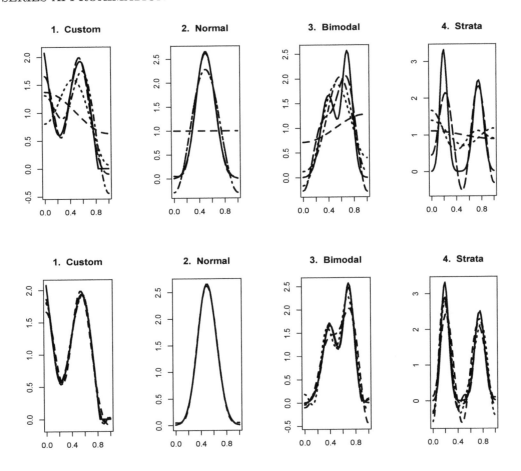

Figure 2.3 *Approximations of the custom-made and three corner functions by cosine series with different cutoffs J. The solid lines are the underlying functions. Short-dashed, dotted, dot-dashed and long-dashed lines correspond to cutoffs $J = 1$ through $J = 4$ in the top diagrams and $J = 5$ through $J = 8$ in the bottom diagrams, respectively. The first function is custom-made, and the explanation of how to construct it follows in the curly brackets. {The optional argument CFUN allows one to substitute a corner function by a custom-made corner function. For instance, the choice $CFUN = list(3, \;"2 * x - 3 * \cos(x)")$ would imply that the third corner function (the Bimodal) is substituted by the positive part of $2x - 3\cos(x)$ divided by its integral over $[0,1]$, i.e., the third corner function will be $(2x - 3\cos(x))_+ / \int_0^1 (2u - 3\cos(u))_+ du$. Any valid R formula in x (use only the lowercase x) may be used to define a custom-made corner function. This option is available for all figures where corner functions are used. Figure 2.4 allows one to check the sequence of curves and their color when the R software is used, and the curves are well recognizable on a color monitor. Try > **ch2(f=4)** to test colors and repeat Figure 2.4. Here and in what follows, arguments of a corresponding function can be found at the end of the caption in square brackets.} $[CFUN = list(1, \;"2 - 2 * x - \sin(8 * x)")]$*

Now let us add some theoretical results to shed light on Fourier coefficients. The beauty of a theoretical approach is that it allows us to analyze simultaneously large classes of functions; in particular recall that we are interested in functions f that are square integrable on $[0, 1]$, i.e., $\int_0^1 f^2(x)dx < \infty$. For square-integrable functions the famous *Parseval* identity

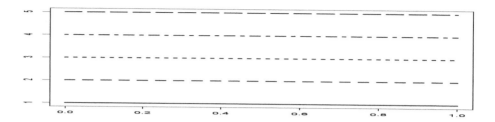

Figure 2.4 *Sequence of curves (lines) used in the book. Five horizontal lines with y-intercepts equal to k where k is the kth curve in the sequence used in all other figures. In a majority of diagrams we are using only the first four curves, and then, to make references simple and short, we refer to them as solid, dashed, dotted and dot-dashed curves (note that the dotted curve may look like short-dashed curve and the dot-dashed curve may look like short-dashed-intermediate-long-dashed curve). If all five curves are used, then we refer to them as solid, short-dashed, dotted, dot-dashed and long-dashed. Depending on your computer and software, the curves may look different and using this figure helps to realize this.* {*To repeat this figure, call after the R prompt >* **ch2(f=4)**}

states that

$$\int_0^1 f^2(x)dx = \sum_{j=0}^{\infty} \theta_j^2 \quad \text{and} \quad \text{ISE}(f, f_J) := \int_0^1 (f(x) - f_J(x))^2 dx = \sum_{j>J} \theta_j^2, \qquad (2.1.4)$$

where f_J is the partial sum (2.1.2) and the ISE stands for the integrated squared error (of the approximation $f_J(x)$).

Thus, the faster Fourier coefficients decrease, the smaller cutoff J is needed to get a good approximation of f by a partial sum $f_J(x)$ in terms of the ISE.

Let us explain the main characteristics of a function f that influence the rate at which its Fourier coefficients decrease. Namely, we would like to understand what determines the rate at which Fourier coefficients of an integrable function f decrease as $j \to \infty$.

To analyze θ_j, let us recall the technique of integration by parts. If $u(x)$ and $v(x)$ are both continuously differentiable functions, then the following equality, called *integration by parts*, holds:

$$\int_0^1 u(x)dv(x) = [u(1)v(1) - u(0)v(0)] - \int_0^1 v(x)du(x). \qquad (2.1.5)$$

Here $du(x) := u^{(1)}(x)dx$ is the differential of $u(x)$, and $u^{(k)}(x)$ denotes the kth derivative of $u(x)$.

Assume that $f(x)$ is differentiable. Using integration by parts and the relations

$$d\cos(\pi j x) = -\pi j \sin(\pi j x)dx, \quad d\sin(\pi j x) = \pi j \cos(\pi j x)dx, \qquad (2.1.6)$$

we may find θ_j for $j \geq 1$,

$$\begin{aligned} \theta_j &= \sqrt{2}\int_0^1 \cos(\pi j x)f(x)dx = \sqrt{2}(\pi j)^{-1}\int_0^1 f(x)d\sin(\pi j x) \\ &= \frac{\sqrt{2}}{(\pi j)}[f(1)\sin(\pi j) - f(0)\sin(0)] - \frac{\sqrt{2}}{(\pi j)}\int_0^1 \sin(\pi j x)f^{(1)}(x)dx. \end{aligned}$$

Recall that $\sin(\pi j) = 0$ for all integers j, so we obtain

$$\theta_j = -\sqrt{2}(\pi j)^{-1}\int_0^1 \sin(\pi j x)f^{(1)}(x)dx. \qquad (2.1.7)$$

Note that $|\int_0^1 \sin(\pi j x) f^{(1)}(x)dx| \leq \int_0^1 |f^{(1)}(x)|dx$, and thus we may conclude the following. *If a function $f(x)$ is differentiable, then for the cosine system,*

$$|\theta_j| \leq \sqrt{2}(\pi j)^{-1} \int_0^1 |f^{(1)}(x)|dx, \quad j \geq 1. \qquad (2.1.8)$$

We established the first rule (regardless of a particular f) about the rate at which the Fourier coefficients decrease. Namely, if f is differentiable and $\int_0^1 |f^{(1)}(x)|dx < \infty$, then $|\theta_j|$ decrease with rate at least j^{-1}.

Let us continue the calculation. Assume that f is twice differentiable. Then using the method of integration by parts on the right-hand side of (2.1.7), we get

$$\theta_j = -\frac{\sqrt{2}}{\pi j} \int_0^1 \sin(\pi j x) f^{(1)}(x)dx = \frac{\sqrt{2}}{(\pi j)^2} \int_0^1 f^{(1)}(x)d\cos(\pi j x)$$

$$= \frac{\sqrt{2}}{(\pi j)^2}[f^{(1)}(1)\cos(\pi j) - f^{(1)}(0)\cos(0)] - \frac{\sqrt{2}}{(\pi j)^2} \int_0^1 \cos(\pi j x) f^{(2)}(x)dx. \qquad (2.1.9)$$

We conclude that *if $f(x)$ is twice differentiable,* then

$$|\theta_j| \leq j^{-2}\Big[|f^{(1)}(1)(-1)^j - f^{(1)}(0)| + \int_0^1 |f^{(2)}(x)|dx\Big], \quad j \geq 1. \qquad (2.1.10)$$

Thus, the Fourier coefficients θ_j of smooth (twice differentiable) functions decrease with a rate not slower than j^{-2}.

So far, boundary conditions (i.e., values of $f(x)$ near boundaries of the unit interval $[0,1]$) have not affected the rate. The situation changes if f is smoother, for instance, it has three derivatives. In this case, integration by parts can be used again. However, now the decrease of θ_j may be defined by boundary conditions, namely by the term $[f^{(1)}(1)(-1)^j - f^{(1)}(0)]$ on the right-hand side of (2.1.9). These terms are equal to zero if $f^{(1)}(1) = f^{(1)}(0) = 0$. *This is the boundary condition that allows θ_j to decrease faster than j^{-2}. Otherwise, if the boundary condition does not hold, then θ_j cannot decrease faster than j^{-2} regardless of how smooth the underlying function f is.*

Now we know two main factors that define the decay of Fourier coefficients of the cosine system and therefore the performance of an orthonormal approximation: smoothness and boundary conditions. These are the fundamentals that we need to know about the series approximation.

The topic of how the decay of Fourier coefficients depends on various properties of an underlying function is a well-developed branch of mathematics, see the Notes. Here we introduce two classical functional classes that we will refer to later. The Sobolev function class (ellipsoid) $\mathcal{S}_{\alpha,Q}$, $0 \leq \alpha$, $0 < Q < \infty$ is

$$\mathcal{S}_{\alpha,Q} := \Big\{ f : \sum_{j=0}^{\infty}(1 + (\pi j)^{2\alpha})\theta_j^2 \leq Q \Big\}. \qquad (2.1.11)$$

For integer α the class is motivated by α-fold differentiable functions. Another important example is the class of *analytic* functions,

$$\mathcal{A}_{r,Q} := \{f : |\theta_j| \leq Qe^{-rj}, j = 0, 1, \ldots\}. \qquad (2.1.12)$$

Analytic functions have the derivatives of any order; this class is the subclass of Sobolev's functions and it is used to describe functions that are smoother than Sobolev's functions. In particular, the class nicely fits mixtures of normal densities and describes spectral densities of ARMA processes discussed in Chapter 8.

Let us present an example which explains why it is convenient to work with the above-introduced classes. The relation (2.1.4) explains that the partial sum $f_J(x) := \sum_{j=0}^{J} \theta_j \varphi_j(x)$ always approximates a square-integrable function $f(x)$, $x \in [0,1]$ in terms of the ISE. Indeed, (2.1.4) shows that the $\mathrm{ISE}(f, f_J) := \int_0^1 (f(x) - f_J(x))^2 dx = \sum_{j>J} \theta_j^2$ vanishes as J increases. At the same time, we cannot quantify decrease of the ISE without additional information about the function f. The theoretical importance of the two function classes is that they allow us to quantify the decrease of the ISE uniformly over all functions from the classes. Namely, for any $J \geq 1$ and f from the Sobolev class $\mathcal{S}_{\alpha,Q}$ we have

$$\mathrm{ISE}(f, f_J) < Q J^{-2\alpha}, \tag{2.1.13}$$

that is the ISE decreases as a power in J, while for any function from the analytic class $\mathcal{A}_{r,Q}$ the ISE decreases exponentially in J,

$$\mathrm{ISE}(f, f_J) < Q^2 (2r)^{-1} e^{-2rJ}. \tag{2.1.14}$$

Further, the function classes allow us to establish that the partial sum $f_J(x)$ approximates $f(x)$ uniformly at all points $x \in [0,1]$. Let us explain this assertion. Using the Cauchy-Schwatz inequality (1.3.43) we can write for functions from the Sobolev class $\mathcal{S}_{\alpha,Q}$ with $\alpha > 1/2$,

$$|f(x) - f_J(x)| = |\sum_{j>J} \theta_j \varphi_j(x)| \leq 2^{1/2} \sum_{j>J} |\theta_j|$$

$$\leq 2^{1/2} [\sum_{j>J} j^{-2\alpha} \sum_{j>J} j^{2\alpha} \theta_j^2]^{1/2} < [Q/(2\alpha - 1)]^{1/2} J^{-\alpha+1/2}. \tag{2.1.15}$$

In particular, (2.1.15) explains how to verify consistency of an estimator based on mimicking a partial sum $f_J(x)$. A similar inequality holds for an analytic function class.

2.2   Density Estimation for Complete Data

Density estimation is one of the most fundamental and practically important problems in statistics. In this section the classical model of probability density estimation is considered when all n independent and identically distributed observations $X_1, X_2, \ldots, X_n$ (the sample of size n) of a continuous random variable X are available. It is supposed that X is distributed according to an unknown probability density $f^X(x)$ supported on $[0,1]$; the latter means that X takes on values within this unit interval. More general settings will be considered later.

The problem is to estimate density $f^X(x)$, $x \in [0,1]$ using a sample from X. Because no assumption about shape of the estimand is made (we do not assume that the density is linear, or unimodal, or has a particular parametric distribution, etc.), the estimation procedure and the corresponding estimate are called nonparametric.

A note on notation. In what follows we use diacritics (e.g., "hat," "tilde," or "bar") above a parameter or a function to indicate that this is an estimator (statistic) of the corresponding parameter or function, for instance $\tilde{f}$ is an estimator of f. $I(A)$ is the indicator function of the event A, that is, the indicator is equal to 1 if A occurs and is 0 otherwise, $E\{\cdot\}$ denotes the expectation (population mean). The used criterion for the quality of nonparametric estimation of a function $f(x)$, $x \in [0,1]$ by an estimator $\tilde{f}$ is the Mean Integrated Squared Error (MISE) defined as

$$\mathrm{MISE}(\tilde{f}, f) := \mathbb{E}\left\{ \int_0^1 [\tilde{f}(x) - f(x)]^2 dx \right\}. \tag{2.2.1}$$

In (2.2.1) and in what follows it is understood that $\mathbb{E}\{\cdot\} := \mathbb{E}_f\{\cdot\}$.

Further, let us briefly recall the method of moments estimator of a parameter which equates a theoretical moment to a sample moment and then solves the equation with respect to the parameter. In what follows we are primarily interested in the first moment when the theoretical mean is equated to the sample mean. For instance, let the parameter of interest be $\kappa := \mathbb{E}\{g(X)\}$, then the method of moments estimator is $\hat{\kappa} := n^{-1}\sum_{l=1}^{n} g(X_l)$. Note that this estimator may be also referred to as the sample mean estimator and this terminology will be used in what follows. Further, if the function $g(x)$ is unknown but can be estimated by an estimator $\hat{g}(x)$ then $\tilde{\kappa} := n^{-1}\sum_{l=1}^{n} \hat{g}(X_l)$ is referred to as the plug-in sample mean estimator (in what follows we may skip the "plug-in" and simply say the sample mean estimator). Let us present one more example of construction of a sample mean estimator. Suppose that we want to estimate the parameter $\gamma := \mathbb{P}(X > c)$ where c is a given constant. Because we can write $\gamma = \mathbb{E}\{I(X > c)\}$ (recall that $I(\cdot)$ is the indicator function) then the sample mean estimator of γ is $\hat{\gamma} = n^{-1}\sum_{l=1}^{n} I(X_l > c))$.

A nice property of a sample mean estimator $n^{-1}\sum_{j=1}^{n} g(X_l)$ is that it is unbiased estimator of $\mathbb{E}\{g(X)\}$ and its variance may be calculated via the following simple formula,

$$\mathbb{V}\Big(n^{-1}\sum_{l=1}^{n} g(X_l)\Big) = n^{-1}\mathbb{V}(g(X)). \tag{2.2.2}$$

Also, the variance $\mathbb{V}(g(X))$ may be estimated by the familiar sample variance estimator $(n-1)^{-1}\sum_{l=1}^{n}[g(X_l) - n^{-1}\sum_{r=1}^{n} g(X_r)]^2$. Then, according to (2.2.2), multiply this estimator by n^{-1} and get the sample variance estimator of the variance of $n^{-1}\sum_{l=1}^{n} g(X_l)$.

Now let us return to the nonparametric estimation. For all statistical problems considered in the book (density estimation, regression, spectral density, etc.) the same three-step approach is used for construction of a nonparametric data-driven series estimator. The corresponding estimator will be referred to as the *E-estimator*.

E-estimator of a function $f(x)$, $x \in [0,1]$:
1. Consider a series expansion $f(x) = \sum_{j=0}^{\infty} \theta_j \varphi_j(x)$ explained in Section 2.1. Suggest a sample mean estimator $\hat{\theta}_j$ of Fourier coefficients $\theta_j := \int_0^1 f(x)\varphi_j(x)dx$. Then calculate a corresponding sample variance estimator $\hat{v}_{jn}$ of the variance $v_{jn} := \mathbb{V}(\hat{\theta}_j)$ of the sample mean estimator.
2. The E-estimator is defined as

$$\hat{f}(x) := \sum_{j=0}^{\hat{J}} \hat{\theta}_j I(\hat{\theta}_j^2 > c_{TH}\hat{v}_{jn})\varphi_j(x). \tag{2.2.3}$$

Here the empirical cutoff is

$$\hat{J} := \operatorname{argmin}_{0 \le J \le c_{J0} + c_{J1}\ln(n)}\Big\{\sum_{j=0}^{J}[2\hat{v}_{jn} - \hat{\theta}_j^2]\Big\}, \tag{2.2.4}$$

and c_{J0}, c_{J1} and c_{TH} are parameters (nonnegative constants).
3. If there are bona fide restrictions on $f(x)$ (for instance, the probability density is nonnegative and integrated to one, or it is known that the function is monotonic) then a projection of $\hat{f}(x)$ on the bona fide function class is performed.

Let us recall that the function $\operatorname{argmin}_{0 \le s \le S}\{a_s\}$, used in (2.2.4), returns the value s^* that is the index of the smallest element among $\{a_0, a_1, \ldots, a_S\}$.

Remark 2.2.1 Steps 2 and 3 in construction of the E-estimator are the same for all nonparametric statistical problems. As a result, as soon as a sample mean estimator of Fourier coefficients is proposed, this Fourier estimator yields the corresponding nonparametric E-estimator. Further, the choice of parameters of E-estimator may depend on a priori information about smoothness of $f(x)$. In this chapter the default values for the parameters are $c_{J0} = 3$, $c_{J1} = 0.8$ and $c_{TH} = 4$. It will be left as an exercise to propose better values for each considered problem.

Let us explain how the universal estimator is constructed for the case of estimation of the density $f^X(x)$ supported on $[0, 1]$ and what is the motivation of the E-estimation methodology. We know from Section 2.1 that a square-integrable density may be approximated by a partial sum

$$f_J^X(x) := \sum_{j=0}^{J} \theta_j \varphi_j(x), \ 0 \le x \le 1, \quad \text{where} \ \theta_j := \int_0^1 \varphi_j(x) f^X(x) dx. \tag{2.2.5}$$

Here $\{\varphi_j(x), j = 0, 1, \ldots\}$ is the cosine basis $\{\varphi_0(x) := 1, \varphi_j(x) := \sqrt{2}\cos(\pi j x), j = 1, 2, \ldots\}$. Recall that J is called the cutoff and θ_j is called the jth Fourier coefficient of $f^X(x)$ corresponding to the jth element $\varphi_j(x)$ of the used basis.

Step 1 of the E-estimation methodology is to estimate θ_j using a sample mean estimator. To find such an estimator, we express the Fourier coefficient as an expectation. Write

$$\theta_j := \int_0^1 \varphi_j(x) f^X(x) dx = \mathbb{E}\{\varphi_j(X)\}. \tag{2.2.6}$$

We can conclude that the parameter of interest θ_j is the expectation (population mean) of the random variable $\varphi_j(X)$. This yields the sample mean estimator

$$\hat{\theta}_j := n^{-1} \sum_{l=1}^{n} \varphi_j(X_l). \tag{2.2.7}$$

Note that $\theta_0 = 1$ and there is no need to estimate it, but regardless $\hat{\theta}_0 = 1$. This is the recommended, according to Step 1, method of moments estimator of Fourier coefficients. (Do you see that $\theta_0 = 1$ for any underlying density?) Let us also stress that all known series density estimators, proposed in the literature, use the recommended Fourier estimator (2.2.7). As soon as the Fourier estimator is chosen, its variance can be estimated. According to (2.2.7), we can use the classical sample variance estimator and set

$$\hat{v}_{jn} := n^{-1}\Big[(n-1)^{-1} \sum_{l=1}^{n} [\varphi_j(X_l) - \hat{\theta}_j]^2\Big]. \tag{2.2.8}$$

This completes Step 1.

Now let us explain Step 2. The step performs *adaptation* to unknown smoothness of an underlying density. We begin with its justification. Consider the projection estimator

$$\tilde{f}_J^X(x) := \sum_{j=0}^{J} \hat{\theta}_j \varphi_j(x) \tag{2.2.9}$$

and calculate its MISE. The Parseval identity and unbiasedness of $\hat{\theta}_j$ allow us to write,

$$\text{MISE}(\tilde{f}_J^X, f^X) = \sum_{j=0}^{J} E\{(\hat{\theta}_j - \theta_j)^2\} + \sum_{j>J} \theta_j^2 = \sum_{j=0}^{J} v_{jn} + \sum_{j>J} \theta_j^2. \tag{2.2.10}$$

Consider the two sums on the right-hand side of (2.2.10). The first sum is the integrated variance of $\tilde{f}_J^X(x)$ which is equal to the sum of $J+1$ variances $\mathbb{V}(\hat{\theta}_j)$. The second sum in (2.2.10) is the integrated squared bias

$$\text{ISB}_J(f^X) := \sum_{j>J} \theta_j^2. \tag{2.2.11}$$

It is impossible to estimate the integrated squared bias directly because it contains infinitely many terms. Instead, let us note that the Parseval identity allows us to rewrite this infinite sum via a finite sum,

$$\text{ISB}_J(f) = \int_0^1 [f^X(x)]^2 dx - \sum_{j=0}^J \theta_j^2. \tag{2.2.12}$$

The term $\int_0^1 [f^X(x)]^2 dx$ is a constant. Thus, the problem of finding a cutoff J that minimizes (2.2.10) is equivalent to finding a cutoff that minimizes $\sum_{j=0}^J (v_{jn} - \theta_j^2)$. Hence, we can rewrite (2.2.10) as

$$\text{MISE}(\tilde{f}_J^X, f^X) = \sum_{j=0}^J [v_{jn} - \theta_j^2] + \int_0^1 [f^X(x)]^2 dx, \tag{2.2.13}$$

and we need to choose J which minimizes the sum in the right-hand side of (2.2.13). Estimation of θ_j^2 is based on the relation which is valid for any $\breve{\theta}_j$ (recall (1.3.4)),

$$[\mathbb{E}\{\breve{\theta}_j\}]^2 = \mathbb{E}\{\breve{\theta}_j^2\} - \mathbb{V}(\breve{\theta}_j). \tag{2.2.14}$$

The Fourier estimator $\hat{\theta}_j$ is unbiased, and this together with (2.2.14) yield

$$\theta_j^2 = \mathbb{E}\{\hat{\theta}_j^2\} - \mathbb{V}(\hat{\theta}_j). \tag{2.2.15}$$

This implies that we can propose the unbiased estimator $\hat{\theta}_j^2 - \hat{v}_{jn}$ of θ_j^2.

Using this result in (2.2.13) we conclude that the search for the cutoff which minimizes the MISE is converted into finding a cutoff J which minimizes $\sum_{j=0}^J [2\hat{v}_{jn} - \hat{\theta}_j^2]$. This is exactly the sum on the right side of (2.2.4). Further, it was explained in Section 2.1 that we do not need to use very large cutoffs J to get a good approximation of a function by its partial sum, and this leads us to the search of the cutoffs over the set defined in (2.2.4).

Now let us explain why the indicator function is used in (2.2.3). For practically all functions some Fourier coefficients are very small or even zero. For instance, the Uniform corner function has all Fourier coefficients, apart of $\theta_0 = 1$, equal to zero. Due to symmetry, the Normal has all even Fourier coefficients equal to zero. As a result, the indicator performs the so-called *thresholding* which allows us to remove small $\hat{\theta}_j$ from the estimator. The default value of the coefficient of thresholding $c_{TH} = 4$, and the R software allows us to change it, as well as parameters c_{J0} and c_{J1}, in each figure.

Finally, step 3 is the projection on a class of bona fide functions. For the density this is the class of nonnegative and integrated to 1 functions. The projection is

$$\bar{f}^X(x) := \max(0, \hat{f}^X(x) - u), \text{ where the constant } u \text{ implies } \int_0^1 \bar{f}^X(x) dx = 1. \tag{2.2.16}$$

This procedure cuts off values of $\hat{f}^X(x)$ that are smaller than u, and this may create unpleasant bumps in $\bar{f}^X(x)$. There is a procedure, described in Section 3.1 of Efromovich (1999a), which removes the bumps, and it is used by the R software.

Let us make one more remark about the following property of the variance v_{jn},

$$nv_{jn} = 1 + o_j(1). \qquad (2.2.17)$$

Here and in what follows $o_s(1)$ is the little-o notation for generic sequences which tend to zero as $s \to \infty$. To prove (2.2.17), we use the trigonometric equality

$$\cos^2(\alpha) = [1 + \cos(2\alpha)]/2, \qquad (2.2.18)$$

and write for $j > 0$

$$\mathbb{V}(\varphi_j(X)) = \mathbb{E}\{[\varphi_j(X)]^2\} - \theta_j^2 = 1 + [\theta_{2j}2^{-1/2} - \theta_j^2]. \qquad (2.2.19)$$

As we know from the previous section, Fourier coefficients θ_j decay as j increases and this, together with (2.2.2), proves (2.2.17). The conclusion is that while we do not know values of individual v_{jn}, we have $n \lim_{j \to \infty} v_{jn} = 1$; actually for many densities only several first nv_{jn} may be far from 1.

This finishes our explanation of the E-estimation methodology and how to construct a density E-estimator. Several questions immediately arise. First, how does E-estimator perform for small sample sizes? Second, is it possible to suggest a better estimator? We are considering these questions in turn.

To evaluate performance of the E-estimator for small samples, we use Monte Carlo simulations where samples are generated according to the corner densities shown in Figure 2.1. E-estimates for sample sizes 100, 200, and 300 are shown in Figure 2.5 whose caption explains the diagrams. This figure exhibits results of 4 times 3 (that is, 12) independent Monte Carlo simulations. Note that the estimates are based on simulations, hence another simulation will yield different estimates. A particular outcome, shown in Figure 2.5 (as well as in all other figures), is chosen primarily with the objective of the discussion of a variety of possible E-estimates.

We begin with the estimates for the Uniform density shown in Diagram 1. As we see, while for $n = 100$ and $n = 200$ the estimates are perfect, the estimate for $n = 300$ is bad. On one hand, this outcome is counterintuitive; on the other, it is a great teachable moment with two issues to discuss. The former is that for a particular simulation a larger sample size may lead to a worse estimate. To understand this phenomenon, let us consider a simple example. Suppose an urn contains 5 chips and we know that 3 of them have one color (the "main" color) and that the other two chips have another color. We know that the colors are red and blue but do not know which one is the main color. We draw a chip from the urn and then want to make a decision about the main color. The natural bet is that the color of the drawn chip is the main color (after all, the chances are $\frac{3}{5}$ that the answer is correct). Now let us draw two more chips. Clearly, a decision based on three drawn chips should only be better. However, there is a possible practical caveat. Assume that the main color is blue and the first chip drawn is also blue. Then the conclusion is correct. On the other hand, if the two next chips are red (and this happens with probability $\frac{1}{6}$), the conclusion will be wrong despite the increased "sample size." Of course, if we repeat this experiment many times, then on average our bet will prevail, but in a particular experiment the proposed "reasonable" solution may imply a wrong answer. The latter issue is that we can realize what is the underlying series estimate for $n = 300$. Please return to Figure 2.2, where the basis functions are exhibited, and think about a formula for the series estimate shown by the dot-dashed line. This estimate is $\hat{f}(x) = 1 + \hat{\theta}_2\varphi_2(x)$ with $\hat{\theta}_2 \approx 0.14$. We know that $\theta_2 = 0$ for the Uniform density, so why was $\hat{\theta}_2$ included? The answer is because $\hat{\theta}_2^2 \approx 0.02 > c_{TH}\hat{v}_{2n} \approx 4/300 \approx 0.013$. Hence, we would need to use $c_{TH} \approx 6$ to threshold this large Fourier estimate, and while this may be good for the Uniform (actually, any larger threshold coefficient c_{TH} will benefit estimation of the Uniform), that can damage

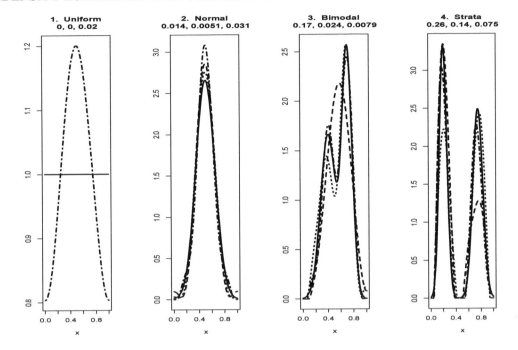

Figure 2.5 *Probability density E-estimates for sample sizes 100, 200 and 300 and 4 underlying corner densities. The corresponding ISEs (integrated squared errors) of the estimates are shown in the title. In a diagram the solid line is an underlying density while dashed, dotted and dot-dashed lines correspond to E-estimates based on samples of sizes $n = 100$, $n = 200$, and $n = 300$, respectively. (Note that the sequence of lines is the same as in Figure 2.4 which may be used to check curves on your computer.) Note that a solid line may "hide" other lines, and this implies a perfect estimation. For instance, in the first diagram the Uniform density (the solid line) hides the dashed and dotted lines and makes them invisible. {Recall that this figure may be repeated (with other simulated datasets) by calling after the R-prompt > the R-function > ch2(f=5). Also, see the caption of Figure 2.3 about a custom-made density. All the arguments, shown below in square brackets, may be changed. Let us review these arguments. The argument set.n allows one to choose three (or fewer) sample sizes. The arguments cJ0, cJ1, and cTH control the parameters c_{J0}, c_{J1}, and c_{TH} used by the E-estimator. Note that R does not recognize subscripts, so we use cJ0 instead of c_{J0}, etc. Also recall that below in the square brackets the default values for these arguments are given. Thus, after the call > ch2(f=5) the estimates will be calculated with these values of the coefficients. If one would like to change them, for instance to use a different threshold level, say $c_{TH} = 3$, make the call > ch2(f=5, cTH=3).} [set.n = c(100,200,300), cJ0 = 3, cJ1 = 0.8, cTH = 4]*

estimation of other densities. Another comment is as follows. The specific of the Uniform is that any deviation of an estimate from the Uniform looks like a "tragic" mistake despite the fact the deviation may be very small in terms of the ISE. It is worthwhile to compare the ISE = 0.02 for the case $n = 300$ with ISEs for other corner functions to appreciate this comment.

In the Diagram 2 for the underlying Normal density, the estimates nicely exhibit the symmetric and unimodal shape of the Normal. Curiously, here again the worst estimate is for the largest sample size $n = 300$. Now, let us compare the visual appeal of the estimates with the corresponding ISEs and ask ourselves the following question. Does the ISE reflect the quality of nonparametric estimation? This is an important question because the expected ISE, which we call the MISE, is our main criterion in finding a good estimator. Overall, if you

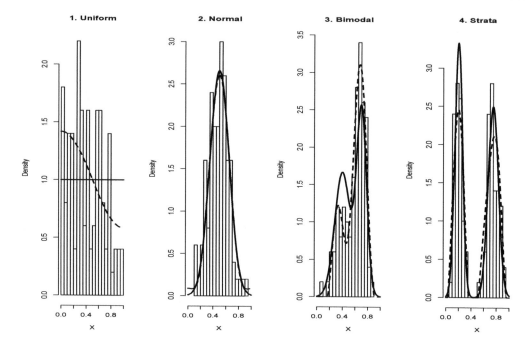

Figure 2.6 *Performance of density E-estimator for simulated samples of size $n = 100$. A sample is shown by the histogram overlaid by the underlying density (solid line) and its E-estimate (dashed line). [n = 100, cJ0 = 3, cJ1 = 0.8, cTH = 4]*

repeat Figure 2.5 several times, it becomes clear that the MISE is a reasonable criterion. Returning to Diagram 2, note that the dashed curve has funny tails. It is impossible to explain them without knowing the underlying sample, and this is what our next Figure 2.6 will allow us to do.

Estimates for the Bimodal, exhibited in Diagram 3, are of a special interest. For the case $n = 100$ the E-estimate (the dashed line) oversmooths the Bimodal and shows a single mode. This is a typical outcome and it indicates that this sample size is too small to indicate the closely located modes of the Bimodal. The larger sample sizes have allowed the E-estimator to exhibit the bimodal shape of the density. Also note how ISEs reflect the quality of estimation. The E-estimates for the Strata exhibit two pronounced strata. This clearly shows the flexibility of the proposed E-estimator because in the Strata we have two high spikes, a pronounced flat valley, and rapidly vanishing tails. The E-estimator is clearly capable to show us these characteristics of the density. Please keep in mind that the E-estimator does not know an underlying density, it has only data, and the densities change from the flat Uniform to the rough Strata. Also, note how well the projection on the class of bona fide densities performs here (compare with the ideal Fourier approximations of the Strata in Figure 2.2).

Figure 2.6 is another tool to understand how the E-estimator performs. Here in each diagram we can see the E-estimate which overlays the frequency histogram of an underlying sample from the density (the solid line) indicated in the title. A frequency histogram (or simply histogram) is a special nonparametric density estimator that uses vertical columns above bins to show frequencies of observations falling within bins. The histogram is a popular statistical method to visualize data; more about the histogram can be found in Section 8.1 of Efromovich (1999a).

For the Uniform we observe the monotonic E-estimate (the dashed line). This is a bad estimate of the flat Uniform density, and also look at the huge deviation of the estimate from the Uniform. But is this the fault of the proposed estimation procedure? The histogram helps us to answer the question because it shows us the data. The answer is "no." Recall that the density estimate describes the underlying sample, and the sample is heavily skewed to the left. Hence the E-estimate is consistent with the data. Two comments are due. First, the studied nonparametric estimation may be viewed upon as a correct smoothing of the histogram; this explains why statisticians working on topics of nonparametric curve estimation are often called the "smoothers." The second comment is that similarly to our analysis of Diagram 1 in Figure 2.5, we can figure out the functional form of the shown series estimate, and here it is $1 + \hat{\theta}_1\varphi_1(x)$. Repeated simulations show that such a skewed sample is a rare appearance but it is a real possibility to be ready for.

An interesting simulation is shown in the Normal diagram. The symmetric bell-shaped density created a sample with curious and asymmetric tails. This is what we also see in the otherwise very good E-estimate.

For the Bimodal density the E-estimate shows two pronounced modes and this fact should be appreciated. On the other hand, the magnitudes of modes are shown incorrectly. Is this the fault of the E-estimator? To answer the question, try to smooth the histogram using your imagination (and forget about the underlying density shown by the solid line); most likely you will end up with a curve resembling the E-estimate. Indeed, there is nothing in the data that may tell us about larger left mode or smaller right mode. Further, let us look at the left tail of the estimate. Here we see how the projection on the bona fide class performs. Indeed, recall that the E-estimator $\hat{f}^X(x)$, defined in (2.2.3), is an extremely smooth function in x. Hence, the left tail shows us that the underlying E-estimate $\hat{f}^X(x)$ takes on negative values to reflect the smallest values in the sample that are separated from the main part of the sample, and then the negative values are cut off.

Finally, consider the Strata in Figure 2.6. The E-estimate is good but not as sharp as one would like it to be. Here again it is worthwhile to put yourself in the shoes of the E-estimator. Look at this particular dataset, compare it with the underlying density (the solid line), and then try to answer the following question: How can the E-estimator perform better for this sample? One possibility here is to consider a larger cutoff (by increasing parameters c_{J0} and c_{J1}) and decreasing the thresholding coefficient c_{TH}. This and other figures allow one to make these changes and then test performance of the E-estimator via repeated simulations. Of course, a change that benefits one density may hurt another one, and this is when simultaneous analysis of different underlying densities becomes beneficial.

Two practical conclusions may be drawn from the analysis of these particular simulations. First, the visualization of a particular estimate is useful and sheds light on the estimator. Second, a conclusion may not be robust if it is based only on the analysis of just several simulations. The reason is that for any estimator one can find a dataset where an estimate is perfect or, conversely, very bad; this is why it is important to complement your reading of the book by repeating figures using the R software and analyzing the results. Every experiment (every figure) should be repeated many times; the rule of thumb, based on the author's experience, is that the results of at least 20 simulations should be analyzed before making a conclusion. Also, try to understand the parameters of each figure and change them to understand their role.

Remark 2.2.2. In the case where the density is estimated over a given support $[a, b]$ (or data are given only for this interval), to convert the problem to the case of the $[0, 1]$ support one first should rescale the data and compute $Y_l := (X_l - a)/(b - a)$. The rescaled observations are distributed according to a density $f^Y(y)$, which is then estimated over the unit interval $[0, 1]$. Let $\tilde{f}^Y(y)$, $y \in [0, 1]$ be the obtained estimate of the density $f^Y(y)$. Then the corresponding estimate of $f^X(x)$ over the interval $[a, b]$ is defined by $\tilde{f}^X(x) := (b-a)^{-1}\tilde{f}^Y((x-a)/(b-a))$, $x \in [a, b]$.

Remark 2.2.3. Consider the setting where one would like to estimate an underlying density over its finite support $[a, b]$, which is unknown. In other words, both the density and its support are of interest. The simplest and a reliable solution is to set $\tilde{a} := X_{(1)}$, $\tilde{b} := X_{(n)}$ and then use the method proposed in the previous remark. Here $X_{(1)} \leq X_{(2)} \leq \cdots \leq X_{(n)}$ are the ordered observations. Let us also explain a more sophisticated solution. It is known that $\mathbb{P}(a < X < X_{(1)}) = \mathbb{P}(X_{(n)} < X < b) = 1/(n+1)$, Thus, $a =: X_{(1)} - d_1$ and $b =: X_{(n)} + d_2$, where both $d_1 > 0$ and $d_2 > 0$ should be estimated. Let us use the following approach. If an underlying density is flat near the boundaries of its support, then for a sufficiently small positive integer s we have $(X_{(1+s)} - X_{(1)})/s \approx X_{(1)} - a = d_1$, and similarly $(X_{(n)} - X_{(n-s)})/s \approx b - X_{(n)} = d_2$. The default value of s is 1. Thus, we may set $\hat{d}_1 := (X_{(1+s)} - X_{(1)})/s$ and $\hat{d}_2 := (X_{(n)} - X_{(n-s)})/s$. More precise estimation of d_1 and d_2 requires estimation of both the density and its derivatives near $X_{(1)}$ and $X_{(n)}$, and this is a complicated problem for the case of small sample sizes.

Remark 2.2.4. We shall see that for different settings, including regression, missing and modified data, the parameter $d := \lim_{j,n \to \infty} nE\{(\hat{\theta}_j - \theta_j)^2\}$, where $\hat{\theta}_j$ is an appropriate sample mean estimator of θ_j, defines the factor in changing a sample size that makes estimation of an underlying curve comparable, in terms of the same precision of estimation, with the density estimation model over the known support $[0, 1]$ when $d = 1$. In other words, the problem of density estimation may be considered as a benchmark for analyzing other models. Thus we shall refer to the coefficient d as the *coefficient of difficulty*. We shall see that it is a valuable tool, which allows us to appreciate the complexity of a problem based on our experience with the density estimation.

We are finishing this section with the asymptotic analysis of the MISE of the recommended projection estimator $\tilde{f}_J^X(x) = \sum_{j=0}^J \hat{\theta}_j \varphi_j(x)$, and then comment on its consistency. The MISE is calculated in (2.2.10) and we would like to evaluate it for the case of Sobolev and analytic function classes defined in (2.1.11) and (2.1.12), respectively. We begin with the Sobolev's densities. Using (2.1.11) and (2.2.17) we can write that

$$\text{MISE}(\tilde{f}_J^X, f^X) \leq c^*[Jn^{-1} + J^{-2\alpha}], \quad f^X \in \mathcal{S}_{\alpha,Q}. \tag{2.2.20}$$

Here c^* is a finite positive constant which depends on (α, Q). Then the cutoff J^*, which minimizes the right-hand side of (2.2.20), is proportional to $n^{1/(2\alpha+1)}$. Further, the optimal cutoff yields the classical result for Sobolev densities: the MISE vanishes with the rate $n^{-2\alpha/(2\alpha+1)}$. This rate is also optimal meaning that no other estimator can improve it for the Sobolev densities. This result is intuitively clear because a single parameter can be estimated with the variance not smaller in the order than n^{-1}, and this together with (2.1.11) gives us a lower bound for the MISE which is, up to a constant, the same as the right-hand side of (2.2.20). We conclude that the proposed E-estimation methodology is rate optimal for Sobolev's functions. Further, it is explained in Efromovich (1999a) that the used thresholding decreases the MISE of E-estimator.

For analytic densities the projection estimator is not only rate but also sharp optimal meaning that its MISE converges with the optimal rate and constant (see more in Efromovich 1999a). Let us shed light on this assertion. Using (2.1.12) and (2.2.17) we can write for any $J_n \to \infty$ as $n \to \infty$,

$$\text{MISE}(\tilde{f}_{J_n}^X, f^X) = \sum_{j=0}^{J_n} v_{jn} + \sum_{j > J_n} \theta_j^2$$

$$= n^{-1} J_n(1 + o_n(1)) + \sum_{j > J_n} \theta_j^2$$

$$\leq [n^{-1}J_n + (Q^2/2r)e^{-2rJ_n}] + o_n(1)J_n n^{-1}, \quad f^X \in \mathcal{A}_{r,Q}. \tag{2.2.21}$$

The cutoff which minimizes the right-hand side of (2.2.21) is $J_n^* = (1/2r)\ln(n)(1 + o_n(1))$. Note that only a logarithmic in n number of Fourier coefficients is necessary to estimate the density, and the latter is the theoretical justification for the set of cutoffs used by the procedure (2.2.4). Using this optimal cutoff we can continue (2.2.21) and get

$$\text{MISE}(\tilde{f}_{J_n^*}^X, f^X) \leq (1/2r)\ln(n)n^{-1}[1 + o_n(1)], \quad f^X \in \mathcal{A}_{r,Q}. \tag{2.2.22}$$

Further, this upper bound is sharp, that is there is no other estimator for which the right-hand side of (2.2.22) may be smaller for the analytic functions.

Now let us comment on consistency of the projection estimator (2.2.9) for functions from a Sobolev class $\mathcal{S}_{\alpha,Q}$, $\alpha > 1/2$. (Recall that for the density $\hat{\theta}_0 = \theta_0 = 1$, but, whenever possible, we are presenting general formulas that may be used for the analysis of any problem like regression or spectral density estimation.) Write,

$$|\tilde{f}_J^X(x) - f^X(x)| = |\sum_{j=0}^{J}(\hat{\theta}_j - \theta_j)\varphi_j(x) + \sum_{j>J}\theta_j\varphi_j(x)|$$

$$\leq 2^{1/2}\sum_{j=0}^{J}|\hat{\theta}_j - \theta_j| + |\sum_{j>J}\theta_j\varphi_j(x)|. \tag{2.2.23}$$

Using (2.1.15), (2.2.19) and $\mathbb{E}\{|\hat{\theta}_j - \theta_j|\} \leq [\mathbb{E}\{(\hat{\theta}_j - \theta_j)^2\}]^{1/2}$ we conclude that for some constant $c_{\alpha,Q} < \infty$

$$\mathbb{E}\{|\tilde{f}_J^X(x) - f^X(x)|\} \leq c_{\alpha,Q}[Jn^{-1/2} + J^{-\alpha+1/2}]. \tag{2.2.24}$$

Inequality (2.2.24), together with the Markov inequality (1.3.26), yields that for any $t > 0$

$$\mathbb{P}(|\tilde{f}_J^X(x) - f^X(x)| > t) \leq c_{\alpha,Q}[Jn^{-1/2} + J^{-\alpha+1/2}]/t. \tag{2.2.25}$$

This inequality allows us to establish consistency of the projection estimator for a wide class of increasing to infinity cutoffs $J = J_n \to \infty$ as $n \to \infty$. A similar calculation can be made for analytic densities.

The presented asymptotical results give theoretical justification of the E-estimation methodology. Another useful conclusion of the asymptotic theory is that the MISE converges slower than the traditional rate n^{-1} known for parametric problems like estimation of the mean or the variance of a random variable. This is what makes nonparametric problems so special and more challenging than classical parametric problems.

2.3 Nonparametric Regression

Nonparametric regression is another classical nonparametric curve estimation problem. Here we are dealing with a pair (X, Y) of random variables with X called the predictor (or covariate) and Y called the response. A sample $(X_1, Y_1), \ldots, (X_n, Y_n)$ from (X, Y) is given, and the aim is to find a relationship between variables X and Y that allows one to quantify the impact of X on Y or to predict Y given X. The joint distribution of the pair gives us a complete description of the relationship, but this is not a simple problem to estimate a corresponding bivariate function. Instead, it may be reasonable to formulate the following univariate problem. Suppose that we would like to find a function $m(x)$ which yields the best prediction of Y in terms of minimization of the conditional mean squared error

$$\text{MSE}(m, x) := \mathbb{E}\{(Y - m(X))^2 | X = x\}. \tag{2.3.1}$$

The function $m(x)$, which minimizes the conditional MSE, is

$$m(x) := \mathbb{E}\{Y|X = x\}. \tag{2.3.2}$$

The last assertion follows from the fact that for any random variable Z with finite second moment and any constant c we have

$$\mathbb{E}\{(Z - c)^2\} = \mathbb{E}\{([Z - \mathbb{E}\{Z\}] + [\mathbb{E}\{Z\} - c])^2\}$$

$$= \mathbb{E}\{(Z - \mathbb{E}\{Z\})^2\} + [\mathbb{E}\{Z\} - c]^2 \geq \mathbb{E}\{(Z - \mathbb{E}\{Z\})^2\}. \tag{2.3.3}$$

Furthermore, the random variable $m(X)$ minimizes the MSE, that is it minimizes $\mathbb{E}\{(Y - \mu(X))^2\}$ over all possible $\mu(x)$. The latter follows from expressing the expectation via the expectation of a conditional expectation, namely we can write $\mathbb{E}\{(Y - \mu(X))^2\} = \mathbb{E}\{\mathbb{E}\{(Y - \mu(X))^2|X\}\}$.

The univariate function (2.3.2) is called the *regression* function, and it represents the best prediction of Y given $X = x$ in terms of the MSE criterion. The problem is to estimate $m(x)$ based on a sample from the pair (X, Y). In what follows, it is always assumed that X is supported on [0,1], otherwise X should be rescaled on $[0, 1]$ as explained in the previous section.

A simpler (than the above-discussed) regression model is also often considered where the response is written as

$$Y := m(X) + \sigma(X)\varepsilon. \tag{2.3.4}$$

Here ε is the zero mean and unit variance random variable which is independent of X, and the function $\sigma(x)$ defines the standard deviation (scale or spread) of the additive error $\sigma(X)\varepsilon$ given $X = x$. Note that (2.3.4) is a particular case of (2.3.2). Predictor X can be either deterministic (equidistant regression with $X_l = l/n$, $l = 1, 2, \ldots, n$ is a particular example) or random (in this case $X_1, \ldots, X_n$ is a sample from a random variable X), and then the regression is referred to as fixed- or random-design regression. The regression model is called *homoscedastic* if $\sigma(x)$ is constant (does not depend on the predictor), otherwise it is *heteroscedastic*.

According to Section 2.2, to construct a regression E-estimator we need to propose a sample mean estimator of Fourier coefficients $\theta_j := \int_0^1 m(x)\varphi_j(x)dx$. We are going to consider the case of the predictor X being a continuous random variable supported on $[0, 1]$ and with the marginal density bounded below from zero, that is $f^X(x) \geq c_* > 0$, $x \in [0, 1]$. The latter excludes cases when there are no observations of X over some subsets of $[0, 1]$ and hence a consistent regression estimation is impossible. Also note that $f^X(x)$ may be referred to as the *design density* of the predictor. We can write,

$$\theta_j = \mathbb{E}\left\{\frac{Y\varphi_j(X)}{f^X(X)}\right\}. \tag{2.3.5}$$

It is easy to see why (2.3.5) holds for the model (2.3.4), and here we are proving this formula for the general model (2.3.2). Write,

$$\mathbb{E}\left\{\frac{Y\varphi_j(X)}{f^X(X)}\right\} = \mathbb{E}\{\varphi_j(X)[f^X(X)]^{-1}\mathbb{E}\{Y|X\}\}$$

$$= \int_0^1 \varphi_j(x)[f^X(x)]^{-1}m(x)f^X(x)dx = \theta_j. \tag{2.3.6}$$

Here in the first equality we used the standard formula $\mathbb{E}\{g(X, Y)\} = \mathbb{E}\{\mathbb{E}\{g(X, Y)|X\}\}$ of writing the expectation of a bivariate function $g(x, y)$ via the expectation of the conditional expectation.

If the design density $f^X(x)$ is known, then the sample mean Fourier estimator of θ_j is

$$\tilde{\theta}_j := n^{-1} \sum_{l=1}^{n} \frac{Y_l \varphi_j(X_l)}{f^X(X_l)}. \tag{2.3.7}$$

If the design density is unknown then its E-estimator $\hat{f}(x)$ of Section 2.2, based on the sample $X_1, \ldots, X_n$, may be used. This yields a plug-in sample mean Fourier estimator

$$\hat{\theta}_j := n^{-1} \sum_{l=1}^{n} \frac{Y_l \varphi_j(X_l)}{\max(\hat{f}^X(X_l), c/\ln(n))}. \tag{2.3.8}$$

In (2.3.8) $\hat{f}^X(x)$ is truncated from below because it is used in the denominator, c is the additional parameter of the regression E-estimator with the default value $c = 1$, and all corresponding figures allow the user to choose any $c > 0$.

The variance of $\hat{\theta}_j$ can be estimated by the sample variance based on statistics $\{Y_l \varphi_j(X_l)/\max(\hat{f}^X(X_l), c/\ln(n)), \ l = 1, \ldots, n\}$. This yields the regression E-estimator $\hat{m}(x)$. Further, if some bona fide restrictions are known (for instance, it is known that the corner functions are nonnegative), then a projection on the bona fide functions is performed.

Remark 2.3.1. Let θ_k be the kth Fourier coefficient of the regression function $m(x)$, and $k \neq j$. Then it is possible to show that

$$\theta_j = \mathbb{E}\{[Y - \theta_k \varphi_k(X)]\varphi_j(X)/f^X(X)\}. \tag{2.3.9}$$

This observation points upon a more accurate procedure of estimation which, theoretically, yields asymptotically efficient estimation. The simplest and most valuable step in implementation of the idea is to calculate $\hat{\theta}_0$ and then subtract it from Y_l, $l = 1, 2, \ldots, n$ and use the differences for estimating Fourier coefficients θ_j, $j \geq 1$. We are utilizing this step in the regression E-estimator. There is also another idea of estimation of the Fourier coefficients based on using formula $\theta_j = \int_0^1 m(x)\varphi_j(x)dx$. Namely, it is possible to replace an unknown $m(X_l)$ by its observation Y_l and then use responses in a numerical integration. This approach is convenient for the case of fixed-design predictors. More about these and other methods can be found in Chapter 4 of Efromovich (1999a).

Figure 2.7 allows us to look at regression data and check how the regression E-estimator performs for small sample sizes. A plot of the pairs (X_l, Y_l) in the xy-plane (so-called *scattergram* or *scatter plot*) is a useful tool to get a first impression about a dataset at hand. Consider Monte Carlo simulations of observations according to (2.3.4) with $n = 100$; underlying experiments are explained in the caption. Four scattergrams are shown by circles. For now let us ignore the lines and concentrate on the data. An appealing nature of the regression problem is that one can easily appreciate its difficulty. To do this, try to use your imagination and draw a line $m(x)$ through the middle of the cloud of circle in a scattergram that, according to your understanding of the regression problem, gives a good fit according to model (2.3.4). Or even simpler, because in the diagrams the underlying regression functions are shown by the solid line, try to recognize them in the cloud of circles. If you are not successful in this imagination and are confused, do not be upset because these particular scattergrams are difficult to read due to a large scale function (just look at the range of responses).

Let us examine the four diagrams in turn where the dashed line shows E-estimates. For the Uniform case (here the regression function is the Uniform) the estimate is good. Can you see that there are more observations near the right tail than the left one? This is because the design density is increasing. The scattergram for larger predictors may also suggest that an underlying regression should have a decreasing right tail, but this is just an illusion created

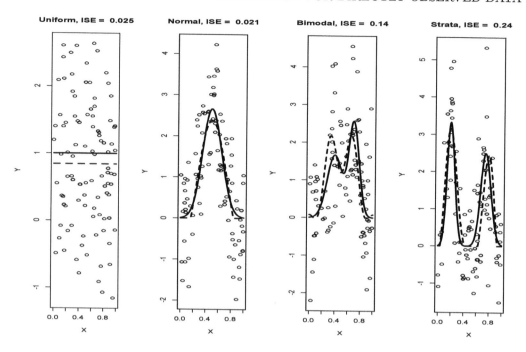

Figure 2.7 *Heteroscedastic nonparametric regression. Observations (the scattergram) are shown by circles overlaid by the underlying regression (the solid line) and its regression E-estimate (the dashed line). The underlying model is $Y = m(X) + \sigma s(x)\varepsilon$ where $m(x)$ is a corner function indicated in the title, ε is standard normal and independent of X and $\sigma s(x)$ is the scale function. ISE is the integrated squared error $\int_0^1 (m(x) - \hat{m}(x))^2 dx$. The E-estimator knows that an underlying regression function is nonnegative. {The arguments are: n controls the sample size, sigma controls σ, the string scalefun defines the shape of a custom function $s(x)$ which is truncated from below by the value dscale and then rescaled into the bona fide density supported on $[0,1]$, the string desden controls the shape of the design density which is then truncated from below by the value dden and then rescaled into a bona fide density. Argument c controls parameter c in (2.3.8). Arguments cJ0, cJ1 and cTH are explained in Figure 2.6.} [n = 100, desden = "1 + 0.5 * x", scalefun = "3 − (x − 0.5) ^2 ", sigma = 1, dscale = 0, dden = 0.2, c = 1, cJ0 = 3, cJ1 = 0.8, cTH = 4]*

by the regression errors. Actually, here one can imagine a number of interesting shapes of the regression. For the Normal regression the estimate is not perfect but it correctly shows the unimodal shape. Because the Normal regression has a pronounced shape and its range is comparable with the regression noise, we can see the correct shape in the scatter plot. The reason why the magnitude of the estimate is smaller than it should be is due to the large noise which does not allow the E-estimator use larger frequency corresponding to smaller Fourier coefficients (recall Figure 2.3). Also note how the E-estimator uses the fact that the underlying regression function is nonnegative by truncating negative values of $\hat{m}(x)$. The E-estimate for the Bimodal regression is respectful keeping in mind the relatively small sample size. Note that it would be tempting to oversmooth the scattergram and indicate just one mode. The Strata case is interesting. The E-estimator does a superb job here. To realize this, just try to draw a regression line through the scattergram; it would be a difficult task to ignore the outliers.

It is a good exercise to repeat this figure with different parameters and get used to the nonparametric regression and the E-estimator.

2.4   Bernoulli Regression

Suppose that we are interested in a relationship between a continuous random variable X (the predictor) and a Bernoulli random variable A. Bernoulli regression (often referred to as binary, probit, and also recall a parametric logistic regression) is an important statistical model that is used in various fields, including actuarial science, machine learning, engineering, most medical fields, and social sciences. For example, the likelihood of an insurable event, as a function of a covariate, is the key topic for insurance industry. The likelihood of admission to a university, as a function of the SAT score, is another example. Clinical trials, whose aim is to understand effectiveness of a new drug, as a function of known covariates, is another classical example. Bernoulli regression is widely used in engineering, especially for predicting the probability of failure of a given process, system or product. Prediction of a customer's propensity to purchase a product or halt a subscription is important for marketing. Furthermore, as we will see shortly, Bernoulli regression often occurs in statistical problems with modified and missing data.

We begin with reviewing the notion of a classical Bernoulli random variable. Suppose that a trial, or an experiment, whose outcome can be classified as either a "success" or as a "failure," is performed. Introduce a random variable A which is equal to 1 when the outcome is the success and $A = 0$ otherwise. Set $\mathbb{P}(A = 1) = w$ and correspondingly $\mathbb{P}(A = 0) = 1 - w$ for some constant $w \in [0,1]$; note that w is the probability of the success. A direct calculation shows that $\mathbb{E}\{A\} = w$ and $\mathbb{V}(A) = w(1 - w)$. If the probability w of success is unknown and a sample $A_1, \ldots, A_n$ from A is available, then the sample mean estimator $\hat{w} := n^{-1} \sum_{l=1}^{n} A_l$ may be used for estimation of the parameter w. This estimator is unbiased and enjoys an array of optimal properties. These are the basic facts that we need to know about a Bernoulli random variable.

Now let us translate the classical parametric Bernoulli setting into a nonparametric one. Assume that the probability of the success w is the function of a predictor X which is a continuous random variable supported on the unit interval $[0,1]$ with the design density $f^X(x) \geq c_* > 0$, $x \in [0,1]$. In other words, we assume that for $x \in [0,1]$ and $a \in \{0,1\}$

$$\mathbb{P}(A = 1 | X = x) = w(x), \quad f^{A,X}(a,x) = f^X(x)[w(x)]^a [1 - w(x)]^{1-a}. \qquad (2.4.1)$$

Note that here $w(x)$ may be considered as either the probability of success given the predictor equal to x or as the conditional expectation of A given the predictor equal to x, that is

$$w(x) := \mathbb{P}(A = 1 | X = x) = \mathbb{E}\{A | X = x\}. \qquad (2.4.2)$$

This explains why the problem of estimation of the conditional probability of success $w(x)$ may be solved using nonparametric regression.

The Bernoulli regression problem is to estimate the function $w(x)$ using a sample of size n from the pair (X, A).

To solve the problem using the E-estimation methodology, we begin with writing down Fourier coefficients θ_j of $w(x)$ as an expectation. Write,

$$\theta_j := \int_0^1 w(x)\varphi_j(x)dx = \mathbb{E}\left\{ \frac{A\varphi_j(X)}{f^X(x)} \right\}. \qquad (2.4.3)$$

If the design density is known, (2.4.3) yields the sample mean estimator

$$\tilde{\theta}_j := n^{-1} \sum_{l=1}^{n} \frac{A_l \varphi_j(X_l)}{f^X(X_l)}. \qquad (2.4.4)$$

Design density may be known in controlled regressions and some other situations, but in general it is unknown. In the latter case we may estimate the design density using the

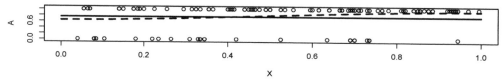

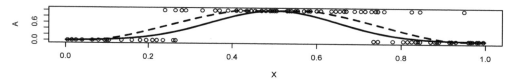

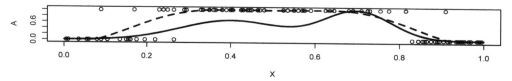

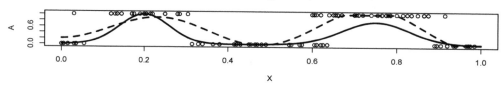

Figure 2.8 *Bernoulli regression. In each diagram a scattergram is overlaid by the underlying regression $w(x)$ (the solid line) and its E-estimate (the dashed line). {n controls the sample size, the string desden defines the shape of the design density which is then truncated from below by dden and then rescaled into a bona fide density.} [n = 100, desden = "1 + 0.5 * x", dden = 0.2, c = 1, cJ0 = 3, cJ1 = 0.8, cTH = 4]*

E-estimator $\hat{f}^X(x)$ of Section 2.2 and plug it in (2.4.4). Because the density E-estimate is used in the denominator, it is prudent to truncate it below from zero. This yields

$$\hat{\theta}_j := n^{-1} \sum_{l=1}^{n} \frac{A_l \varphi_j(X_l)}{\max(\hat{f}^X(X_l), c/\ln(n))}. \qquad (2.4.5)$$

Recall that c is the parameter of E-estimator with the default value $c = 1$.

The Fourier estimator yields the Bernoulli regression E-estimator $\hat{w}(x)$, $x \in [0,1]$. Further, because the function $w(x) \in [0,1]$ (it is the conditional probability), we project the estimator onto the unit interval by replacing all negative values by zero and replacing all values larger than one by one.

Let us check, using Figure 2.8, how the Bernoulli regression E-estimator performs for the case of the corner functions being the probability of success. Here all corner functions, shown in Figure 2.1, with the exception of the Uniform, are divided by their maximal value, and in place of the Uniform the function $w(x) = 2/3$ is used (in what follows we still refer to it as the Uniform). For the Uniform, the estimate wrongly shows a positive slope implying that the probability of a success $w(x)$ increases with x. Is this a mistake? Of course, if one knows the underlying Uniform regression function, the answer is "yes." On the other hand, let us look at the simulated observations. It is clear from the scattergram that the number of successes, with respect to the corresponding number of failures, increases as x increases. This is what the data tell us and this is what the E-estimator shows us. In other words, the E-estimator knows only data and it performs correspondingly. The solid line, showing the underlying regression function, serves as a reference, and allows us to understand that the sample size $n = 100$ may not yield a scattergram corresponding to an underlying regression function. In other words, it is always important to keep in mind that we are dealing with n random pairs of realizations of (X, A) that may create a pattern different from the one that can be expected from an underlying regression function.

For the Normal regression the estimate is also far from the underlying regression function, but again it correctly describes the data at hand. Namely, please note that there are no failures within interval $[0.3, 0.7]$, and this is why the E-estimate is larger than the underlying regression. A similar outcome we see for the Bimodal regression, and again there is nothing in the data that may tell us (and the E-estimator) about the two modes of the underlying regression. E-estimate for the Strata is also poor, but again it correctly describes the data at hand. For instance, look at observations corresponding to the right stratum. There is not a single failure for responses from the interval $[0.65, 0.85]$, and this explains why the E-estimate indicates larger $\hat{w}(x)$ than the underlying $w(x)$.

The particular simulations, shown in Figure 2.8, allow us to understand that the sample size $n = 100$ is relatively small for the considered nonparametric problem and it may create a scattergram which does not reflect an underlying regression function. The reader is advised to repeat Figure 2.8 and get used to the model and possible outcomes.

Finally, have you noticed that the presentation of diagrams in Figure 2.8 is different from those in Figure 2.7? Which one do you prefer?

2.5 Multivariate Series Estimation

Suppose that we would like to estimate a joint density of several random variables or a regression on several covariates. As we shall see shortly, both the series approach and the E-estimation are effortlessly extended to multivariate models. At the same time, the outcome of estimation becomes worse or, to state this differently, to get an accuracy of estimation similar to a univariate case, larger sample sizes are needed. This fact is often referred to as the curse of dimensionality, and it will be explained shortly.

We begin with discussion of the bivariate series approach and the coresponding E-estimation; general case will follow shortly. Consider a square integrable on $[0, 1]^2$ bivariate function $f(x_1, x_2)$ (such that $\int_{[0,1]^2} f^2(x_1, x_2)dx_1dx_2 < \infty$) and any two univariate bases $\{\phi_j(x), j = 0, 1, \ldots\}$ and $\{\eta_j(x), j = 0, 1, \ldots\}$ on the unit interval $[0, 1]$. Then products of elements from these two bases,

$$\{\varphi_{j_1 j_2}(x_1, x_2) := \phi_{j_1}(x_1)\eta_{j_2}(x_2), \quad j_1, j_2 = 0, 1, \ldots\}, \tag{2.5.1}$$

create a basis on $[0, 1]^2$ which is called a *tensor-product* basis. This useful mathematical fact implies a great flexibility in creating convenient bivariate bases. In the previous section, the cosine basis on $[0, 1]$ was used; for bivariate functions we will use the cosine tensor-product basis with elements $\varphi_{j_1 j_2}(x_1, x_2) := \varphi_{j_1}(x_1)\varphi_{j_2}(x_2)$. Recall that $\varphi_0(x) := 1$ and $\varphi_j(x) := 2^{1/2}\cos(\pi j x)$, $j = 1, 2, \ldots$

If a function $f(x_1, x_2)$ is square integrable on $[0, 1]$, then its partial sum approximation with cutoffs J_1 and J_2 is

$$f_{J_1 J_2}(x_1, x_2) = \sum_{j_1=0}^{J_1} \sum_{j_2=0}^{J_2} \theta_{j_1 j_2} \varphi_{j_1 j_2}(x_1, x_2), \tag{2.5.2}$$

and the Fourier coefficients $\theta_{j_1 j_2}$ are defined by the formula

$$\theta_{j_1 j_2} := \int_{[0,1]^2} f(x_1, x_2) \varphi_{j_1 j_2}(x_1, x_2) dx_1 dx_2. \qquad (2.5.3)$$

This formula immediately yields the sample mean estimator of Fourier coefficients for the case of a bivariate density $f^{X_1 X_2}(x_1, x_2)$ supported on $[0,1]^2$. Denote by $(X_{11}, X_{21}), \dots, (X_{1n}, X_{2n})$ a sample of size n from (X_1, X_2) and define the sample mean Fourier estimator

$$\hat{\theta}_{j_1 j_2} := n^{-1} \sum_{l=1}^{n} \varphi_{j_1 j_2}(X_{1l}, X_{2l}). \qquad (2.5.4)$$

The Fourier estimator allows us to use the E-estimation methodology of construction of the bivariate density E-estimator of Section 2.2; the only necessary comment to make is that now a pair of cutoffs $(\hat{J}_1, \hat{J}_2)$, which minimizes the empirical risk of the projection estimator, is calculated.

Figure 2.9 illustrates performance of the bivariate density E-estimator. This figure allows one to choose two bivariate densities that are the products of any two corner densities. The sample size $n = 100$ is considered small for bivariate functions but it allows us to visualize data in the top diagrams. These diagrams are classical scattergrams only now the estimand is the bivariate density and not the univariate regression. Let us look at the data. We know that the joint density is the product of two corner densities. Hence, the random variables X_1 and X_2 are independent. Do we see this in the scattergrams? This is not a simple question due to a very small number of observations in some areas; for instance, in the right scattergram we have just one observation with X_2 smaller than 0.25. A training in reading the scattergrams is needed, the interested reader may repeat this figure a number of times to get a correct "feeling" of data. For instance, to realize the independence, you may move a horizontal or vertical line along the range of a corresponding variable and then convince yourself that the univariate shape of the density along the moving line remains the same. Then compare your "feeling" with an underlying density shown in a middle diagram. Or even simpler, check that the regressions of X_1 on X_2 and vice versa are constant (this approach, of course, can tell you only about dependence but cannot prove independence).

The middle diagrams show underlying bivariate densities, and they allow us to evaluate the simulations. Note that in an ideal simulation the larger the crowdedness of points in a scattergram, the larger the underlying bivariate density. The bottom diagrams show us E-estimates. The left estimate is very good. It correctly shows the independence and the overall shape. The right estimate is also good for such a small sample size and the complicated shape of the bivariate density. In particular, note that, relative to each other, magnitudes of the two peaks are shown correctly.

Now let us comment on the asymptotic theory of the bivariate projection estimation. If a function $f(x_1, x_2)$ has β bounded partial derivatives in x_1 and x_2, then, under a mild additional assumption, it is known that the integrated squared biased can be bounded from above as follows,

$$\text{ISB}(J_1, J_2) := \int_{[0,1]^2} [f(x_1, x_2) - \sum_{j_1=0}^{J_1} \sum_{j_2=0}^{J_2} \theta_{j_1 j_2} \varphi_{j_1 j_2}(x_1, x_2)]^2 dx_1 dx_2$$

$$= \sum_{j_1 > J_1} \sum_{j_2 > J_2} \theta_{j_1 j_2}^2 \leq c^* [J_1^{-2\beta} + J_2^{-2\beta}], \quad c^* < \infty. \qquad (2.5.5)$$

Note that the equality in the bottom line of (2.5.5) is due to the Parseval identity. The decrease in the ISB is similar to what we have in the case of univariate differentiable functions, and this is the good news. The bad news is that now $(J_1 + 1)(J_2 + 1)$ Fourier coefficients

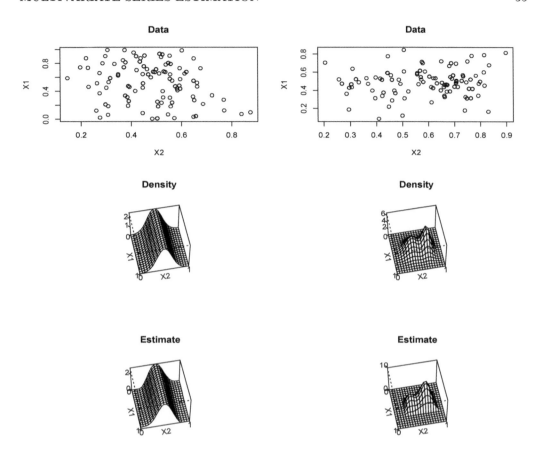

Figure 2.9 *E-estimation of the bivariate density. Two samples of size $n = 100$ are generated and the corresponding estimation is exhibited in the two columns of diagrams. The first row of diagrams shows the data (scattergrams). The second row shows the underlying bivariate densities. The third row shows corresponding bivariate density E-estimates. {n controls the sample size. An underlying bivariate density is the product of two corner densities defined by parameters c11 and c12 for the left column of diagrams and c21 and c22 for the right column.} [n = 100, c11 = 1, c12 = 2, c21 = 2, c22 = 3, cJ0 = 3, cJ1 = 0.8, cTH = 4]*

must be estimated, and this yields the variance of the projection bivariate estimator to be of order $n^{-1}J_1J_2$. If we optimize the MISE with respect to J_1 and J_2, then we get that optimal cutoffs and the corresponding MISE are

$$J_1^* = J_2^* = O_n(1)n^{1/(2\beta+2)}, \quad \text{MISE} = O_n(1)n^{-2\beta/(2\beta+2)}. \tag{2.5.6}$$

Here $O_n(1)$ are generic sequences in n such that $0 < c_* < O_n(1) < c^* < \infty$.

Recall that in a univariate case the MISE decreases faster with the rate $n^{-2\beta/(2\beta+1)}$. In a case of a d-variate density, the MISE slows down to $n^{-2\beta/(2\beta+d)}$.

What is the conclusion for the case of multivariate functions? We have seen that using tensor-product bases makes the problem of series approximation of multivariate functions similar to the univariate case. Nonetheless, several technical difficulties arise. First, apart from bivariate functions (surfaces), there is no simple tool to visualize a multidimensional curve. Second, we have seen in Section 2.1 that to approximate fairly well a smooth univariate function, about 5 to 10 Fourier coefficients are needed. For the case of a d-dimensional curve this translates into 5^d to 10^d Fourier coefficients. Since all these coefficients must be estimated, the rate of the MISE convergence slows down as d increases. Third, suppose that $n = 100$ points are uniformly distributed over the five-dimensional unit cube $[0, 1]^5$. What is the probability of having some points in a neighborhood of reasonable size, say a cube with side 0.2? Since the volume of such a cube is $0.2^5 = 0.00032$, the expected number of points in this neighborhood is n times 0.2^5, i.e., 0.032. As a result, no averaging over that neighborhood can be performed. For this example, to get on average of 5 points in a cube, its side should be 0.55, that is, more than a half of the range along each coordinate. This shows how sparse multivariate observations are. Fourth, the notion of a small sample for multivariate problems mutates. Suppose that for a univariate regression a grid of 50 points is considered sufficient. Then this translates into 50 points along each axis, i.e., into 50^d data points. These complications present a challenging problem, which is customarily referred to as the curse of dimensionality. However, in no way the curse implies that the situation is always hopeless as we have seen for the case of bivariate functions.

2.6   Confidence Bands

Consider the problem of estimation of the probability density $f^X(x)$ supported on $[0, 1]$. Based on a direct sample $X_1, X_2, \ldots, X_n$ from X, a data-driven E-estimator has been suggested in Section 2.2. A corresponding E-estimate is a curve which allows us to visualize an underlying density. It would be nice to complement this curve by showing intervals/bands around it which would point upon variance of the E-estimator.

In parametric statistics, confidence intervals are traditionally used to complement a point estimator. Suppose that we are interested in estimation of the mean of a random variable X supported on $[0, 1]$. Then the parameter $\mu := \int_0^1 x f^X(x) dx = \mathbb{E}\{X\}$ is the mean, and we would like to evaluate variability of the sample mean estimator $\hat{\mu} = n^{-1} \sum_{l=1}^{n} X_l$. The variance of the estimator is $\mathbb{V}(\hat{\mu}) = n^{-1}\mathbb{V}(X)$ (recall that variance of the sum of independent random variables is the sum of their variances), and the unknown variance $\mathbb{V}(X)$ can be estimated by the sample variance estimator $\hat{\sigma}^2 := (n-1)^{-1} \sum_{l=1}^{n}(X_l - \hat{\mu})^2$. Then, if the sample size is sufficiently large to use the central limit theorem ($n \geq 30$ is often recommended), then the classical $1 - \alpha$ confidence interval is

$$[\hat{\mu} - z_{\alpha/2}\hat{\sigma}n^{-1/2}, \hat{\mu} + z_{\alpha/2}\hat{\sigma}n^{-1/2}]. \qquad (2.6.1)$$

Here $\alpha \in (0, 1)$ and its typical values are 0.01, 0.02, 0.05 and 0.1, $z_{\alpha/2}$ is a function in α such that $\int_{z_{\alpha/2}}^{\infty} (2\pi)^{-1/2} e^{-x^2/2} dx = \alpha/2$ and $\hat{\sigma} := \sqrt{\hat{\sigma}^2}$. Note that the confidence interval is random (it is a statistic of observations) and its main property is the following relation

$$\mathbb{P}\Big([\hat{\mu} - z_{\alpha/2}\hat{\sigma}n^{-1/2} \leq \mu \leq \hat{\mu} + z_{\alpha/2}\hat{\sigma}n^{-1/2}] \Big| \mu \Big) = 1 - \alpha + o_n(1). \qquad (2.6.2)$$

Recall that $o_n(1)$ denotes a vanishing sequence in n, and $\mathbb{P}(D|\mu)$ or $\mathbb{P}(D|f)$ denote the probabilities of event D given an estimated parameter μ or an estimated function f, respectively.

Relation (2.6.2) sheds light on the notion of the $1 - \alpha$ confidence interval (2.6.1), namely, this confidence interval covers an underlying parameter μ with the probability $1 - \alpha$. Note how the half-length of the confidence interval, called the margin of error, sheds light on the used sample mean estimator $\hat{\mu}$, namely, it tells us how far the estimator can be from the underlying parameter of interest. Finally, let us stress that while a confidence interval is based on the estimated variance, its length is proportional to the standard deviation $\hat{\sigma}$.

Now let us return to our problem of nonparametric estimation of the density. Suppose that for a particular sample of size n we get an E-estimate $\hat{f}^X(x) = 1 + \hat{\theta}_j\varphi_j(x)$; recall that $\theta_0 = 1$ for any density supported on $[0, 1]$. This estimate allows us to conjecture that an underlying density is $f_j^X(x) := 1 + \theta_j\varphi_j(x)$. Assume that the conjecture is correct. Then (2.6.1) and (2.6.2) yield that (we use notation $\varphi_j^2(x) := [\varphi_j(x)]^2$)

$$\mathbb{P}\Big(|f_j^X(x) - \hat{f}^X(x)| \le z_{\alpha/2}[n^{-1}\mathbb{V}(\varphi_j(X))\varphi_j^2(x)]^{1/2}\Big|f_j^X\Big) = 1 - \alpha + o_n(1), \quad x \in [0,1], \quad (2.6.3)$$

and

$$\mathbb{P}\Big(\max_{x \in [0,1]} |f_j^X(x) - \hat{f}^X(x)|[(n^{-1}\mathbb{V}(\varphi_j(X))\varphi_j^2(x)]^{-1/2} \le z_{\alpha/2}\Big|f_j^X\Big) = 1 - \alpha + o_n(1). \quad (2.6.4)$$

In what follows we are referring to (2.6.3) as a pointwise confidence variance-band (or simply pointwise band) and to (2.6.4) as a simultaneous confidence variance-band (or simply simultaneous band). The reason for this terminology is as follows. In (2.6.3) the estimand is the value of the density at point x, and this is a parameter. This is why the band is called pointwise, that is the band is for a point $x \in [0,1]$. In (2.6.2) the largest deviation of the density estimator from the density over all $x \in [0,1]$ is under control, and this explains the terminology of simultaneous band. The notion of the variance-band is motivated by the following consideration. In (2.6.3) and (2.6.4) it is assumed that an underlying density is of a special parametric type $f_j^X(x) = 1 + \theta_j\varphi_j(x)$. In general, an underlying density is

$$f^X(x) = f_j^X(x) + \sum_{k \in \{1,2,\dots\}\backslash j} \theta_k\varphi_k(x). \quad (2.6.5)$$

Then, using the Parseval identity, the MISE (the mean integrated squared error) of the estimator $\hat{f}^X(x) = 1 + \hat{\theta}_j\varphi_j(x)$ can be written as the sum of the variance and the integrated squared bias (ISB) of the estimator, namely

$$\mathbb{E}\{\int_0^1 (\hat{f}^X(x) - f^X(x))^2 dx\} = \mathbb{V}(\hat{\theta}_j) + \sum_{k \in \{1,2,\dots\}\backslash j} \theta_k^2. \quad (2.6.6)$$

In (2.6.3)-(2.6.4) the term ISB is ignored, and this explains the variance-band terminology. There are several traditional justifications for using the variance-band approach. The first one is that in nonparametric inference it is difficult (even theoretically) to take into account the bias term because it involves an infinite number of unknown Fourier coefficients. Unfortunately, any approach that takes into account the bias may lessen effect of the bias but cannot eliminate it. The second justification is that for many densities and cases of small to moderate sample sizes, the bias of the E-estimate is relatively small, and simulations show that the variance-band approach implies reliable confidence bands. Finally, as we will see in the following chapters, in a number of applications the variance-band is of interest on its own. This is why the variance-band approach is recommended for nonparametric estimation. Furthermore, knowing flaws of the variance-band approach allows us to make reasonable inferences about E-estimates.

Now let us expand the considered example of a single Fourier coefficient in an E-estimate of the density to a general case. Suppose that an E-estimate is $\hat{f}^X(x) = 1 + \sum_{j \in \mathcal{N}} \hat{\theta}_j\varphi_j(x)$. The estimate allows us to conjecture that an underlying density is $f_{\mathcal{N}}^X(x) = 1 + \sum_{j \in \mathcal{N}} \theta_j\varphi_j(x)$. Then we can write that $\hat{f}^X(x) - f_{\mathcal{N}}^X(x) = \sum_{j \in \mathcal{N}} (\hat{\theta}_j - \theta_j)\varphi_j(x)$. The expectation of this difference is zero and the variance is (below a general formula for calculation of the variance of a sum of random variables is used)

$$\sigma_{\mathcal{N}}^2(x) := \mathbb{V}\Big(\sum_{j \in \mathcal{N}} (\hat{\theta}_j - \theta_j)\varphi_j(x)\Big) = \sum_{i,j \in \mathcal{N}} \varphi_i(x)\varphi_j(x)\text{Cov}(\hat{\theta}_i, \hat{\theta}_j), \quad (2.6.7)$$

where the covariance $\text{Cov}(\hat{\theta}_i, \hat{\theta}_j) := \mathbb{E}\{(\hat{\theta}_i - \mathbb{E}\{\hat{\theta}_i\})(\hat{\theta}_j - \mathbb{E}\{\hat{\theta}_j\})\} := \sigma_{ij}(n) =: \sigma_{ij}$. This

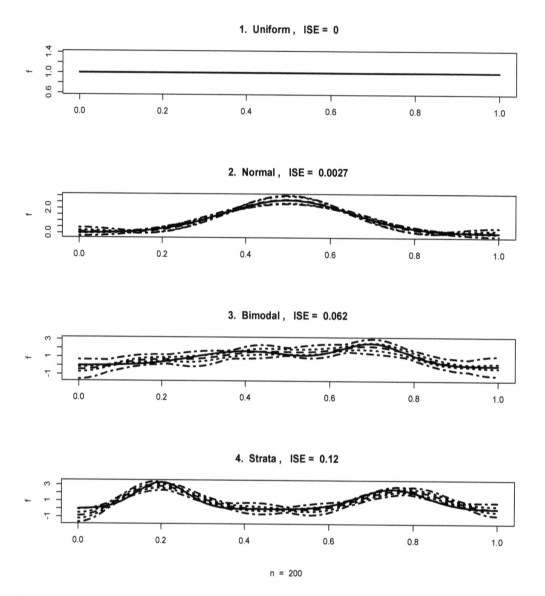

Figure 2.10 *Pointwise and simultaneous confidence bands for the density E-estimator. Four diagrams correspond to the four underlying corner densities. The solid and dashed lines show the underlying density and its E-estimate, respectively. Two dotted lines show a pointwise band and two dot-dashed lines show a simultaneous band. Each title shows the integrated squared error of the E-estimate which is $\int_0^1 (\hat{f}^X(x) - f^X(x))^2 dx$. {Argument alpha controls α.} [n = 200, alpha = 0.05, cJ0 = 3, cJ1 = 0.8, cTH = 4]*

formula allows us to calculate sample mean estimates $\hat{\sigma}_{ij}$ of σ_{ij} and then plug them in (2.6.7) and get an estimate $\hat{\sigma}_{\mathcal{N}}(x)$ for $\sigma_{\mathcal{N}}(x)$. This, together with the central limit theorem, implies the following pointwise variance-band (or in short pointwise band)

$$\mathbb{P}\Big(|f_{\mathcal{N}}^X(x) - \hat{f}^X(x)| \leq z_{\alpha/2}\hat{\sigma}_{\mathcal{N}}(x)\Big| f_{\mathcal{N}}^X \Big) = 1 - \alpha + o_n(1), \quad x \in [0,1]. \qquad (2.6.8)$$

Developing a simultaneous confidence band is a more challenging problem, and here a simple (but not necessarily optimal) solution is proposed. Consider a set of positive numbers

$\{c_{j,\alpha/2,n}, j \in \mathcal{N}\}$ such that a vector $\{Z_j, j \in \mathcal{N}\}$ of normal random variables with the zero mean and covariance function of the vector $\{(\hat{\theta}_j - \theta_j), j \in \mathcal{N}\}$ satisfies

$$\mathbb{P}\Big(\cup_{j \in \mathcal{N}} \{|Z_j| \le c_{j,\alpha/2,n}\} \Big) = 1 - \alpha + o_n(1). \qquad (2.6.9)$$

This allows us to conclude that

$$\mathbb{P}\Big(\sup_{x \in [0,1]} \frac{|f_{\mathcal{N}}^X(x) - \hat{f}^X(x)|}{\sum_{j \in \mathcal{N}} c_{j,\alpha/2,n}|\varphi_j(x)|} \le 1 \Big| f_{\mathcal{N}}^X \Big) = 1 - \alpha + o_n(1). \qquad (2.6.10)$$

Formula (2.6.10) defines the $1 - \alpha$ simultaneous confidence band.

Figure 2.10 helps us to understand how the bands perform. Each diagram corresponds to one of our corner densities. Here we are suppressing the projection of both the E-estimator and the bands on the class of densities; we do this to highlight the proposed bands (and the same will be done throughout the book). The top diagram shows an outcome for the Uniform. This is a very interesting case where the underlying density is $f^X(x) = 1$ and the E-estimate is the perfect $\hat{f}^X(x) = \hat{\theta}_0 = 1$, $x \in [0,1]$. Note that $\mathbb{P}(\hat{\theta}_0 = 1) = 1$ and $\mathbb{V}(\hat{\theta}_0) = 0$. This is why we see that the widths of the pointwise and simultaneous bands are zero, and this is why we see just one line. This is a nice example which sheds light on the notion of a variance-band. For the Normal density we can observe a typical relationship between the more narrow pointwise band and the wider simultaneous band. Note that the bands are wider near boundaries. Now let us look at the Bimodal. As we already know, this density is more complicated for estimation than the Normal, and the bands clearly exhibit this. For the Strata, the interesting outcome is the relatively wide bands for the flat valley between the strata.

Overall, repeated simulations indicate that the proposed bands are a feasible and useful inference tool.

2.7   Exercises

2.1.1 Describe the four corner functions. What are their special characteristics and differences?

2.1.2 How can a sample from the Bimodal density be generated?

2.1.3 Prove that if function $f(x)$ is square integrable on $[0,1]$ and $\{\varphi_j(x), j = 0,1,\ldots\}$ is the basis on this interval, then $\int_0^1 f^2(x)dx = \sum_{j=0}^{\infty} \theta_j^2$.

2.1.4 Use Figures 2.1 and 2.2 and suggest a reasonable Fourier approximation of the Normal corner function using a minimal number of elements of the cosine basis.

2.1.5 Use Figure 2.3 to choose minimal cutoffs that imply a reasonable approximation of corner functions. Explain why the cutoffs are different.

2.1.6 Check tails of approximations in Figure 2.3 and explain why they are poor for some functions and good for others.

2.1.7 Explain how the integration by parts formula (2.1.5) can be verified.

2.1.8 Verify (2.1.7).

2.1.9* Assume that a function $f(x)$, $x \in [0,1]$ has k bounded derivatives on $[0,1]$. Explain how its Fourier coefficients θ_j decrease. Do they decrease with the rate j^{-k} or faster?

2.1.10* How do boundary conditions affect the decrease of Fourier coefficients?

2.1.11 Define Sobolev and Analytic function classes.

2.1.12 Do Fourier coefficients from a Sobolev class decrease faster than from an Analytic class?

2.1.13 Verify (2.1.13). Can a sharper inequality be proposed?

2.1.14 Verify (2.1.14). Explain how it may be used for the analysis of the MISE.

2.1.15 Prove validity of (2.1.5) or correct it.

2.1.16 Consider an analytic class $\mathcal{A}_{r,Q}$ and evaluate $|f(x) - f_J(x)|$. Hint: Follow (2.1.15).

2.1.17* Let $r := r(\alpha)$ denote the largest integer strictly less than a positive number α, and $L > 0$ be another positive number. Introduce a Hölder class $H_{\alpha,L}$ of functions $f(x)$, $x \in [0,1]$ whose rth derivative $f^{(r)}(x)$ satisfies

$$|f^{(r)}(x) - f^{(r)}(y)| \le L|x-y|^{\alpha-r}, \quad (x,y) \in [0,1]^2. \tag{2.7.1}$$

Further, for a positive integer k introduce a (Sobolev) class $D_{k,L}$ of functions $f(x)$, $x \in [0,1]$ such that $f^{(k-1)}(x)$ is absolutely continuous and

$$\int_0^1 [f^{(k)}(x)]^2 dx \le L^2. \tag{2.7.2}$$

Prove that the function class $D_{k,L}$ contains the Hölder function class $H_{k,L}$.

2.1.18* Using notation of Exercise 2.1.17*, introduce a periodic function class

$$D_{k,L}^* := \{f : f \in D_{k,L}, \ f^{(r)}(0) = f^{(r)}(1), \ r = 0,1,\ldots,k-1\}. \tag{2.7.3}$$

Further, introduce a classical trigonometric basis $\{\eta_0(x) = 1, \eta_{2j-1}(x) = 2^{1/2}\sin(2\pi j x),$ $\eta_{2j}(x) = 2^{1/2}\cos(2\pi j x), j = 1,2,\ldots\}$ on $[0,1]$. Show that if $f \in D_{k,L}^*$ then

$$f(x) = \sum_{j=0}^{\infty} \nu_j \eta_j(x), \tag{2.7.4}$$

where $\nu_j := \int_0^1 f(x)\eta_j(x)$ are Fourier coefficients and

$$\sum_{j=1}^{\infty} (2\pi j)^{2k} [\nu_{2j-1}^2 + \nu_{2j}^2] \le L^2. \tag{2.7.5}$$

2.1.19* Using notation of Exercise 2.1.18*, prove that if (2.7.4)-(2.7.5) holds, then $f \in D_{k,L}^*$.

2.1.20 Combine assertions of the previous three exercises and make a conclusion about the introduced function classes.

2.1.21* Let a function $f(x)$, $x \in [0,1]$ be r-times differentiable on $[0,1]$, $\int_0^1 [f^{(r)}(x)]^2 dx < \infty$, and for all positive and odd integers $k < r$

$$f^{(k)}(0) = f^{(k)}(1) = 0. \tag{2.7.6}$$

Further, let $\varphi_j(x)$, $j = 0,1,\ldots$ be elements of the cosine basis (2.1.3) on $[0,1]$. Prove that Fourier coefficients $\theta_j := \int_0^1 f(x)\varphi_j(x)dx$ of the function $f(x)$ satisfy the Parseval-type identity

$$\sum_{j=1}^{\infty} (\pi j)^{2r} \theta_j^2 = \int_0^1 [f^{(r)}(x)]^2 dx. \tag{2.7.7}$$

Hint: Use integration by parts.

2.1.22* Consider a positive integer r and set

$$r' = \begin{cases} r & \text{if } r \text{ is even,} \\ r-1 & \text{if } r \text{ is odd.} \end{cases} \tag{2.7.8}$$

Let a function $f(x)$ be r-times differentiable on $[0,1]$ and $\int_0^1 [f^{(r)}(x)]^2 dx < \infty$. Then there exists a unique polynomial–cosine expansion

$$f(x) = \sum_{i=1}^{r'} c_i x^i + \sum_{j=0}^{\infty} \nu_j \varphi_j(x), \quad x \in [0,1], \tag{2.7.9}$$

where the coefficients c_i are defined from the following relations for the polynomial term $p(x) := \sum_{i=1}^{r'} c_i x^i$,

$$p^{(k)}(0) = f^{(k)}(0), \quad p^{(k)}(1) = f^{(k)}(1), \tag{2.7.10}$$

that hold for all positive and odd $k < r$, and $\varphi_j(x)$ are elements of the cosine basis (2.1.3). In (2.7.9) it is understood that for $r = 1$ (and correspondingly $r' = 0$) the term $\sum_{i=1}^{0} c_i x^i = 0$. Moreover,

$$\sum_{j=1}^{\infty} \pi^{2r} v_j^2 = \int_0^1 [f^{(r)}(x)]^2 dx - (1 - r + r')[r'!c_{r'}]^2. \tag{2.7.11}$$

2.1.23* Explain how the result of the previous exercise may help to accelerate convergence of the cosine Fourier series, or in other words may help to avoid the boundary problem discussed in Section 2.1.

2.1.24* Consider the cosine basis $\{\varphi_j(x), j = 0, 1, \ldots\}$ on $[0, 1]$, and introduce an analytic (exponential) ellipsoid

$$\mathcal{E}_{\alpha,Q} := \{f(x): \sum_{j=0}^{\infty} e^{2\alpha j} \theta_j^2 \leq Q, \ \theta_j := \int_0^1 f(u)\varphi_j(u)du, \ x \in [0, 1]\}. \tag{2.7.12}$$

Consider a series approximation $f_J(x) := \sum_{l=0}^{J} \theta_j \varphi_j(x)$ for functions f from the ellipsoid, and then explore its ISB and $|f_J(x) - f(x)|$.

2.2.1 Give definitions of a discrete, continuous and mixed random variables.

2.2.2 For a continuous random variable, what is the relationship between the cumulative distribution function, the survivor function and the probability density?

2.2.3 Suppose that the density of a random variable X is known. Define the expectation and variance of a function $g(X)$.

2.2.4 What is the relationship between the variance, first and second moments of a random variable?

2.2.5 Is the variance of a random variable less, equal or larger than the second moment?

2.2.6 Consider a sample $X_1, \ldots, X_n$ from a random variable X. Is the sample mean $n^{-1}\sum_{l=1}^{n} g(X_l)$ unbiased estimator of $\mathbb{E}\{g(X)\}$? Do you need any assumption to support your conclusion?

2.2.7 Consider a sample $X_1, \ldots, X_n$ from a random variable X. What is the variance of statistic $n^{-1}\sum_{l=1}^{n} g(X_l)$? Do you need any assumption to support your conclusion?

2.2.8 Propose a sample mean estimator of a Fourier coefficient $\theta_j := \int_0^1 f^X(x)\varphi_j(x)dx$ based on a sample of size n from the random variable X supported on $[0, 1]$.

2.2.9* Consider the problem of Exercise 2.2.8 where now it is no longer assumed that $[0, 1]$ is the support of X. Suggest the sample mean estimate for θ_j. Hint: Write down θ_j as the expectation and recall the use of the indicator function $I(\cdot)$, for instance, we can write $\int_0^1 g(x)dx = \int_{-\infty}^{\infty} I(x \in [0, 1])g(x)dx$.

2.2.10 Explain the first step of the E-estimator.

2.2.11* Explain the second step of the E-estimator.

2.2.12 Explain the third step of the E-estimator.

2.2.13 What is the reason for having the indicator function in (2.2.3)?

2.2.14* Why do we have a factor 2 in the right side of (2.2.4)?

2.2.15 Is the Fourier estimator (2.2.7) unbiased?

2.2.16* Why is the sum in (2.2.8) divided by $(n-1)$ despite the fact that there are n terms in the sum?

2.2.17* Verify formula (2.2.10).

2.2.18 What is the definition of the integrated squared bias of a density estimate?

2.2.19 What is the definition of the MISE? Comment on this criterion.

2.2.20 Verify relation (2.2.13).

2.2.21* Prove that (2.2.16) is the projection on the class of densities under the ISE criterion.

2.2.22 Formula (2.2.16) defines the projection of a function on a class of densities. Draw a graphic of a continuous density estimate that takes on both positive and negative values and is integrated to one. Then explain how projection (2.2.16) will look like. Hint: Think about a horizontal line such that the curve above it is the projection.

2.2.23 Verify formula (2.2.19). What can one conclude from the formula?

2.2.24 Explain all parameters of the E-estimator used in Figure 2.5.

2.2.25* Use Figure 2.5 and choose an optimal argument $cJ0$ for each combination of sample sizes 100, 200, 300 and four corner functions. Comment on your recommendations.

2.2.26* Use Figure 2.5 and choose an optimal argument $cJ1$ for each combination of sample sizes 100, 200, 300 and all four corner functions. Comment on your recommendations.

2.2.27* Use Figure 2.5 and choose an optimal argument cTH for each combination of sample sizes 100, 200, 300 and all four corner functions. Comment on your recommendations.

2.2.28* Use Figure 2.6 and repeat Exercises 2.2.25-27.

2.2.29 Suppose that a random variable of interest Y is supported on a given interval $[a, b]$. How can the software, developed for the unit interval $[0, 1]$, be used for estimation of the density $f^Y(y)$?

2.2.30* Consider the question of Exercise (2.2.29) only now the support is unknown.

2.2.31* Explain and prove formulas (2.2.20), (2.2.21) and (2.2.22).

2.2.32 What is the definition of a consistent density estimator?

2.2.33* Prove (2.2.25) for the case of Sobolev densities. Is it possible to establish the consistency of a projection estimator for densities from a Sobolev class of order $\alpha \leq 1/2$?

2.2.34* For a class of analytic densities, establish an inequality similar to (2.2.25).

2.2.35* Prove consistency of the E-estimator for a Sobolev class of order $\alpha > 1/2$.

2.3.1 Explain the problem of a nonparametric regression. What is the difference between the classical linear regression and nonparametric one?

2.3.2 Why is the regression function $\mathbb{E}\{Y|X = x\}$ a reasonable predictor of Y given $X = x$?

2.3.3 Verify (2.3.3). What does the inequality tell us about the variance? Does this inequality shed light on the MSE in (2.3.1)?

2.3.4* Is (2.3.4) a more general regression model than (2.3.2)? Justify your answer. Can these two models be equivalent?

2.3.5 Define homoscedastic and heteroscedastic regressions. Which one, in your opinion, is more challenging?

2.3.6 Explain every step in establishing (2.3.6).

2.3.7 Is (2.3.7) a sample mean estimator? Explain your answer.

2.3.8* What is the variance of the estimator (2.3.7)? What is the coefficient of difficulty?

2.3.9 Explain the motivation of the estimator (2.3.8) and why it uses truncation from the zero of the design density estimator.

2.3.10* Evaluate the mean, variance and the coefficient of difficulty of Fourier estimator (2.3.8). Hint: Begin with the case of known design density.

2.3.11* Explain the modification of the E-estimate proposed in Remark 2.3.1. Then suggest a procedure which uses not only $\hat{\theta}_0$ but also other Fourier estimates.

2.3.12* Propose a procedure for estimation of θ_j using the idea of numerical integration. Then explore its mean and variance. Hint: Use (2.3.6) and then replace $m(X_l)$ by Y_l.

2.3.13 Explain arguments of Figure 2.7.

2.3.14* For each corner function and each sample size, find optimal parameters of the E-estimator used in Figure 2.7. Explain the results.

2.3.15 Repeat Figure 2.7 with different design densities that make estimation of the Bimodal more and less complicated.

2.3.16 Using Figure 2.7, explain how the scale function affects estimation of each corner function.

2.4.1 What is the definition of a Bernoulli random variable?

2.4.2 What is the mean and variance of a Bernoulli random variable?

2.4.3 Consider the sum of n identically distributed and independent Bernoulli random variables. What is the mean and variance of the sum?

2.4.4 Present several examples of the Bernoulli regression.

2.4.5 Why is the Bernoulli regression problem equivalent to estimation of the conditional probability of the success given the predictor?

2.4.6 Describe the E-estimator for Bernoulli regression.

2.4.7* Explain the motivation behind the Fourier estimator (2.4.5). Then find its mean, variance and the coefficient of difficulty.

2.4.8 Estimates in Figure 2.8 are not satisfactory. Can they be improved by some innovations or there is nothing that can be done better for the data at hand?

2.4.9* Repeat Figure 2.8 for different sample sizes and for each corner function choose a minimal sample size that implies a reliable regression estimation.

2.4.10 If our main concern is the shape of curves, which of the corner functions is more difficult and simpler for estimation?

2.4.11 Find better values for parameters of the E-estimator used in Figure 2.8.

2.4.12 Compare presentation of diagrams in Figures 2.7 and Figure 2.8. Which one do you prefer? Explain your choice.

2.5.1 How to construct a tensor-product basis?

2.5.2 Write down Fourier approximation of a three-dimensional density, supported on $[0,1]^3$, using a tensor-product cosine basis.

2.5.3* Explain the sample mean Fourier estimator (2.5.4). Find the mean, variance, and the coefficient of difficulty.

2.5.4 Consider the problem of estimation of a bivariate density supported on $[0,1]^2$. Explain constriction of the E-estimator.

2.5.5 Repeat Figure 2.9 several times, and for each simulation compare your understanding of the underlying density, exhibited by data, with the E-estimate.

2.5.6 Using Figure 2.9, suggest optimal parameters for the E-estimator.

2.5.7 Repeat Figure 2.9 with underlying densities generated by different densities for the variables X_1 and X_2. Does this make estimation of the bivariate density more complicated or simpler?

2.5.8 For the setting of Exercise 2.5.7, would you recommend to use the same or different parameters of the E-estimator?

2.5.9 Verify relations (2.5.5).

2.5.10* Verify expressions for the optimal cutoffs and the MISE presented in (2.5.6).

2.5.11 What is the curse of multidimensionality?

2.5.12 How can the curse of multidimensionality be overcome?

2.5.13* Figure 2.9 allows us to estimate a density supported on $[0,1]^2$. How can a density with unknown support be estimated using the software? Will this change the coefficient of difficulty of the E-estimator?

2.5.14* Consider a bivariate regression function $m(x_1, x_2) := \mathbb{E}\{Y | X_1 = x_1, X_2 = x_2\}$. A sample of size n from a triplet (X_1, X_2, Y) is available. Suggest a bivariate regression E-estimate. Hint: Propose convenient assumptions.

2.6.1 Consider estimation of the mean of a bounded function $g(X)$ based on a sample of size n from X. Propose an estimator and its $(1 - \alpha)$ confidence interval.

2.6.2* Suppose that we have a sample of size n from a normal random variable X with unknown mean θ. Let $g(x)$ be a bounded and differentiable function. Propose an estimator of $g(\theta)$ and its $1 - \alpha$ confidence interval.

2.6.3 How is the Central Limit Theorem used in construction of confidence intervals?

2.6.4 Explain the approach leading to (2.6.3).

2.6.5 Prove (2.6.4).

2.6.6 What is the definition of a nonparametric pointwise confidence band?

2.6.7 What is the definition of a nonparametric simultaneous confidence band?

2.6.8 Explain how the relation (2.6.6) is obtained.

2.6.9 Define the variance and the ISB (integrated squared bias) of a nonparametric estimator.

2.6.10 Why an unknown ISB may prevent us from construction of a reliable confidence band for small samples?

2.6.11 Using Figure 2.3 explain why a variance-band, which ignores the ISB, may be a reasonable approach for the corner densities.

2.6.12* An analytic class of densities was introduced in (2.1.12). Consider an analytic density, write down the MISE of a projection estimate, and then evaluate the ISB using (2.1.12). Using the obtained expression and the Central Limit Theorem, suggest a reasonable confidence band. Hint: Note that the inference for analytic functions justifies the variance-band approach.

2.6.13* Verify (2.6.7). Hint: Recall the rule of calculating the variance of a sum of dependent random variables

2.6.14* Write down a reasonable estimator for the covariance σ_{ij} defined below line (2.6.7). Then explore its statistical properties like the mean and variance.

2.6.15* Verify (2.6.8). Hint: Use the Central Limit Theorem. Further, note that the set $\mathcal{N}$ is random, so first consider the conditional expectation given the set.

2.6.16 Explain the underlying idea of the proposed simultaneous confidence band defined in (2.6.10).

2.6.17* Propose an algorithm for calculating the simultaneous confidence band introduced in (2.6.10). In other words, think about a program that will do this. Hint: You need to find coefficients $c_{j,\alpha/2,n}$ satisfying (2.6.9). Note that Z_j in (2.6.9) are components of a specific normal random vector defined above line (2.6.9).

2.6.18 Repeat Figure 2.10. Find a simulation when the E-estimate for the Normal is not perfect, and explain the shape of the bands.

2.6.19 If the parameter α is decreased, then will a corresponding band be wider or narrower? Check your answer using Figure 2.10.

2.6.20* If you look at a band in Figure 2.10, then it is plain to notice that its width is not constant. Why?

2.6.21 Titles of diagrams in Figure 2.10 show values of the ISE. What is the definition of the ISE and why is it used to quantify quality of a nonparametric estimate?

2.6.22 How the ISE of an E-estimate is related to a band? Answer this question heuristically and using mathematical analysis, and then check your answers using Figure 2.10.

2.6.23* Fourier estimators $\hat{\theta}_j$, $j = 1, 2, \ldots$ of the density E-estimator are dependent random variables. In particular, verify the following formula or find a correct expression,

$$\text{Cov}(\hat{\theta}_j, \hat{\theta}_i) = n^{-1}[2\mathbb{E}\{\cos(j\pi X)\cos(i\pi X)\} - \theta_j\theta_i] = n^{-1}[\theta_{j-i} + \theta_{j+i} - \theta_j\theta_i]. \quad (2.7.13)$$

Hint: Trigonometric formula

$$\cos(\gamma)\cos(\beta) = [\cos(\gamma + \beta) + \cos(\gamma - \beta)]/2 \quad (2.7.14)$$

may be useful.

2.6.24* Explore the problem of Exercise 2.6.23 for a regression model.

2.8 Notes

In what follows the subsections correspond to sections in the chapter.

2.1 Fourier series approximation is one of the cornerstones of the mathematical science. The presented material is based on Chapter 2 in Efromovich (1999a). The basic idea of Fourier series is that a periodic function may be expressed as a sum of sines and cosines. This idea was known to the Babylonians, who used it for the prediction of celestial events. The history of the subject in more recent times begins with d'Alembert, who in the eighteenth century studied the vibrations of a violin string. Fourier's contributions began in 1807 with his studies of the problem of heat flow. He made a serious attempt to prove that any function may be expanded into a trigonometric sum. A satisfactory proof was found later by Dirichlet. These and other historical remarks may be found in the book by Dym and McKean (1972). Also, Section 1.1 of that book gives an excellent explanation of the Lebesgue integral, which should be used by readers with advanced mathematical background. The mathematical books by Krylov (1955), Bary (1964) and Kolmogorov and Fomin (1957) give a relatively simple discussion (with rigorous proofs) of Fourier series. There are many good books on approximation theory. Butzer and Nessel (1971) and Nikolskii (1975) are the classical references, and DeVore and Lorentz (1993), Temlyakov (1993), and Lorentz, Golitschek and Makovoz (1996) may be recommended as solid mathematical references. An interesting topic of wavelet bases and series wavelet estimators are discussed in Walter (1994), Donoho and Johnstone (1995), Efromovich (1997c; 1999a; 2000a,c; 2004e; 2007a,b; 2009b), Härdle et al. (1998), Mallat (1998), Vidakovic (1999), Efromovich et al. (2004), Efromovich et al. (2008), Nason (2008), Efromovich and Valdez-Jasso (2010), and Efromovich and Smirnova (2014a,b).

2.2 This section is based on Chapter 3 of Efromovich (1999a). The first result about optimality of Fourier series estimation of nonparametric densities is due to Chentsov (1962). Professor Chentsov was not satisfied with the fact that this estimate could take on negative values, and later proposed to estimate $g(x) := \log(f(x))$ by a series estimator $\hat{g}(x)$ and then set $\hat{f}(x) := e^{\hat{g}(x)}$; see Chentsov (1980) and also Efron and Tibshirani (1996). Clearly, the last estimator is nonnegative. Recall that we dealt with this issue by using the projection (2.2.17), see also a discussion in Efromovich (1999a) and Glad, Hjort and Ushakov (2003).

An interesting discussion of the influence of Kolmogorov's results and ideas in approximation theory on density estimation may be found in Chentsov (1980) and Ibragimov and Khasminskii (1981). Series density estimators are discussed in the books by Devroye and Györfi (1985), Devroye (1987), Thompson and Tapia (1990), Tarter and Lock (1993), Hart (1997), Wasserman (2006), Tsybakov (2009), Hollander, Wolfe and Chicken (2013). Asymptotic justification of using adaptive Fourier series density estimators is given in Efromovich (1985; 1989; 1996b; 1998a; 1999d; 2000b; 2008b; 2009a; 2010a,c; 2011c) and Efromovich and Pinsker (1982). A discussion of plug-in estimation may be found in Bickel and Doksum (2007). Projection procedures are discussed in Birgé and Massart (1997), Efromovich (1999a, 2010c), and Tsybakov (2009). There is also a rich literature on other approaches to density estimation, see a discussion in the classical book by Silverman (1986) as well as in Efromovich (1999a), Scott (2015), and more recent results in Sakhanenko (2015, 2017). See also an interesting practical application in Efromovich and Salter-Kubatko (2008).

Let us specifically stress that the used default values of parameters of the E-estimator are not necessarily "optimal," and even using the word optimal is questionable because there is no feasible risk for the analysis of E-estimator for small samples. It is up to the reader to choose parameters that imply better fit for the reader's favorite corner densities and sample sizes. And this is where the R package becomes so handy because it allows one to find custom-chosen parameters for every considered setting.

For the more recent development in the sharp minimax estimation theory, including the superefficiency, see Efromovich (2014a, 2016a, 2017).

2.3 The discussed nonparametric regression is based on Chapter 4 of Efromovich (1999a). There are also many good books where different applied and theoretical aspects of non-parametric regression are discussed. These books include Carroll and Ruppert (1988), Eubank (1988), Müller (1988), Härdle (1990), Wahba (1990), Green and Silverman (1994), Wand and Jones (1995), Simonoff (1996), Nemirovskii (1999), Györfi et al. (2002), Takezawa (2005), Li and Racine (2009), Berk (2016), Faraway (2016), Matloff (2017) and Yang (2017) among others. A chapter-length treatment of orthogonal series estimates may be found in Eubank (1988, Chapter 3). Asymptotic justification of orthogonal series estimation for the regression model is given in Efromovich (1986; 1992; 1994b; 1996a; 2001c; 2002; 2005a; 2007d,e), where it is established that for smooth functions a data-driven series estimator outperforms all other possible data-driven estimators. Practical applications are also discussed. The heteroscedastic regression was studied in Efromovich (1992, 2013a) and Efromovich and Pinsker (1996), where it is established that asymptotically a data-driven orthogonal series estimator outperforms any other possible data-driven estimators whenever an underlying regression function is smooth. Also, Efromovich and Pinsker (1996) present results of numerical comparison between series and local linear kernel estimators. In some applications it is known that an underlying regression function satisfies some restrictions on its shape, like monotonicity or unimodality. Nonparametric estimation under the shape restrictions is discussed in Efromovich (1999a, 2001a), Horowitz and Lee (2017), and in the book Groeneboom and Jongbloed (2014) solely devoted to the estimation under shape restrictions.

2.4 Bernoulli regression is considered in Chapter 4 of Efromovich (1999a), see also a discussion and further references in Mukherjee and Sen (2018). It also may be referred to as binary regression. The asymptotic justification of the E-estimator is given in Efromovich (1996a) and Efromovich and Thomas (1996) where also an interesting application to the analysis of the sensitivity can be found. As we will see shortly, this regression is a pivot in solving a number of statistical problems with missing data.

2.5 This section is based on Chapter 6 of Efromovich (1999a) which contains discussion of a number of multivariate problems arising in nonparametric estimation. Asymptotic theory can be found in Efromovich (2000b; 2002; 20010c,d). Although generalization of most of the univariate series estimators to multivariate series estimators appears to be feasible, we have seen that serious problems arise due to the curse of multidimensionality, as it was termed by Bellman (1961). The curse is discussed in the books by Silverman (1986) and Hastie and Tibshirani (1990). Many approaches have been suggested aimed at a simplification and overcoming the curse: additive and partially linear modeling, principal components analysis, projection pursuit regression, classification and regression trees (CART), multivariate adaptive regression splines, etc. A discussion may be found in the book Izenman (2008). Many of these methods are supported by a number of specialized R packages, see also Everitt and Hothorn (2011). Approximation theory is discussed in the books by Nikolskii (1975), Temlyakov (1993), and Lorentz, Golitschek, and Makovoz (1996). A book-length discussion of multivariate density estimators (with a particular emphasis on kernel estimators) is given in Scott (2015).

2.6 Inference about nonparametric estimators, based on the analysis of the MISE and its decomposition into variance and integrated squared bias, can be found in the books Efromovich (1999a) and Wasserman (2006). The latter contains an interesting overview of different approaches to construction of confidence bands as well as a justification of the variance-band approach. Theoretical analysis of nonparametric confidence bands, including proofs of impossibility of constructing efficient data-driven bands, can be found in (mathematically involved) papers Cai and Low (2004), Genovese and Wasserman (2008), Giné and Nickl (2010), Hoffmann and Nickl (2011), Cai and Guo (2017), and Efromovich and Chu (2018a,b). Nonparametric hypotheses testing is explored in the book by Ingster and Suslina (2003). Bayesian approach is discussed in Yoo and Ghosal (2016).

Chapter 3

Estimation for Basic Models of Modified Data

This chapter considers basic models where underlying observations are modified and the process of modification is known. The studied topics serve as a bridge between the case of direct data and the case of missing, truncated and censored data considered in the following chapters. Section 3.1 is devoted to density estimation based on biased data. Biased data is a classical statistical example of modified data. The interesting aspect of the presentation is that biased sampling is explained via a missing mechanism. Namely, there are underlying hidden realizations of a random variable of interest X^*, and then a hidden realization may be observed or skipped with the likelihood depending on the value of the realization. This sampling mechanism creates biased data because the distribution of the observed X is different from the distribution of the hidden X^*. As we will see in the following chapters, missing, truncation, censoring and other modifications typically imply biased data. Section 3.2 considers regression with biased responses. Section 3.3 explores regression with biased predictors and responses. Other sections are devoted to special topics. Among them, results of Section 3.7, where Bernoulli regression with unavailable failures is considered, will be often used in the next chapters.

3.1 Density Estimation for Biased Data

We are interested in estimation of the density $f^{X^*}(x)$ of the random variable of interest X^*. If a sample from X^* is available (the case of direct observations), then we know from Section 2.2 how to construct E-estimator of the density. In many applications a direct sampling is impossible and instead the following biased sampling may be possible. There is a hidden sequential sampling from pair (X^*, A) where A is a Bernoulli random variable such that $\mathbb{P}(A = 1|X^* = x) =: B(x)$ and $B(x)$ is called the biasing function. (Recall that a Bernoulli random variable A takes on values 0 or 1). If (X_1^*, A_1) is the first hidden realization of the pair, then we observe $X_1 = X_1^*$ if $A_1 = 1$ and skip the hidden realization otherwise. Suppose that we are interested in a sample of size n. Then the hidden sequential simulation continues until n observations $X_1, \ldots, X_n$ of X are obtained.

Let us present a particular example of biased data. Suppose that a researcher would like to know the distribution of the ratio of alcohol X^* in the blood of liquor-intoxicated drivers traveling along a particular highway. A sample $X_1, \ldots, X_n$ of the ratios of alcohol is available from routine police reports on arrested drivers charged with driving under the influence of alcohol (a routine report means that there are no special police operations to reveal all intoxicated drivers). Because a drunker driver has a larger chance of attracting the attention of the police, it is clear that the available observations are biased (and we may say length-biased) toward higher ratios of alcohol in the blood. Thus, the researcher should make an appropriate adjustment in a method of estimation of the distribution of the ratio of alcohol in the blood of all intoxicated drivers. There are many other similar examples in different sciences where a likelihood for an observation to appear in a sample depends on its value.

To estimate the density $f^{X^*}(x)$, we need to understand how it is related to the density $f^X(x)$ of biased observations. In what follows, we assume that X^* is supported on $[0, 1]$ and the biasing function $B(x) := \mathbb{P}(A = 1|X^* = x)$ is known and

$$B(x) \geq c_0 > 0. \tag{3.1.1}$$

The joint density of the pair (X^*, A) can be written as

$$f^{X^*,A}(x, 1) = f^{X^*}(x)\mathbb{P}(A = 1|X^* = x) = f^{X^*}(x)B(x). \tag{3.1.2}$$

Further, we observe $X = X^*$ given $A = 1$. This, together with (3.1.2), allows us to write,

$$f^X(x) = f^{X^*|A}(x|1) = \frac{f^{X^*,A}(x, 1)}{\mathbb{P}(A = 1)} = \frac{f^{X^*}(x)B(x)}{\mathbb{P}(A = 1)}. \tag{3.1.3}$$

Further, (3.1.2) yields

$$\mathbb{P}(A = 1) = \int_0^1 f^{X^*,A}(x, 1)dx = \int_0^1 f^{X^*}(x)B(x)dx = \mathbb{E}\{B(X^*)\}. \tag{3.1.4}$$

Finally, (3.1.3) allows us to get the following formula

$$\mathbb{E}\left\{\frac{1}{B(X)}\right\} = \int_0^1 \frac{f^X(x)}{B(x)}dx = \int_0^1 \frac{f^{X^*}(x)B(x)}{B(x)\mathbb{P}(A = 1)}dx = \frac{1}{\mathbb{P}(A = 1)}. \tag{3.1.5}$$

This formula points upon a simple sample mean estimator of $\mathbb{P}(A = 1)$.

Note that the above-presented formulas are based on the sequential model of creating biased data. In general, instead of exploring the process of collecting biased data, the problem of estimation of $f^{X^*}(x)$ based on a biased sample from X is formulated via the following relation:

$$f^X(x) = \frac{f^{X^*}(x)B(x)}{\int_0^1 f^{X^*}(u)B(u)du}. \tag{3.1.6}$$

In this case $B(x)$ is not necessarily the probability and may take on values larger than 1. On the other hand, according to (3.1.6), the biasing function can be always rescaled to make it not larger than 1, and this rescaling does not change the probability model.

In what follows we will use model (3.1.3)-(3.1.4) rather than (3.1.6) to stress the fact that biased data may be explained via a sequential missing mechanism. As it was already explained, this does not affect the generality of considered model. Also, we will use notation $B^{-1}(x) := 1/B(x)$.

Now, given known $B(x)$, (3.1.1) and (3.1.3)-(3.1.4), we are in a position to explore the problem of estimation of the density f^{X^*} based on a biased sample $X_1, \ldots, X_n$ from X.

First of all, let us stress that if the biasing function $B(x)$ is unknown, then no consistent estimation of the density of interest is possible. This immediately follows from (3.1.6) which shows that only the product $f^{X^*}(x)B(x)$ is estimable. Unless the nuisance function is known (or may be estimated), we cannot factor out $f^{X^*}(x)$ from the product $f^{X^*}(x)B(x)$. This is the pivotal moment in our understanding of the biasing modification of data, and we will observed it in many particular examples of missing and modified data. Only if the biasing function is known (for instance from previous experiments) or may be estimated based on auxiliary data, the formulated estimation problem becomes feasible and consistent estimation of $f^{X^*}(x)$ becomes possible.

The E-estimation methodology of Section 2.2 tells us that we need to propose a sample mean estimator of Fourier coefficients θ_j of the density of interest $f^{X^*}(x)$. To do this, we should first write down θ_j as an expectation and then mimic the expectation by a sample mean estimator. To make the first step, using (3.1.3) let us write down Fourier coefficients θ_j as follows:

$$\theta_j := \int_0^1 \varphi_j(x) f^{X^*}(x) dx$$

$$= \mathbb{P}(A = 1) \int_0^1 \varphi_j(x) f^X(x) B^{-1}(x) dx = \mathbb{P}(A = 1) \mathbb{E}\{\varphi_j(X) B^{-1}(X)\}. \tag{3.1.7}$$

Here $\varphi_j(x)$ are elements of the cosine basis on $[0, 1]$ (the definition and discussion can be found in Section 2.1). Note that $\theta_0 = \int_0^1 \varphi_0(x) f^{X^*}(x) dx = \int_0^1 f^{X^*}(x) dx = 1$, and hence we need to estimate only Fourier coefficients θ_j, $j \geq 1$. Formula (3.1.7) implies the following plug-in sample mean estimator of Fourier coefficients:

$$\hat{\theta}_j := \hat{P} n^{-1} \sum_{l=1}^n \varphi_j(X_l) B^{-1}(X_l), \tag{3.1.8}$$

where according to (3.1.5)

$$\hat{P} := \frac{1}{n^{-1} \sum_{l=1}^n B^{-1}(X_l)} \tag{3.1.9}$$

is the plug-in sample mean estimator of $\mathbb{P}(A = 1)$.

Fourier estimator (3.1.8) yields the density E-estimator $f^{X^*}(x)$ of Section 2.2, and the coefficient of difficulty is

$$d := \lim_{n,j \to \infty} n \mathbb{E}\{(\hat{\theta}_j - \theta_j)^2\}$$

$$= [\mathbb{P}(A = 1)]^2 \mathbb{E}\{B^{-2}(X)\} = \mathbb{P}(A = 1) \mathbb{E}\{B^{-1}(X^*)\}. \tag{3.1.10}$$

Recall that the coefficient of difficulty of estimation of density f^{X^*}, based on direct observations of X^*, is 1 (see Section 2.2). Is the coefficient (3.1.10) larger? In other words, do biased data make the estimation problem more complicated or some biasing schemes may improve the estimation? To answer the question, let us use the Cauchy-Schwarz inequality (1.3.33) and write,

$$1 = [\int_0^1 f^{X^*}(x) dx]^2 \leq [\int_0^1 f^{X^*}(x) B(x) dx][\int_0^1 f^{X^*}(x) B^{-1}(x) dx]$$

$$= \mathbb{P}(A = 1) \mathbb{E}\{B^{-1}(X^*)\} = d, \tag{3.1.11}$$

with the equality only for the case of direct observations when $B(x) \equiv 1$. We conclude that biasing makes estimation more complicated and requires larger sample sizes for the same quality of estimation.

Figure 3.1 allows us to both understand the problem of biased data and test performance of the E-estimator of the underlying density f^{X^*}. Its caption explains the simulation and the four diagrams with a corner function being the underlying density. The histograms show simulated biased data. The biasing function is linear $B(x) = a + bx$, it is often referred to as length-biasing and frequently occurs in applications. The particular values of a and b, used in the simulations, are 0.2 and 0.8, respectively. This biasing favors larger values of X^* to be observed and hinders observing smaller values. The latter skews observed data to the right, and this is clearly seen in the left diagram. Here the underlying density f^{X^*} is the Uniform, and the histogram is dramatically skewed to the right. The proposed E-estimator does a perfect recovering of the underlying Uniform density. Let us also stress that if we do not suspect that the data is biased, or suspect that the data may be biased but do not know the biasing function, it is impossible to restore the underlying density. The latter is an important observation that we will repeatedly see in cases of modified and missing data.

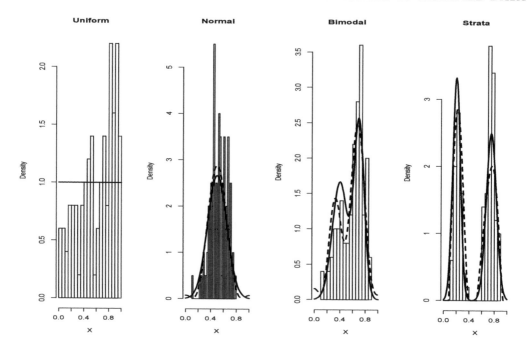

Figure 3.1 *Density estimation for biased data with the biasing function* $B(x) = a + bx$. *Four diagrams correspond to different underlying densities* $f^{X^*}(x)$ *indicated in the title and shown by the solid line. The biased observations are shown by the histogram and the density E-estimate* $\hat{f}^{X^*}(x)$ *by the dashed line.* {*Parameters of the biasing function are controlled by set.B* $= c(a,b)$} *[n = 100, set.B = c(0.2,0.8), cJ0 = 4, cJ1 = .5, cTH = 4]*

The Normal diagram in Figure 3.1 is another interesting example of the length-biased data. Again we see the skewed to the right histogram of the biased data, and the E-estimator correctly restores the symmetric shape of the Normal density. The funny tails are due to outliers. The Bimodal is another teachable example. This density is difficult for estimation even for the case of direct observations, and here look at how the E-estimator correctly increases (with respect to the histogram) the left mode and decreases the right mode. A similar outcome is observed in the Strata diagram where the E-estimator also corrects the histogram. Let us also shed light on underlying coefficients of difficulty. For instance, for the case of $B(x) = 0.1 + 0.9x$, coefficients of difficulty for our 4 corner densities are 1.4, 1.1, 1.1 and 1.3, respectively.

Figure 3.2 allows us to zoom on biased data and quantify the quality of estimation. The diagrams are similar to those in Figure 3.1, only here confidence bands are added (note that they are cut below the bottom line of the histogram). In the top diagram the underlying density is the Bimodal shown by the solid line. The histogram clearly exhibits the biased data that are skewed to the right. The E-estimate (the dashed line) fairly well exhibits the underlying density, and the 0.95-level confidence bands shed additional light on the quality of estimation. The subtitle shows the integrated squared error (ISE) of the E-estimate, the estimated coefficient of difficulty and the sample size. The bottom diagram exhibits a simulation with the underlying density of interest being the Strata. The left stratum is estimated worse than the right one, and we see that the underlying left strata is beyond the pointwise band but still within the simultaneous one. While the E-estimate is far from being perfect, it does indicate the two pronounced strata and even shows that the left mode is larger than the right one despite the heavily right-skewed biased data.

Histogram of X, E-estimate and Confidence Bands

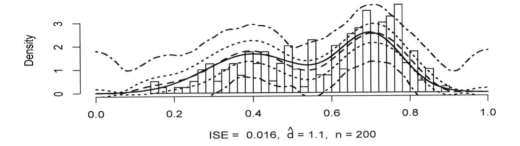

ISE = 0.016, $\hat{d}$ = 1.1, n = 200

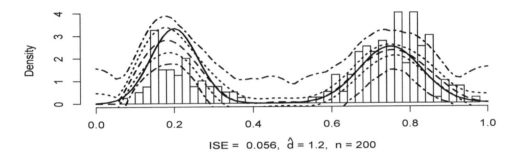

ISE = 0.056, $\hat{d}$ = 1.2, n = 200

Figure 3.2 *Density estimation for biased data. Results of two simulations are exhibited for the Bimodal and the Strata underlying densities $f^{X^*}(x)$. Simulations and the structure of diagrams are similar to Figure 3.1 only here $1-\alpha$ confidence bands, explained in Section 2.6, are added. The $1-\alpha$ pointwise and simultaneous bands are shown by the dotted and dot-dashed lines, respectively. The exhibited confidence bands are truncated from below by the bottom line of the histogram. {Parameters of the biasing function are controlled by set.B =c(a,b), underlying densities are chosen by set.corn, and α is controlled by the argument alpha.} [n = 200, set.B = c(0.2,0.8), set.corn = c(3,4), alpha = 0.05, cJ0 = 3, cJ1 = 0.8, cTH = 4]*

One theoretical remark is due about the proposed E-estimator. Its natural competitor is the ratio estimator which is based on formula (3.1.3). The ratio estimator is defined as the E-estimator $\hat{f}^X(x)$, based a biased sample, divided by $B(x)/\hat{P}$. It is possible to show that the proposed E-estimator is more efficient than the ratio estimator, and this is why it is recommended. On the other hand, the appealing feature of the ratio estimator is in its simplicity.

We are finishing this section with a remark about the relation between the biased and missing data. Recall that a biased sample may be generated by a sequential sampling from X^* when some of the realizations are missed. Further, the sample size n of biased observations corresponds to a larger random sample size N (stopping time) of the hidden sample from X^*. One may think that $N-n$ observations in the hidden sample from X^* are missed. This thinking bridges the biased data and the missing data. Furthermore, there is an important lesson that may be learned from the duality between the biasing and missing. We know that unless the biasing function is known, consistent estimation of the density of X^* is impossible. Hence, missing data may preclude us from consistent estimation of an underlying density f^{X^*} unless some additional information about the missing mechanism is available.

In other words, missing may completely destroy information about density contained in a hidden sample. In the next chapters we will consider such missing mechanisms and refer to them as destructive missing.

3.2 Regression with Biased Responses

The aim is to estimate the regression function $m(x) := \mathbb{E}\{Y^*|X = x\}$ for the pair of continuous random variables (X, Y^*) where X is the predictor and Y^* is the response. For the case of a direct sample from (X, Y^*) this problem was discussed in Section 2.3 where a regression E-estimator was proposed. Here we are considering a more complicated setting when a sample $(X_1, Y_1), \ldots, (X_n, Y_n)$ from (X, Y) is available, and

$$f^{X,Y}(x,y) = f^X(x)f^{Y|X}(y|x) = f^X(x)[f^{Y^*|X}(y|x)B(x,y)D(x)]. \qquad (3.2.1)$$

Here $B(x, y)$ is a known biasing function and

$$B(x,y) \geq c_0 > 0, \qquad (3.2.2)$$

the function $D(x)$ makes the expression in the square brackets a bona fide conditional density and it is defined as

$$D(x) := \frac{1}{\mathbb{E}\{B(X, Y^*)|X = x\}} = \mathbb{E}\{[1/B(X,Y)]|X = x\}, \qquad (3.2.3)$$

and $f^X(x)$ is the design density of the predictor X supported on $[0, 1]$, and it is assumed that $f^X(x) \geq c_* > 0$.

Let us explain how a sample with biased responses, satisfying (3.2.1), may be generated. First, a sample $X_1, \ldots, X_n$ from X is generated according to the design density $f^X(x)$. Then for each X_l, a single biased observation Y_l is generated according to the algorithm of Section 3.1 with (notation of that section is used) the density $f^{X^*}(y) = f^{Y^*|X}(y|X_l)$ and the biasing function $B(y) := \mathbb{P}(A = 1|X^* = y, X_l) = B(X_l, y)$, where A is the Bernoulli variable. Note that the difference between this sampling and sampling in Section 3.1 is that here biased responses are generated n times with different underlying densities and different biasing functions.

It is a nice exercise to check that the above-described biasing mechanism implies (3.2.1). Write,

$$f^{Y|X}(y|x) = f^{Y^*|A,X}(y|1,x)$$

$$= \frac{f^{Y^*,A|X}(y, 1|x)}{\mathbb{P}(A = 1|X = x)} = \frac{f^{Y^*|X}(y|x)\mathbb{P}(A = 1|Y^* = y, X = x)}{\mathbb{P}(A = 1|X = x)}. \qquad (3.2.4)$$

According to the above-described biasing algorithm, $P(A = 1|Y^* = y, X = x) = B(x, y)$, and we also get that

$$\mathbb{P}(A = 1|X = x) = \mathbb{E}\{\mathbb{P}(A = 1|Y^*, X)|X = x\}$$

$$= \mathbb{E}\{B(X, Y^*)|X = x\} = 1/D(x). \qquad (3.2.5)$$

Now we plug (3.2.5) in the right side of (3.2.4) and get (3.2.1) with $B(x, y) = P(A = 1|Y^* = y, X = x)$ and $D(x) = 1/\mathbb{P}(A = 1|X = x)$, as we wished to show. Further, the sampling mechanism and formulas (3.2.4)-(3.2.5) shed a new light on the model (3.2.1).

Now we are ready to propose an E-estimator for the regression with biased responses. The good news is that the biasing function $B(x, y)$ is assumed to be known and positive. The bad news is that function $D(x)$, $x \in [0, 1]$ is unknown. Hence our first task is to estimate this function.

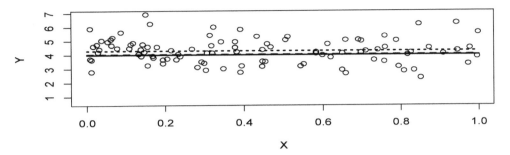

Biased Responses, n = 100 , ISE = 0.00079 , ISEN = 0.083

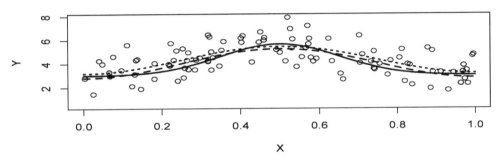

Biased Responses, n = 100 , ISE = 0.06 , ISEN = 0.17

Figure 3.3 *Regression with biased responses. The biasing function is $B(x, y) = b_1 + b_2 x + b_3 y$, the design density is the Uniform, and the hidden regression is $Y^* = [m(X) + 3\sigma] + \sigma\xi$ where $m(x)$ is a corner function, ξ is a standard normal regression error which is independent of the predictor X, and σ is a parameter. The two diagrams correspond to the Uniform and the Normal functions $m(x)$, and the regression functions $m(x) + 3\sigma$ are shown by the solid lines. Biased data are shown by circles. Regression E-estimate (the dashed line) and the naïve regression E-estimate (the dotted line), based on biased data, have integrated squared errors indicated as ISE and ISEN, respectively. {Parameters of the biasing function are controlled by set.B = c(b_1, b_2, b_3) and note that the biasing function must be positive, underlying regressions are chosen by the argument set.corn, σ is controlled by the argument sigma.} [n = 100, sigma = 1, set.B = c(0.3,0.5,2), set.corn = c(1,2), c = 1, cJ0 = 3, cJ1 = 0.8, cTH = 4]*

As usual, we use E-estimation methodology for estimating $D(x)$. Using (3.2.3) we can write for its Fourier coefficients

$$\kappa_j := \int_0^1 D(x)\varphi_j(x)dx$$

$$= \int_0^1 \mathbb{E}\{[1/B(x, y)]|X = x\}\varphi_j(x)dx = \mathbb{E}\Big\{\frac{\varphi_j(X)}{f^X(X)B(X, Y)}\Big\}. \qquad (3.2.6)$$

This implies the plug-in sample mean estimator

$$\hat{\kappa}_j := n^{-1}\sum_{l=1}^{n}\frac{\varphi_j(X_l)}{\max(\hat{f}^X(X_l), c/\ln(n))B(X_l, Y_l)}, \qquad (3.2.7)$$

where $\hat{f}^X(x)$, $x \in [0, 1]$ is the E-estimator of the density $f^X(x)$ based on $X_1, \ldots, X_n$ (recall that the predictor is not biased).

The Fourier estimator (3.2.7) yields the E-estimator $\hat{D}(x)$, $x \in [0,1]$.

Now we are ready to explain how we can estimate Fourier coefficients of the regression function $m(x) := \mathbb{E}\{Y^*|X = x\}$. Using (3.2.1), Fourier coefficient of $m(x)$, $x \in [0,1]$ can be written as follows,

$$\theta_j := \int_0^1 m(x)\varphi_j(x)dx = \int_0^1 \left[\int_{-\infty}^{\infty} yf^{Y^*|X}(y|x)dy \right]\varphi_j(x)dx$$

$$= \int_0^1 \left[\int_{-\infty}^{\infty} yf^{Y|X}(y|x)[B(x,y)D(x)]^{-1}dy \right]\varphi_j(x)dx$$

$$= \mathbb{E}\left\{ \frac{Y\varphi_j(X)}{f^X(X)B(X,Y)D(X)} \right\}. \tag{3.2.8}$$

This yields the following plug-in sample mean estimator of θ_j,

$$\hat{\theta}_j := \int_0^1 \frac{Y_l\varphi_j(X_l)}{\max(\hat{f}^X(X_l), c/\ln(n))B(X_l, Y_l)\hat{D}(X_l)}. \tag{3.2.9}$$

There is one useful remark about the plug-in $\hat{D}(X_l)$. It follows from (3.2.2) and (3.2.3) that if $B(x,y) \leq c_B < \infty$, and recall that the biasing function is known, we get $D(x) \geq 1/c_B$. Then $\hat{D}(X_l)$, used in (3.2.9), may be truncated from below by $1/c_B$, that is we may plug in $\max(\hat{D}(X_l), 1/c_B)$.

Fourier estimator (3.2.9) yields the regression E-estimator $\hat{m}(x)$, $x \in [0,1]$ defined in Section 3.3.

As we see, the E-estimator for the regression with biased response is more complicated than the one for the regular regression proposed in Section 2.3 because now we need to estimate the nuisance function $D(x)$.

Figure 3.3 allows us to test the proposed estimator, and its caption explains the simulation and the diagrams. The top diagram shows the scattergram of regression with biased responses when the underlying regression is the Uniform plus 3. Note the high volatility of the biased data. The biasing clearly skews data up, and this is highlighted by the naïve regression E-estimate of Section 2.3 (the dotted line) based solely on the biased data. As we know, without information about biasing, a regression estimator cannot be consistent. The proposed regression estimator (the dashed line) is almost perfect, and this is highlighted by the small ISE. The bottom diagram shows a similar simulation for the Normal plus 3 underlying regression function. Here we have an interesting divergence between the two estimates. The naïve one is better near the mode and worse otherwise. The integrated squared errors quantify the quality of estimation.

It is worthwhile to repeat Figure 3.3 with different parameters and learn to read scattergrams with biased responses.

3.3   Regression with Biased Predictors and Responses

We are interested in the regression $m(x) := \mathbb{E}\{Y^*|X^* = x\}$ of the response Y^* on the predictor X^*. Realizations of the pair of continuous random variables (X^*, Y^*) are hidden and instead a sample of size n from a biased pair (X, Y) is available. It is known that joint density of the biased pair (X, Y) is

$$f^{X,Y}(x,y) = f^{X^*,Y^*}(x,y)B(x,y)D, \tag{3.3.1}$$

where

$$D = \frac{1}{\mathbb{E}\{B(X^*, Y^*)\}} = \mathbb{E}\{1/B(X,Y)\} \tag{3.3.2}$$

is a constant that makes the joint density bona fide. Further, the biasing function $B(x,y)$ is known and is bounded below from zero,

$$B(x,y) \geq c_0 > 0, \tag{3.3.3}$$

and the design density $f^{X^*}(x)$ of the hidden predictor X^* is supported on $[0,1]$ and $f^{X^*}(x) \geq c_* > 0$, $x \in [0,1]$.

Model (3.3.1) looks similar to the model (3.2.1) for the biased response. The difference is that now the predictor X may be also biased. Indeed, from (3.3.1) we get that

$$f^X(x) = f^{X^*}(x)[D\mathbb{E}\{B(x,Y^*)|X^* = x\}]. \tag{3.3.4}$$

As a result, unless $\mathbb{E}\{B(x,Y^*)|X^* = x\}$ is a constant (an example of the latter is $B(x,y) = B(y)$), the observed predictor is also biased. This is what differentiates models (3.3.1) and (3.2.1).

Let us present a particular example and then explain how biased data may be generated via a sequential missing algorithm. Recall the example of Section 3.1 about the distribution of the ratio of alcohol in the blood of liquor-intoxicated drivers based on routine police reports on arrested drivers. It was explained that the data in reports was biased given that a drunker driver was more likely to be stopped by the police. Suppose that now we are interested in the relationship between the level of alcohol and the age (or income level) of the driver. If it is reasonable to assume that both the level of alcohol and the age (income level) are the factors defining the likelihood of the driver to be stopped (as the thinking goes, your wheels give clues to your age, gender, income level and marital status) then both the level of alcohol and age (income) in the reports are biased.

A possible method of simulation of the biased data is based on a sequential missing. There is an underlying hidden sequential sampling from triplet (X^*, Y^*, A) where A is a Bernoulli random variable such that $\mathbb{P}(A = 1|X^* = x, Y^* = y) = B(x,y)$ satisfying (3.3.3). If (X_1^*, Y_1^*, A_1) is the first hidden realization of the pair, then we observe $(X_1, Y_1) := (X_1^*, Y_1^*)$ if $A_1 = 1$ and skip the hidden realization otherwise. Then the hidden simulation continues until n observations of (X, Y) are available.

Let us check that the simulated sample satisfies (3.1.3). For the joint density of the observed pair (X, Y) we can write,

$$f^{X,Y}(x,y) = f^{X^*,Y^*|A}(x,y|1) = \frac{f^{X^*,Y^*,A}(x,y,1)}{\mathbb{P}(A = 1)}$$

$$= \frac{f^{X^*,Y^*}(x,y)\mathbb{P}(A = 1|X^* = x, Y^* = y)}{\mathbb{P}(A = 1)}. \tag{3.3.5}$$

If we compare (3.3.5) with (3.3.1), then we can conclude that the formulas are identical because $B(x,y) = \mathbb{P}(A = 1|X^* = x, Y^* = y)$ and $D = 1/\mathbb{P}(A = 1)$.

Now let us explain how an underlying regression function $m(x) := \mathbb{E}\{Y^*|X^* = x\}$ can be estimated by the regression E-estimator of Section 2.3. Following the E-estimation methodology, we need to understand how to estimate Fourier coefficients θ_j of the regression function. The approach is to write down Fourier coefficients as an expectation and then mimic the expectation by a sample mean estimator. Write,

$$\theta_j := \int_0^1 m(x)\varphi_j(x)dx = \int_0^1 \left[\int_{-\infty}^\infty y[f^{X^*,Y^*}(x,y)/f^{X^*}(x)]dy\right]\varphi_j(x)dx. \tag{3.3.6}$$

Using (3.3.1) we get the following expression for the marginal density $f^{X^*}(x)$ (compare with (3.3.4))

$$f^{X^*}(x) = f^X(x)\mathbb{E}\{[1/B(X,Y)]|X = x\}/D. \tag{3.3.7}$$

Using this formula in (3.3.6), together with (3.3.1), we continue,

$$\theta_j = \int_0^1 \left[\int_{-\infty}^{\infty} \frac{y f^{X,Y}(x,y)}{DB(X,Y) f^{X^*}(x)} dy \right] \varphi_j(x) dx$$

$$= \mathbb{E}\left\{ \frac{Y \varphi_j(X)}{B(X,Y) f^X(X) \mathbb{E}\{[1/B(X,Y)]|X\}} \right\}. \tag{3.3.8}$$

This is a pivotal formula that sheds light on the possibility to estimate θ_j. First, we need to estimate two nuisance functions $f^X(x)$, $x \in [0,1]$ and

$$D(x) := \mathbb{E}\{[1/B(X,Y)]|X = x\}, \quad x \in [0,1], \tag{3.3.9}$$

that are used in the denominator of (3.3.8).

The density $f^X(x)$ is estimated by the E-estimator $\hat{f}^X(x)$ of Section 2.2 using the available sample $X_1, \ldots, X_n$. Estimation of $D(x)$ requires developing its own E-estimator. Fourier coefficients of $D(x)$ can be written as

$$\kappa_j := \int_0^1 D(x) \varphi_j(x) dx = \int_0^1 \mathbb{E}\{[1/B(X,Y)]|X = x\} \varphi_j(x) dx$$

$$= \mathbb{E}\{\varphi_j(X)/[f^X(X)B(X,Y)]\}. \tag{3.3.10}$$

This implies the plug-in sample mean estimator of Fourier coefficients,

$$\hat{\kappa}_j := n^{-1} \sum_{l=1}^n \frac{\varphi_j(X_l)}{\max(\hat{f}^X(X_l), c/\ln(n)) B(X_l, Y_l)}. \tag{3.3.11}$$

In its turn, Fourier estimator (3.3.11) yields the E-estimator $\hat{D}(x)$, $x \in [0,1]$.

The two nuisance functions are estimated, and then (3.3.8) yields the following plug-in sample mean estimator of Fourier coefficients θ_j of the regression function $m(x)$,

$$\hat{\theta}_j := n^{-1} \sum_{l=1}^n \frac{Y_l \varphi_j(X_l)}{B(X_l, Y_l) \max(\hat{f}^X(X_l), c/\ln(n)) \hat{D}(X_l)}. \tag{3.3.12}$$

One remark about (3.3.12) is due. If the biased data is created by a missing mechanism, then $D(x) \geq 1$, and then its E-estimator may be truncated from below by 1.

Apart of estimation of the regression, in applied examples it may be of interest to estimate the marginal densities f^{X^*} and f^{Y^*} of the hidden predictor X^* and the hidden response Y^*. We estimate these densities in turn.

Estimation of the hidden design density $f^{X^*}(x)$ is based on the following useful probability formula. We divide both sides of (3.3.1) by $DB(X,Y)$, then integrate both sides with respect to y, use (3.3.2), (3.3.9) and get

$$f^X(x) = f^{X^*}(x) \frac{D}{\mathbb{E}\{[1/B(X,Y)]|X = x\}} = f^{X^*}(x) \frac{D}{D(x)}. \tag{3.3.13}$$

Formula (3.3.13) tells us that the density $f^X(x)$ of the observable variable X is biased with respect to the density of interest $f^{X^*}(x)$ with the biasing function $1/D(x)$. Because we already constructed the E-estimator $\hat{D}(x)$, it can be used in place of unknown $D(x)$, and then $f^{X^*}(x)$, $x \in [0,1]$ be estimated by the plug-in density E-estimator of Section 3.1.

For the density of the hidden response Y^*, again using (3.3.1) we obtain the following useful formula (compare with (3.3.13))

$$f^Y(y) = f^{Y^*}(y) \frac{D}{\mathbb{E}\{[1/B(X,Y)]|Y = y\}}. \tag{3.3.14}$$

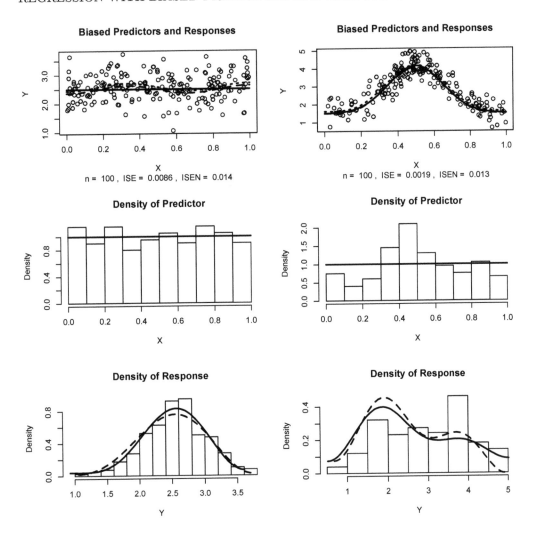

Figure 3.4 *Regression with biased predictors and responses. The underlying regression and the biasing function $B(x, y)$ are the same as in Figure 3.3 whose caption explains the parameters. Two columns of diagrams correspond to simulations with different underlying regression functions. A top diagram shows the scattergram of biased data by circles overlaid by the underlying regression (the solid line), the proposed regression E-estimate (the dashed line) and the naïve regression E-estimate of Section 2.3 based on biased data and not taking into account the biasing. The corresponding integrated squared errors of the two estimates are shown as ISE and ISEN. The middle and bottom diagrams show histograms of biased predictors and responses overlaid by the underlying marginal densities (the solid line) and their E-estimates (the dashed line). These densities are shown over the empirical range of biased observations. [n = 100, sigma = 0.5, set.B = c(0.2,0.5,1), set.corn = c(1,2), c = 1, cJ0 = 3, cJ1 = 0.8, cTH = 4]*

Sure enough, density $f^Y(y)$ is biased with respect to the density of interest $f^{Y^*}(y)$ with the biasing function $1/\mathbb{E}\{[1/B(X,y)]|Y = y\}$. The conditional expectation may be estimated similarly to how $D(x)$ was estimated, and then the density E-estimator of Section 3.1 may be used.

Let us stress one more time that we need to know the biasing function $B(x, y)$ to solve the regression and density estimation problems.

Figure 3.4 allows us to test performance of the proposed regression estimator and the marginal density estimators. Its diagrams and curves are explained in the caption, and note that while the underlying regression and even the biasing function are the same as in Figure 3.3, the biased data are generated according to (3.3.1) and hence the data are different from those generated according to formula (3.2.1) in Section 3.2.

Now let us look at the scattergrams. We begin with the left column where the underlying regression is created by the Uniform and it is shown by the solid line in the left-top diagram. First of all, note that we are dealing with a large volatility in the biased data. It looks like there are several modes in an underlying regression, but we do know that this is not the case. The dotted line shows the naïve regression estimate of Section 2.3 which ignores the known information that the data is biased. It does indicate modes, and recall that the E-estimator follows data while we know the underlying simulation and the hidden regression function. The proposed regression E-estimator, which takes into account the biasing nature of predictors and responses, also shows the same modes but it is much closer to the underlying regression. The latter is also supported by the indicated in the subtitle ISE = 0.0086 and ISEN = 0.014. Overall, despite a not perfect shape, the performance of the E-estimator is impressive keeping in mind the large volatility in the biased data. The middle and bottom diagrams are devoted to estimation of the hidden marginal densities. The density E-estimate of predictor is perfect despite the histogram of biased predictors exhibiting several modes. The bottom diagram is even more interesting because it allows us to look at the elusive marginal density of the response. Note that the histogram is asymmetric, and the density E-estimate is also skewed to the left, but overall it is a good estimate.

The right column shows us results of a simulation with the regression function created by the Normal. Again, the top diagram highlights the large volatility of data. Further, the scattergram and the underlying regression (the solid line) highlight the biased nature of the data (just notice that the data are skewed up). The latter is highlighted by the dotted line of the naïve regression estimate which goes above the underlying regression (the large volatility attenuates the difference). The proposed estimator (the dashed line) practically coincides with the underlying regression. Further, look at the corresponding integrated squared errors and note that taking into account the biased nature of data yields almost a seven-fold decrease in the integrated squared error. The middle diagram is of a special interest on its own. Here we can visualize the histogram of biased predictors, note the influence of the regression function on the distribution of observed biased predictors. It may be a good exercise to write down the density $f^X(x)$ for this simulation and then analyze it. The proposed marginal density E-estimator does a perfect job for this heavily biased data, and recall that this estimator is rather involved and based on estimation of the nuisance function $D(x)$. The bottom diagram is even more interesting because here we are dealing with the elusive marginal density of responses. Note that the hidden marginal density $f^{Y^*}(y)$ has a peculiar asymmetric shape with two modes. Further, look at how "disturbed" the histogram is, and how it magnifies the tiny right mode of the underlying density and, at the same time, diminishes the main left mode. Keeping in mind complexity of the marginal density estimation, which involves estimation of a nuisance conditional expectation, the E-estimator does an impressive job in exhibiting the shape of the underlying marginal density of the response Y^*.

The studied nonparametric regression problem with biased predictors and responses is a complicated one, both in terms of the model and the solution. It is highly advisable to repeat Figure 3.4, with both default and new arguments, and learn more about the biased regression and its consequences.

A remark about a regression with biased predictors is due. It is worthwhile to explain the problem via a particular example of the corresponding data modification. There is a hidden simulation from pair (X^*, A) where X^* is a continuous variable (predictor) supported on

$[0,1]$ and $f^{X^*}(x) \geq c_* > 0$, and A is a Bernoulli random variable generated according to the conditional density $\mathbb{P}(A = 1|X^* = x) =: B'(x) \geq c_0 > 0$. Denote the first realization of the pair as (X_1^*, A_1). If $A_1 = 1$, then $X_1 := X_1^*$ is observed, next the response Y_1 is generated according to the conditional density $f^{Y|X^*}(y|X_1)$, and the first realization (X_1, Y_1) of a regression sample with biased predictors is obtained. If $A_1 = 0$, then the realization (X_1^*, A_1) is skipped. Then the next realization of the pair (X^*, A) occurs. The sequential sampling stops whenever n realizations $(X_1, Y_1), \ldots, (X_n, Y_n)$ are collected. The problem is to estimate the regression of the response Y on the hidden predictor X^*, that is, we want to estimate $m(x) := \mathbb{E}\{Y|X^* = x\} = \int_{-\infty}^{\infty} y f^{Y|X^*}(y|x) dy$.

Let us explore the regression function for the considered regression model with biased predictors. To do this, it suffices to find a convenient expression for the conditional density $f^{Y|X}(y|x)$. We begin with the corresponding joint density,

$$f^{Y,X}(y,x) = f^{Y,X^*|A}(y,x|1) = \frac{f^{Y,X^*,A}(y,x,1)}{\mathbb{P}(A = 1)}$$

$$= \frac{f^{Y,X^*}(y,x)\mathbb{P}(A = 1|Y = y, X^* = x)}{\mathbb{P}(A = 1)}. \tag{3.3.15}$$

According to the considered biased sampling, the equality $\mathbb{P}(A = 1|Y = y, X^* = x) = \mathbb{P}(A = 1|X^* = x)$ holds. This equality, together with the relation $f^{Y,X}(y,x) = f^{Y|X}(y|x)f^X(x)$, the inequality $f^X(x) > 0$, $x \in [0,1]$ and (3.3.15), yield

$$f^{Y|X}(y|x) = \frac{f^{Y|X^*}(y|x)f^{X^*}(x)\mathbb{P}(A = 1|X^* = x)}{f^X(x)\mathbb{P}(A = 1)}. \tag{3.3.16}$$

Next, we note that

$$f^X(x) = \frac{f^{X^*}(x)\mathbb{P}(A = 1|X^* = x)}{\mathbb{P}(A = 1)}, \tag{3.3.17}$$

and this formula quantifies the biased modification of the predictors.

Using (3.3.17) in (3.3.16) we conclude that

$$f^{Y|X}(y|x) = f^{Y|X^*}(y|x). \tag{3.3.18}$$

Equality (3.3.18) sheds a light on the case of biased predictors in the regression setting. Here, despite the fact that X is biased, we have the equality (3.3.17) which implies that the regressions of Y on X and Y on X^* are the same. If you think about this outcome, it may seem either plain or confusing. If the latter is the feeling, then think about the fact that X is equal to X^* whenever X^* is observed, and then Y is generated according to $f^{Y|X^*}$. Of course, the same conclusion can be made from our general formula (3.3.1) when the biasing function $B(x,y) = B(x)$.

Finally, let us note that there is a special (and quite different) notion of unbiased predictors in finance theory, namely that forward exchange rates are unbiased predictors of future spot rates. In general, forward exchange rates are widely expected as a good predictor of future spot rates. For instance, any international transaction involving foreign exchange is risky due to unexpected change in currency exchange rates. Forward contract can be used to lower such risk, and as a result, the relation between the forward exchange rate and the corresponding future spot rate is of great concern for investors, portfolio managers, and policy makers. Forward rates are often expected to be unbiased estimator of corresponding future spot rates. It is possible to explore this problem, using our nonparametric technique, via statistical analysis of the joint distribution of the forward and spot rates.

3.4   Ordered Grouped Responses

So far we have explored problems where an underlying data was modified by a biasing mechanism caused, for instance, by observing a realization of a hidden sampling only if a specific event occurs. In this section we are considering another type of modification when it is only known that an underlying observation belongs to a specific group of possible observations. A group, depending on a situation and tradition, can be referred to by many names, for instance the stratum, category, cluster, etc.

Let us present several motivating examples. Strata, categories or clusters may define the socioeconomic statuses of a population: (i) Lowest 25 percent of wage earners; (ii) Middle 50 percent of wage earners; and (iii) Highest 25 percent of wage earners. A car may be driven with speed below 25, between 25 and 45, or above 45 miles per hour. A patient may have no pain, mild pain, moderate pain, severe pain, or acute pain. A patient in a study drinks no beer a day, 1 beer a day, more than 1 but fewer than 2 beers a day, and at least 2 beers a day. The overall rating of a proposal can be poor, fair, good, very good, or excellent.

In the above-presented examples, there is a logical ordering of the groups and hence they may be referred to as *ordinal* responses. To finish with the terminology, *nominal* responses have no natural logical ordering; examples are the color of eyes or the place of birth of a respondent to a survey.

Classical examples of nonparametric regression with grouped regression are the prediction of how a dosage of this or that medicine affects pain, or how the length of a rehabilitation program affects drug addiction, or how the quality of published papers affects the rating of a proposal.

To shed light on grouped (categorical, strata, cluster) nonparametric regression, let us consider the numerically simulated data shown in Figure 3.5. The left diagram shows an example of simulated classical additive regression $Y^* = m(X) + \sigma\eta$ which is explained in the caption. The small sample size $n = 30$ is chosen to improve visualization of each observation. The scatter plot is overlaid by boundaries for 4 ordered groups: $Y^* < -1$, $-1 \leq Y^* < 1$, $1 \leq Y^* < 3$, and $3 \leq Y^*$. Then the data are modified by combining the responses into the above-highlighted groups (categories) shown in the right diagram. Thus, instead of the hidden underlying pairs (X_l, Y_l^*), where $Y_l^* = m(X_l) + \sigma\eta_l$, we observe modified pairs (X_l, Y_l) where Y_l is the number of a group (cell, category, stratum, etc.) for an unobserved Y_l^*. Figure 3.5 visually stresses the loss of information about the underlying regression function, because grouped data give no information on how underlying unobserved responses are spread out over cells. Please look at the right diagram and imagine that you need to visualize an underlying regression function. Further, the fact that heights of cells may be different, make the setting even more complicated.

The interesting (and probably unexpected) feature of the grouped regression is that the regression noise may help to recover an underlying regression. Indeed, consider a case where a regression function is $m(x) = 0$, $\sigma = 0$ and cells are as shown in Figure 3.5. Then the available observations are $(X_l, 2)$, $l = 1, 2, \ldots, n$ and there is no way to estimate the underlying regression function. Further, even if there are additive errors but their range is not large enough, for instance $\sigma\eta_l$ are uniform $U(-0.99, 0.99)$, then the modified observations are again $(X_l, 2)$, $l = 1, \ldots, n$.

It is a good exercise to repeat Figure 3.5 with different arguments and get used to this special type of data modification.

Now we are ready to explain how an underlying regression function may be estimated based on observed grouped responses.

In what follows it is assumed that the underlying regression model is

$$Y = m(X) + \varepsilon, \tag{3.4.1}$$

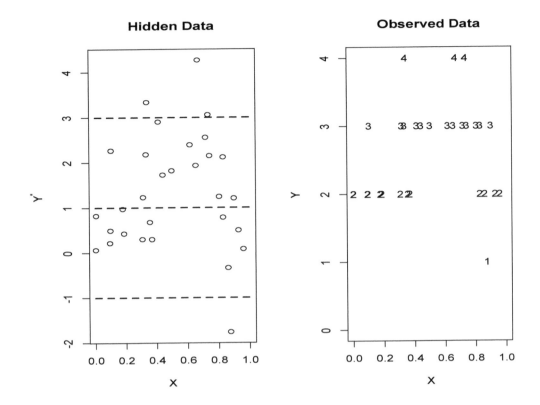

Figure 3.5 *Example illustrating grouped responses in nonparametric regression. A scattergram of underlying observations is shown in the left diagram, and it is generated by the model $Y^* = m(X) + \sigma\eta$ where X is the Uniform, η is standard normal and independent of X, and σ is a parameter controlling the standard deviation of the regression error. The sample size is $n = 30$. Responses are grouped according to 4 groups separated by dashed horizontal lines. The right diagram shows the corresponding grouped data. {Horizontal lines are controlled by the argument bound.set, regression errors are independent additive Normal with zero mean and standard deviation σ controlled by argument sigma.} [n = 30, set.corn = 3, sigma = 1, bound.set = c(-50,-1,1,3,50), cJ0 = 3, cJ1 = 0.8, cTH = 4]*

where X is supported on $[0, 1]$ and $f^X(x) \geq c_* > 0$, and the regression error ε is a continuous random variable with zero mean, finite variance and independent of the predictor X.

We begin with the parametric case $m(x) = \theta$ and the model of grouped data shown in Figure 3.5. Let $\bar{p}$ be the proportion of observations that have categories 3 or 4. Then the probability $\mathbb{P}(\theta + \varepsilon \geq 1) =: p$, which is the theoretical proportion of observations in the third and fourth categories, is

$$p = \mathbb{P}(\varepsilon \geq 1 - \theta) = 1 - F^\varepsilon(1 - \theta). \tag{3.4.2}$$

By solving this equation we get a natural estimate of θ,

$$\bar{\theta} = 1 - Q^\varepsilon(1 - \bar{p}), \tag{3.4.3}$$

where $Q^\varepsilon(\alpha)$ is the quantile function, that is, $\mathbb{P}(\varepsilon \leq Q^\varepsilon(\alpha)) = \alpha$.

Note that we converted the problem of grouped regression into Bernoulli regression discussed in Section 2.4. The latter is the underlying idea of the proposed solution.

There are three steps in the proposed regression estimator for regression with grouped responses.

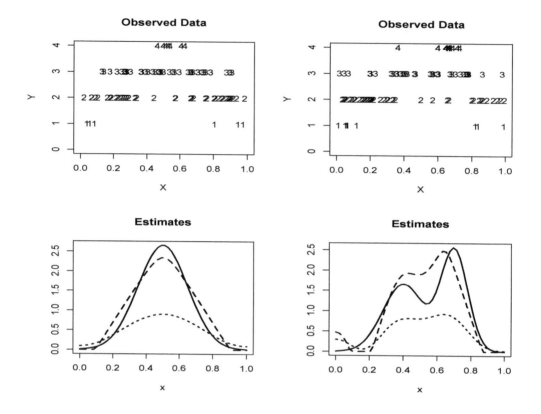

Figure 3.6 *E–estimates for nonparametric regression with grouped responses. The underlying simulation is the same as in Figure 3.5. Dotted and dashed lines show estimates $\hat{p}$ and $\hat{m}$ of the binary
probabilities and the regression function, respectively. The solid line is the underlying regression
function. [n = 100, set.corn = c(2,3), sigma = 1, bound.set = c(-50,-1,1,3,50),a = 0.005,b =
0.995, cJ0 = 3, cJ1 = 0.8, cTH = 4]*

Step 1. Combine the ordered groups into two groups of "successes" and "failures." Ide-
ally, the boundary in responses that separates these two groups should be such that both
successes and failures spread over the domain of predictors. For instance, for the example
shown in Figure 3.5, the only reasonable grouping is $\{(1, 2), (3, 4)\}$.

Step 2. Use the Bernoulli regression E-estimator $\hat{p}(x)$ of Section 2.4 to estimate the
probability of a success as a function in x. If no information about the regression error ε is
given, this is the last step. If the distribution of ε is given, then go to step 3.

Step 3. This step is based on the assumption that the distribution of ε is known. Assume
that an observed Y_l belongs to the success group iff $Y_l \geq c^*$ where c^* is a constant. Then

$$\hat{m}(x) = c^* - Q^\varepsilon\big(1 - [\hat{p}(x)]_a^b\big). \qquad (3.4.4)$$

Here $[z]_a^b = \max(a, \min(z, b))$ is the truncation (or we can say projection) of z onto interval
$[a, b]$. The truncation allows us to avoid infinite values for $\hat{m}$. The "default" values of a and
b are 0.005 and 0.995.

Let us check how the proposed estimator performs. Figure 3.6 exhibits results of two
simulations in two columns of diagrams. Underlying regression functions are the Normal
and the Bimodal shown by the solid lines in the bottom diagrams. The regression errors
are standard normal. The estimates $\hat{p}(x)$ and $\hat{m}(x)$ are shown by dotted and dashed lines,
respectively. The datasets are simulated according to Figure 3.5, only here the sample size

$n = 100$. The estimates $\hat{p}(x)$ (the dotted lines) look not too impressive but not too bad either keeping in mind the complexity of the grouped data. After all, we could observe similar shapes in estimates based on $n = 100$ direct observations. The estimate for the Bimodal (see the right-bottom diagram) has a wrong and confusing left tail, but it corresponds to the left tail of the grouped data exhibited in the right-top diagram.

Knowing the distribution of regression error ε dramatically improves the visual appeal of estimates $\hat{m}(x)$ shown in the bottom diagrams of Figure 3.6 by the dashed lines. The estimate for the Normal is truly impressive keeping in mind both complexity of the setting and the small sample size. The estimate for the Bimodal is also a significant improvement both in terms of the two pronounced modes and their magnitudes (just compare with the dotted line which shows the estimate $\hat{p}(x)$).

The reader is advised to repeat this figure with different arguments and get used to this particular data modification and the proposed estimates.

The proposed estimator is not optimal because it is based on creating just two groups from existing groups. Nevertheless, asymptotically the suggested estimator is rate optimal, it is a good choice for the case of small sample sizes where typically only several groups contain a majority of responses, and its simplicity is appealing.

3.5 Mixture

This section presents a new type of data modification that occurs in a number of practical applications, and it will be explained via a regression example.

There is an underlying sample of size n from pair (X, Y) where X is the predictor and Y is the response. It is known that X is a continuous random variable supported on $[0, 1]$, Y is Bernoulli and $\mathbb{P}(Y = 1|X = x) = m(x)$. The problem is to estimate the conditional probability $m(x)$. As we know from Section 2.4, the problem may be treated as a Bernoulli regression because

$$m(x) := \mathbb{E}\{Y|X = x\}. \tag{3.5.1}$$

If the sample from (X, Y) is available, then the E-estimator of Section 2.4 can be used. In the considered mixture model, the responses are hidden and instead we observe realizations from (X, Z) where

$$Z = Y\zeta + (1 - Y)\xi. \tag{3.5.2}$$

Here ζ and ξ are random variables with known and different mean values μ_ζ and μ_ξ.

As we can see, the mixture (3.5.2) is a special modification of an underlying variable of interest Y.

One of the classical practical examples of the mixture is a *change-point* problem in observed time series where $X_l = l/n$ is time and $Y = 1$ if an object functions normally and $Y = 0$ if the object functions abnormally. Then Equation (3.5.2) tells us that while we do not observe Y directly, observations of ζ correspond to the case where the object functions normally and observations of ξ correspond to the case where it functions abnormally. Then changing the regression $m(X)$ from 0 to 1 implies that the object recovers from abnormal functioning.

Now let us propose an E-estimator for the underlying regression function (3.5.1). In what follows it is assumed that in model (3.5.2) $\mu_\zeta \neq \mu_\xi$ and that X is independent of ζ and ξ. Introduce a scaled version of the observed Z defined as

$$Z' := (Z - \mu_\xi)/(\mu_\zeta - \mu_\xi). \tag{3.5.3}$$

The underlying idea of the new random variable Z' is based on the following relation:

$$\mathbb{E}\{Z'|X = x\} = \mathbb{E}\{\frac{Z - \mu_\xi}{\mu_\zeta - \mu_\xi}|X = x\} = \mathbb{E}\{\frac{Y\zeta + (1 - Y)\xi - \mu_\xi}{\mu_\zeta - \mu_\xi}|X = x\}$$

Mixtures Regression,   n = 100

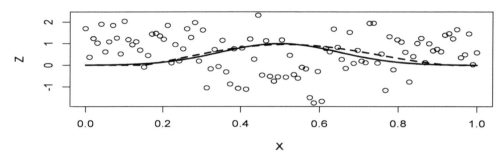

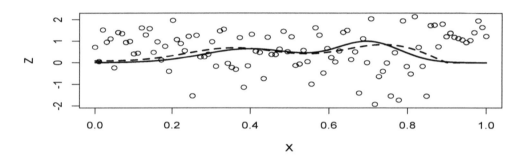

Figure 3.7 *E-estimation for mixtures regression. Underlying regression (the solid line) and its E-estimate (the dashed line) overlaid the scattergram shown by circles. In the top and bottom diagrams the underlying regressions are the Normal and the Bimodal divided by their maximum value. If the Uniform is chosen, then it is equal to 3/4. Realizations of X are equidistant and hence observations imitate a time series. {Random variable ξ is Normal(muxi,(sdxi)²) and ζ is Normal(muzeta,(sdzeta)²).} [n = 100, set.corn = c(2,3), muxi = 1, muzeta = 0, sdxi = 0.6, sdzeta = 0.9, cJ0 = 4, cJ1 = 0.8, cTH = 4]*

$$= \frac{\mathbb{E}\{Y|X = x\}\mathbb{E}\{\zeta\} + (1 - \mathbb{E}\{Y|X = x\})\mathbb{E}\{\xi\} - \mu_\xi}{\mu_\zeta - \mu_\xi} = \mathbb{E}\{Y|X = x\} = m(x). \quad (3.5.4)$$

We conclude that

$$m(x) = \mathbb{E}\{Z'|X = x\}, \quad (3.5.5)$$

and the problem of the mixtures regression is converted into the traditional regression problem for a sample from (X, Z') for which we can use the E-estimator of Section 2.3. Let us also note that (3.5.1) implies the relation $0 \le m(x) \le 1$ which can be used by the E-estimator for bona fide estimation.

Figure 3.7 allows us to look at a mixtures regression and how the proposed E-estimator performs, and the caption explains the diagrams. We begin our discussion with the top diagram where the underlying regression function is the Normal divided by its maximum value (because the regression should be between zero and one). The regression function is shown by the solid line. Let us look at the scattergram of observations, shown by the circles, and the solid curve which is the estimand. It is clear that in no way the scattergram resembles the regression. This is a very interesting observation because it tells us that a mixture scattergram should be visualized differently than a scattergram for classical regression discussed in Section 2.3. We can realize that larger values of the regression function correspond to smaller values of Z because the mean of ζ is smaller than the mean of ξ. Further, even with this fact taken into consideration, we can realize the shape of the regression but not its values. This is why we should appreciate performance of the E-estimator with the E-estimate shown by the dashed line. Of course, the estimate is too large around $x = 0.75$. But is this the fault of the E-estimator or do the mixture data indicate the larger tail? This is a teachable issue to explore because it may hint on how to read mixture data. Let us note that there are three relatively small observations of Z around $x = 0.8$. This is what causes the regression estimate to increase its value around this area. Keeping in mind the relatively small sample size $n = 100$, the outcome is impressive.

Now let us look at the bottom diagram where the underlying hidden regression is the Bimodal divided by its maximum value. We already know that estimation of the Bimodal regression is a challenging task even for the case of direct observations. (The reader is advised to return to Section 2.3 and check estimation of the Bimodal regression using Figure 2.7.) After the discussion of the top diagram, it is more clear why the scattergram corresponds to the underlying Bimodal regression. Further, the mixtures shed light on the fact why the main mode shifted to the right while the smaller one shifted to the left. To see this, just look at the smallest realizations of Z. Overall, for this particular simulation the E-estimator performed well.

Of course, the regression setting with mixtures is complicated and another simulation may produce a worse outcome. This is why it is instructive and useful to make more simulations, find a poor outcome, and then try to understand why the E-estimator performed in such a way.

3.6   Nuisance Functions

In many practical situations direct observations of a nonparametric function of interest are not available. A classical example of such a setting, considered in this section, is the case of a heteroscedastic regression with independent observations $(X_1, Y_1), \ldots, (X_n, Y_n)$ of the pair of continuous variables (X, Y) where

$$Y = m(X) + \sigma(X)\varepsilon. \tag{3.6.1}$$

Here ε is zero mean, unit variance and independent of X random variable, the nonnegative function $\sigma(x)$ is called the scale (spread or volatility) function, and the predictor X is supported on $[0, 1]$ and $f^X(x) \geq c_* > 0$.

Traditional regression problem is to estimate the function $m(x)$, and then the design density $f^X(x)$ and the scale function $\sigma(x)$ become nuisance ones. These two functions may be of interest on their own. We know from Section 2.2 how to estimate $f^X(x)$, $x \in [0, 1]$ based on the observed predictors. In this section our task is to estimate the scale $\sigma(x)$, $x \in [0, 1]$.

In the statistical literature the same problem of estimating the scale function may be referred to as either estimation of a nuisance function in a regression problem or as estimation based on data modified by a nuisance regression function.

Let us explain the latter formulation of the problem of scale estimation. There are hidden observations $Z_l = \sigma(X_l)\varepsilon_l$ of the scale function. If they would be available, we could convert estimation of the scale into a regression problem. To do this, we write

$$Z^2 = \sigma^2(X) + \sigma^2(X)(\varepsilon^2 - 1), \tag{3.6.2}$$

and note that because ε is independent of X and $\mathbb{E}\{\varepsilon^2\} = 1$ we get

$$\mathbb{E}\{\sigma^2(X)(\varepsilon^2 - 1)|X\} = 0. \tag{3.6.3}$$

Estimation of Scale Function, n = 100 , s = 0.3

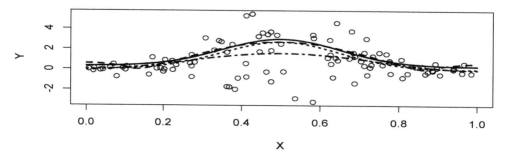

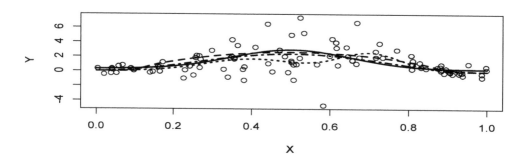

Figure 3.8 *Estimation of the scale function. Each diagram exhibits observations (scattergram) of a heteroscedastic regression generated by the Uniform design density, a regression function (the Normal and the Bimodal in the top and bottom diagrams), and the scale function $\sigma(x) = s + \sigma f(x)$ where $f(x)$ is the Normal, and s and σ are positive constants. In each diagram an underlying regression function is shown by the dotted line and its E-estimate by the dot-dashed line, while an underlying scale function and its E-estimate are shown by the solid and dashed lines, respectively. {Underlying regression functions are controlled by the argument set.corn, parameter σ by sigma, parameter s by s, and the choice of f is controlled by argument scalefun.} [n = 100, set.corn = c(2,3), sigma = 1, s = 0.3, scalefun = 2, cJ0 = 3, cJ1 = 0.8, cTH = 4]*

As a result, (3.6.2) is the regression model discussed in Section 2.3 and we can use the regression E-estimator for estimation of the regression $\sigma^2(x)$. Further, if the E-estimator takes on nonnegative values, then they are replaced by zero. Taking the square root of the above-defined estimator yields the estimator $\tilde{\sigma}(x, Z_1^n)$ of the scale function, and here $Z_1^n := (Z_1, Z_2, \ldots, Z_n)$. Of course, in the regression model (3.6.1) realizations of Z are hidden, instead we observe pairs (X_l, Y_l) such that

$$Y_l = m(X_l) + Z_l, \quad l = 1, 2, \ldots, n. \tag{3.6.4}$$

Equation (3.6.4) explains how the hidden observations Z_l are modified by the nuisance and unknown function $m(X_l)$.

A natural possible solution of a problem with data modified by nuisance functions is to first estimate them and then plug them in. In our case we can estimate the regression function $m(x)$ by the regression E-estimator $\hat{m}(x)$, and then replace unknown Z_l by

$$\hat{Z}_l := Y_l - \hat{m}(X_l). \tag{3.6.5}$$

Finally, we plug obtained $\hat{Z}_l$ in the above-defined estimator $\tilde{\sigma}(x, Z_1^n)$ and get the wished plug-in E-estimator $\hat{\sigma}(x) := \tilde{\sigma}(x, \hat{Z}_1^n)$.

Figure 3.8 allows us to appreciate complications that may be created by nuisance functions and how the proposed estimator performs, and its caption explains the simulation and the diagrams. We begin with the top diagram where the underlying regression function is the Normal, the scale function is the Normal plus $s = 0.3$, and the design density (the density of predictor) is the Uniform. Let us look at the scattergram (available pairs of observations) shown by circles. Can you visualize an underlying regression function that goes through the middle of the cloud of circles? It is not a simple task due to the large volatility of data in the middle of the unit interval. The regression E-estimate (the dot-dashed line) is poor, but it does correctly indicate the unimodal and symmetric nature of the Normal. Further, the regression E-estimate correctly describes the scattergram where we see a large number of negative responses in the middle of the unit interval. Nonetheless, the scale estimate (the dashed line) is impressively good.

The bottom diagram shows a similar simulation only with the underlying regression function being the Bimodal (the dotted line). Again, due to the strong volatility caused by the Normal scale function, it is practically impossible to visualize the underlying regression. The regression estimate (the dot-dashed line) shows a unimodal regression which barely catches characteristics of the Bimodal. This yields the overall poor scale estimate (the dashed line) which, nonetheless, correctly shows the unimodal and symmetric about 0.5 character of the Normal scale.

The presented simulations show the complexity of the studied modification, and having larger samples is the remedy. It is highly advisable to repeat Figure 3.8 and to get better understanding of this complicated problem.

More statistical examples of estimation of nuisance functions and estimation of data modified by nuisance functions will be considered in Chapters 4 and 9.

3.7 Bernoulli Regression with Unavailable Failures

In this section we are considering an important modification of data in Bernoulli regression when only cases with successes are observed while all cases with failures are unavailable. As we will see in the following chapters, this modification occurs in many applied problems and it is the pivot for solving many problems with missing data.

We begin with reviewing classical Bernoulli regression discussed in Section 2.4. Let us briefly recall the problem and the Bernoulli regression E-estimator. We are interested in a relationship between a continuous random variable X^* (the predictor) and a Bernoulli random variable A^*. A Bernoulli random variable takes on only two values 0 and 1, and traditionally the outcome 0 is classified as a "failure" and the outcome 1 as a "success." For instance, every day the level of pollution in a city can be below or above some threshold level, and the outcome is a Bernoulli random variable. Bernoulli random variable is completely defined by the probability of success $w := \mathbb{P}(A^* = 1)$, and then we have the formulae $\mathbb{E}\{A^*\} = w$ and $\mathbb{V}(A^*) = w(1 - w)$. These are the basic facts that we need to know about a Bernoulli random variable.

Now let us recall the model of Bernoulli regression considered in Section 2.4. Consider a situation when the probability of the success w is the function of a predictor X^*, which is a continuous random variable with the density $f^{X^*}(x)$ supported on the unit interval $[0, 1]$ and $f^{X^*}(x) \geq c_* > 0$, $x \in [0, 1]$. Introduce the regression function

$$w(x) := \mathbb{P}(A^* = 1 | X^* = x) = \mathbb{E}\{A^* | X^* = x\}, \tag{3.7.1}$$

and note that the joint mixed density of the pair (X^*, A^*) is

$$f^{X^*, A^*}(x, 1) = f^{X^*}(x)\mathbb{P}(A^* = 1 | X^* = x) = f^{X^*}(x)w(x),$$

$$f^{X^*,A^*}(x,0) = f^{X^*}(x)(1 - w(x)). \tag{3.7.2}$$

Using a directly observed sample $(X_1^*, A_1^*), \ldots, (X_n^*, A_n^*)$ from (X^*, A^*), the aim is to estimate the regression function $w(x)$. The proposed E-estimator is based on the E-estimation methodology of constructing a sample mean estimator of Fourier coefficients of an underlying regression function. Following the methodology, a Fourier coefficient $\theta_j := \int_0^1 w(x)\varphi_j(x)dx$ of the regression function $w(x)$, $x \in [0,1]$ can be written as

$$\theta_j = \int_0^1 \mathbb{E}\{A^*|X^* = x\}\varphi_j(x)dx = \mathbb{E}\left\{\frac{A^*\varphi_j(X^*)}{f^{X^*}(X^*)}\right\}. \tag{3.7.3}$$

Hence the corresponding sample mean estimator of θ_j is

$$\tilde{\theta}_j := n^{-1}\sum_{l=1}^{n}A_l^*\frac{\varphi_j(X_l^*)}{f^{X^*}(X_l^*)}. \tag{3.7.4}$$

If the design density $f^{X^*}(x)$ is unknown, then it is replaced by its E-estimator $\hat{f}^{X^*}(x)$ of Section 2.2 truncated below from zero by $c/\ln(n)$. In its turn, the plug-in Fourier estimator (3.7.4) yields the regression E-estimator $\tilde{w}(x)$ of Section 2.4.

Now we are ready to consider the Bernoulli regression problem with unavailable failures. The aim is still to estimate the regression function $w(x)$, $x \in [0,1]$ defined in (3.7.1), but now the sample $(X_1^*, A_1^*), \ldots, (X_n^*, A_n^*)$ is hidden. Instead, a subsample $X_1, \ldots, X_N$ of the predictors $X_1^*, \ldots, X_n^*$, corresponding to successes, is available and also the sample size n of the hidden sample is known. The subsampling is done as follows. If $A_1^* = 1$ then $X_1 := X_1^*$, and otherwise X_1^* is skipped. Then this subsampling continues, and finally if $A_n^* = 1$ then $X_N := X_n^*$ and otherwise X_n^* is skipped. Note that the number N of available predictors in the subsample is

$$N := \sum_{l=1}^{n}A_l^*. \tag{3.7.5}$$

Further, as usual we do not consider settings with $N = 0$ because there are no data, and in general we also exclude cases with relatively small N that are not feasible for nonparametric estimation. Let us also note that the available data may be equivalently written as $A_1^*X_1^*, \ldots, A_n^*X_n^*$ or as $(A_1^*X_1^*, A_1^*), \ldots, (A_n^*X_n^*, A_n^*)$.

It is convenient to use a different notation X for the observed predictor in a success case because the distribution of X is different from the distribution of the underlying predictor X^*. Indeed,

$$f^X(x) := f^{X^*|A^*}(x|1) = \frac{f^{X^*,A^*=1}(x)}{\mathbb{P}(A^* = 1)} = f^{X^*}(x)\frac{w(x)}{\mathbb{P}(A^* = 1)}. \tag{3.7.6}$$

This result implies that the observed predictor X has a biased distribution with respect to the hidden predictor X^*, and the biasing function is equal to the regression function $w(x)$.

Recall that biased distributions and biased data were discussed in Section 3.1. As we know from that section (and this also follows from (3.7.6)), based on the biased data we can consistently estimate only the product $f^{X^*}(x)w(x)$. The pivotal conclusion is that we need to know the design density $f^{X^*}(x)$ or its estimate for consistent estimation of $w(x)$.

As a result, we are exploring the following path for solving the problem. Formulas (3.7.3) and (3.7.4) tell us that to estimate Fourier coefficients of the regression function $w(x)$ (and hence to construct a regression E-estimator), it is sufficient to know only predictors X_l^* corresponding to $A_l^* = 1$. As a result, it is sufficient to know only the observed predictors $X_1, \ldots, X_N$. This is good news. The bad news is that we need to know the underlying density $f^{X^*}(x)$ which, as we already know, cannot be estimated based on the available data.

Suppose that we know values $f^{X^*}(X_l)$, $l = 1, \ldots, N$. Then the regression function may be estimated solely on available predictors corresponding to successes in the hidden Bernoulli sample. Indeed, we may rewrite (3.7.4) as

$$\tilde{\theta}_j = n^{-1} \sum_{l=1}^{N} \frac{\varphi_j(X_l)}{f^{X^*}(X_l)}. \tag{3.7.7}$$

This Fourier estimator yields the regression E-estimator $\tilde{w}(x)$, $x \in [0,1]$. In some practical applications, when design of predictors is controlled, this conclusion allows us to use this regression E-estimator. Further, theoretically this E-methodology implies asymptotically (in n) optimal regression estimation.

If the design density $f^{X^*}(x)$ is unknown, then in some situations it may be possible to get an extra sample $X_{E1}^*, \ldots, X_{Ek}^*$ of size $k \ll n$ from X^*; here $\ll$ means "significantly smaller." Then we may use the extra sample to calculate the density E-estimator $\hat{f}^{X^*}(x)$ and plug it in (3.7.7). Because the density estimator is used in the denominator, it is prudent to truncate it from below by $c/\ln(n)$ where c is the new parameter of the E-estimator. Then the (plug-in) sample mean estimator of Fourier coefficients of $w(x)$, $x \in [0,1]$ is

$$\hat{\theta}_j := n^{-1} \sum_{l=1}^{N} \frac{\varphi_j(X_l)}{\max(\hat{f}^{X^*}(X_l), c/\ln(n))}. \tag{3.7.8}$$

This Fourier estimator yields the regression estimator $\hat{w}(x)$, $x \in [0,1]$. The asymptotic theory shows that, under a mild assumption, this approach is consistent and implies optimal MISE (mean integrated squared error) convergence.

One more remark is due. In all future applications of the Bernoulli regression with unavailable failures, we need to know $w(x)$ only for $x \in \{X_1, \ldots, X_n\}$. This is important information to know because it means that the range of observations in the E-sample should be close to the range of available observations $X_1, \ldots, X_n$.

Let us test the proposed E-estimator on several simulated examples. Figure 3.9 presents the first set of four simulations, its caption explains the diagrams and the simulation. Here a left diagram shows the histogram of an extra sample of size k from X^*; the extra sample is referred to as an E-sample. An E-sample is used to estimate $f^{X^*}(X_l)$, $l = 1, \ldots, N$. Values of an underlying design density $f^{X^*}(X_l)$ and its E-estimate $\hat{f}^{X^*}(X_l)$ are shown by circles and crosses, respectively. A corresponding right diagram shows via circles observed pairs $(X_l, 1)$. The size of a hidden sample n and the number of available predictors $N = \sum_{l=1}^{n} A_l^*$ are shown in the title. Further, the solid and dashed lines show the underlying regression $w(x)$ and its oracle-estimate based on all n hidden realizations of (X^*, A^*). Crosses show values of $\hat{w}(X_l)$.

Now we can look at specific simulations and outcomes shown in Figure 3.9. The top row shows the case of the constant regression $w(x) = 3/4$. The extra E-sample is tiny ($k = 30$) for a nonparametric estimation of the density. The default histogram stresses complexity of the density estimation which is a linear function shown by the circles. The E-estimate is surprisingly good here (look at the crosses). It is fair to say that visualization of data (the histogram) does not help us to recognize the density, and this is why the density E-estimate is impressive. Then the estimated values of the underlying design density are plugged in the regression E-estimator (3.7.8), and results are shown in the right-top diagram. First of all, let us compare the solid line (the underlying regression) and crosses showing values $\hat{w}(X_l)$. The regression estimate is perfect, and this is despite the fact that only $N = 75$ observations are available. Interestingly, the oracle's E-estimate, based on hidden

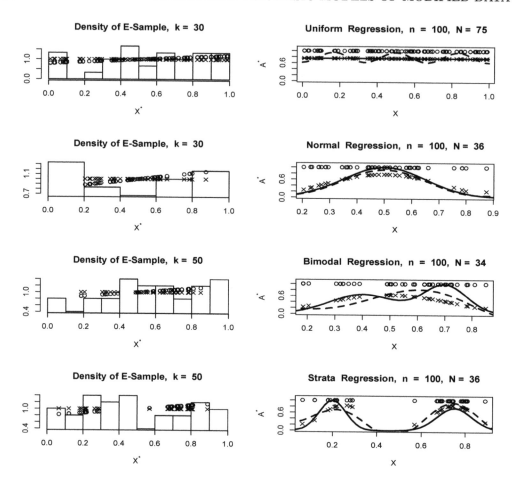

Figure 3.9 *Bernoulli regression with unavailable failures and an extra sample (E-sample) of pre-dictors. Four rows of diagrams exhibit results of simulations with different underlying regression functions $w(x)$ shown by the solid line and named in the title of a right diagram. The sizes of a hidden Bernoulli regression and E-sample are n and k, respectively. A left diagram shows the histogram of E-sample and values of the design density $f^{X^*}(X_l)$ and its E-estimate $\hat{f}^{X^*}(X_l)$, $l = 1, \ldots, N$ by circles and crosses, respectively. A right diagram shows by circles available observations in the Bernoulli regression, the underlying regression function $w(x)$ (the solid line), oracle's regression E-estimate (the dashed line) based on n hidden observations, and by crosses values of the proposed E-estimator $\hat{w}(X_l)$, $l = 1, \ldots, N$. {The figure allows to choose different parameters of E-estimator for the design density (they are controlled by traditional arguments) and the regression E-estimator. Further, for the regression E-estimator, parameters c_{J0} and c_{J1} can be specified for each of the 4 experiments. The latter is done by arguments setw.cJ0 and setw.cJ1. The argument desden controls the shape of the design density which is then truncated from below by the value dden and rescaled into a bona fide density. The argument st.k controls sample sizes of E-samples for each row.} [n = 100, set.k = c(30,30,50,50), desden = "1 + 0.5 * x", dden = 0.2, c=1, cJ0 = 3, cJ1 = 0.8, cTH = 4, setw.cJ0 = c(3,3,3,3), setw.cJ1 = c(0.3,0.3,0.3,0.3)]*

pairs (X_l^*, A_l^*), $l = 1, 2, \ldots, n$, is much worse (look at the oscillating dashed line). This is an interesting outcome but it is rare, in general the oracle-estimate is much better.

The second (from the top) row of diagrams in Figure 3.9 considers the same setting only with the Normal regression function. Here again only $k = 30$ extra observations of X^* are available for estimating the design density f^{X^*}. Note that the histogram clearly deviates from the underlying linear design density. However, fortunately for us, we need values of the density E-estimator only for $X_l \in [0.2, 0.9]$ interval, and within this interval the E-estimate is satisfactory. As a result, the right diagram shows us a fair regression estimate despite the fact that only $N = 36$ observations from hidden $n = 100$ are available. The oracle estimate of the regression (the dashed line) is good. In the third row of diagrams the case of the Bimodal regression is considered. Here the larger sample size $k = 50$ of E-sample is used and the design density estimate is fair. Unfortunately, this cannot help the regression E-estimator because the size $N = 34$ of available observations is too small and the regression function is too complicated (recall our simulations in Section 2.3). The poor oracle estimate, based on $n = 100$ hidden observations, sheds additional light on the difficult task. In the bottom diagram we explore the case of the Strata regression function. The design density estimate is fair, and the regression E-estimate is truly impressive given that only $N = 36$ observations are available. Further, this estimate is on par with the oracle estimate.

It is advisable to repeat Figure 3.9 with different parameters and get used to this challenging problem. Further, it is of interest to explore the relation between k and n that implies a reliable estimation comparable with the oracle's estimation. Further, Figure 3.10 allows us to use different parameters for the density estimator and regression E-estimators used in each row. The latter is a nice feature if we want to take into account different sample sizes and shapes of the underlying curves.

In many applications the support of the predictor may not be known. We have discussed this situation in Chapter 2, and let us continue it here because this case may imply some additional complications for our regression E-estimator. Namely, so far it has been explicitly assumed that the design density is bounded below from zero (recall that in Figure 3.9 the design density is not smaller than the argument $dden$). Let us relax these two assumptions and explain how this setting may be converted into the above-considered one.

Suppose that the hidden predictor X^* is a continuous random variable supported on a real line. Then our methodology of E-estimation is as follows. First, we combine N available predictors X_l and k extra observations X_{El} and find among these $N + k$ observations the smallest and largest values X_S and X_L, respectively. Then, using the transformation $(X - X_S)/(X_L - X_S)$ we rescale onto $[0, 1]$ the two available samples, and repeat all steps of the above-proposed regression E-estimation. The only new element here is that the obtained $\hat{f}^{X^*}(x)$ should be divided by $(X_L - X_S)$ to restore its values to the original interval.

Figure 3.10 illustrates this setting and the proposed solution. Its structure is similar to Figure 3.9, only here the regression function is the same in all 4 experiments, it is a custom-made function, and other differences are explained in the caption. Let us look at the top row of diagrams. Here the density E-estimate is fair, keeping in mind the small sample size $k = 30$, and its deviation from the underlying one is explained by the histogram. Please pay attention to the fact that 30 observations from a normal density may not be representative of an underlying density (as we see from the heavily skewed histogram). The deficiency in the density estimate is inherited by the regression estimate. Namely, note that the regression E-estimate (shown by crosses) is significantly smaller for positive values of X, and this is due to larger values of the density E-estimate. In the second row of diagrams results of an identical simulation are shown. Here the density estimate, at the required values X_l, is almost perfect. Of course, recall that the smallest values are truncated from below to avoid almost zero values in the denominator. The corresponding regression E-estimate is better. Overall, keeping in mind the small sample sizes $N = 61$ and $N = 68$ of available observations, the two regression estimates are fairly good and correctly indicate the sigmoid shape of the regression.

Simulations in the two bottom rows in Figure 3.10 use larger size $k = 50$ of E-samples. The second from the bottom row of diagrams exhibits a teachable outcome which stresses the fact that outcomes of small random samples may present surprises. Here we observe the worst density and regression estimates despite the largest $N = 82$. Note how the shortcom-

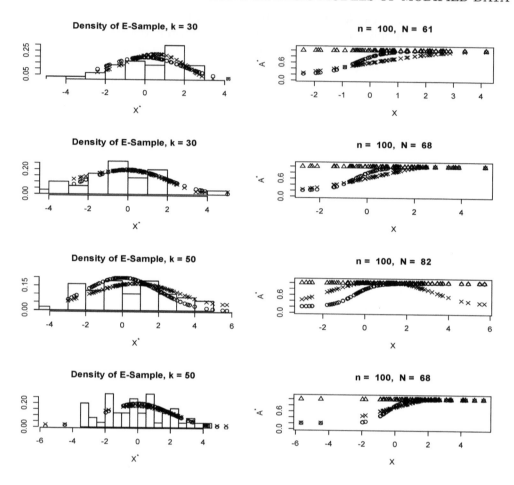

Figure 3.10 *Bernoulli regression with unavailable failures. The structure of the diagrams is the same as in Figure 3.9. The difference is that the design density is Normal$(0, \sigma^2)$, in all rows the same underlying regression function $w(x)$ is used and it is controlled by the string w, and in a right diagram triangles show available observations while circles and crosses show values of the underlying regression function $w(X_l)$ and its E-estimate $\hat{w}(X_l)$, $l = 1, \ldots, N$. {The argument sigma controls the standard deviation σ of the normal design density. The string w defines the regression function $w(x)$, and note that $w(x) \in [0, 1]$. All other arguments are the same as in Figure 3.9.} [n = 100, set.k = c(30,30,50,50), sigma = 2, w = "0.2 + 0.8 * exp(1 + 2 * x)/(1 + exp(1 + 2 * x))", c = 1, cJ0 = 3, cJ1 = 0.8, cTH = 4, setw.cJ0 = c(3,3,3,3), setw.cJ1 = c(0.3,0.3,0.3,0.3)]*

ings in the density estimate are inherited by the regression E-estimate. The bottom row of diagrams exhibits another outcome of the same simulation, and here both the density and the regression E-estimates are very good.

The simulations indicate that the issue that we should be aware of is that the range of the E-sample should be close to the range of available predictors. This remark may be useful if sequential E-sampling is possible.

Overall, we may conclude that if using a relatively small extra sample of hidden predictors is possible, then the proposed regression E-estimator is a feasible solution of the otherwise unsolvable problem of Bernoulli regression with unavailable failures.

It is highly advisable to repeat Figure 3.10 with different parameters and learn more about this important problem that will play a key role in statistical analysis of missing data.

3.8   Exercises

3.1.1 What is the definition of biased data?

3.1.2 Present an example of biased data.

3.1.3 Suppose that an underlying random variable X^* is observed only if it is larger than another independent random variable T. Are the observed realizations of X^* biased?

3.1.4 Verify (3.1.2) and (3.1.3).

3.1.5 Explain all components of formula (3.1.5).

3.1.6* For the setting of Exercise 3.1.3, write down a formula that relates the density of X^* with the density of X.

3.1.7 Is (3.1.8) a reasonable estimator of the Fourier coefficient (3.1.7)?

3.1.8 Find the mean of $\hat\theta_j$ defined in (3.1.8). Hint: Begin with the case when $P := \mathbb{P}(A = 1)$ is given, and then look at how using the plug-in estimate $\hat P$ affects the mean.

3.1.9* Evaluate the variance of $\hat\theta_j$ defined in (3.1.8). Hint: Prove formula (3.1.10).

3.1.10 Verify inequality (3.1.11).

3.1.11 Repeat Figure 3.1 for different biasing functions and explain outcomes.

3.1.12 Repeat Figure 3.2 with different underlying corner densities and biasing functions. What combinations imply worse and better estimates?

3.1.13 How parameters of the biasing function, used in Figure 3.2, affect the coefficient of difficulty for the four corner densities?

3.1.14 A naive estimate for biased data first estimates the density of the observed random variable X and then corrects its using a known biasing function $B(x)$. Write down a formula for this estimator. Hint: Density E-estimator can be used for estimating f^X, then use formula

$$f^X(x) = f^{X^*}(x)B(x)/B \tag{3.8.1}$$

where B is a constant which makes the density $f^X(x)$ bona fide (integrated to 1).

3.2.1 Explain a regression setting with biased responses.

3.2.2 Is the predictor, the response, or both biased under the model (3.2.1)?

3.2.3 Explain all functions in (3.2.1).

3.2.4 How is formula (3.2.3) obtained?

3.2.5 Explain how formula (3.2.4) is obtained. What is its relation to (3.2.1)?

3.2.6 How can a simulation of regression with biased responses be designed?

3.2.7 Consider a model where an underlying response Y^* is observed only if $Y^* > T$ where T is an independent random variable. Is this a sampling with biased responses?

3.2.8 For the setting of the previous exercise, what is the formula for the joint density of the observed pair of random variables (predictor and response)?

3.2.9 Suppose that in the setting of Exercise 3.2.7 the random variable T depends on predictor X. Does this information make a difference in your conclusions about the biased data? Is this a response-biased sampling?

3.2.10 Verify formula (3.2.8).

3.2.11 What is the underlying idea of the estimator (3.2.9)?

3.2.12 Evaluate the bias of estimator (3.2.9).

3.2.13 What is the underlying idea of the estimator $\hat D(x)$?

3.2.14 What is the bias of the estimator $\hat D(x)$?

3.2.15* The corresponding coefficient of difficulty of the proposed regression E-estimator is

$$d := \mathbb{E}\{[1/(f^X(X)B(X,Y)D(X))]^2\} = \int_0^1 \int_{-\infty}^\infty \frac{f^{Y^*|X}(y|x)}{f^X(x)B(x,y)D(x)} dy dx. \tag{3.8.2}$$

Prove this assertion, or show that it is wrong and then suggest a correct formula. Hint: Begin with the case when all nuisance functions (like $f^X(x)$ or $D(x)$) are known.

3.2.16 Explain all arguments used by Figure 3.3.

3.2.17 Repeat Figure 3.3 a number of times using different regression functions. Which one is more difficult for estimation? Hint: Use both visual analysis and ISEs to make a conclusion.

3.2.18* Use Figure 3.3 to answer the following question. For each underlying regression function, what are the parameters of the biasing function that make estimation less and more challenging? Confirm your observations using theoretical analysis based on the coefficient of difficulty.

3.2.19* Consider the case $B(x, y) = B^*(y)$ when the biasing is defined solely by the value of the underlying response. Present all related probability formulas for this case, and propose an E-estimator.

3.2.20* In the literature, statisticians often consider a model where

$$f^{X|Y}(x|y) = f^{X|Y^*}(x|y). \tag{3.8.3}$$

Explore this case.

3.3.1 Explain the model of regression with both predictors and responses being biased.

3.3.2 Present an example of regression where both predictors and responses are biased.

3.3.3 What is the relationship between models discussed in Sections 3.2 and 3.3?

3.3.4 Why do we use in formula (3.3.1) a constant factor B and not a function?

3.3.5 Explain a sampling procedure corresponding to (3.3.4).

3.3.6 In formula (3.3.4), how can constant D be estimated?

3.3.7 Explain relations in (3.3.5).

3.3.8 Verify formula (3.3.6).

3.3.9 How can the right-hand side of (3.3.6) be written as expectation?

3.3.10 Explain how formula (3.3.8) is obtained.

3.3.11 Propose an estimator of the function $D(x)$.

3.3.12 Explain the motivation behind the Fourier estimator (3.3.11).

3.3.13 Explain how the marginal density of the underlying predictor X^* may be estimated.

3.3.14* Evaluate the mean, the variance and the coefficient of difficulty of the Fourier estimator (3.3.12).

3.3.15* Consider the model of regression with biased predictors and explain how the regression function can be estimated. Prove your assertion.

3.3.16 Explain the notion of unbiased predictor used in financial literature. Suggest a statistical analysis of this setting.

3.3.17 Repeat Figure 3.4 several times and explain simulated data and the E-estimates.

3.3.18 Using Figure 3.4 explain how parameters of the biasing function affect observations and the estimates.

3.3.19 Which of the underlying regression functions that may be used by Figure 3.4 is more challenging for estimation?

3.3.20 How does parameter σ affect estimation in Figure 3.4?

3.3.21 Using Figure 3.4, propose optimal arguments for the E-estimator for each corner function.

3.3.22 Prove equality (3.3.18), and then explain how it may be used for estimation of the regression function.

3.4.1 Present several examples of grouped observations. Explain the used terminology.

3.4.2 Use Google and find definitions and applications of grouped observations referred to as strata, categories and clusters.

3.4.3 What is the difference between ordinal and nominal categories?

3.4.4 Consider a regression with grouped responses. Can smaller regression errors improve regression estimation?

3.4.5 Suppose that an underlying regression function is a constant θ. Explain why the regression noise can help in estimation of θ.

3.4.6 For the setting of Exercise 3.4.5, explain how one can estimate parameter θ.

3.4.7 Explain the three-step procedure of regression estimation for the case of grouped responses.

3.4.8 How are the grouped observations shown in the left column of the diagrams in Figure 3.6 obtained?

3.4.9 Explain the two types of estimates shown in Figure 3.6.

3.4.10 Using Figure 3.6, create several different groups that make estimation less and more complicated. Explain your results.

3.4.11 Explain how the noise affects the estimation. Use the argument *sigma* of Figure 3.6 to support your conclusion.

3.4.12 Explain the estimator (3.4.3).

3.4.13* Find the expectation and the variance of the estimator (3.4.3).

3.4.14* Find the expectation and the variance of the estimator (3.4.4).

3.4.15 What is the role of parameters a and b in the estimator (3.4.4)?

3.4.16 Using Figure 3.6, explore the effect of parameters a and b on the proposed estimator.

3.4.17* Propose better parameters of the E-estimator used in Figure 3.6. Then explore how the underlying regression function affects the choice. Present empirical and theoretical justifications.

3.5.1 Explain the notion of data modification via mixture.

3.5.2 Present an example which corresponds to model (3.5.2).

3.5.3 How is the model (3.5.2) related to the change-point problem?

3.5.4 Consider a setting where the underlying regression is constant. Write down the corresponding model and propose a procedure of estimation of that constant.

3.5.5* Verify formula (3.5.5). Formulate all necessary assumptions.

3.5.6* Explain the proposed regression E-estimator. Then calculate its coefficient of difficulty.

3.5.7 Knowing the underlying sampling mechanism, explain the scattergrams shown in the diagrams of Figure 3.7. Then repeat the simulation and compare outcomes.

3.5.8* Typically one can visualize (guess about) an underlying regression function as a curve going through the middle of a scattergram. Why is this no longer the case for the mixture regression? Hint: Diagrams in Figure 3.7 may be helpful.

3.5.9 Repeat Figure 3.7 with different corner functions and different sample sizes. What is your conclusion about quality of estimation?

3.5.10* How do distributions of ζ and ξ affect the estimation? Suggest a theoretical explanation and then use Figure 3.7 to check the answer.

3.5.11* It is assumed that means of the random variables ζ and ξ are different. Suppose that they are the same but the corresponding variances are different. Propose an E-estimator for this case.

3.6.1 Explain the heteroscedastic regression model (3.6.1). Presenting several practical examples when using this model may be appropriate.

3.6.2 Assume that in model (3.6.1) the function of interest is the scale $\sigma(x)$. What are the nuisance functions?

3.6.3 Verify equality (3.6.3). Formulate used assumptions.

3.6.4 Does (3.6.3) hold if X and ε are dependent?

3.6.5* Explain how the E-estimator of $\sigma^2(x)$ can be constructed. Then find its coefficient of difficulty.

3.6.6 Explain model (3.6.4).

3.6.7 Explain all steps in construction of the E-estimator $\hat{\sigma}(x)$. Then calculate its coefficient of difficulty. Does it depend on the underlying regression function? Explain your answer.

3.6.8* Consider a sampling from hidden X^* where a realization of X^* is observed only if $X^* \leq C$ and C is a continuous random variable. We are interested in estimation of the density of X^*. Do we have nuisance functions here?

3.6.9* Consider a hidden sampling (X_l^*, Y_l^*, T^*), $l = 1, 2, \ldots$ where its lth realization is observed only if $Y_l^* > T_l^*$. We are interested in estimation of the regression of Y^* on X^*. Develop the probability model for observed triplets and explore nuisance functions needed for construction a regression E-estimator.

3.6.10 Explain plots shown in Figure 3.8.

3.6.11 In the top diagram of Figure 3.8 the regression estimate (the dot-dashed line) is clearly bad. Nonetheless, the scale estimate is reasonable. Can you explain this outcome?

3.6.12* In a regression estimation problem, larger regression errors typically imply worse estimation. Is this also the case for the scale estimation?

3.6.13 In the bottom diagram of Figure 3.8 the regression estimate (the dot-dashed line) is far from the underlying Bimodal regression function shown by the dotted line. Nonetheless the scale estimate is relatively good. This is due to relatively large scale function. Repeat Figure 3.8 with smaller values of parameter σ and write a report on how this parameter affects estimation of the scale function. Hint: Pay attention to the fact that smaller scale functions improve regression estimation and, at the same time, may make estimation of the scale function more complicated.

3.6.14* Suppose that X and ε in model (3.6.1) are dependent. Propose an E-estimator of the scale function. Hint: Make necessary assumptions.

3.7.1 Explain the model of Bernoulli regression with unavailable failures.

3.7.2 Explain the equality in (3.7.1).

3.7.3 Verify relations (3.7.2).

3.7.4 Explain formula (3.7.3). How can it be used for estimation of the Fourier coefficient θ_j?

3.7.5* Consider a density $f^Z(z)$ of a random variable Z. Is it possible that $f^Z(Z) = 0$? Then use your answer to explain when the design density $f^{X^*}(X_l)$ can be used in the denominator of (3.7.4).

3.7.6* Evaluate the mean and the variance of the Fourier estimator (3.7.4). Explain the used assumptions.

3.7.7 What is an assumption needed for consistency of the plug-in estimator (3.7.4)?

3.7.8 The number N of available predictors is introduced in (3.7.5). What is the distribution of N?

3.7.9 Calculate the mean and standard deviation of the number N of available predictors.

3.7.10* Write down an exponential inequality for the probability $\mathbb{P}(|N/n - \mathbb{P}(A^* = 1)| > t)$ where t is a positive constant. Further, what is the probability of the event $N = 0$?

3.7.11 Explain each equality in (3.7.6).

3.7.12 Are the observed predictors biased with respect to hidden predictors? Explain your answer and, if it is positive, point upon a biasing function.

3.7.13 Is it always possible to propose consistent estimation of the regression function in the Bernoulli regression with missing failures? Explain your answer.

3.7.14 Explain the idea of using an additional (extra) sample from the hidden predictor to estimate the regression function.

3.7.15 Why may it be important to bound the design density E-estimator from zero?

3.7.16 Repeat Figure 3.9 several times and explain shapes of the estimates via analysis of available data.

3.7.17 Why is, in Figure 3.9, the available number N of predictors small with respect to n?

3.7.18 What changes in the experiment of Figure 3.9 will increase the number N of available predictors?

3.7.19 What type of design densities make the regression estimation simpler or more complicated? Check your answers using Figure 3.9.

3.7.20 Figure 3.9 allows us to use different parameters of the regression E-estimator in each row. Use this feature to propose optimal parameters of the estimator.

3.7.21 How does the argument *dden* in Figure 3.9 affect the estimation? Present a heuristic answer and then test it via simulations.

3.7.22* Explain the E-estimator used in Figure 3.10 for the case of a design density with unknown support. Then calculate its coefficient of difficulty.

3.7.23 Repeat Figure 3.10 and then present analysis of obtained estimates.

3.7.24 Test the effect of the argument c of Figure 3.10 on the E-estimator.

3.7.25 How does parameter σ of the normal design density, used in Figure 3.10, affect estimation of the regression function?

3.7.26 Using Figure 3.10, what minimal size k of the extra E-sample would you recommend for a practitioner? Consider several n.

3.7.27 Write down a formula for the rescaled onto $[0,1]$ design density with unknown support. Hint: Recall the scale-location transformation and how it affects the density. It also may be helpful to recall that the classical z-scoring is used to transfer a normal random variable into a standard normal variable.

3.7.28 In the proposed methodology of rescaling predictors onto the unit interval, the density estimate is divided by $(X_L - X_S)$. On the other hand, no transformation of the regression estimate is mentioned. Is this a mistake and should it be also rescaled? Explain your answer.

3.7.29* Let us complement the proposed approach of regression estimation by another idea of estimation of the Bernoulli regression function. According to (3.7.6) we can write down the regression function as

$$w(x) = \mathbb{P}(A^* = 1)\frac{f^X(x)}{f^{X^*}(x)}. \tag{3.8.4}$$

We know how to estimate the two densities on the right side of (3.8.4). Then suggest a sample mean estimator of $\mathbb{P}(A^* = 1)$, and propose a new estimator of the regression function.

3.9 Notes

3.1 Biased data is a familiar topic in statistical literature, see a discussion in Efromovich (1999a), Comte and Rebafka (2016) and Borrajo et al. (2017). A review of possible estimators may be found in the book by Wand and Jones (1995). Efromovich (2004a) has proved that E-estimation methodology is asymptotically efficient, and then a plug-in estimator of the cumulative distribution function is even second-order efficient, see a discussion and the proof in Efromovich (2004c). A combination of biased and other modifications is also popular in the literature, see Luo and Tsai (2009), Brunel et al. (2009), Ning et al. (2010) and Chan (2013).

3.2-3.3 Biased responses commonly occur in technological, actuarial, biomedical, epidemiological, financial and social studies. In a response-biased sampling, observations are taken according to the values of the responses. For instance, in a study of possible dependence of levels of hypertension (response) on intake of a new medicine (covariate), sampling from patients in a hospital is response-biased with respect to a general population of people with hypertension. Another familiar example of sampling with selection bias in economic and social studies is that the wage is only observed for the employed people. An interesting discussion may be found in Gill, Vardi and Wellner (1988), Bickel and Ritov (1991), Wang (1995), Lawless et al. (1999), Luo and Tsai (2009), Tsai (2009), Ning, Qin and Shen (2010), Chaubey et al. (2017), Kou and Liu (2017), Qin (2017) and Shen et al. (2017).

3.4 Nonparametric estimation for ordered categorical data is considered in the books Simonoff (1996) and Efromovich (1999a). Efromovich (1996a) presents asymptotic justification of E-estimation.

3.5 A discussion of parametric mixture models can be found in the book by Lehmann and Casella (1998). Nonparametric models are discussed in books by Prakasa Rao (1983) and Efromovich (1999a). The asymptotic justification of the E-estimation is given in Efromovich (1996a). For possible further developments see Chen et al. (2016).

3.6 Asymptotic justification of using E-estimation for nuisance functions may be found in Efromovich (1996a; 2004b; 2007a,f).

Chapter 4

Nondestructive Missing

In this chapter nonparametric problems with missing data that allow a consistent estimation are considered. By missing we mean that some cases in data (you may think about rows in a matrix) are incomplete and instead of numbers some elements in cases are missed (empty). In R language missed elements are denoted by a logical flag "NA" which stands for "Not Available," and this is why we are saying that some elements in a case are available (not missed) and others not available (missed); we may similarly say that a case is complete (all elements are available) or incomplete (when some elements are not available).

With missing data, on the top of all earlier discussed issues with nonparametric estimation, we must address the new issue of dealing with incomplete cases. Of course, using the E-estimation methodology converts the new problem into proposing a sample mean Fourier estimator based on missing data. As we will see shortly, for some settings incomplete cases can be ignored (and this may be also the best solution) and for others a special statistical procedure, which takes into account the missing, is necessary for consistent estimation.

Let us recall that if for a missing data a consistent estimation is possible, then we refer to the missing as nondestructive. The meaning of this definition is that while a nondestructive missing may affect accuracy (quality) of estimation via increasing the MISE and other nonparametric risks, at least it allows us to propose a consistent estimator. Some types of missing may destroy all useful information contained in an underlying (hidden) data and hence make a consistent estimation impossible. In this case the missing is called destructive and then some extra information is needed for a consistent estimation; destructive missing is discussed in the next chapter. Traditional examples of nondestructive missing are settings where observations are missed completely at random (MCAR) when the probability of an observation to be available (not missed) does not depend on its value. Another example is settings with missing at random (MAR) when the probability of an observation to be available (not missed) depends only on value of another always observed (never missed) random variable. In some special cases missing not at random (MNAR), when the probability of a variable to be available depends on its value, also implies a nondestructive missing.

In this chapter we often encounter Bernoulli and Binomial random variables. Let us briefly review basic facts about these random variables (more can be found in Section 1.3). A Bernoulli random variable A may be equal to zero (often coded as a "failure") with the probability $1 - w$ or 1 (often coded as a "success") with the probability w. Parameter w, the probability that A is equal to 1, describes this random variable. In short, we can say that A is Bernoulli(w). The mean of A is equal to w, that is $\mathbb{E}\{A\} = w$, and the variance of A is equal to $w(1-w)$. The sum $N := \sum_{l=1}^{n} A_l$ of n independent and identically distributed Bernoulli(w) random variables has a binomial distribution with $\mathbb{P}(N = k) = [n!/(k!(n-k)!)]w^k(1-w)^{n-k}$, $k = 0, 1, \ldots, n$. In short, we can write that N is Binomial(n, w). The mean value of N is nw (indeed, $\mathbb{E}\{N\} = \mathbb{E}\{\sum_{l=1}^{n} A_l\} = \sum_{l=1}^{n} \mathbb{E}\{A_l\} = n\mathbb{E}\{A\} = nw$), and the variance of N is $nw(1-w)$ (it is the sum of variances of A_l). Another useful result about a Binomial distribution is that for large n it can be approximated by a Normal distribution with the same mean and variance. The rule of thumb is that if $\min(nw, n(1-w)) \geq 5$ or

$n \geq 30$, then the distribution of N is Normal with the mean nw and the variance $nw(1-w)$. Hoeffding's inequality states that for any positive constant t,

$$\mathbb{P}\Big(\frac{N}{n} - w < -t\Big) \leq e^{-2nt^2}, \quad \mathbb{P}\Big(|\frac{N}{n} - w| > t\Big) \leq 2e^{-2nt^2}. \tag{4.0.1}$$

In its turn, Hoeffding's inequality yields that for any constant $\delta \in (0,1]$,

$$\mathbb{P}\Big(N/n < w - \sqrt{\ln(1/\delta)/(2n)}\Big) \leq \delta. \tag{4.0.2}$$

In this and the following chapters, a Bernoulli(w) random variable A describes an underlying missing mechanism such that if $A = 1$ (the success) then a hidden observation is available (this is why we use the letter A which stands for "Availability") and if $A = 0$ (the failure), then the hidden observation is not available. Also recall our explanation that in R and some other statistical softwares the logical flag NA is used to indicate a missed (not available) value. The probability w of the success may be referred to as the *availability likelihood*. If there are n realizations in a hidden sample of interest, the number of *complete cases* is $N := \sum_{l=1}^{n} A_l$ which has a Binomial(n, w) distribution. These facts explain why Bernoulli and Binomial distributions are pivotal in statistical analysis of missing data.

In what follows we refer to an underlying and hidden sample as H-sample, and to a sample with missing observations as M-sample. Typically an M-sample is created from a corresponding H-sample by an underlying missing mechanism implying that H- and M-samples may be dependent. Let us also comment about notation used in this and the next sections. Suppose that X is a continuous random variable of interest and $X_1, \ldots, X_n$ is a sample from X. Suppose that A is the availability (Bernoulli random variable) and $A_1, \ldots, A_n$ is a sample from A. Then the sample from X is the H-sample, and $(A_1X_1, A_1), \ldots, (A_nX_n, A_n)$ is the M-sample. Further, because $\mathbb{P}(A = I(AX \neq 0)) = 1$, sample $(A_1X_1), \ldots, (A_nX_n)$ is also M-sample and the two M-samples are equivalent. In graphics, we may use AX and $X[A == 1]$ as axis labels. The label AX means that all observations in the M-sample are considered, while $X[A == 1]$ means that only not missed observations in the M-sample are considered. The latter notation corresponds to R operator of extracting elements of a vector X corresponding to unit elements of vector A.

Let us make one important remark. In the previous chapters we have learned that a feasible sample size n of a sample, used to solve a nonparametric curve estimation problem, cannot be small. In missing data the number N of complete cases mimics the sample size n, and hence an M-sample with relatively small N should not be taken lightly even if the size n of the hidden H-sample is relatively large. In other words, even if n is large, for missing data it is prudent to look at the number of complete cases N and only then decide on feasibility of using a nonparametric estimator. If the sample size n should be chosen a priori, then (4.0.1), (4.0.2) and probability inequalities of Section 1.3 may help us to understand how large the size n should be to avoid a prohibitively small N. Further, numerical simulations (and in particular those in this and the following chapters) become an important tool in gaining a necessary experience in the statistical analysis of missing data.

The context of this chapter is as follows. Section 4.1 considers the classical problem of a univariate density estimation where elements of an H-sample may be missed purely at random meaning that the missing mechanism does not depend on values of observations in an underlying H-sample. This is the case of MCAR and it is not difficult to understand that a complete-case approach implies a consistent density estimation. Nonetheless, because it is our first problem with missing data, everything is thoroughly explained. In particular, it is explained how to deal with the random number N of available observations. Section 4.2 considers a case of missing responses in nonparametric regression when the probability of missing may depend on predictor. The main conclusion is that the simplest procedure of estimation based on complete cases (and ignoring incomplete ones) dominates all other possible approaches. The case of missed predictors, considered in Section 4.3, is more involved. The latter is not surprising because in regression analysis a deviation from basic assumptions about predictors typically causes major statistical complications. It is explained that a multi-step statistical methodology of regression estimation, involving estimation of nuisance functions, is required. Further, no longer a complete-case approach implies a consistent estimation. Estimation of the conditional density, which is a bivariate estimation problem, is discussed in Section 4.4. A special topic of regression with discrete responses is discussed in Section 4.5 via the classical and practically important example of Poisson regression. Scale (volatility) estimation in a regression setting is explored in Section 4.6. Multivariate regression is discussed in Sections 4.7 and 4.8.

4.1 Density Estimation with MCAR Data

The following model is considered. There is a hidden sample (H-sample) $X_1, X_2, \ldots, X_n$ from a continuous random variable of interest X supported on $[0, 1]$. This sample is not observed because some realizations are missed and the fact of missing is known. For instance, in R missing values are represented by the symbol NA (not available). There may be different underlying missing mechanisms, and in this section we are considering the MCAR (missing completely at random) mechanism when a realization is missed at random and the probability of missing does not depend on the value of missed variable.

Probabilistically, the MCAR mechanism may be described by a Bernoulli(w) random variable A referred to as the *availability*. If $A = 1$ then X is available (not missed), $\mathbb{P}(A = 1|X) = \mathbb{P}(A = 1) =: w \geq c_0 > 0$, and w is called the *availability likelihood*. If $A = 0$ then X is not available (missed) and $\mathbb{P}(A = 0) = 1 - w$. Note that $A = 0$ is equivalent to the logical flag NA (not available) in R language.

As a result, under the MCAR, instead of the H-sample of interest $X_1, \ldots, X_n$ we observe a sample with missing realizations (M-sample) $(A_1 X_1, A_1), \ldots, (A_n X_n, A_n)$ generated by the pair (AX, A).

Two comments about the missing are due. First, because X is a continuous random variable, the probability formula $\mathbb{P}(A = I(AX \neq 0)) = 1$ holds. This formula implies that the above-defined M-sample is equivalent to the sample $(A_1 X_1), \ldots, (A_n X_n)$ from AX. As a result, a sample from AX also may be referred to as M-sample. The second comment is about the MCAR. Under the MCAR, the variable of interest X and the availability A are independent.

The aim is identical to the one considered in Section 2.2. We are interested in estimation of the probability density $f^X(x)$ of the random variable X, only here M-sample is available while in Section 2.2 density estimation was based on H-sample.

To propose a density estimator, we begin with the probability formula for the joint mixed (because X is the continuos variable and A is the discrete variable) density of the observed pair (AX, A),

$$f^{AX,A}(ax, a) = [f^X(x)w]^a [1 - w]^{1-a}, \quad x \in [0, 1], \ a \in \{0, 1\}. \tag{4.1.1}$$

Let us comment on (4.1.1). The data from (AX, A) is collected as follows. First an observation of X is generated according to the density f^X, then independently of X an observation of A is generated according to the Bernoulli distribution with $\mathbb{P}(A = 1) = w$. If $A = 1$ then the observation of X is available (not missed) and the joint density is $f(AX, A)(x, 1) = f^X(x)\mathbb{P}(A = 1|X = x) = f^X(x)w$. Otherwise, if $A = 0$ then the observation of X is not available (missed) and $f^{AX,A}(0, 0) = \mathbb{P}(X \in [0, 1], A = 0) = \int_0^1 f^X(x)\mathbb{P}(A = 0|X = x)dx = \int_0^1 f^X(x)\mathbb{P}(A = 0)dx = 1 - w$.

It is important to stress that the MCAR yields a random number

$$N := \sum_{l=1}^{n} A_l \tag{4.1.2}$$

of available (not missed) observations of X in the M-sample, or equivalently we may say that we have only N complete cases in the M-sample. The distribution of N is Binomial(n, w), $\mathbb{E}\{N\} = nw$, $\mathbb{V}(N) = nw(1 - w)$. In a particular simulation the number of available observations N can be small with respect to n; this is the main complication of the MCAR. Inequalities (4.0.1) and (4.0.2) may be used to evaluate the probability of small N, and also recall that, according to the Central Limit Theorem, if n is sufficiently large then the distribution of N is close to the Normal$(nw, nw(1 - w))$.

Now we are ready to explain how to use our E-estimation methodology for construction of E-estimator of the density $f^X(x)$, $x \in [0, 1]$ for a MCAR sample.

First, we need to propose a sample mean estimator of Fourier coefficients $\theta_j := \int_0^1 f^X(x)\varphi_j(x)dx$, $j \geq 1$. Recall that $\varphi_0(x) = 1$, $\varphi_j(x) = 2^{1/2}\cos(\pi jx)$, $j = 1, 2, \ldots$ are elements of the cosine basis on $[0, 1]$, and Fourier coefficient $\theta_0 = \int_0^1 f^X(x)dx = 1$ is always known for a density supported on $[0, 1]$.

The idea of construction of a feasible sample mean estimator of θ_j is based on the assertion that the distribution of available (not missed) observations in M-sample is the same as the distribution of the variable of interest X. This assertion may look plain due to the independence between X and A, but because this is our first example of missing data, let us prove it. Consider the cumulative distribution function of an available observation of X in M-sample, that is an observation of (AX, A) given $A = 1$. Write,

$$F^{AX,A|A}(x, 1|1) := \mathbb{P}(AX \leq x|A = 1) = \frac{\mathbb{P}(A = 1, X \leq x)}{\mathbb{P}(A = 1)}$$

$$= \frac{\mathbb{P}(A = 1)\mathbb{P}(X \leq x)}{\mathbb{P}(A = 1)} = \mathbb{P}(X \leq x) = F^X(x). \tag{4.1.3}$$

Taking the derivative of both sides in (4.1.3) yields the important equality between the involved densities,

$$f^{AX,A|A}(x, 1|1) = f^X(x). \tag{4.1.4}$$

This is what was wished to prove. Of course, we may also get (4.1.4) from (4.1.1) using definition of the conditional density.

Note that $f^{AX,A|A}(x, 1|1) = f^{AX|A}(x|1)$, and hence we may use the latter density in (4.1.4), and also recall that a sample from AX is equivalent to the sample from the pair (AX, A).

We conclude that observations in a complete-case subsample have the distribution of the underlying random variable of interest X. This allows us to write down a Fourier coefficient of $f^X(x)$ as

$$\theta_j := \int_0^1 f^X(x)\varphi_j(x)dx = \int_0^1 f^{AX,A|A}(x, 1|1)\varphi_j(x)dx = \mathbb{E}\{A\varphi_j(AX)|A = 1\}. \tag{4.1.5}$$

Assume that the availability likelihood w is known. Then (4.1.5) implies a classical sample mean estimator of θ_j,

$$\check{\theta}_j := \frac{n^{-1}\sum_{l=1}^{n} A_l\varphi_j(A_lX_l)}{w}. \tag{4.1.6}$$

To polish our technique of dealing with missing data, let us formally check that estimator (4.1.6) is unbiased. Write,

$$\mathbb{E}\{\breve{\theta}_j\} = \mathbb{E}\left\{A\frac{\varphi_j(AX)}{w}\right\} = \frac{\mathbb{P}(A=1)\mathbb{E}\{\varphi_j(X)\}}{\mathbb{P}(A=1)} = \theta_j. \tag{4.1.7}$$

This proves that the estimator is unbiased.

If the availability likelihood w is unknown, then it can be estimated by the sample mean estimator $\hat{w} := N/n$. It is tempting to plug it in (4.1.6), but note that N may be equal to zero and $\mathbb{P}(N=0) = (1-w)^n = e^{-n\ln(1/(1-w))} > 0$. Hence we need to propose a remedy. First of all, it is a good idea to shed numerical light on the probability $\mathbb{P}(N=0)$. In nonparametric we typically work with sample sizes larger than 30. Then, if we may assume that $w \geq 0.5$, that is the likelihood of missing an observation does not exceed 50%, then $\mathbb{P}(N=0) < 10^{-9}$. This is the probability of the event that may be neglected. Nonetheless, the pure theoretical issue with $N=0$ still holds.

There are three feasible approaches to deal with the plugging $\hat{w}$ in (4.1.6). The first one is to note that if $N=0$ then the numerator in (4.1.6) is also equal to zero. As a result, if we formally set $0/0 := 0$, then we may introduce a plug-in sample mean Fourier estimator

$$\hat{\theta}_j := \frac{n^{-1}\sum_{l=1}^n A_l\varphi_j(A_lX_l)}{\hat{w}}$$

$$= \frac{\sum_{l=1}^n A_l\varphi_j(A_lX_l)}{\sum_{l=1}^n A_l} = \frac{\sum_{l=1}^n A_l\varphi_j(A_lX_l)}{N}, \quad j \geq 1. \tag{4.1.8}$$

To shed additional light on this approach, note that if $N=0$ then $\theta_0=1$ and $\hat{\theta}_j=0$, $j \geq 1$, and this yields that the density E-estimator is equal to the Uniform density. In other words, with no information about an underlying H-sample, and this is exactly the case when $N=0$, the E-estimator assumes that the underlying density is the Uniform.

Another feasible approach is to ignore M-samples with no available observations. After all, in any practical situation an M-sample, with all observations being missed, will be ignored and dismissed. As a result, it is reasonable to restrict our attention to M-samples with $N > 0$. Further, even in the parametric statistics it is rare to work with really small samples, and in nonparametric statistics sample sizes smaller than 30 would be considered insufficient. Then the following remark sheds light on this issue. If $n > 30$ then the normal approximation of the distribution of N is applicable. For the often used rule of two standard deviations, this yields that $\mathbb{P}(N < nw - 2[nw(1-w)]^{1/2}) \leq 0.023$. If the likelihood of 2.3% can be ignored, and the latter is the underlying idea of the above-mentioned rule, then we get a simple empirical rule for the smallest size N of an MCAR sample. For instance, if $w = 0.9$, that is, on average 10% of observations are missed, than in an MCAR sample of size $n = 100$ we will have at least $100(0.9) - 2[100(0.9)(1-0.9)]^{1/2} = 84$ observations of X. At the same time, if $w = 0.7$ then this "two standard deviations" minimal number of observations in M-sample is still larger than 56.

As a result, another remedy is to consider M-samples with $N > k$ for some $k \geq 0$. Then the only issue to explain is how to theoretically analyze these M-samples. The probability theory teaches us that in this case we are dealing with conditional distributions given the event $N > k$. As an example, let us evaluate the conditional expectation of the plug-in estimator (4.1.8). First of all, because it is given that $N > k \geq 0$, we have $\hat{w} := N/n > 0$ and the plug-in estimator (4.1.8) is well defined. Next, for the conditional expectation we can write,

$$\mathbb{E}\{\hat{\theta}_j|N>k\} = \mathbb{E}\{N^{-1}\sum_{l=1}^n A_l\varphi_j(A_lX_l)|N>k\}$$

$$= \mathbb{E}\Big\{E\Big\{\frac{\sum_{l=1}^{n} A_l \varphi_j(A_l X_l)}{\sum_{l=1}^{n} A_l}\Big|A_1, A_2, \ldots, A_n, N > k\Big\}\Big|N > k\Big\}. \qquad (4.1.9)$$

Using independence between A and X, as well as $\theta_j = \mathbb{E}\{\varphi_j(X)\}$ and that A takes on values 0 or 1, we continue,

$$\mathbb{E}\{\hat{\theta}_j|N > k\} = \mathbb{E}\Big\{\frac{\sum_{l=1}^{n} A_l \mathbb{E}\{\varphi_j(A_l X_l)|A_l\}}{\sum_{l=1}^{n} A_l}\Big|N > k\Big\}$$

$$= \mathbb{E}\Big\{\frac{\sum_{l=1}^{n} A_l \mathbb{E}\{\varphi_j(X_l)\}}{\sum_{l=1}^{n} A_l}\Big|N > k\Big\} = \theta_j \mathbb{E}\Big\{\frac{\sum_{l=1}^{n} A_l}{\sum_{l=1}^{n} A_l}\Big|N > k\Big\} = \theta_j. \qquad (4.1.10)$$

We conclude that if we restrict our attention to M-samples with $N > k$, the plug-in sample mean estimator (4.1.8) is unbiased. This is a nice and encouraging theoretical result for missing data.

The third remedy that may be used in dealing with zero $\hat{w}$ is to plug in $\max(\hat{w}, c/\ln(n))$ with some positive constant c. Recall that we used a similar remedy in dealing with small estimates of the design density in regression problems. Then a theoretical justification of this remedy is based on the assumption $w \geq c_0 > 0$ and that $\mathbb{P}(\hat{w} < c/\ln(n))$ decreases exponentially in n according to (4.0.1).

While these three remedies are different, in applications they yield similar outcomes according to (4.0.1) and (4.0.2), and we will be able to check this via simulations shortly.

The Fourier estimator (4.1.8) yields the density E-estimator $\hat{f}(x)$ of Section 2.2.

Now let us make one more comment about the Fourier estimator (4.1.8) and, correspondingly, about the density E-estimator. These are complete-case estimators that are based on N available observations of the random variable X in an M-sample. Note that even the sample size n of M-sample is not used. As a result, the random number N of available observations plays the role of fixed n in the proposed density E-estimator of Section 2.2. We know from Section 2.2 that the sample size n should be relatively large for a chance to get a feasible nonparametric density estimation. The same is true for the missing data, and only M-samples with large N must be considered. In practical applications N is known, but in a simulation, according to (4.0.1), this means that both n and w should be relatively large.

We may conclude that the MCAR does not change the procedure of E-estimation which simply uses available observations and ignores missed ones. Interestingly, a complete-case approach is the default approach in all major statistical softwares including R. The recommended complete-case approach, as the asymptotic theory confirms, is optimal and cannot be improved by any other method.

Now let us check how the recommended complete-case approach performs in simulations. Note that no new software is needed because we are using the E-estimator of Section 2.2. Figure 4.1 allows us to look at two simulations with the underlying densities being the Normal and the Bimodal. The caption explains the simulation and diagrams. The corresponding sample sizes are chosen to be small so we can observe the underlying hidden H-sample from X and the M-sample from (AX, A). The two top diagrams exhibit a simulation and E-estimates for the Normal density based on H-sample and M-sample. The sample size $n = 50$ is small and the size $N = 34$ of the M-sample is almost "parametric," but still it is far from being in single digits. It is an important teachable lesson to repeat this figure with different parameters and pay attention to N. H-sample and M-sample, shown in the top diagram, are reasonable, but we may see some issues with the tails (there are more large realizations than small ones). Nonetheless, the corresponding E-estimates, shown in the second (from the top) diagram, are very reasonable. Visually the estimates are too close to see any difference, and this is when their empirical ISEs help us to recognize the effect of the MCAR on estimation. The two bottom diagrams exhibit results for the Bimodal density. Here a larger sample size $n = 75$ is used because otherwise it is difficult to find a

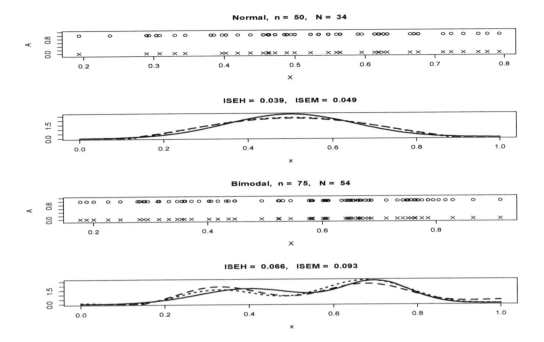

Figure 4.1 *MCAR data and density E-estimation. The two top and two bottom diagrams correspond to samples from the Normal and the Bimodal densities, and the missing is created by Bernoulli availability A with the availability likelihood* $\mathbb{P}(A = 1|X = x) = w = 0.7$. *The top diagram shows both the H-sample and M-sample via the scattergram* $(X_1, A_1), \ldots, (X_n, A_n)$ *where available cases are shown by circles and missing cases are shown by crosses.* $N := \sum_{l=1}^{n} A_l$ *is the number of available observations in the M-sample. In the second from the top diagram the underlying density, E-estimate based on the M-sample, and E-estimate based on the H-sample are shown by the solid, dashed and dotted lines, respectively. The ISEs of E-estimates, based on the H-sample and M-sample, are denoted as ISEH and ISEM, respectively. The two bottom diagrams have the same structure. {Recall that the simulation may be repeated by calling (after the R prompt >) the R function >* **ch4(f=1)**. *All default arguments, shown below in the square brackets, may be changed. Let us review these arguments. The argument set.c controls the choice of underlying corner densities; recall that the caption of Figure 2.3 explains how to use a custom-made density. The argument set.n allows one to choose 2 different sample sizes. The argument w controls the availability likelihood* $P(A = 1)$, *that is the likelihood of a hidden realization of X to be observed. The arguments cJ0, cJ1 and cTH control the parameters* c_{J0}, c_{J1}, *and* c_{TH} *used by the E-estimator defined in Section 2.2. Note that R language does not recognize subscripts, so we use cJ0 instead of* c_{J0}, *etc. To repeat this figure with the defaults arguments, make the call >* **ch4(f=1)**. *If one would like to change arguments, for instance to use a different threshold level, say* $c_{TH} = 3$, *make the call >* **ch4(f=1, cTH=3)**. *To change sample sizes to 100 and 150, make the call >* **ch4(f=1, set.n=c(100,150))**.} *[set.c = c(2,3), set.n = c(50,75), w = 0.7, cJ0 = 3, cJ1 = 0.8, cTH = 4]*

sample which indicates two modes of the Bimodal. Note that the M-sample contains only $N = 54$ observations, and this makes E-estimation of the Bimodal challenging. The two E-estimates, based on H-sample and M-sample, indicate two pronounced modes, but the former more accurately shows the main mode and overall is closer to the Bimodal, and the latter is stressed by the ISEs. We can also note that the Bimodal density is more difficult for estimation than the Normal.

Let us complement the discussion of outcomes of the two simulations in Figure 4.1 by a theoretical analysis of the Fourier estimator (4.1.6). The estimator is unbiased and for its

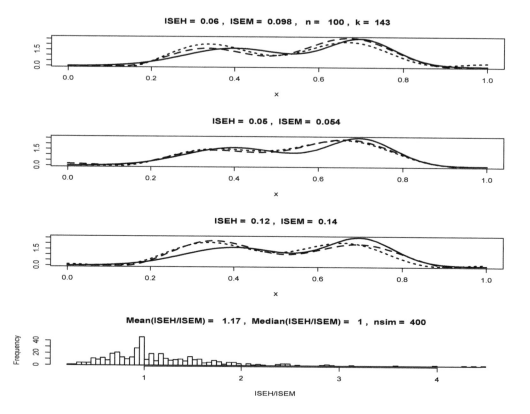

Figure 4.2 *Testing the theoretical conclusion that using M-sample of size k, equal to rounded up ratio n/w, allows us to estimate an underlying density with the same MISE as using H-sample of size n. Simulations are the same as in Figure 4.1, only now extra k − n observations of X are combined with H-sample and then are used to generate M-sample of size k. The first three diagrams show results of particular simulations, and their structure is the same as in the second from the bottom diagram in Figure 4.1. The bottom diagram shows the histogram of ratios ISEH/ISEM for 400 repeated simulations. Its title also shows the sample mean ratio and the sample median ratio. {The argument nsim controls the number of simulations, and corn controls an underlying corner density.} [n = 100, corn = 3, w = 0.7, nsim = 400, cJ0 = 3, cJ1 = 0.8, cTH = 4]*

variance we can write

$$\mathbb{V}(\check{\theta}_j) = n^{-1}\mathbb{V}(Aw^{-1}\varphi_j(AX))$$
$$= n^{-1}[\mathbb{E}\{(A\varphi_j(AX)/w)^2\} - \theta_j^2] = w^{-1}n^{-1}(1 + o_j(1)). \tag{4.1.11}$$

If w is unknown then the estimator $\hat{\theta}_j$, defined in (4.1.8), is used. Recall that this is a plug-in estimator with $\hat{w} = N/n$ used in place of w. Relation (4.0.1), together with some straightforward algebra, shows that

$$\mathbb{V}(\hat{\theta}_j) = w^{-1}n^{-1}(1 + o_j(1) + o_n(1)). \tag{4.1.12}$$

Further, recall the notion of the coefficient of difficulty $d = \lim_{n,j\to\infty} n\mathbb{V}(\hat{\theta}_j)$ introduced in Chapter 2. For the case of an H-sample we have $d = 1$, and for the MCAR $d = 1/w = 1/\mathbb{P}(A = 1) \geq 1$ with the equality only if $\mathbb{P}(A = 1) = 1$ (no missing). Hence, for an MCAR sample of size k we need to have $k \geq n/w$ to compete, in terms of the MISE, with H-samples of size n. This relation quantifies the effect of the MCAR on nonparametric density estimation.

We are finishing the section by exploring an interesting question motivated by the above-presented theory. Is it possible, under any scenario, to prefer an MCAR sampling to the traditional sampling without missing? The question may be confusing because the missing data literature always considers missing as a nuisance which should be avoided if possible. On the other hand, let us consider the following situation. Assume that the price of a single observation in an M-sample (a sample that allows missing observations) is P_M while in the corresponding H-sample (a sample without missing) the price is P_H. As we know from the asymptotic theory, to get the same MISE the sample size k of the M-sample should be equal to n/w where n is the size of the H-sample. This yields that the M-sampling becomes more cost efficient if $P_M < wP_H$. In other words, at least theoretically, if the price of the MCAR sampling is low with respect to the price of sampling without missing (which may require more diligent bookkeeping or collecting meteorological data even during bad weather conditions), then the MCAR sampling can be more cost efficient. Of course, we have $k > n$, and if the total time of sampling is important, then this issue should be taken into account, but if only the price of sampling and accuracy (MISE) are important, then at least theoretically MCAR missing may have the edge.

Can the above-presented asymptotic theory be applied to small samples? Figure 4.2 allows us to explore this question via an intensive numerical study. The underlying experiment is similar to the one in Figure 4.1. Namely, first a direct H-sample of size k, equal to the rounded up ratio n/w, is generated from a corner function, here it is the Bimodal. Then an M-sample is obtained from the H-sample with $w = 0.7$ for which E-estimate $\hat{f}_M^X$ is calculated. Also, based on the first n observations of the H-sample, the E-estimate $\hat{f}_H^X$ is calculated. Then for these two estimates their integrated squared errors are calculated, namely ISEH $:= \int_0^1 (\hat{f}_H(x) - f(x))^2 dx$ and ISEM $:= \int_0^1 (\hat{f}_M(x) - f(x))^2 dx$. Finally, the ratio ISEH/ISEM of the integrated squared errors is calculated. Then the simulation is repeated 400 times (the number of simulations is controlled by the argument $nsim$). The ratio ISEH/ISEM should be close to one if the theory may be applied to small samples. The three top diagrams in Figure 4.2 show us results of particular simulations. As we see, theoretically extra 43 observations are needed to make chances of M-sampling and H-sampling be equal in terms of obtaining the same MISE. The particular three simulations do not support the theory. But we need a larger number of simulations to test the theory. The bottom diagram summarizes results of 400 simulations. Its histogram exhibits 400 ratios ISEH/ISEM and the title shows the sample mean and sample median of the ratios. The histogram indicates a large range of ratios. This range and the sample mean ratio 1.14 are not encouraging. On the other hand, the sample median ratio is equal to 1, and this result supports the theory. The conclusion is that for small samples there is a large variability in outcomes but overall the theory sheds light on small samples.

The reader is advised to repeat Figure 4.2 with the same and other parameters to get a better feeling of the studied nonparametric problem.

4.2   Nonparametric Regression with MAR Responses

Nonparametric regression problem was considered in Section 2.3. Let us briefly recall its setting and aim. There is a pair of continuous random variables (X, Y), where X is the predictor supported on $[0, 1]$ and $f^X(x) \geq c_* > 0$, $x \in [0, 1]$, and Y is the response. A sample $(X_1, Y_1), \ldots, (X_n, Y_n)$ from (X, Y) is available, and then the aim is to estimate the regression function

$$m(x) := E\{Y | X = x\}, \quad x \in [0, 1]. \tag{4.2.1}$$

Also recall that the regression model may be written as

$$Y = m(X) + \sigma(X)\varepsilon, \tag{4.2.2}$$

where ε is a random variable (the error) which may depend on X, $E\{\varepsilon|X\} = 0$, $E\{\varepsilon^2|X\} = 1$, and $\sigma(x)$ is the scale (volatility) function.

In this section the above-defined direct sample is no longer available because it is hidden, this is why we will refer to that sample as H-sample (hidden sample). Instead, the hidden responses may be missed while all predictors are still available. Namely, we observe an M-sample $(X_1, A_1Y_1, A_1), \ldots, (X_n, A_nY_n, A_n)$ of size n from the triplet (X, AY, A). Here A is a Bernoulli random variable, called the availability, which is the indicator that the response is available (not missed). It is assumed that

$$\mathbb{P}(A = 1|X = x, Y = y) = \mathbb{P}(A = 1|X = x) =: w(x) \geq c_0 > 0. \tag{4.2.3}$$

Function $w(x)$ is called the availability likelihood, and (4.2.3) implies that the missing is MAR (missing at random) because the probability of missing the response is defined by the always observed predictor.

To use our E-estimation methodology for the MAR sample, we need to propose a sample mean estimator of Fourier coefficients of the regression function (4.2.1). To understand its construction, we begin with the formula for the joint mixed density of the observed triplet (X, AY, A). For $x \in [0, 1]$, $y \in (-\infty, \infty)$ and $a \in \{0, 1\}$ we can write down the joint mixed density as

$$f^{X,AY,A}(x, ay, a) = \mathbb{P}(A = a|X = x)f^{X,AY}(x, ay)$$
$$= [w(x)f^X(x)f^{Y|X}(y|x)]^a[(1 - w(x))f^X(x)]^{1-a}. \tag{4.2.4}$$

This formula allows us to write down Fourier coefficients of $m(x)$, $x \in [0, 1]$ as follows,

$$\theta_j = \int_0^1 m(x)\varphi_j(x)dx = E\{AY\varphi_j(X)/[f^X(X)w(X)]\}, \quad j = 0, 1, \ldots \tag{4.2.5}$$

Assume for a moment that functions $w(x)$ and $f^X(x)$ are known. Then the sample mean estimator of Fourier coefficients is

$$\bar{\theta}_j := n^{-1}\sum_{l=1}^{n}\frac{A_lY_l\varphi_j(X_l)}{f^X(X_l)w(X_l)}. \tag{4.2.6}$$

If $f^X(x)$ and $w(x)$ are unknown, and this is a typical situation, then they can be estimated based on the M-sample. Indeed, the design density can be estimated using E-estimator of Section 2.2 and the availability likelihood can be estimated by the Bernoulli regression estimator of Section 2.4.

There is also another attractive possibility to deal with the case of unknown $f^X(x)$ and $w(x)$. First, we rewrite the estimate $\bar{\theta}_j$ as

$$\bar{\theta}_j = [n\mathbb{P}(A = 1)]^{-1}\sum_{l=1}^{n}\frac{A_lY_l\varphi_j(X_l)}{f^X(X_l)w(X_l)/\mathbb{P}(A = 1)}. \tag{4.2.7}$$

Next, we note that the marginal density of the predictor in a complete case is

$$f^{X|A}(x|1) = \frac{f^{X,A}(x, 1)}{\mathbb{P}(A = 1)} = \frac{f^X(x)\mathbb{P}(A = 1|X = x)}{\mathbb{P}(A = 1)} = \frac{f^X(x)w(x)}{\mathbb{P}(A = 1)}. \tag{4.2.8}$$

Using formula (4.2.8) in (4.2.7) we get

$$\bar{\theta}_j = [n\mathbb{P}(A = 1)]^{-1}\sum_{l=1}^{n}\frac{A_lY_l\varphi_j(X_l)}{f^{X|A}(X_l|1)}. \tag{4.2.9}$$

In (4.2.10) the probability $\mathbb{P}(A = 1)$ is unknown and it is estimated by the sample mean estimator N/n where $N := \sum_{l=1}^{n} A_l$ is the number of complete cases in M-sample. The density $f^{X|A}(x|1)$ is the density of predictors in complete cases and hence it can be estimated by the E-estimator $\hat{f}^{X|A}(x|1)$ of Section 2.2. Recall our discussion in Section 4.1 about remedy for the case $N = 0$, and define the plug-in sample mean Fourier estimator

$$\hat{\theta}_j := N^{-1} \sum_{l=1}^{n} \frac{A_l Y_l \varphi_j(X_l)}{\max(\hat{f}^{X|A}(X_l|1), c/\ln(n))}. \tag{4.2.10}$$

Further, since in practice we always deal with N comparable with n, we can replace $c/\ln(n)$ in (4.2.10) by $c/\ln(N)$. Then the Fourier estimator is based only on complete-case observations (when $A_l = 1$), and even the sample size n is not used. In other words, this is a complete-case Fourier estimator.

We can make the following conclusion. The E-estimator, proposed for the case of classical regression with no missing data, can be used here for the subsample of complete cases $(X_l, A_l Y_l)$ corresponding to $A_l = 1$, $l = 1, \ldots, n$. This approach, when only complete cases are used in estimation and incomplete cases are ignored, is called a *complete-case approach*. The asymptotic theory supports this approach and asserts that no other estimator may outperform a complete-case regression estimation for the case of MAR responses.

Let us check how the proposed complete-case E-estimator performs for small sample sizes. Two simulations are shown in the two columns of Figure 4.3. Top diagrams show scattergrams of the underlying H-samples (unavailable samples from (X, Y)) overlaid by the corresponding regression functions and their E-estimates. As we already know from Section 2.3, the estimates may be good for the sample size $n = 100$, and this relatively small sample size is chosen to better visualize data. In both cases the E-estimates are fair despite the heteroscedasticity. The bottom diagrams show us MAR samples from (X, AY) that are referred to as M-samples. Note that here complete pairs are shown by circles and incomplete by crosses. In both simulations the same availability likelihood function $w(x)$ is used, and we can see from the M-samples that the function is decreasing in x. Let us look more closely at the left column of diagrams with the Normal regression function. The M-sample is a teachable example of what missing may do to the data. First, the number N of complete pairs is only 77, that is almost a quarter of responses are lost. Second, while for the H-sample the E-estimate overestimates the mode, in M-sample it underestimates it, and we can realize why from the scattergrams. It is also of interest to compare the ISEs which describe how the MAR affects the quality of estimation. For the Bimodal regression, the MAR affects the quality of estimation rather dramatically. The Bimodal is a difficult regression to deal with, and here the heteroscedasticity makes its estimation even more complicated by hiding the two modes. Nonetheless, the E-estimator does a very good job for the H-sample. Magnitudes of the two modes are shown correctly, both modes are slightly shifted to the left, but overall we get a good picture of a bimodal regression function. The MAR, highlighted in the right-bottom diagram, modifies the H-sample in such a way that while the E-estimate shows two modes, its left mode is higher than the right one. It is a teachable moment to analyze the scattergram and, keeping in mind the underlying availability likelihood, to figure out why such a dramatic change has occurred.

Now let us make several theoretical comments about the recommended estimator. First, let us check that (4.2.6) is unbiased estimator of Fourier coefficients θ_j, that is

$$\mathbb{E}\{\bar{\theta}_j\} = \theta_j. \tag{4.2.11}$$

To prove this assertion, we first note that the assumption (4.2.3) implies independence of A and Y given X, and in particular that $\mathbb{E}\{AY|X\} = \mathbb{E}\{A|X\}\mathbb{E}\{Y|X\}$. This equality, together with the rule of calculation of the expectation via conditional expectation and

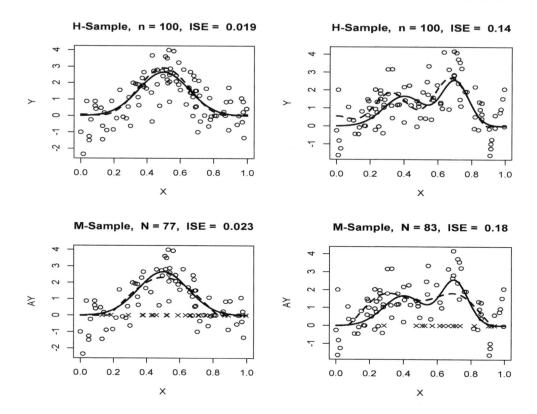

Figure 4.3 *Performance of the complete-case E-estimator for regression with MAR responses. Results for two simulations with the Normal and the Bimodal regression functions are shown in the two columns of diagrams. The underlying model is $Y = m(X) + \sigma S(x)\epsilon$ where ϵ is standard normal and independent of X. The top diagrams show underlying hidden samples (H-samples), the bottom diagrams show the corresponding M-samples with missing responses. Observations in H-samples are shown by circles. Available complete pairs are shown by circles while incomplete pairs $(A_l X_l, X_l)$ with $A_l = 0$ are shown by crosses. The titles show corresponding integrated squared errors (ISE). An underlying regression function and its E-estimate are shown by the solid and dashed lines, respectively. Because it is known that estimated regression functions are nonnegative, a projection on nonnegative functions is used. {The arguments are: set.c controls two underlying regression functions for the left and right columns; n controls the sample size; sigma controls the parameter σ; the string scalefun defines a custom function $S(x)$ which is truncated from below by the value dscale and then it is rescaled to get a bona fide density supported on [0,1]; the string desden controls shape of the design density $f^X(x)$ which is then truncated from below by the value dden and then it is rescaled to get a bona fide density; the availability likelihood function $w(x)$ is defined by the string w and then truncated from below by dwL and from above by dwU.} [n = 100, set.c = c(2,3), sigma = 1, scalefun = "3-(x-0.5)^2", desden = "1+0.5*x", dscale = 0, dden = 0.2, w = "1-0.4*x", dwL = 0.5, dwU = 0.9, c = 1, cJ0 = 3, cJ1 = 0.8, cTH = 4]*

(4.2.3), allows us to write,

$$\mathbb{E}\{\bar{\theta}_j\} = \mathbb{E}\left\{\frac{AY\varphi_j(X)}{f^X(X)w(X)}\right\} = E\left\{\frac{E\{AY|X\}\varphi_j(X)}{f^X(X)w(X)}\right\}$$

$$= E\left\{\frac{\mathbb{E}\{A|X\}\mathbb{E}\{Y|X\}\varphi_j(X)}{f^X(X)w(X)}\right\} = E\left\{\frac{w(X)m(X)\varphi_j(X)}{f^X(X)w(X)}\right\}$$

$$= E\Big\{\frac{m(X)\varphi_j(X)}{f^X(X)}\Big\} = \int_0^1 m(x)\varphi_j(x)dx = \theta_j. \tag{4.2.12}$$

This is what was wished to show.

Second, let us evaluate the variance of $\bar{\theta}_j$ and, using notation of model (4.2.2), show that

$$\mathbb{V}(\bar{\theta}_j) = n^{-1}\big[\int_0^1 \frac{[m^2(x) + \sigma^2(x)]\varphi_j^2(x)}{f^X(x)w(x)}dx - \theta_j^2\big]. \tag{4.2.13}$$

To prove (4.2.13) we recall that the variance of a sum of independent random variables is equal to the sum of the variances of the variables. Using this property we can write,

$$\mathbb{V}(\bar{\theta}_j) = n^{-2}\sum_{l=1}^n \mathbb{V}\Big(\frac{A_l Y_l \varphi_j(X_l)}{f^X(X_l)w(X_l)}\Big) = n^{-1}\mathbb{V}\Big(\frac{AY\varphi_j(X)}{f^X(X)w(X)}\Big). \tag{4.2.14}$$

Using formula

$$\mathbb{V}(Z) = \mathbb{E}\{Z^2\} - [\mathbb{E}\{Z\}]^2, \tag{4.2.15}$$

we continue with calculation of the second moment,

$$\mathbb{E}\Big\{\Big(\frac{AY\varphi_j(X)}{f^X(X)w(X)}\Big)^2\Big\} = \mathbb{E}\Big\{\Big(\frac{Y\varphi_j(X)}{f^X(X)w(X}\Big)^2 E\{A^2|X,Y\}\Big\}$$

$$= \mathbb{E}\Big\{\frac{(m^2(X) + 2m(X)\sigma(X)\varepsilon + \sigma^2(X)\varepsilon^2)\varphi_j^2(X)w(X)}{[f^X(X)w(X)]^2}\Big\}$$

$$= \mathbb{E}\Big\{\mathbb{E}\Big\{\frac{(m^2(X) + 2m(X)\sigma(X)\varepsilon + \sigma^2(X)\varepsilon^2)\varphi_j^2(X)w(X)}{[f^X(X)w(X)]^2}|X\Big\}\Big\}. \tag{4.2.16}$$

Using the assumed $\mathbb{E}\{\varepsilon|X\} = 0$ and $\mathbb{E}\{\varepsilon^2|X\} = 1$, we may continue (4.2.16),

$$\mathbb{E}\Big\{\Big(\frac{AY\varphi_j(X)}{f^X(X)w(X)}\Big)^2\Big\} = \mathbb{E}\Big\{\frac{[m^2(X) + \sigma^2(X)]\varphi_j^2(X)}{[f^X(X)]^2 w(X)}\Big\}$$

$$= \int_0^1 \frac{[m^2(x) + \sigma^2(x)]\varphi_j^2(x)}{f^X(x)w(x)}dx. \tag{4.2.17}$$

This, (4.2.12) and (4.2.15) prove (4.2.13).

Third, in Section 2.3 it was explained how to reduce the variance (3.2.13) by subtracting an appropriate partial sum

$$\tilde{m}_{-j}(X_l) := \sum_{s\in\{0,1,\ldots,b_n\}\backslash j} \bar{\theta}_s \varphi_s(X_l) \tag{4.2.18}$$

from Y_l, where b_n is the rounded $\ln(n)$. This yields the estimator

$$\tilde{\theta}_j = N^{-1}\sum_{l=1}^n \frac{A_l[Y_l - \tilde{m}_{-j}(X_l)]\varphi_j(X_l)}{\max(\hat{f}^{X|A}(X_l|1), c/\ln(n))}. \tag{4.2.19}$$

A direct calculation shows that, under a mild assumption on smoothness of functions $m(x)$, $f^X(x)$, $\sigma(x)$ and $w(x)$, we get

$$\mathbb{V}(\tilde{\theta}_j) = n^{-1}d(f^X, \sigma, w)[1 + o_j(1) + o_n(1)], \tag{4.2.20}$$

where the coefficient of difficulty is

$$d(f^X, \sigma, w) := \int_0^1 \frac{\sigma^2(x)}{f^X(x)w(x)}dx. \tag{4.2.21}$$

Fourth, the asymptotic theory asserts that it is impossible to propose another estimator with a smaller coefficient of difficulty. Further, using the complete-case approach and our plug-in methodology for the case of unknown $f^X(x)$ and $w(x)$ yields the same coefficient of difficulty (4.2.21).

Fifth, based on the above-presented results, to get the same asymptotic MISE convergence as for a hidden H-sample of size n, the M-sample should have the sample size $n^* = nd(f^X, \sigma, w)$. The latter sheds light on the effect of missing responses on the quality of estimation.

Sixth, using the Cauchy-Schwarz inequality we get that the design density $f_*^X(x)$ which minimizes the coefficient of difficulty is

$$f_*^X(x) := \frac{\sigma(x)[w(x)]^{-1/2}}{\int_0^1 \sigma(u)[w(u)]^{-1/2}du}. \tag{4.2.22}$$

If a controlled design of regression with missing responses is possible, then this is the design to use.

Finally, it is a straightforward exercise to explore a problem, formulated in the previous section, about choosing between H-sampling and M-sampling when prices of single observations in each sampling are given.

4.3 Nonparametric Regression with MAR Predictors

This section continues discussion of regression estimation with missing data, only now the case of MAR predictors is considered. Let us describe the regression model. There is a hidden sample $(X_1, Y_1, A_1), \ldots, (X_n, Y_n, A_n)$ from the triplet (X, Y, A). Continuous variable X is the predictor supported on $[0, 1]$ and $f^X(x) \geq c_* > 0$, $x \in [0, 1]$, continuous variable Y is the response, and A is a Bernoulli random variable. The three random variables may be dependent. If $A_l = 1$ then the corresponding predictor X_l is available and the case (X_l, Y_l) is complete, otherwise X_l is missed and the lth case is incomplete because only the response Y_l is available. Formally, we may say that the available M-sample (missing sample) of size n from the triplet (AX, Y, A) is $(A_1 X_1, Y_1, A_1), \ldots, (A_n X_n, Y_n, A_n)$. Also, because X is a continuous random variable, we have $\mathbb{P}(A = I(AX \neq 0)) = 1$, and hence the M-sample is equivalent to the sample $(A_1 X_1, Y_1), \ldots, (A_n X_n, Y_n)$.

The main assumption about the missing mechanism is that it is MAR and

$$\mathbb{P}(A = 1|Y = y, X = x) = \mathbb{P}(A = 1|Y = y) =: w(y) \geq c_0 > 0. \tag{4.3.1}$$

Note that while in general X and A may be dependent random variables, according to (4.3.1) they are conditionally independent given the response Y.

The joint mixed density of the triplet (AX, Y, A) is

$$f^{AX,Y,A}(ax, y, a) = [f^X(x)f^{Y|X}(y|x)w(y)]^a [f^Y(y)(1 - w(y))]^{1-a}, \tag{4.3.2}$$

where $x \in [0, 1]$, $y \in (-\infty, \infty)$ and $a \in \{0, 1\}$.

The problem is to estimate the regression function for the hidden pair (X, Y), namely to estimate the function

$$m(x) := E\{Y|X = x\}, \quad x \in [0, 1]. \tag{4.3.3}$$

Of course, we can rewrite (4.3.3) in a more familiar way,

$$Y = m(X) + \sigma(X)\varepsilon, \tag{4.3.4}$$

where ε is a random variable which may depend on X, $E\{\varepsilon|X\} = 0$ and $E\{\varepsilon^2|X\} = 1$ almost sure, and $\sigma(x)$ is the scale function.

The good news about the regression model with MAR predictors is that the availability likelihood (4.3.1) depends only on the always observed value of the response. The bad news is that in complete cases of M-sample the conditional density $f^{Y|X,A}(y|x,1)$ of the response given the predictor is biased with respect to the conditional density of interest $f^{Y|X}(y|x)$ (discussion of biased densities may be found in Section 3.1). Let us prove the last assertion. Write,

$$f^{Y|X,A}(y|x,1) = \frac{w(y)f^{Y|X}(y|x)f^X(x)}{f^X(x)\int_0^1 w(u)f^{Y|X}(u|x)du}$$

$$= \frac{w(y)}{\int_{-\infty}^{\infty} w(u)f^{Y|X}(u|x)du} f^{Y|X}(y|x). \qquad (4.3.5)$$

This is what was wished to prove, and note that $w(y)$ is the biasing function.

The biased conditional density implies that in general, when only M-sample is available, a complete-case approach cannot be used for estimation of the conditional density $f^{Y|X}(y|x)$ and hence for estimation of the regression function $m(x) := \mathbb{E}\{Y|X = x\}$. This is what sets apart regressions with MAR responses and MAR predictors, because for the former the complete-case approach is both consistent and optimal.

To propose a regression estimator, we are going to consider in turn three scenarios when a consistent estimation is possible (furthermore, the asymptotic theory asserts that the proposed solutions yield optimal rates of the MISE convergence). The first one is when the design density $f^X(x)$ and the availability likelihood $w(y)$ are known. Under this scenario a complete-case approach yields consistent estimation. The second one is when these two nuisance functions are unknown but the marginal density $f^Y(y)$ of the response is known. Under the second scenario a complete-case approach is also consistent. Finally, if only M-sample is available, a consistent E-estimator uses both complete and incomplete cases.

Scenario 1. Functions $f^X(x)$ and $w(y)$ are known. To employ the E-estimation methodology, we need to suggest a sample mean estimator of Fourier coefficients

$$\theta_j := \int_0^1 m(x)\varphi_j(x)dx, \quad j = 0, 1, \ldots \qquad (4.3.6)$$

of the regression function of interest (4.3.3). To do this, we need to write down the coefficients as an expectation. Using (4.3.2) and (4.3.3) we can write,

$$\theta_j = \mathbb{E}\{m(X)\varphi_j(X)[f^X(X)]^{-1}\}$$

$$= \mathbb{E}\{\mathbb{E}\{(Y|X)\}\varphi_j(X)[f^X(X)]^{-1}\}\} = \mathbb{E}\{AY\varphi_j(AX)[f^X(AX)w(Y)]^{-1}\}. \qquad (4.3.7)$$

This implies the following sample mean estimator of the Fourier coefficients,

$$\bar{\theta}_j := n^{-1}\sum_{l=1}^n A_l Y_l \varphi_j(A_l X_l)[f^X(A_l X_l)w(Y_l)]^{-1}. \qquad (4.3.8)$$

The Fourier estimator $\bar{\theta}_j$ yields the regression E-estimator of Section 2.3. Note that the E-estimator is based only on complete cases, that is the regression estimate is a complete-case estimator. Further, the asymptotic theory asserts that a complete-case approach is optimal.

Scenario 2. Known marginal density of responses $f^Y(y)$. This is an interesting case both on its own and because it will explain how to solve the problem of regression estimation when only M-sample is available.

There are two steps in the proposed regression estimation. The first step is to use the known density $f^Y(y)$ of responses for estimation of nuisance functions $w(y)$ and $f^X(x)$. This step may be done using only complete cases. The second step is to utilize the E-estimator proposed under the above-considered first scenario.

Let us show how the two nuisance functions can be estimated. We begin with estimation of $w(y)$. Recall that $w(y) := \mathbb{P}(A = 1|Y = y) = \mathbb{E}\{A|Y = y\}$. This implies that estimation of $w(y)$ is the Bernoulli regression for which E-estimator $\hat{w}(y)$ was proposed in Section 2.4. Also recall that only Y_l in complete cases are needed whenever the density of Y is known.

With the estimator $\hat{w}(y)$ at hand, we can estimate the density $f^X(x)$. The idea of estimation is as follows. Suppose that $w(y)$ is known, and denote by $\kappa_j := \int_0^1 f^X(x)\varphi_j(x)dx$ the jth Fourier coefficient of the density. Then, as we will show shortly, the sample mean estimator of κ_j is

$$\tilde{\kappa}_j := n^{-1} \sum_{l=1}^{n} A_l\varphi_j(A_lX_l)/w(Y_l). \tag{4.3.9}$$

Then we plug $\max(\hat{w}(Y_l), c/\ln(n))$ in place of unknown $w(Y_l)$. This Fourier estimator yields the density E-estimator $\hat{f}^X(x)$, $x \in 0, 1]$ of Section 2.2.

The teachable moment here is that if $f^Y(y)$ is known, then the complete-case approach is still consistent.

To finish our discussion of Scenario 2, we need to prove that (4.3.9) is the sample mean estimator. Using the rule of calculation of the expectation via a conditional expectation, we can write,

$$\mathbb{E}\{\tilde{\kappa}_j\} = \mathbb{E}\{A\varphi_j(AX)/w(Y)\} = \mathbb{E}\{\varphi_j(X)\mathbb{E}\{A|X,Y\}/w(Y)\}$$

$$= \mathbb{E}\{\varphi_j(X)w(Y)/w(Y)\} = \mathbb{E}\{\varphi_j(X)\} = \int_0^1 f^X(x)\varphi_j(x)dx = \kappa_j. \tag{4.3.10}$$

This is what was wished to show.

Scenario 3. Only M-sample is available. The available M-sample is $(A_1X_1, Y_1, A_1), \ldots,$ (A_nX_n, Y_n, A_n) from the triplet (AX, Y, A). No other information is available.

The proposed solution is to convert this scenario into the previous one. To do this, we note that all n observations of Y are available and hence the density $f^Y(y)$ may be estimated by the density E-estimator $\hat{f}^Y(y)$ of Section 2.2. Then the estimator proposed for the second scenario can be utilized with the density f^Y replaced by its E-estimate. Let us also stress that values of all nuisance functions are needed only at (A_lX_l, Y_l) from complete cases.

Let us comment on the proposed data-driven regression estimator from the point of view of a complete-case approach. All observations, including responses in incomplete cases, are needed only for estimation of the density $f^Y(y)$, furthermore, this density is needed only at points A_lY_l corresponding to $A_l = 1$. As a result, if an extra sample of responses is available to estimate $f^Y(Y_l)$, then only complete cases can be used. This remark is important whenever the reliability of data in incomplete cases is in doubt.

Figure 4.4 allows us to explain the setting and each step in construction of the regression estimator. The caption explains all diagrams.

We begin with the left-top diagram exhibiting the underlying H-sample (the hidden scattergram). Pairs of observations are shown by circles, the solid and dashed lines show the underlying regression and the E-estimate based on the H-sample, respectively. The regression is $Y = m(X) + \sigma S(X)\varepsilon$, where $m(x)$ and $S(x)$ are two custom-made functions, σ is the parameter and ε is an independent from X standard normal random variable. For the particular simulation $Y = 2X + (3 - (X - 0.5)^2)\varepsilon$. The distribution of X is also custom-made, here it is uniform on $[0, 1]$. The title indicates the sample size n and the ISE of the regression E-estimate. Let us note that the linear regression, as a function, is not an element of the cosine basis, and it is a challenging nonparametric function. Overall, the estimate correctly shows the underlying regression.

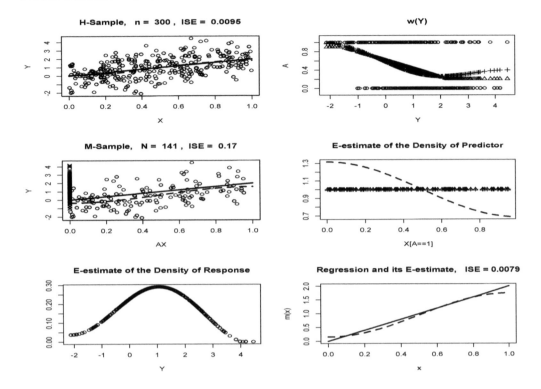

Figure 4.4 *Nonparametric regression with MAR predictors. The underlying model is $Y = m(X) + \sigma S(x)\varepsilon$ where ε is standard normal and independent of X. The left-top diagram shows the hidden H-sample, regression $m(x)$ (the solid line) and its E-estimate (the dashed line). The left-middle diagram shows the corresponding M-sample, $m(x)$ by the solid line and E-estimate, based on complete cases, by the dashed line. The left-bottom diagram shows E-estimate of the marginal density $f^Y(Y_l)$ at observed responses. The right-top diagram shows the scattergram of (Y_l, A_l) by circles, the underlying availability likelihood function $w(Y_l)$ by triangles and its E-estimate $\hat{w}(Y_l)$ by crosses. The right-middle diagram shows by triangles $f^X(A_l X_l)$ for complete cases (when $A_l = 1$), and by crosses its E-estimate (here they coincide). The dashed line shows density E-estimate based only on predictors in complete cases (available predictors). The axis label $X[A == 1]$ indicates that $f^X(X_l)$ and $\hat{f}^X(X_l)$ are shown only for X_l corresponding to $A_l = 1$, that is only for complete cases. The right-bottom diagram shows the underlying regression function $m(x)$ (the solid line) and the regression E-estimate (the dashed line) proposed for the case of MAR predictors. {The arguments are: mx defines a custom-made underlying regression function, n controls the size of H-sample, sigma controls the parameter σ; scalefun defines shape of function $S(x)$ which is truncated from below by the value dscale and then rescaled into a bona fide density supported on $[0,1]$; desden allows to choose the shape of design density $f^X(x)$ which is then truncated from below by dden and then rescaled into a bona fide density. The shape of the availability likelihood function $w(y)$ is defined by the argument-function w which is then truncated from below by dwL and from above by dwU.} [mx = "2*x", n = 300, sigma = 1, scalefun = "3-(x-0.5)^2", desden = "1+0*x", dscale = 0, dden = 0.2, w = "1-1.2*y", dwL = 0.2, dwU = 0.9, c = 1, cJ0 = 3, cJ1 = 0.8, cTH = 4]*

The left-middle diagram shows us the observed missing data (M-sample). The missing is created by the Bernoulli random variable A with an availability likelihood function $\mathbb{P}(A = 1|X = x, Y = y) = w(y)$, in this particular simulation $w(y) = \min(0.9, \max(0.2, 1 - 1.2y))$; the caption explains how to change this function. The diagram shows n realizations of (Y, AX), and its title also shows the number $N := \sum_{l=1}^{n} A_l$ of complete pairs. As we see, from $n = 300$ of hidden complete pairs, the missing mechanism allows us to have only

$N = 141$ complete pairs, that is more than a half of predictors are missed. The naïve complete-case regression E-estimator is shown by the dashed line; its visualization and the $ISE = 0.17$ supports the above-made theoretical opinion about inconsistency of complete-case approach for the case of MAR predictors. The complete-case E-estimate indicates smaller y-intercept and changing slope. To understand this outcome, let us briefly look at the right-top diagram. Here the circles show the scattergram of (Y_l, A_l). As we see, all smaller underlying responses belong to complete cases, while many larger responses belong to incomplete cases. This is what explains the complete-case E-estimate in the left-middle diagram. To finish the discussion of the diagram, let us comment on how the M-sample was generated. First the random variable X (the predictor) is generated. Second, the response Y is generated according to the custom-made regression formula. Then the Bernoulli($w(Y)$) random variable A is generated. This creates the M-sample. A particular sample is shown in the diagram. Note that if in the H-sample the predictors are distributed uniformly, available predictors in the M-sample are no longer uniform; it is clear that the distribution is skewed to the left. Can you support this conclusion theoretically via a formal analysis of $f^{X|A}(x|1)$? We will return to this density shortly.

The first two diagrams described the data. All the following diagrams are devoted to the above-outlined regression estimation based solely on the M-sample.

The left-bottom diagram illustrates the first step: estimation of the response density $f^Y(y)$. Because all responses are available, we may use the density E-estimator of Section 2.2. The only issue to explain is how we are dealing with an unknown support of Y. The density E-estimator of Section 2.2 assumes that the support is $[0,1]$. This is why we use the rescaling approach explained in Section 2.2. Namely, we do the following: (i) Rescale all Y_l onto the unit interval by introducing $Y_l' := (Y_l - Y_{(1)})/(Y_{(n)} - Y_{(1)})$; (ii) Calculate density E-estimate $\hat{f}^{Y'}(y')$ of $f^{Y'}(y')$, $y' \in [0,1]$; (iii) Rescale back the calculated estimate and get the wished $\hat{f}^Y(Y_l) := [Y_{(n)} - Y_{(1)}]^{-1}\hat{f}^{Y'}(Y_l')$. The diagram shows that the calculated E-estimate is unimodal with the mode at $Y = 1$. Another informative aspect of the plot is that majority of responses are located around the mode and just few create tails (as should be expected from the curve).

Let us stress that the left-bottom diagram illustrates the only step where all n observations of Y are used. Hence, if an additional sample from Y is available, we can estimate $f^Y(y)$ and then use it in the complete-case regression E-estimator described below. This is a useful remark because it explains what can be done if only complete cases in an M-sample are available.

The right-top diagram shows us the scattergram of A versus Y, it is shown by circles. The regression of A on Y is the searching after the availability likelihood function $w(y) = \mathbb{P}(A = 1|Y = y)$. This is a classical Bernoulli regression discussed in Section 2.4. Recall that if $f^Y(y)$ is known then the regression E-estimator is based only on values of Y_l in available complete cases (when $A_l = 1$). In the diagram triangles and crosses show the underlying $w(Y_l)$ and its E-estimate $\hat{w}(Y_l)$ at all observed values of Y (note that Figure 4.4 allows us to chose any custom-made availability likelihood function $w(y)$). The E-estimate is not perfect due to the wrong right tail. At the same time, only a small proportion (from $n = 300$) of responses belong to that tail. Furthermore, recall that $w(y)$ is a nuisance function whose values are needed only at values Y_l corresponding to $A_l = 1$. Hence, a bad estimation of the right tail may not ruin the regression E-estimate.

The right-middle diagram illustrates estimation of the design density $f^X(x)$ at values $x = X_l$ of available predictors in the M-sample. The density is estimated with the help of the estimates $\hat{w}(Y_l)$ for l such that $A_l = 1$. The diagram shows us by the triangles and crosses the underlying and estimated densities (here they coincide). Note that the E-estimate is perfect despite the imperfect estimate of $w(y)$. The dashed line shows us the density E-estimate $\check{f}^X(x)$ based on $N = 141$ available realizations of X. This estimate correctly shows the biased (by the MAR) character of predictors in complete cases. The reader may look theoretically at the density $f^{X|A}(x|1)$ and get a formula for the biasing function. This is an interesting example of creating a biased data when the underlying density can be restored without knowing the biasing function.

Finally, with the help of the estimated $w(y)$ and $f^X(x)$, we can estimate the regression function $m(x)$. In the right-bottom diagram, the solid and dashed lines show the underlying regression and the proposed E-estimate, respectively. The E-estimate nicely shows the monotone character of the regression. Further, if we compare its ISE = 0.0079 with the ISE = 0.0095 of the E-estimate based on the H-sample, then the outcome is surprisingly good especially keeping in mind the number $N = 141$ of complete cases and the imperfect estimate of the availability likelihood. Of course, this is an atypical outcome. It is advisable to repeat Figure 4.4 and learn more about this interesting statistical problem.

The asymptotic theory asserts that under a mild assumption the proposed methodology is optimal. Further, we may conclude that the case of MAR predictors is dramatically more complicated than the case of MAR responses where a complete-case approach may be used.

4.4   Conditional Density Estimation

Consider two continuous random variables X and Y with a joint density $f^{XY}(x,y)$. If the marginal density $f^X(x_0) := \int_{-\infty}^{\infty} f^{X,Y}(x_0, y)dy$ for some fixed value x_0 of X is positive (in other words, x_0 belongs to the support of X), then the conditional density of Y given $X = x_0$ is defined as

$$f^{Y|X}(y|x_0) := \frac{f^{X,Y}(x_0, y)}{f^X(x_0)}. \tag{4.4.1}$$

Note that the conditional density $f^{Y|X}(y|x_0)$ is the probability density in y. Indeed, according to (4.4.1) the conditional density is nonnegative and

$$\int_{-\infty}^{\infty} f^{Y|X}(y|x_0)dy = \int_{-\infty}^{\infty} [f^{X,Y}(x_0, y)dy/f^X(x_0)]dy = f^X(x_0)/f^X(x_0) = 1. \tag{4.4.2}$$

We conclude that, for each value x from the support of X, the conditional density describes a new random variable V_x whose distribution depends on x. At the same time, if X and Y are independent, then for any x the random variables V_x and Y have the same distribution.

As any other random variable, V_x has its own descriptive characteristics. For instance, its mean is $m(x) := \mathbb{E}\{V_x\} = \int_{-\infty}^{\infty} y f^{Y|X}(y|x)dy = \mathbb{E}\{Y|X = x\}$, and from the two previous sections we know the name of this mean - it is the nonparametric regression. Hence, it is natural to refer to Y as the response and to X as the predictor. The standard deviation of V_x is another important characteristic which is called the scale or volatility. Quantiles of V_x are also of the statistical interest and they are referred to as quantile regressions.

Note that the above-introduced univariate functions are functionals of the conditional density. The latter means that if the conditional density is known then the regression, scale and quantile regressions are known but not vice versa. If this is the case, then why is there a vast literature about estimation of all these univariate functions instead of concentrating on estimation of the conditional density? The reason is that the conditional density is a bivariate function. Indeed, $f^{Y|X}(y|x)$ is the function in y and x, and as we know from Section 2.5, this makes the problem of conditional density estimation more complicated than estimation of the univariate functions.

Following the two previous regression sections, we are considering cases of missed responses and missed predictors in turn. Recall that we are referring to X and Y as the predictor and the response, respectively, and here this is simply a convenient terminology to use. The aim is to estimate the conditional density of the response Y given the predictor X. Similarly to the previous sections, it is assumed that X is supported on $[0,1]$ and $f^X(x) \geq c_* > 0$, $x \in [0,1]$.

MAR responses. The model is the same as in Section 4.2. There is a hidden H-sample $(X_1, Y_1, A_1), \ldots, (X_n, Y_n, A_n)$ of size n from the triplet (X, Y, A) where A is a Bernoulli random variable, called the availability, which defines an underlying missing mechanism. Namely, a realization Y_l is available when $A_l = 1$ and otherwise the realization is missed. This creates the observed M-sample $(X_1, A_1 Y_1, A_1), \ldots, (X_n, A_n Y_n, A_n)$ with missing responses. Note that $\mathbb{P}(A = I(AY \neq 0)) = 1$, and hence the M-sample is equivalent to the sample $(X_1, A_1 Y_1), \ldots, (X_n, A_n Y_n)$. Further, it is assumed that the missing mechanism is MAR and the availability likelihood is

$$\mathbb{P}(A = 1 | X = x, Y = y) = \mathbb{P}(A = 1 | X = x) =: w(x) \geq c_0 > 0. \qquad (4.4.3)$$

The latter means that we always have a chance of observing (not missing) Y given X. The problem is to estimate conditional density $f^{Y|X}(y|x)$ using M-sample.

It was shown in Section 4.2 that a complete-case approach is consistent for estimation of the conditional expectation $\mathbb{E}\{Y|X\}$. Can this approach also shine in estimation of the conditional density? Let us explore this question theoretically. Consider the conditional density of Y given X in complete cases of an M-sample. Using (4.4.3) we can write,

$$f^{Y|X,A}(y|x, 1) = \frac{f^{X,Y,A}(x, y, 1)}{f^{X,A}(x, 1)} = \frac{f^{X,Y}(x, y)\mathbb{P}(A = 1 | X = x, Y = y)}{f^X(x)\mathbb{P}(A = 1 | X = x)}$$

$$= \frac{f^{X,Y}(x, y)}{f^X(x)} = f^{Y|X}(y|x). \qquad (4.4.4)$$

We conclude that the conditional density in the complete cases of an M-sample is the same as the conditional density of interest $f^{Y|X}$ in the underlying H-sample. Of course, the number $N = \sum_{l=1}^n A_l$ of complete cases is a Binomial$(n, \mathbb{P}(A = 1))$ random variable, and for some M-samples N may be dramatically smaller than n. On the other hand, we know from the inequality (4.0.1) that the likelihood of relatively small N is negligible for large n. In other words, the more serious issue is that here we are dealing with estimation of a bivariate function, recall the discussion in Section 2.5. As a result, similarly to our discussion in Section 4.1, only M-samples with relatively large $N > k$ should be considered, and for a bivariate problem a reasonable k is in the hundreds.

Now we are in a position to propose an E-estimator of the conditional density (4.4.1). First, let us explain how we can solve the problem when an H-sample of size n from (X, Y) is available. Following our methodology of constructing an E-estimator, we need to propose a sample mean estimator of Fourier coefficients of the conditional density. Suppose that a pair (X, Y) is supported on $[0, 1]^2$ and recall the corresponding tensor-product basis with elements $\varphi_{j_1 j_2}(x, y) := \varphi_{j_1}(x)\varphi_{j_2}(y)$ (recall Section 2.5). Using this basis, we may write down Fourier coefficients of the conditional density,

$$\theta_{j_1 j_2} = \int_{[0,1]^2} f^{Y|X}(y|x)\varphi_{j_1 j_2}(x, y)dxdy$$

$$= \int_{[0,1]^2} \frac{f^{X,Y}(x, y)\varphi_{j_1 j_2}(x, y)}{f^X(x)}dxdy = \mathbb{E}\left\{\frac{\varphi_{j_1 j_2}(X, Y)}{f^X(X)}\right\}. \qquad (4.4.5)$$

Using (4.4.5) could immediately yield a sample mean estimator if we knew the marginal density $f^X(x)$ (recall that in a regression setting it would be called the design density). We do not know this marginal density, but can estimate it by the density E-estimator $\hat{f}^X(x)$ of Section 2.2. This implies the sample mean estimator of $\theta_{j_1 j_2}$,

$$\bar{\theta}_{j_1 j_2} := n^{-1} \sum_{l=1}^n \frac{\varphi_{j_1 j_2}(X_l, Y_l)}{\max(\hat{f}^X(X_l), c/\ln(n))}. \qquad (4.4.6)$$

The Fourier estimator yields the bivariate E-estimator $\bar{f}^{Y|X}(y|x)$, $(x,y) \in [0,1]^2$ of Section 2.5. The asymptotic theory supports the proposed estimation methodology.

For the case of M-sample we use the above-defined conditional density estimator only now it is based on $N := \sum_{l=1}^{n} A_l$ complete cases of the M-sample. Namely, denote the density E-estimator of $f^{X|A}(x|1)$ as $\hat{f}^{X|A}(x|1)$, and then introduce the plug-in sample mean Fourier estimator

$$\hat{\theta}_{j_1 j_2} := N^{-1} \sum_{l=1}^{n} \frac{A_l \varphi_{j_1 j_2}(X_l, A_l Y_l)}{\max(\hat{f}^{X|A}(X_l|1), c/\ln(n))}. \tag{4.4.7}$$

This Fourier estimator yields the conditional density E-estimator $\hat{f}^{Y|X}(y|x)$, $(x,y) \in [0,1]^2$ of Section 2.5.

In a general case of an unknown support of the pair (X,Y), we use our standard procedure of rescaling available observations on the unit square.

Let us check how the E-estimator performs for a sample with MAR responses and also add an explanation why estimation of the conditional density may shed an important light on relationship between two random variables which cannot be gained from the analysis of regression functions.

Figure 4.5 helps us to understand the MAR setting and how the E-estimator performs. We postpone for now explanation of the underlying simulation, and this will allow us to use our imagination and guess about an underlying distribution. The left-top diagram shows a scattergram of a hidden H-sample; for now do not look at the other diagrams. For us, after the analysis of so many scattergrams in the previous sections, this one is not a difficult one. It is clear that the underlying regression is a unimodal and symmetric around 0.5 function which resembles the Normal function. And this is a reasonable answer. Keeping in mind that we have $n = 500$ observations (see the title), it was not a difficult task to visualize a Normal-like regression. At the same time, if you are still confused with the regression and the scattergram, your feeling is correct because the scattergram is not as simple as it looks. We return to this diagram shortly.

Now let us look at the left-bottom diagram. It shows E-estimate of the conditional density $f^{Y|X}(y|x)$ based on the H-sample. Let us explain how to analyze the estimate. Use a vertical slice with a constant $x = x_0$, and then the cut along the shown surface exhibits the estimate $f^{Y|X}(y|x_0)$ as a function in y. The surface indicates two pronounced ridges with a valley between, and this yields a conditional density which, as a function in y, has two pronounced modes.

Now let us return to the scattergram for the H-sample. Look one more time at the scattergram and please pay attention to the following detail. Do you see a pronounced gap between two clusters of circles? Actually, it looks like we have two unimodal regressions shown in the same diagram. This is what the E-estimator sees in the H-sample and shows us via the conditional density. By the way, here we have 500 observations, does it look like that you see so many circles? If the answer is "no," then you are not alone.

To appreciate the conditional density estimate and get a better understanding of the scattergram, let us explain how the H-sample is generated. The response is $Y = u(X) + \sigma\eta + \sigma_N\varepsilon$ where X is the Uniform random variable, η is the Strata random variable, ε is the Normal random variable and these three variables are mutually independent, and $u(x)$ is the Normal function. The variable η creates the special shape of the data (the two ridges). Note that this type of data also may be created by a mixture of two underlying regressions.

After this discussion it becomes clear that for the H-sample at hand its regression function is not a very informative tool because it does not explain the data. The conditional density sheds a better light on the data. The downside is that a larger sample size and paying attention to arguments of the E-estimator is critical here. For instance, note that the argument cTH is increased to 10 with the aim of removing relatively small Fourier coefficients. A rule of thumb, often recommended, is to choose cTH close to $2\ln(n)$. For

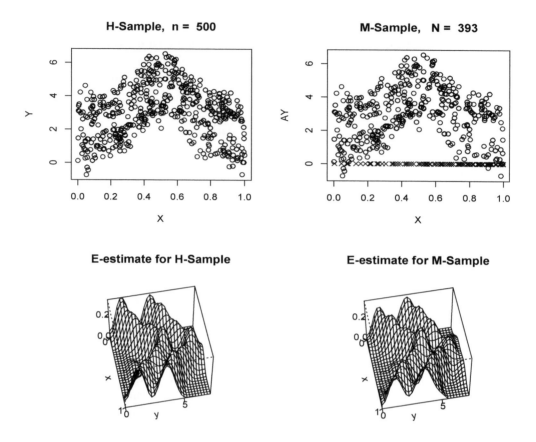

Figure 4.5 *Estimation of conditional density $f^{Y|X}(y|x)$ for H-sample and M-sample with MAR responses. The simulation is explained in the text. In the right-top diagram circles and crosses show complete and incomplete cases, respectively. {The argument corn controls function $u(x)$, cS controls the distribution of η, sigmaN controls σ_N, the string w controls the availability likelihood function while dwL and dwU control its lower and upper bounds.} [n = 500, corn = 2, sigma = 4, sigmaN = 0.5, cS = 4, w = "1-0.4*x", dwL = 0.5 ,dwU = 0.9, c = 1, cJ0 = 3, cJ1 = 0.8, cTH = 10]*

the reference, when n increases from 100 to 500, this argument changes from 9 to 12. Some experience is also needed in choosing other arguments (they are the same as in the previous figures). Recall that figures in the book allow the reader to change their arguments, and then via simulations to learn how to choose better parameters of E-estimators for different statistical models.

The right column of the diagrams in Figure 4.5 shows us an M-sample produced by MAR responses with $P(A = 1|Y = y, X = x) = w(x) = \max(d_{wL}, \min(d_{wU}, 1 - 0.4x))$. Note that only $N = 393$ complete pairs, from $n = 500$ hidden ones, are available (see the title); here we lost a bit more than one-fifth of the responses. Further, note that responses with larger predictors are more likely to be missed, and this creates the biased scattergram. Nonetheless, in the scattergram of the M-sample we still can see the two pronounced ridges and a valley between them. The reader is advised to compare the two scattergrams and analyze the effect of the missing responses. The E-estimate for the M-sample, shown in the right-bottom diagram, is very impressive. The two ridges and the valley between them are clearly exhibited and, despite the heavy missing for larger values of the predictor, the estimate for larger x is as good as the one in the left-bottom diagram for the H-sample.

We may conclude that for the considered problem of MAR responses, the proposed complete-case approach has worked out very nicely for the particular simulation. This may not be the case in another simulation, some adjustments in arguments for other classes of conditional densities may be beneficial, etc. Figure 4.5 allows us to explore all these issues and gain necessary experience in dealing with the complicated problem of estimating a conditional density with MAR responses.

MAR predictors. Let us describe the problem of estimation of a conditional density of response Y given predictor X based on data with missing predictors. There is a hidden sample $(X_1, Y_1, A_1), \ldots, (X_n, Y_n, A_n)$ from the triplet (X, Y, A) of dependent random variables where A is a Bernoulli random variable which controls availability (not missing) of X. If $A_l = 1$ then the corresponding predictor X_l is available and the case (X, Y) is complete, otherwise X_l is missed and the case is incomplete because only the response Y_l is available. Formally, we may say that an available M-sample of size n from the triplet (AX, Y, A) is $(A_1 X_1, Y_1, A_1), \ldots, (A_n X_n, Y_n, A_n)$. The considered missing mechanism is MAR and

$$\mathbb{P}(A = 1 | Y = y, X = x) = \mathbb{P}(A = 1 | Y = y) =: w(y) \geq c_0 > 0. \tag{4.4.8}$$

Two remarks are due about the model. First, while in general X and A are dependent random variables, according to (4.4.8) they are conditionally independent given the response Y. Second, the mechanism of generating the H-sample and M-sample is the same as in Section 4.3 only here the problem is to estimate the conditional density $f^{Y|X}(y|x)$ of the response given the predictor.

We begin our discussion of an appropriate E-estimator of the conditional density for the case of continuous (X, Y) supported on $[0, 1]^2$, and it is also assumed that the design density $f^X(x) \geq c_* > 0$.

The following formula is valid for the joint mixed density of the triplet (AX, Y, A),

$$f^{AX,Y,A}(ax, y, a) = [f^X(x) f^{Y|X}(y|x) w(y)]^a [f^Y(y)(1 - w(y))]^{1-a}, \tag{4.4.9}$$

where $(x, y) \in [0, 1]^2$ and $a \in \{0, 1\}$.

To construct a conditional density E-estimator, we need to understand how to estimate Fourier coefficients

$$\theta_{j_1 j_2} := \int_{[0,1]^2} f^{Y|X}(y|x) \varphi_{j_1 j_2}(x, y) dx dy. \tag{4.4.10}$$

Here $\{\varphi_{j_1 j_2}(x, y)\}$ is the cosine tensor-product basis on $[0, 1]^2$ defined in Section 2.5. If the marginal density $f^X(x)$ and the availability likelihood function $w(y)$ are known, then according to (4.4.9) the following sample mean estimator of Fourier coefficients may be recommended,

$$\bar{\theta}_{j_1 j_2} := n^{-1} \sum_{l=1}^{n} \frac{A_l \varphi_{j_1 j_2}(A_l X_l, Y_l)}{f^X(A_l X_l) w(Y_l)}. \tag{4.4.11}$$

Let us check that the Fourier estimator is unbiased. Using (4.4.9) we can write,

$$\mathbb{E}\{\bar{\theta}_{j_1 j_2}\} = \mathbb{E}\left\{ \frac{A \varphi_{j_1 j_2}(AX, Y)}{f^X(AX) w(Y)} \right\}$$

$$= \int_{[0,1]^2} \frac{f^X(x) f^{Y|X}(y|x) w(y) \varphi_{j_1 j_2}(x, y)}{f^X(x) w(y)} dx dy = \theta_{j_1 j_2}. \tag{4.4.12}$$

This is what was wished to check.

The Fourier estimator (4.4.11) yields the conditional density E-estimator defined in Section 2.5. Further, the coefficient of difficulty is

$$\lim_{n,j_1,j_2\to\infty}[n\mathbb{V}(\bar{\theta}_{j_1j_2})] = \int_{[0,1]^2} \frac{f^{Y|X}(y|x)}{f^X(x)w(y)}dxdy. \qquad (4.4.13)$$

This formula shows how the availability likelihood (the missing mechanism) affects estimation of the conditional density.

In general functions $f^X(x)$ and $w(x)$ are unknown and should be estimated. The availability likelihood $w(y)$ is estimated by the Bernoulli regression E-estimator $\hat{w}(y)$ of Section 2.4 based on n realizations of (Y,A). To estimate the design density $f^X(x)$, we note that its Fourier coefficients are

$$\kappa_j := \int_0^1 f^X(x)\varphi_j(x)dx = \mathbb{E}\{A\varphi_j(AX)/w(Y)\}, \qquad (4.4.14)$$

where the last equality holds due to (4.4.9). This yields the plug-in sample mean estimator of κ_j,

$$\hat{\kappa}_j := n^{-1}\sum_{l=1}^n \frac{A_l\varphi_j(A_lX_l)}{\max(\hat{w}(Y_l), c/\ln(n))}. \qquad (4.4.15)$$

This Fourier estimator yields the density E-estimator $\hat{f}^X(x)$, $x \in [0,1]$ of Section 2.2.

Now we can plug the obtained estimators in (4.4.11) and get the plug-in sample mean estimator of Fourier coefficients of the conditional density,

$$\hat{\theta}_{j_1j_2} := n^{-1}\sum_{l=1}^n \frac{A_l\varphi_{j_1j_2}(A_lX_l,Y_l)}{[\max(\hat{f}^X(A_lX_l), c/\ln(n))][\max(\hat{w}(Y_l), c/\ln(n))]}. \qquad (4.4.16)$$

This Fourier estimator yields the conditional density E-estimator for the considered model of MAR predictors.

If the support of (X,Y) is unknown, then we use our traditional rescaling onto the unit square.

Figure 4.6 helps us to understand the model and how the E-estimator performs. The simulation of an H-sample is the same as in Figure 4.5, and a particular H-sample is shown in the left-top diagram. Further, the right-top diagram shows the conditional density E-estimate based on the H-sample. After our analysis of Figure 4.5, we can plainly realize the two ridges and the valley between them.

The left-middle diagram shows the M-sample generated by MAR predictors with the availability likelihood function shown in the right-middle diagram by triangles. Note that in the left-middle diagram incomplete cases are shown by crosses. The title shows that from n hidden cases only $N = 408$ are complete in the M-sample.

These two diagrams show us the data. Now let us explain how the proposed E-estimator performs. The right-middle diagram shows us by circles the Bernoulli scattergram of n realizations of (Y,A), and then triangles and crosses show the underlying values of $w(Y_l)$ and their E-estimates $\hat{w}(Y_l)$, $l = 1, 2, \ldots, n$. As we see, the E-estimate is almost perfect. The left-bottom diagram shows us by triangles and crosses the design density $f^X(X_l)$ and its E-estimate $\hat{f}^X(X_l)$, respectively. The estimate is perfect. Finally, the right-bottom diagram shows us the proposed E-estimate based on the M-sample. Overall, the E-estimate is on par with (but not as good as) the reference E-estimate based on the H-sample and exhibited in the right-top diagram.

We may conclude that: (i) A data-driven estimation of a conditional density for a model with MAR predictors is a feasible task; (ii) E-estimate requires estimation of several nuisance

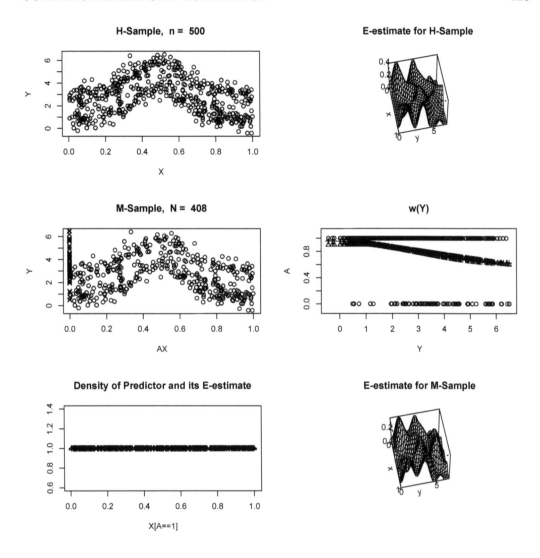

Figure 4.6 *Estimation of conditional density $f^{Y|X}(y|x)$ based on data with MAR predictors. Underlying simulation and the structure of diagrams exhibiting scattergrams and estimates of conditional densities are the same as in Figure 4.5, only here the missing is defined by the response. The left-bottom and right-middle diagrams show by triangles and crosses the underlying functions and their estimates, respectively. {The string w controls $w(y)$ where y is the value of response rescaled onto [0,1].}* [n = 500, corn = 2, sigma = 4, sigmaN = 0.5, cS = 4, w = "1-0.4*y", dwL = 0.2, dwU = 0.9, c = 1, cJ0 = 3, cJ1 = 0.8, cTH = 10]

functions (the design density and the availability likelihood) that may be of interest on their own; (iii) Due to the curse of multidimensionality, a relatively large number N of complete cases is required for a reliable estimation. In its turn, this implies a large sample size n; (iv) Arguments of E-estimator require an adjustment to take into account the bivariate nature of a conditional density.

It is highly recommended to repeat Figures 4.5 and 4.6 with different parameters and gain necessary experience in dealing with the complicated problem of estimation of the conditional density.

4.5 Poisson Regression with MAR Data

So far in this chapter we have considered regression models with continuous responses. These models are very popular in applied statistics. At the same time, there are many applications where responses are discrete random variables. For instance, we considered a Bernoulli regression in Section 2.4 when response takes only two values.

In this section we are considering a so-called Poisson regression when the conditional distribution of the response given the predictor is Poisson. First, let us recall the notion of a Poisson random variable. A random variable Y taking on one of the values 0, 1, 2, ... is said to be a Poisson random variable with parameter $m > 0$ if $\mathbb{P}(Y = k) = e^{-m}m^k/k!$, $k = 0, 1, \ldots$ For the Poisson random variable we have $\mathbb{E}\{Y\} = \mathbb{V}(Y) = m$. The reader may recall from a standard probability class several customary examples of random variables that obey the Poisson probability law: The number of misprints on a page of a book; the number of wrong telephone numbers that are dialed in a day; the number of customers entering a shopping mall on a given day; the number of α-particles discharged in a fixed period of time from some radioactive material; and the number of earthquakes occurring during some fixed time span.

The classical probability considers m as a constant. At the same time, in all the above-mentioned examples the parameter m may depend on another given variable (predictor) X. In this case the function $m(x)$ may be considered as a regression function because

$$m(x) := \mathbb{E}\{Y|X = x\}. \tag{4.5.1}$$

In other words, given $X = x$ the response Y has Poisson distribution with parameter $m(x)$. This definition immediately implies that

$$\mathbb{V}(Y|X = x) = m(x). \tag{4.5.2}$$

Before explaining the case of missing data, let us propose an E-estimator for the Poisson regression based on a hidden H-sample $(X_1, Y_1), \ldots, (X_n, Y_n)$ from the pair (X, Y) where X is the predictor which is a continuous random variable supported on $[0, 1]$ and $f^X(x) \geq c_* > 0$, $x \in [0, 1]$, and given $X = x$ the response Y is a Poisson random variable with the mean $m(x)$. The aim is to estimate the regression function $m(x)$.

Using our methodology of construction of a regression E-estimator, we need to find a (possibly plug-in) sample mean estimator of Fourier coefficients

$$\theta_j := \int_0^1 m(x)\varphi_j(x)dx. \tag{4.5.3}$$

To do this, we are rewriting θ_j as an expectation of observed variables. Using (4.5.1) we can write,

$$\theta_j = \int_0^1 \mathbb{E}\{Y|X = x\}\varphi_j(x)dx = \int_0^1 f^X(x)[\mathbb{E}\{Y|X = x\}\varphi_j(x)/f^X(x)]dx$$

$$= \mathbb{E}\{\mathbb{E}\{[Y\varphi_j(X)/f^X(X)]|X\}\} = \mathbb{E}\left\{\frac{Y\varphi_j(X)}{f^X(X)}\right\}. \tag{4.5.4}$$

If the design density f^X is known, then we immediately get the sample mean estimator of θ_j,

$$\bar{\theta}_j := n^{-1}\sum_{l=1}^{n}\frac{Y_l\varphi_j(X_l)}{f^X(X_l)}. \tag{4.5.5}$$

If the design density f^X is unknown, then it may be estimated by the density E-estimator

$\hat{f}^X$ of Section 2.2 based on observations $X_1, \ldots, X_n$. This yields the plug-in sample mean estimator of Fourier coefficient θ_j,

$$\tilde{\theta}_j := n^{-1} \sum_{l=1}^{n} \frac{Y_l \varphi_j(X_l)}{\max(\hat{f}^X(X_l), c/\ln(n))}. \tag{4.5.6}$$

The Fourier estimator yields the regression E-estimator $\tilde{m}(x)$, $x \in [0,1]$ for the case of a known H-sample.

Now we are considering a setting with missing realizations of a Poisson random variable. The model is as follows. There are n hidden realizations of $(X_1, Y_1, A_1), \ldots, (X_n, Y_n, A_n)$ from the triplet (X, Y, A) where A is a Bernoulli random variable, called the availability, and the availability likelihood is

$$\mathbb{P}(A = 1|X = x, Y = y) = \mathbb{P}(A = 1|X = x) = w(x) \geq c_0 > 0. \tag{4.5.7}$$

Then the observed M-sample is a sample $(X_1, A_1 Y_1, A_1), \ldots, (X_n, A_n Y_n, A_n)$ from the triplet (X, AY, A).

Note that the Poisson Y_l is not available if $A_l = 0$ and we observe a complete case (X_l, Y_l) if $A_l = 1$. Furthermore, (4.5.7) implies that the missing is MAR (missing at random) because the probability of missing Y depends on the value of always observed predictor X.

To propose a sample mean estimator of Fourier coefficients (4.5.3) of the regression function of interest $m(x)$, $x \in [0,1]$, we begin with a formula for the joint density of pair (X, Y) in a complete case of an M-sample. This joint density is the conditional density of the pair given $A = 1$,

$$f^{X,Y|A}(x, y|1) = \frac{f^{X,Y,A}(x, y, 1)}{\mathbb{P}(A = 1)}$$

$$= \frac{f^{X,Y}(x, y)\mathbb{P}(A = 1|X = x, Y = y)}{\mathbb{P}(A = 1)} = f^{X,Y}(x, y) \frac{w(x)}{\mathbb{P}(A = 1)}. \tag{4.5.8}$$

Note that the density is biased with the biasing function being $w(x)$.

Integrating (4.5.8) with respect to y we get a formula for the marginal density of X in a complete case of M-sample,

$$f^{X|A=1}(x|1) = f^X(x) \frac{w(x)}{\mathbb{P}(A = 1)}. \tag{4.5.9}$$

This marginal density is also biased with the same biasing function $w(x)$. However, combining (4.5.8) and (4.5.9) we get the following pivotal result,

$$f^{Y|X,A}(y|x, 1) = \frac{f^{X,Y|A}(x, y|1)}{f^{X|A}(x|1)} = f^{Y|X}(y|x). \tag{4.5.10}$$

We conclude that the conditional density of the response given the predictor in a complete case of M-sample is equal to the underlying conditional density of the predictor given the response in the underlying H-sample.

This conclusion immediately implies that the above-defined regression estimator $\tilde{m}(x)$, developed for an H-sample, also may be used for consistent regression estimation for the considered case of missing data whenever the estimator is based on complete cases in an M-sample.

Figure 4.7 illustrates the setting of Poisson regression and how the E-estimator performs. The left column of diagrams illustrates the case of the Normal being the regression function $m(x) = \mathbb{E}\{Y|X = x\}$. First of all, let us look at the scattergram for the H-sample. Note how special the scattergram for the discrete Poisson response is. This is because Poisson

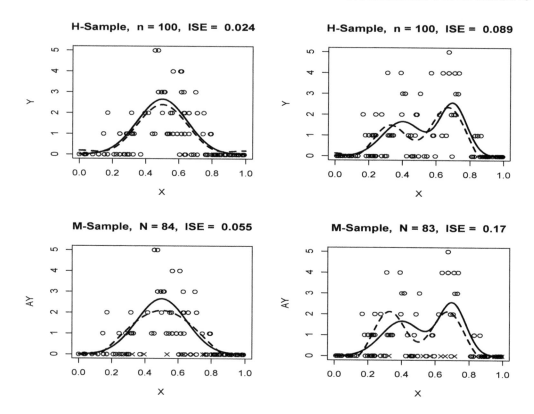

Figure 4.7 *Poisson regression with MAR responses. Two simulations with different regression functions are shown in the two columns of diagrams. Scattergrams are shown by circles, while for M-samples incomplete cases are shown by crosses. Underlying regressions and their E-estimates are shown by solid and dashed lines, respectively. Titles show the sample size n, the number $N = \sum_{l=1}^{n} A_l$ of complete cases, and integrated squared errors (ISE). Because it is known that estimated regression functions are nonnegative (Poisson random variables are nonnegative), a projection on nonnegative functions is used. {The arguments are: set.c controls two underlying regression functions for the left and right columns; n controls the sample size; desden controls the shape of the design density $f^X(x)$ which is then truncated from below by the value dden and rescaled to get a bona fide density; the availability likelihood function w(x) is chosen by the string w and then truncated from below by dwL and from above by dwU.} [n = 100, set.c = c(2,3), dden = 0.2, desden = "0.7+0.4*x", w = "1-0.4*x", dwL = 0.5, dwU=0.9, c = 1, cJ0 = 3, cJ1 = 0.8, cTH = 4]*

random variable takes on only nonnegative integer values. The E-estimate is relatively good here. It correctly shows location of the mode and the symmetric shape of the regression. Now, please look at the circles one more time and answer the following question. Should the estimate be shifted to the right for better fitting the data? The answer is likely "yes," but this is the illusion created by the increasing design density. Indeed, please pay attention that there are more observations in the right half of the diagram than in the left one. The left-bottom diagram shows the M-sample and the corresponding E-estimate based only on circles (complete cases). Note that the number of complete cases is $N = 84$. The estimate is definitely worse, and the reader is advised to look at the 16 missed responses and explain why they so dramatically affected the estimate.

The right column of the diagrams shows two similar diagrams for the case of the underlying regression being the Bimodal. Here again it is of interest to compare the two scattergrams and understand why the E-estimate for the M-sample is worse. The obvious possible reason is that the sample size decreased from 100 to 83, but this alone cannot cause almost doubled ISE. The reason is in missing several "strategic" responses, and the reader is asked to find them.

It is recommended to repeat Figure 4.7 with different arguments to get better understanding of the Poisson regression with MAR responses.

4.6   Estimation of the Scale Function with MAR Responses

In Section 3.6 we considered the model of a heteroscedastic regression with observations of the pair (X, Y) of continuous random variables such that

$$Y = m(X) + \sigma(X)\varepsilon, \ \ m(X) = \mathbb{E}\{Y|X\}, \ \ \mathbb{V}(\varepsilon) = 1. \tag{4.6.1}$$

Here ε is a zero mean and unit variance random variable (regression error) which is independent of the predictor X. The predictor X is supported on $[0, 1]$ and $f^X(x) \geq c_* > 0$. The nonnegative function $\sigma(x)$ is called the scale (spread or volatility) function. The problem of Section 3.6 was to estimate the scale function $\sigma(x)$ based on a sample from (X, Y). Using our terminology for missing data, we can say that in Section 3.6 estimation for the case of H-sample was considered.

Let us describe an M-sample with MAR responses. We observe an M-sample $(X_1, A_1Y_1, A_1), \ldots, (X_n, A_nY_n, A_n)$ of size n from the triplet (X, AY, A). Here A is a Bernoulli random variable which is the indicator that the response is available (not missed), and the availability likelihood is

$$\mathbb{P}(A = 1|X, Y) = \mathbb{P}(A = 1|X) =: w(X) \geq c_0 > 0. \tag{4.6.2}$$

To propose an E-estimator of the scale function, we begin with several probability formulas. The joint mixed density of the triplet (X, AY, A) is

$$f^{X,AY,A}(x, ay, a) = [w(x)f^X(x)f^{Y|X}(y|x)]^a [(1 - w(x))f^X(x)]^{1-a}, \tag{4.6.3}$$

where $x \in [0, 1]$, $y \in (-\infty, \infty)$ and $a \in \{0, 1\}$.

The marginal density of the predictor in a complete case is

$$f^{X|A}(x|1) = \frac{f^{X,A}(x, 1)}{\mathbb{P}(A = 1)} = \frac{f^X(x)\mathbb{P}(A = 1|X = x)}{\mathbb{P}(A = 1)} = \frac{f^X(x)w(x)}{\mathbb{P}(A = 1)}, \tag{4.6.4}$$

where

$$\mathbb{P}(A = 1) = \int_0^1 f^{X,A}(x, 1)dx = \int_0^1 f^X(x)w(x)dx = \mathbb{E}\{w(X)\}. \tag{4.6.5}$$

Now we are ready to propose an E-estimator of the scale function. In Section 4.2 the regression E-estimator $\hat{m}(x)$, based on complete cases, was proposed. Using this estimator, we calculate residuals $A_lY_l - \hat{m}(X_l)$ for complete cases $A_l = 1$. Further, following the solution proposed in Section 3.6, we introduce Fourier coefficients of the squared scale function $\sigma^2(x)$, $x \in [0, 1]$,

$$\theta_j := \int_0^1 \sigma^2(x)\varphi_j(x)dx. \tag{4.6.6}$$

Suppose that nuisance functions $f^{X|A}(x|1)$, $w(x)$ and $m(x)$ are known. Then we can propose the following sample-mean estimator of θ_j,

$$\bar{\theta}_j := n^{-1}\sum_{l=1}^{n} \frac{A_l(A_lY_l - m(X_l))^2\varphi_j(X_l)}{f^X(X_l)w(X_l)}. \tag{4.6.7}$$

Let us show that (4.6.7) is unbiased estimator of θ_j. Using (4.6.3), the rule of calculation of the expectation via conditional expectation, and the assumed conditional independence of A and Y given X, we write,

$$\mathbb{E}\{\bar{\theta}_j\} = \mathbb{E}\left\{\frac{A(AY - m(X))^2\varphi_j(X)}{f^X(X)w(X)}\right\}$$

$$= E\left\{\frac{E\{A(Y - m(X))^2|X\}\varphi_j(X)}{f^X(X)w(X)}\right\} = E\left\{\frac{\mathbb{E}\{A|X\}\mathbb{E}\{(Y - m(X))^2|X\}\varphi_j(X)}{f^X(X)w(X)}\right\}$$

$$= E\left\{\frac{w(X)\sigma^2(X)\varphi_j(X)}{f^X(X)w(X)}\right\} = \int_0^1 \frac{f^X(x)\sigma^2(x)\varphi_j(x)}{f^X(x)}dx = \theta_j. \qquad (4.6.8)$$

This is what was wished to show.

Next, assume that $f^{X|A}(x|1)$ is given, then using (4.6.4) we can rewrite (4.6.7) as

$$\bar{\theta}_j = [n\mathbb{P}(A = 1)]^{-1} \sum_{l=1}^n \frac{A_l(A_lY_l - m(X_l))^2\varphi_j(X_l)}{f^{X|A}(X_l|1)}. \qquad (4.6.9)$$

This is the pivotal result that will allow us to propose an estimator based exclusively on complete cases in an M-sample. Indeed, we know from Section 4.2 that the regression function $m(x)$ and the density $f^{X|A}(x|1)$ may be estimated by corresponding E-estimators based on complete cases. Further, the product $n\mathbb{P}(A = 1)$ can be estimated by the sample mean estimator $N := \sum_{l=1}^n A_l$, which is the number of complete cases, and recall our discussion in Section 4.1 of the remedy for the event $N = 0$. Then we plug these estimators in (4.6.9) and get the plug-in sample-mean Fourier estimator based on the subsample of complete cases,

$$\hat{\theta}_j = N^{-1} \sum_{\{l:\, A_l=1,\, l=1,\dots,n\}} \frac{A_l(A_lY_l - \hat{m}(X_l))^2\varphi_j(X_l)}{\max(\hat{f}^{X|A}(X_l|1), c/\ln(n))}. \qquad (4.6.10)$$

The Fourier estimator yields the E-estimator of the squared scale, and then taking the square root gives us the wished scale E-estimator $\hat{\sigma}(x)$.

Figure 4.8 explains the setting, sheds light on performance of the complete-case scale E-estimator, and its caption explains the diagrams. We begin with the top diagram. For now, ignore titles and curves, and pay attention to the observations. Circles show $N = 77$ complete cases and crosses show $n - N = 23$ incomplete cases. Probably the first impression from the scattergram that there is a large volatility in the middle of the support of X and furthermore many missed responses have predictors from this set. The large volatility makes the problem of visualization of an underlying regression function complicated. Now let us check how the E-estimator solved the problem. The diagram shows the underlying regression and scale functions by the dotted and solid lines, respectively. The regression E-estimate is shown by the dot-dashed line. The estimate is significantly lower in the middle of the support (and this could be predicted from the data) but it does exhibit the unimodal and symmetric character of the Normal. The scale estimate (the dashed line), despite the poor regression estimate, is surprisingly fair estimate with respect to its regression counterpart.

The bottom diagram exhibits a similar simulation only here the more complicated Bimodal defines the scale function. Here we have a large volatility within the interval $[0.6, 0.8]$. The regression E-estimate is good, and the scale estimate is also relatively good apart of the left tail which reflects the observed volatility of responses with smallest predictors.

It is recommended to repeat Figure 4.8 with different parameters and learn more about this interesting and practically important problem. Overall, repeated simulations conducted by Figure 4.8 indicate that the proposed scale E-estimator for regression with MAR responses is feasible. The asymptotic theory also supports the proposed complete-case approach.

M-Sample for Normal Scale, N = 77 , n = 100

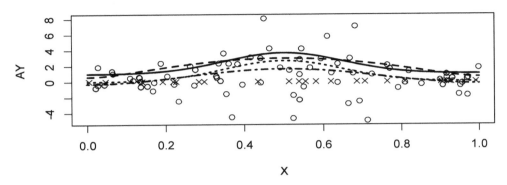

M-Sample for Bimodal Scale, N = 83 , n = 100

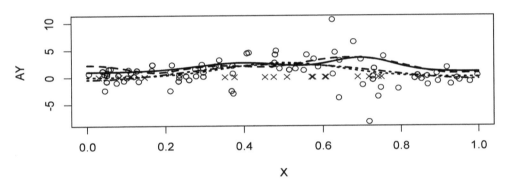

Figure 4.8 *Estimation of the scale function for regression with MAR responses. Each diagram exhibits an M-sample generated by the Uniform design density, the Normal regression function, and the scale function equal to 1 plus a corner function, indicated in the title, and then the total is multiplied by parameter σ. In a scattergram, incomplete cases are shown by crosses and complete cases by circles, respectively. Each diagram also shows the underlying regression function (the dotted line), the regression E-estimate (the dot-dashed line), the underlying scale function (the solid line), and the scale E-estimate (the dashed line). {Underlying regression functions are controlled by the argument set.corn, scale functions are controlled by arguments set.scalefun and sigma, availability likelihood is $\max(\min(w(x), dwU), dwL)$.} [n = 100, set.corn = c(2,2), sigma = 1, set.scalefun = c(2,3), w = "1-0.4*x", dwL = 0.5, dwU = 0.9, cJ0 = 4, cJ1 = .5, cTH = 4]*

4.7 Bivariate Regression with MAR Responses

We begin with a classical bivariate regression model which, using our terminology for missing data, produces an H-sample. Consider the bivariate regression

$$Y = m(X_1, X_2) + \sigma(X_1, X_2)\,\varepsilon. \tag{4.7.1}$$

Here Y is the response, $\sigma(x_1, x_2)$ is the bivariate scale function, and ε is a zero mean and unit variance regression error that is independent of (X_1, X_2). The pair (X_1, X_2) is the vector-predictor (the two random variables may be dependent, and in what follows we may refer to them as *covariates*) with the design density $f^{X_1 X_2}(x_1, x_2)$ supported on the unit square $[0,1]^2$ and $f^{X_1 X_2}(x_1, x_2) \geq c_* > 0$, $(x_1, x_2) \in [0,1]^2$.

Then a classical regression problem is to estimate the bivariate regression function (surface) $m(x_1, x_2)$ based on a sample $(Y_l, X_{1l}, X_{2l}), \ldots, (Y_n, X_{1n}, X_{2n})$ from the triplet (Y, X_1, X_2).

Using the approach of Section 2.5 we need to propose a sample mean estimator of Fourier coefficients

$$\theta_{j_1 j_2} := \int_{[0,1]^2} m(x_1, x_2)\varphi_{j_1 j_2}(x_1, x_2)dx_1\, dx_2, \qquad (4.7.2)$$

where $\{\varphi_{j_1 j_2}(x_1, x_2)\}$ is the cosine tensor-product basis on $[0, 1]^2$.

For the model (4.7.1) we can rewrite (4.7.2) as

$$\theta_{j_1 j_2} = \mathbb{E}\left\{ \frac{Y\varphi_{j_1 j_2}(X_1, X_2)}{f^{X_1 X_2}(X_1, X_2)} \right\}. \qquad (4.7.3)$$

If the design density $f^{X_1, X_2}(x_1, x_2)$ is known, then the sample mean estimator of $\theta_{j_1 j_2}$ is

$$\breve{\theta}_{j_1 j_2} := n^{-1} \sum_{l=1}^{n} \frac{Y_l \varphi_{j_1 j_2}(X_{1l}, X_{2l})}{f^{X_1 X_2}(X_{1l}, X_{2l})}. \qquad (4.7.4)$$

If the design density is unknown, then its E-estimator $\hat{f}^{X_1 X_2}(x_1, x_2)$, defined in Section 2.5, may be plugged in (4.7.4). This Fourier estimator yields the bivariate regression E-estimator for the case of H-sample. Further, the case of k covariates is considered absolutely similarly only a corresponding tensor-product basis should be used in place of the bivariate basis.

Now we are in a position to describe a sampling with MAR responses. We observe an M-sample $(X_{11}, X_{21}, A_1 Y_1, A_1), \ldots, (X_{1n}, X_{2n}, A_n Y_n, A_n)$ of size n from the quartet (X_1, X_2, AY, A). The triplet (X_1, X_2, Y) is the same as in the above-described regression (4.7.1), A is a Bernoulli random variable which is the indicator that the response is available (not missed), and the availability likelihood is

$$\mathbb{P}(A = 1 | X_1 = x_1, X_2 = x_2, Y = y)$$

$$= \mathbb{P}(A = 1 | X_1 = x_1, X_2 = x_2) =: w(x_1, x_2) \geq c_0 > 0. \qquad (4.7.5)$$

The problem is to estimate the regression function $m(x_1, x_2)$.

To propose a regression E-estimator, we begin with several probability formulas that will help us in developing the estimator. The joint mixed density of the quartet (X_1, X_2, AY, A) is

$$f^{X_1, X_2, AY, A}(x_1, x_2, ay, a)$$

$$= [w(x_1, x_2)f^{X_1 X_2}(x_1, x_2)f^{Y|X_1, X_2}(y|x_1, x_2)]^a [(1 - w(x_1, x_2))f^{X_1, X_2}(x_1, x_2)]^{1-a}. \qquad (4.7.6)$$

Further, the marginal joint density of the vector-predictor in a complete case is

$$f^{X_1, X_2 | A}(x_1, x_2 | 1) = \frac{f^{X_1, X_2, A}(x_1, x_2, 1)}{\mathbb{P}(A = 1)}$$

$$= \frac{f^{X_1, X_2}(x_1, x_2)\mathbb{P}(A = 1 | X_1 = x_1, X_2 = x_2)}{\mathbb{P}(A = 1)} = \frac{f^{X_1, X_2}(x_1, x_2)w(x_1, x_2)}{\mathbb{P}(A = 1)}, \qquad (4.7.7)$$

where

$$\mathbb{P}(A = 1) = \int_{[0,1]^2} f^{X_1, X_2, A}(x_1, x_2, 1)dx_1 dx_2$$

$$= \int_{[0,1]^2} f^{X_1, X_2}(x_1, x_2)w(x_1, x_2)dx_1 dx_2 = \mathbb{E}\{w(X_1, X_2)\}. \qquad (4.7.8)$$

Now we are ready to propose an E-estimator of the bivariate regression function $m(x_1, x_2)$. Following the E-estimation methodology, we need to estimate Fourier coefficients (4.7.2) by a sample mean estimator. To propose a sample mean Fourier estimator, we first need to write down Fourier coefficients as an expectation. Let us show that

$$\theta_{j_1 j_2} = \mathbb{E}\Big\{ \frac{AY\varphi_{j_1 j_2}(X_1, X_2)}{w(X_1, X_2) f^{X_1 X_2}(X_1, X_2)} \Big\}. \tag{4.7.9}$$

Indeed, using (4.7.7) we can write,

$$\mathbb{E}\Big\{ \mathbb{E}\Big\{ \frac{AY\varphi_{j_1 j_2}(X_1, X_2)}{w(X_1, X_2) f^{X_1 X_2}(X_1, X_2)} | X_1, X_2 \Big\} \Big\} = \int_{[0,1]^2} m(x_1, x_2)\varphi_{j_1 j_2}(x_1 x_2). \tag{4.7.10}$$

This and (4.7.2) prove (4.7.9).

Given that the design density and the availability likelihood are known, (4.7.9) yields the sample mean estimator

$$\bar{\theta}_{j_1 j_2} := n^{-1} \sum_{l=1}^{n} \frac{A_l Y_l \varphi_{j_1 j_2}(X_{1l}, X_{2l})}{w(X_{1l}, X_{2l}) f^{X_1 X_2}(X_{1l}, X_{2l})}. \tag{4.7.11}$$

Now consider the case of unknown design density and availability likelihood. Using (4.7.7) we note that the denominator in (4.7.11) is

$$w(X_{1l}, X_{2l}) f^{X_1 X_2}(X_{1l}, X_{2l}) = \mathbb{P}(A = 1) f^{X_1 X_2 | A}(X_{1l}, X_{2l} | 1). \tag{4.7.12}$$

Set $N := \sum_{l=1}^{n} A_l$ and note that this is the total number of complete cases in the M-sample. Recall our discussion in Section 4.2 about remedy for the event $N = 0$, and also note that we must restrict our attention to samples with $N > k$ and k being a relatively large integer for a feasible estimation of a bivariate regression function. The sample mean estimator of $\mathbb{P}(A = 1)$ is N/n and this, together with (4.7.11) and (4.7.12), yields the following plug-in sample mean estimator of $\theta_{j_1 j_2}$,

$$\hat{\theta}_{j_1 j_2} := N^{-1} \sum_{l=1}^{n} \frac{A_l Y_l \varphi_{j_1 j_2}(X_{1l}, X_{2l})}{\max(\hat{f}^{X_1 X_2 | A}(X_{1l}, X_{2l} | 1), c/\ln(n))}. \tag{4.7.13}$$

Here $\hat{f}^{X_1 X_2 | A}(X_{1l}, X_{2l} | 1)$ is the density E-estimator of Section 2.5 based on complete cases in the M-sample.

Now, if we look at the estimator (4.7.13) one more time, then it is plain to realize that it is based solely on complete cases, and even the sample size n does not need to be known if we replace $c/\ln(n)$ by $c/\ln(N)$.

The Fourier estimator (4.7.13) yields the regression E-estimator $\hat{m}(x_1, x_2)$ for the case of MAR responses.

Let us check how the proposed complete-case regression E-estimator performs for a particular simulation.

The top diagram in Figure 4.9 exhibits both the H-sample and M-sample. Dots and stars show 100 realizations of the hidden triplet (X_1, X_2, Y) with predictors shown as $X1$ and $X2$. The sample size is small and this allows us to visualize all predictors and responses. Due to the missing mechanism, some responses in the H-sample are missed, and those observations are shown by stars. In other words, for realizations shown by stars only their $(X1, X2)$ coordinates are available in the M-sample. The title indicates that the M-sample has $N = 85$ complete cases and 15 cases are incomplete. Now let us look more closely at the scattergram. First, here we have $n = 100$ observations, but do you have a feeling that there are 100 vertical lines? It is difficult to believe that we see so many vertical lines. Also, there

Scattergram, n = 100 , N = 85

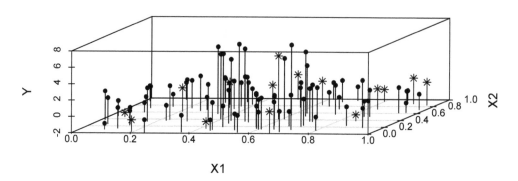

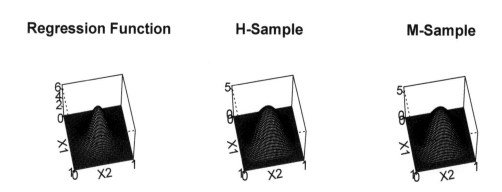

Figure 4.9 *Bivariate regression with MAR responses. The top diagram is a 3-dimensional scatter-plot that shows us H- and M-samples. The dots show complete cases and the stars show hidden cases in which the responses are missed. The underlying bivariate regression function* $m(X1, X2)$ *and its E-estimates based on H-sample and M-sample are exhibited in the second row of diagrams. The H-sample is generated as* $Y = m(X1, X2) + \sigma\varepsilon$ *where* ε *is standard normal,* $X1$ *and* $X2$ *are the Uniform. {Underlying regression function* $m(x1, x2)$ *is the product of the two corner functions whose choice is controlled by the argument set.corn, the availability likelihood is* $\max(\min(w(x1, x2), dwU), dwL)$ *and* $w(x1, x2)$ *is defined by the string w.}* [n = 100, set.corn = c(2,2), sigma = 1, w = "1-0.6*x1*x2", dwL = 0.5, dwU = 0.9, cJ0 = 4, cJ1 = 0.5, cTH = 4]

are relatively large empty spaces on the square with no observations at all, and therefore no information about an underlying regression surface for those spots is available. Further, for instance in the subsquare $[.8, 1] \times [0, 0.2]$ there is just one predictor (look at the vertical line with coordinates around $(.95, 0.2)$). This is what makes the bivariate regression problem so complicated. Note that the scattergram explains complications of a multivariate setting better than words and theorems.

The underlying bivariate regression function is shown in the left-bottom diagram, it is the Normal by the Normal. Its E-estimates, based on the H-sample and the M-sample, are shown in the middle-bottom and right-bottom diagrams. We may notice that the estimates are wider than the underlying regression surface, but overall they do show the symmetric and bell-type shape of the underlying bivariate regression function.

It takes time and practice to get used to the multivariate regression problem, and Figure 4.9 is a good tool to get this experience. The advice is not to use large sample sizes because then it will be difficult to analyze 3-dimensional scattergrams.

4.8   Additive Regression with MAR Responses

The classical linear regression model for the case of a d-dimensional predictor (covariate) assumes that the regression function is $m_L(x_1, \ldots, x_d) := \beta_0 + \sum_{k=1}^d \beta_k x_k$. This regression function is both linear and additive in covariates $x_1, \ldots, x_d$. If we drop the assumption on the linearity and preserve the additivity, then we get an *additive nonparametric model*,

$$Y := \beta + \sum_{k=1}^d m_k(X_k) + \sigma\varepsilon. \tag{4.8.1}$$

Here Y is a response, $X_1, \ldots, X_d$ are possibly dependent covariates with a joint d-variate design density $f^{X_1, \ldots, X_d}(x_1, \ldots, x_d)$, $m_k(x)$, $k = 1, \ldots, d$ are unknown univariate functions, ε is a random variable that is independent of the vector-predictor and has zero mean and unit variance, and σ is the scale.

The classical problem of additive regression is to estimate additive univariate components $m_k(x_k)$ based on the sample $\{(Y_l, X_{1l}, \ldots, X_{dl}), \ l = 1, 2, \ldots, n\}$. Note that we may refer to that sample as the H-sample.

Let us explain how to construct E-estimator for the case of a known H-sample. To simplify notation, set $\mathbf{Z} := (X_1, \ldots, X_d)$. Suppose that the design density $f^{\mathbf{Z}}(\mathbf{z})$ is known, supported on the d-dimensional unit cube $[0, 1]^d$, and $f^{\mathbf{Z}}(\mathbf{z}) \geq c_* > 0$, $\mathbf{z} \in [0, 1]^d$. Also, to make the additive functions $m_k(x)$ unique, it is assumed that

$$\int_0^1 m_k(x)dx = 0, \quad k = 1, 2, \ldots, d. \tag{4.8.2}$$

To suggest a regression E-estimator for a kth additive component $m_k(x_k)$, we need to suggest a sample mean estimator of its Fourier coefficients

$$\theta_{kj} := \int_0^1 m_k(x)\varphi_j(x)dx, \quad j = 1, 2, \ldots \tag{4.8.3}$$

Due to (4.8.2) we have $\theta_{k0} = 0$, and this explains why we need to estimate only Fourier coefficients with $j \geq 1$.

As usual, to propose a sample mean estimator for θ_{jk} we need to write down the Fourier coefficient as the expectation of a function of observed random variables. Let us look at the following expectation,

$$E\{Y\varphi_j(X_k)/f^{\mathbf{Z}}(\mathbf{Z})\} = E\Big\{ [\beta + \sum_{k=1}^d m_k(X_k) + \sigma\varepsilon]\varphi_j(X_k)/f^{\mathbf{Z}}(\mathbf{Z}) \Big\}$$

$$= \int_{[0,1]^d} [\beta + \sum_{k=1}^d m_k(x_k)]\varphi_j(x_k)dx_1 \cdots dx_d. \tag{4.8.4}$$

Recall that $\{\varphi_j(x)\}$ is the basis on $[0,1]$, and hence for any $j \geq 1$ and $s \neq k$ we have

$$\int_0^1 m_s(x_s)\varphi_j(x_k)dx_k = m_s(x_s)\int_0^1 \varphi_j(x_k)dx_k = 0. \qquad (4.8.5)$$

This is the place where the assumed $j \geq 1$ is critical because $\int_0^1 \varphi_0(x)dx = 1$.

Using (4.8.5) we continue (4.8.4),

$$E\{Y\varphi_j(X_k)/f^{\mathbf{Z}}(\mathbf{Z})\} = \int_0^1 m_k(x_k)\varphi_j(x_k)dx_k = \theta_{kj}. \qquad (4.8.6)$$

We obtained the desired expression for θ_{kj}, and it immediately implies the sample mean estimator

$$\bar{\theta}_{kj} = n^{-1}\sum_{l=1}^n \frac{Y_l\varphi_j(X_{kl})}{f^{\mathbf{Z}}(\mathbf{Z}_l)}. \qquad (4.8.7)$$

Now we need to explain how to estimate the constant β in (4.8.1). Again we are thinking about a sample mean estimator. Write,

$$\beta = \int_{[0,1]^d}[\beta + \sum_{k=1}^n m_k(x_k)]dx_1\cdots dx_d = E\{Y/f^{\mathbf{Z}}(\mathbf{Z})\}. \qquad (4.8.8)$$

This implies a sample mean estimator

$$\bar{\beta} := n^{-1}\sum_{l=1}^n \frac{Y_l}{f^{\mathbf{Z}}(\mathbf{Z}_l)}. \qquad (4.8.9)$$

If the design density is unknown, then its estimator is plugged in.

The proposed estimators of Fourier coefficients yield the corresponding univariate regression E-estimators for H-sample.

Let us stress that the advantage of additive regression with respect to multivariate regression, discussed in the previous Section 4.7, is in the faster rate of the MISE convergence because only univariate functions are estimated. On the other hand, it is a serious statistical problem to verify that an underlying regression is additive.

Now we are in a position to describe a sampling with MAR responses. We observe an M-sample $(\mathbf{Z}_1, A_1Y_1, A_1), \ldots, (\mathbf{Z}_n, A_nY_n, A_n)$ of size n from $(\mathbf{Z}, AY, A)$. Here A is a Bernoulli random variable which is the indicator that the response is available (not missed), and the pair $(\mathbf{Z}, Y)$ is described by (4.8.1). It is assumed that the availability likelihood is

$$\mathbb{P}(A = 1|\mathbf{Z}, Y) = \mathbb{P}(A = 1|\mathbf{Z}) =: w(\mathbf{Z}) \geq c_0 > 0. \qquad (4.8.10)$$

The problem is to estimate components β and $m_k(x_k)$, $k = 1, 2, \ldots, d$ of the additive regression (4.8.1).

To solve the estimation problem, we begin with the joint mixed density of $(\mathbf{Z}, AY, A)$. Write,

$$f^{\mathbf{Z},AY,A}(\mathbf{z}, ay, a) = [w(\mathbf{z})f^{\mathbf{Z}}(\mathbf{z})f^{Y|\mathbf{Z}}(y|\mathbf{z})]^a[(1-w(\mathbf{z}))f^{\mathbf{Z}}(\mathbf{z})]^{1-a}. \qquad (4.8.11)$$

The marginal density of covariates in a complete case is

$$f^{\mathbf{Z}|A}(\mathbf{z}|1) = \frac{f^{\mathbf{Z},A}(\mathbf{z},1)}{\mathbb{P}(A=1)} = \frac{f^{\mathbf{Z}}(\mathbf{z})\mathbb{P}(A=1|\mathbf{Z}=\mathbf{z})}{\mathbb{P}(A=1)} = \frac{f^{\mathbf{Z}}(\mathbf{z})w(\mathbf{z})}{\mathbb{P}(A=1)}, \qquad (4.8.12)$$

where

$$\mathbb{P}(A=1) = \int_{[0,1]^d} f^{\mathbf{Z}}(\mathbf{z})w(\mathbf{z})d\mathbf{z} = \mathbb{E}\{w(\mathbf{Z})\}. \qquad (4.8.13)$$

These formulas allow us to suggest a sample mean estimator for Fourier coefficients (4.8.3) of additive components $m_k(x_k)$. Write for $k = 1, \ldots, d$ and $j \geq 1$,

$$\theta_{kj} = \mathbb{E}\left\{\frac{AY\varphi_j(\mathbf{Z})}{w(\mathbf{Z})f^{\mathbf{Z}}(\mathbf{Z})}\right\}. \tag{4.8.14}$$

This yields the sample mean Fourier estimator

$$\bar{\theta}_{kj} := n^{-1}\sum_{l=1}^{n}\frac{A_l Y_l \varphi_j(\mathbf{Z}_l)}{w(\mathbf{Z}_l)f^{\mathbf{Z}}(\mathbf{Z}_l)}. \tag{4.8.15}$$

If the nuisance functions are unknown, then using the density E-estimator $\hat{f}^{\mathbf{Z}|A}(\mathbf{z}|1)$, (4.8.12) and the sample mean estimator $n^{-1}\sum_{l=1}^{n}A_l$ for $\mathbb{P}(A = 1)$, we get the plug-in sample mean estimator

$$\hat{\theta}_{kj} := N^{-1}\sum_{l=1}^{n}\frac{A_l Y_l \varphi_j(\mathbf{Z}_l)}{\hat{f}^{\mathbf{Z}|A}(\mathbf{Z}_l|1)}. \tag{4.8.16}$$

Here $N := \sum_{l=1}^{n}A_l$ is the number of complete cases in the M-sample, and recall our discussion in Section 4.1 about remedy for the case $N = 0$.

Absolutely similarly we estimate parameter β,

$$\hat{\beta} := N^{-1}\sum_{l=1}^{n}\frac{A_l Y_l}{\hat{f}^{\mathbf{Z}|A}(\mathbf{Z}_l|1)}. \tag{4.8.17}$$

The Fourier estimators yield E-estimators of the additive components in the regression. Note that the proposed regression E-estimator is based on the complete-case approach.

Figure 4.10 sheds light on the setting and shows how the additive regression E-estimator recovers additive components. Here $d = 2$ so we can visualize data via the 3-dimensional scattergram. The main title shows that the sample size is $n = 100$ (a larger size would overcrowd the scattergram), the number of complete cases is $N = \sum_{l=1}^{n}A_l = 76$, and the underlying $\beta = 1$. The interesting feature of the design is the dependence between X_1 and X_2 explained in the caption, and note that in the diagrams we use $X1$ and $X2$ in place of X_1 and X_2, respectively. The underlying additive components are the Normal and the Strata (note that 1 is always subtracted from these functions to satisfy (4.8.2)).

Let us look at the 3-dimensional scattergram. First, can you realize the underlying additive components from this scattergram? It is apparently not a simple question even if you know the components. Second, can you see that the predictors are dependent? To do this, we need to compare the distributions of X_2 for cases $X_1 < 0.5$ and cases $X_1 \geq 0.5$. In the first case the distribution of X_2 is the Uniform, while in the second it is the Normal. This is not a simple task for just $n = 100$ observations, but overall we can see that there are less realizations of X_2 near boundaries for the second case than for the first one.

Now let us look at how the proposed E-estimator performs. The estimate of β for the H-sample (denoted as H-beta.est) is shown in the subtitle of the left-bottom diagram and it is 0.99. This is an excellent outcome because the underlying $\beta = 1$ (see the main title). For the M-sample the estimate is 0.91 and it is, of course, worse but not bad for the sample of size 76 and dependent covariates. Estimates of the first component are not perfect and worse than we could get for the same sample sizes in univariate regressions. On the other hand, they correctly describe the bell-shaped and symmetric around 0.5 component. Further, can you visualize this component in the scattergram? The answer is probably "no." The second component is estimated relatively well and we do see two strata.

It is highly recommended to repeat Figure 4.10 with different parameters and learn to read scattergrams created by additive regressions. It is also a good exercise to compare scattergrams produced by Figures 4.9 and 4.10.

Scattergram, n = 100 , N = 76 , beta = 1

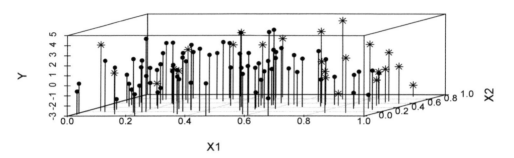

First Component   **Second Component**

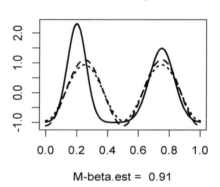

H-beta.est = 0.99   M-beta.est = 0.91

Figure 4.10 *Additive regression with MAR responses and dependent covariates. The top diagram is a 3-dimensional scattergram which, as explained in the caption of Figure 4.9, allows us to visualize the H- and M-samples. The underlying additive regression function is $m(X1, X2) = \beta + m_1(X1) + m_2(X2)$ and $Y = m(X1, X2) + \sigma\varepsilon$ where ε is standard normal. Function $m_1(x)$ is the Normal minus 1, and function $m_2(x)$ is the Strata minus 1. The marginal density of X1 is the Uniform while the conditional density of X2 is the mixture $f^{X2|X1}(x_2|x_1) = I(x_1 < t)f^U(x_2) + I(x_1 \geq t)f^N(x_2)$ where $t \in [0, 1]$, U is the Uniform and N is the Normal. In the bottom diagrams, the solid, dashed and dotted lines show the underlying component and its E-estimates based on H-sample and M-sample, respectively, and the corresponding estimates of β are shown in the subtitles as H-beta.est and M-beta.est. {Two additive components of the regression function are controlled by the argument set.corn, parameter β by argument beta, argument t controls the conditional density of X2 given X1, all other arguments are the same as in Figure 4.9.} [n = 100, set.corn = c(2,4), t = 0.5, sigma = 1, w = "1-0.6*x1*x2", dwL = 0.5, dwU = 0.9, cJ0 = 4, cJ1 = 0.5, cTH = 4]*

We may conclude that if an underlying regression model is additive, then its components can be fairly well estimated even for the case of relatively small sample sizes. Also, even if an underlying regression function is not additive, the additive model may shed light on the data at hand. This explains why additive regression is often used by statisticians as a first look at data.

4.9 Exercises

4.1.1 Explain heuristically (using words) the statistical model of a MCAR sample.

4.1.2 Describe the probability model of a MCAR sample via a hidden sample (H-sample) and a corresponding sample with missing observations (M-sample).

4.1.3 Prove formula (4.1.1) and then comment on the underlying sampling procedure.

4.1.4 Suppose that the joint mixed density of pair (AX, A) is defined by (4.1.1) and the parameter w depends on the value of x. Is this still the case of MCAR?

4.1.5* Consider the random variable $N := \sum_{l=1}^{n} A_l$ and describe its statistical properties. Develop inequalities that evaluate the probability of event $N \leq \gamma n$ for $\gamma \in [0, 1]$.

4.1.6 Suppose that your colleague sent you R-data and you discovered that it contains both numbers and logical flags NA. What does the NA stand for?

4.1.7 Explain the meaning of the probability $\mathbb{P}(AX \leq x | A = 1)$. Then prove that this conditional probability is equal to $\mathbb{P}(X \leq x)$.

4.1.8 Prove that estimator $\check{\theta}_j$, defined in (4.1.6), is unbiased estimator of Fourier coefficient θ_j of an underlying density of interest $f^X(x)$.

4.1.9 Is it correct to say that the estimator (4.1.8) is based solely on complete cases? Explain your answer.

4.1.10 Do we need to know the sample size n to calculate the estimate (4.1.8)?

4.1.11 Why is the availability likelihood w used in the denominator of (4.1.6)?

4.1.12* If w is unknown, what can be done?

4.1.13 How small can the number N of complete cases be? Further, what to do if $N = 0$? Is it reasonable to consider M-samples with $N < 10$?

4.1.14 Explain the diagrams in Figure 4.1. Then repeat this figure 20 times and find the sample mean and sample median of the ratios ISEH/ISEM. Does it match the theoretical ratio?

4.1.15 Can one expect good estimation of the tails of the Normal for samples with size $n = 50$?

4.1.16* Verify and explain all relations in (4.1.9).

4.1.17* Verify and explain all relations in (4.1.10).

4.1.18 Explain possible remedies for density estimation when N is small.

4.1.19 Discuss the possibility of a MCAR sampling being more cost efficient than a corresponding direct sampling without missing.

4.1.20 Explain all diagrams in Figure 4.2.

4.1.21 Repeat Figure 4.2 with different parameters and write a report about obtained results.

4.1.22* Let us consider the following two propositions, and the aim of the exercise is to check their validity. First, show that formula (4.1.5) yields the following sample mean estimator of θ_j,

$$\hat{\theta}_j := \frac{\sum_{\{l: \, A_l = 1, \, l \in \{1,...,n\}\}} A_l \varphi_j(A_l X_l)}{\sum_{l=1}^{n} A_l}$$

$$= \frac{\sum_{l=1}^{n} A_l \varphi_j(A_l X_l)}{\sum_{l=1}^{n} A_l} = N^{-1} \sum_{l=1}^{n} A_l \varphi_j(A_l X_l). \tag{4.9.1}$$

Second, let us verify that Fourier estimator (4.9.1) is unbiased. Write,

$$\mathbb{E}\{\hat{\theta}_j\} = \mathbb{E}\left\{ \frac{\sum_{l=1}^{n} A_l \mathbb{E}\{\varphi_j(X_l) | A_l\}}{\sum_{l=1}^{n} A_l} \right\} = \theta_j \mathbb{E}\left\{ \frac{\sum_{l=1}^{n} A_l}{\sum_{l=1}^{n} A_l} \right\} = \theta_j. \tag{4.9.2}$$

Check correctness of (4.9.1) and (4.9.2), and then write down your comments. Hint: Pay attention to the case $N = 0$.

4.1.23* Can collecting missing data be attractive? Assume that there is an extra cost for avoiding missing data, and then propose a procedure of a controlled sampling that minimizes the cost of sampling given a guaranteed value of the MISE.

4.2.1 Describe the statistical model of nonparametric regression with MAR responses. Give definition of the MAR.

4.2.2 What is the difference between MCAR and MAR responses?

4.2.3 Assume that (4.2.3) holds. Are A and Y dependent? Given X, are A and Y dependent?

4.2.4 Which regression model is more general: model (4.2.1) or (4.2.2)? Explain your answer and present examples.

4.2.5 Verify formula (4.2.4) for the joint (mixed) density.

4.2.6 Find the probability that the response is missed, that is find the probability $\mathbb{P}(A = 1)$.

4.2.7 Verify expression (4.2.5) for Fourier coefficients of the regression function.

4.2.8 Based on (4.2.5) and assuming that the design density is known, propose a sample mean estimator of Fourier coefficients of the regression function. Then compare your answer with (4.2.6).

4.2.9 Explain motivation behind the estimator (4.2.7).

4.2.10 Explain how the estimator (4.2.9) is obtained.

4.2.11 Is the estimator (4.2.9) based solely on complete cases? If the answer is "yes," then can incomplete cases be ignored? Further, do we need to know the sample size n?

4.2.12* Explain estimator (4.2.10). If $N = 0$, then what is the meaning of this estimator? Further, will you use it for $N = 10$? Hint: Recall remedies discussed in Section 4.1.

4.2.13 Using Figure 4.3, for each corner function propose less and more favorable availability likelihood functions $w(x)$ that imply the same probability $\mathbb{P}(A = 1)$. Then verify your recommendation via comparison of ISEs.

4.2.14 For each corner function, propose optimal parameters $cJ0$ and $cJ1$ and verify your conclusion using Figure 4.3.

4.2.15 For each corner function, propose optimal parameter cTH and verify your conclusion using Figure 4.3.

4.2.16 Prove equality (4.2.11).

4.2.17 Explain how (4.2.13) is established.

4.2.18 Verify expression (4.2.14) for the variance of the Fourier estimator $\bar{\theta}_j$.

4.2.19* Explain estimator (4.2.19) and prove (4.2.20).

4.2.20 Using expression (4.2.21) for the coefficient of difficulty, explain how the missing mechanism and characteristics of the regression define the difficulty of regression problem with MAR responses.

4.2.21 Suppose that we can choose the design density. Which one should we recommend?

4.2.22 The optimal design density (4.2.22) depends on the scale and availability likelihood. These functions are typically unknown but may be estimated based on the MAR sample. Propose a sequential estimation plan that allows to adapt the design density to the data at hand.

4.2.23* Can working with missing data be attractive? Assume that there is an extra cost for avoiding missing data, and then propose a procedure of a controlled sampling that minimizes the cost of sampling given a guaranteed value of the MISE.

4.3.1 Explain the model of nonparametric regression with MAR predictors. Present several examples.

4.3.2 Explain the model (4.3.1). Is it MAR? When does this model become MCAR?

4.3.3 Prove (4.3.2).

4.3.4 What is the difference, if any, between models (4.3.3) and (4.3.4)?

4.3.5 Write down the conditional density of the response given the predictor in a complete case. Hint: Check with (4.3.5).

4.3.6 Can the regression problem with MAR predictors be solved using only complete cases?

4.3.7 Suppose that the design density and the availability likelihood functions are given. Propose a regression E-estimator.

4.3.8 Suppose that the marginal density of responses is known. Propose a regression E-estimator.

4.3.9 Propose a data-driven regression E-estimator.

4.3.10 Can a complete-case regression estimator be consistent?

4.3.11 Explain diagrams in Figure 4.4.

4.3.12 Find optimal parameters of the E-estimator using simulations produced by Figure 4.4.

4.3.13 Explore how the shape of the availability likelihood affects regression estimation given $\mathbb{P}(A = 1) = 0.7$.

4.3.14 How does the shape of the scale function affect the estimation? Verify your conclusion: (i) Using Figure 4.4; (ii) Theoretically via analysis of the coefficient of difficulty.

4.3.15 Prove (4.3.5) and explain the assumptions.

4.3.16 Explain regression estimator under scenario 1.

4.3.17 Verify each equality in (4.3.7).

4.3.18* Find the mean and the variance of Fourier estimator (4.3.8). What is the corresponding coefficient of difficulty?

4.3.19 Explain how to construct a regression estimator under scenario 2.

4.3.20 What is the motivation of the Fourier estimator (4.3.9)?

4.3.21* Find the mean and the variance of estimator (4.3.9).

4.3.22 Propose a regression estimator for scenario 3.

4.3.23* Evaluate the MISE of regression estimator based on an M-sample.

4.3.24 Consider the case $N = 0$. How can E-estimator treat this case?

4.3.25* Suppose that we consider an M-sample only if $N := \sum_{l=1}^{n} A_l \geq k > 0$. Calculate the MISE of E-estimator.

4.3.26* Why is the regression with MAR responses less complicated than the regression with MAR predictors?

4.3.27* Can collecting missing data be attractive? Assume that there is an extra cost for avoiding missing data, and then propose a procedure of a controlled sampling that minimizes the cost of sampling given a guaranteed value of the MISE.

4.4.1 What is the definition of a conditional density?

4.4.2 Verify (4.4.2).

4.4.3 Consider $f^{Y|X}(y|x)$. Given x, is this a regular density in y or not?

4.4.4 Consider a regression model with predictor X and response Y. What is the more general description of the relationship between X and Y, the conditional density or the regression function?

4.4.5 If the conditional density $f^{Y|X}(y|x)$ is a more general characteristic of relationship between the predictor X and the response Y than the regression function $m(x) = \mathbb{E}\{Y|X = x\}$, then why do we study the regression function?

4.4.6 Describe a model of conditional density estimation with MAR responses.

4.4.7 Describe a model of conditional density estimation with MAR predictors.

4.4.8 Explain each equality in (4.4.4).

4.4.9 What conclusion can be made from (4.4.4)?

4.4.10 What are the difference and similarity between H-sample and M-sample?

4.4.11 Describe the distribution of the number N of complete cases in M-sample.

4.4.12* Explain the methodology of construction of the estimator (4.4.7). What should be done if $N = 0$?

4.4.13 Repeat Figure 4.5 and explain all diagrams.

4.4.14 Will a regression model be useful for description of data in Figure 4.5?

4.4.15 Explain a model with MAR predictors.

4.4.16 Verify formula (4.4.9) for the joint density.

4.4.17 Can a complete-case estimator imply a consistent estimation for the case of MAR predictors?

4.4.18 Is the estimator (4.4.11) of Fourier coefficients unbiased?

4.4.19* Verify formula (4.4.13).

4.4.20* Write down a formula for the coefficient of difficulty for the case of MAR predictors. Then find the optimal design density.

4.4.21* Find the mean and the variance of Fourier estimator (4.4.16).

4.4.22 How can the design density f^X be estimated? Hint: Note that the density of predictors in complete cases is biased.

4.4.23 How can the availability likelihood function $w(y)$ be estimated? Hint: Pay attention to the fact that the support of Y is unknown.

4.4.24 Repeat Figure 4.6 and explain all diagrams.

4.4.25 Use Figure 4.6 to understand which nuisance function has the least and largest effect on the regression estimation.

4.4.26 Consider a setting when information in incomplete cases cannot be considered as reliable. In other words, consider the satiation when cases $(A_l X, Y_l, A_l)$ with $A_l = 0$ may have corrupted values of Y_l. What can be done in this case? Hint: Consider using an additional sample of responses $(Y_1', Y_2', \ldots, Y_{n'}')$ from Y (without observing predictors).

4.4.27 Find the coefficient of difficulty of the proposed regression E-estimator.

4.4.28* Can collecting missing data be attractive? Assume that there is an extra cost for avoiding missing data, and then propose a procedure of a controlled sampling that minimizes the cost of sampling given a guaranteed value of the MISE.

4.5.1 What is the Poisson distribution?

4.5.2 What are the mean and the variance of a Poisson random variable?

4.5.3 Define a Poisson regression.

4.5.4 Explain how to construct an E-estimator for Poisson regression with no missing data.

4.5.5 Describe a Poisson regression model with MAR responses.

4.5.6 Verify (4.5.8).

4.5.7 What conclusion can be made from (4.5.10)?

4.5.8 Suggest an E-estimator for a Poisson regression with MAR responses.

4.5.9 Repeat Figure 4.7 and explain all diagrams.

4.5.10 What type of shape of the availability likelihood function makes the estimation worse and better for the corner functions? Check your conclusion using Figure 4.7.

4.5.11 Using Figure 4.7, try to find optimal parameters of the E-estimator for each corner function.

4.5.12 Find the coefficient of difficulty of the proposed regression E-estimator.

4.5.13* Suppose that the availability likelihood depends on Y. Propose a consistent regression estimator. Hint: It is possible to ask about additional information.

4.5.14* Can collecting missing data be attractive? Assume that there is an extra cost for avoiding missing data, and then propose a procedure of a controlled sampling that minimizes the cost of sampling given a guaranteed value of the MISE.

4.6.1 Explain the problem of scale estimation for the case of direct observations (H-sample).

4.6.2 Verify formula (4.6.3) for the joint density.

4.6.3 Propose the methodology of estimation of the scale function. Hint: Convert it into a regression problem with several nuisance functions.

4.6.4 Is the estimator (4.6.7) unbiased? Hint: You may follow (4.6.8).

4.6.5 Explain the underlying idea of the estimator (4.6.10).

4.6.6 Repeat Figure 4.8 and explain diagrams.

4.6.7 Repeat Figure 4.8 and comment on the estimates.

4.6.8 How does the shape of regression function affect estimation of the scale?

4.6.9 How does the shape of availability likelihood function affect estimation of the scale?

4.6.10* Find the coefficient of difficulty of Fourier estimator (4.6.7).

4.6.11 Verify each equality in (4.6.8).

4.6.12* Evaluate the mean and variance of Fourier estimator (4.6.8).

4.6.13* Explain how the nuisance functions $\hat{m}(x)$ and $\hat{f}^{X|A}(x|1)$, used in (4.6.10), are constructed. Then evaluate their MISEs.

4.6.14* Explain how the parameter c of estimator (4.6.10) affects its coefficient of difficulty.

4.6.15* Can collecting missing data be attractive? Assume that there is an extra cost for avoiding missing data, and then propose a procedure of a controlled sampling that minimizes the cost of sampling given a guaranteed value of the MISE.

4.7.1 Describe the problem of bivariate regression.

4.7.2 Propose an E-estimator for bivariate regression based on H-sample.

4.7.3 Suppose that we are considering a bivariate regression with MAR responses. What is the availability likelihood function in this case?

4.7.4 Explain how to construct a basis for bivariate functions with domain $[0,1]^2$.

4.7.5 Write down and then prove the Parseval identity for a bivariate function.

4.7.6 Prove equality (4.7.3). What is the used assumption?

4.7.7* Find the mean and the variance of the estimator (4.7.4).

4.7.8 Explain the assumption (4.7.5).

4.7.9 Suppose that (4.7.5) holds. Are A and Y dependent given X_1?

4.7.10 Verify (4.7.6).

4.7.11 Prove every equality in (4.7.7).

4.7.12 Is (4.7.8) correct? Prove or disprove it.

4.7.13 Verify (4.7.9) Formulate all necessary assumptions.

4.7.14 Prove (4.7.10).

4.7.15 Is the estimator (4.7.11) unbiased? Prove your assertion.

4.7.16 Verify (4.7.12). Formulate necessary assumptions.

4.7.17 Explain the estimator (4.7.13). What can be done when $N = 0$?

4.7.18* Find the mean and the variance of Fourier estimator (4.7.13).

4.7.19* Explain how plug-in estimators of nuisance functions, used in (4.7.13), are constructed.

4.7.20 Explain the simulation used in Figure 4.9.

4.7.21 Using Figure 4.9, suggest a minimal sample size which yields a reliable estimation of bivariate regressions.

4.7.22 Explain diagrams in Figure 4.9.

4.7.23 Propose optimal parameters of the E-estimator for two different underlying regressions. Hint: Use Figure 4.9.

4.7.24 Using Figure 4.9, explain how the availability likelihood affects the regression estimation.

4.7.25* Consider a problem of bivariate regression with MAR predictors. Propose an E-estimator.

4.7.26* Can collecting missing data be attractive? Assume that there is an extra cost for avoiding missing data, and then propose a procedure of a controlled sampling that minimizes the cost of sampling given a guaranteed value of the MISE.

4.8.1 Describe a regression model with additive components.

4.8.2 Consider the case of two predictors. What is the difference between bivariate and additive regressions? Which one would you suggest to use?

4.8.3 Why do we need the restriction (4.8.2)?

4.8.4 Propose a sample mean estimator of Fourier coefficients (4.8.3).

4.8.5 Prove (4.8.4). What is the used assumption?

4.8.6 Why is (4.8.5) correct?

4.8.7 Prove (4.8.6).

4.8.8 Is (4.8.6) correct for $j = 0$?

4.8.9 Is the estimator (4.8.7) unbiased?

4.8.10 Explain how parameter β can be estimated.

4.8.11* Evaluate the variance of Fourier estimator (4.8.7).

4.8.12 Describe an additive regression model with MAR responses.

4.8.13 Explain the assumption (4.8.10). What will be if the availability likelihood depends on Y?

4.8.14 Verify (4.8.11).

4.8.15 Prove (4.8.12), and formulate the used assumption.

4.8.16 Verify (4.8.13).

4.8.17 Explain how formula (4.8.14) is established.

4.8.18* Is the estimator (4.8.15) unbiased? Find its variance.

4.8.19* Explain how estimators (4.8.16) and (4.8.17) were suggested. Then find their variances.

4.8.20 Explain all diagrams in Figure 4.10.

4.8.21 Use Figure 4.10 and explain the so-called "curse of multidimensionality."

4.8.22 Repeat Figure 4.10 and try to explain E-estimates via analysis of the scattergram.

4.8.23 Suggest better parameters of the E-estimator used in Figure 4.10.

4.8.24* Can collecting missing data be attractive? Assume that there is an extra cost for avoiding missing data, and then propose a procedure of a controlled sampling that minimizes the cost of sampling given a guaranteed value of the MISE.

4.8.25* Consider additive regression with MAR predictors. Propose an E-estimator.

4.10 Notes

In many practical applications, and almost inevitably in those dealing with activities of human beings, some entries in the data matrix may be missed. A number of interesting practical examples and a thorough discussion of missing data can be found in books by Rubin (1987), Allison (2002), Little and Rubin (2002), Groves et al. (2002), Tsiatis (2006), Molenberghs and Kenward (2007), Daniels and Hogan (2008), Davey and Salva (2009), Enders (2010), Graham (2012), Carpenter and Kenward (2013), Molenberghs et al. (2014), and Raghunathan (2016). There is a strong opinion in the statistical community that researchers often play down the presence of missing data in their studies. To change this attitude, there has been an explosion of reviews/primaries on missing data for different sciences, see, for instance, Bodner (2006), Enders (2006, 2010), Baraldi and Enders (2010), Honaker and King (2010), Young, Weckman and Holland (2011), Cheema (2014), Zhou et al. (2014), Nakagawa (2015), Newgard and Lewis (2015), Lang and Little (2016), Little et al. (2016), Efromovich (2017), Sullivan et al. (2017). The available literature is primarily devoted to parametric models.

The MCAR, MAR and MNAR terminology is due to Rubin (1976).

A number of methods for dealing with missing data is proposed in the literature. Among the more popular are the following:

1. The most common approach is referred to as a complete-case approach (case-wise or list-wise deletion is another name often used in the literature). It involves eliminating all records with missing values on any variable. Popular statistical softwares, like SAS, S-PLUS, SPSS and R, by default ignore cases of observations with missing values, and this tells us how popular and well accepted the complete-case methodology is. Why? Because this method is simple, intuitively appealing and optimal for some settings. A disadvantage of the approach is two-fold. First, in some applications it is too wasteful. Second, in many

important settings a complete-case approach yields inconsistent estimation, and this fact should be known and taken into account.

2. Imputation is a common alternative to the deletion. It is used to "fill in" a value for a missing value using the other information in the database. A simple procedure for imputation is to replace the missing value with the mean or median of that variable. Another common procedure is to use simulation to replace the missing value with a value randomly drawn from the records having values for the variable. It is important to stress that imputation is only the first step in any estimation and inference procedures. The second step is to propose an estimator and then complement it by statistical inference about properties of the estimator, and these are not trivial steps because imputation creates dependence between observations. Warning: the fact that a complete data is created by imputation should be always clearly cited because otherwise a wrong decision can be made by a data analyst who is not aware about this fact. As an example, suppose that we have a sample from a Normal(μ, σ^2) distribution of which some observations are missed. Suppose that an analyst is interested in estimation of the mean μ, and an oracle helps by imputing the underlying μ in place of the lost observations. This is an ideal imputation, and then the analyst correctly uses the sample mean to estimate μ. However, if then the analyst (or someone else) will use this imputed (and hence complete) sample to estimate the variance σ^2, the classical sample variance estimator will yield a biased estimate (to see this just imagine that all observations were missed and then calculate the sample variance estimator which should be equal to zero). An interesting discussion, examples and references can be found in Little and Rubin (2002), Davidian, Tsiatis and Leon (2005), Enders (2010), Graham (2012) and Molenberghs et al. (2014).

3. Multiple imputation is another popular method. It is based on repeated imputation–estimation steps and then aggregation (for instance via averaging) of the estimates. Multiple imputation is a flexible but complicated statistical procedure which requires a rigorous statistical inference. Books by van Buuren (2012) and Berglund and Heeringa (2014) are devoted to this method.

4. Maximum likelihood method, and its numerically friendly EM (expectation-maximization) algorithm, is convenient for parametric models with missing data. Little and Rubin (2002) is a classical reference.

5. A number of weighting methods, like the adjustment cell method, inverse probability weighting, response propensity model, post-stratification, and survey weights, are proposed in the literature. These methods are well documented and supported by a number of statistical packages. See a discussion in Little and Rubin (2002), Molenberghs et al. (2014) and Raghunathan (2016) where further references may be found.

6. Finally, let us mention that prevention of the missing data, if possible, is probably the most powerful method of dealing with missing; see McKnight et al. (2007). Prevention is discussed in practically all of the above-mentioned books. On the other hand, if collecting of missing data is significantly cheaper, then working with missing data may be beneficial, see Efromovich (2017). Efficiency and robustness are discussed in Cao, Tsiatis and Davidian (2009).

4.1 Asymptotic justification of the proposed E-estimation methodology is presented in Efromovich (2013c). Sequential estimation is considered in Efromovich (2015). An interesting possible extension is to propose a second-order efficient estimate of the cumulative distribution function, see Efromovich (2001b).

4.2 The theory of efficient nonparametric regression with MAR responses may be found in Efromovich (2011b). It is shown that the E-estimation methodology, based on the complete case approach, yields efficient estimation of the regression function, and more discussion can be found in Efromovich (2012a; 2014c,e; 2016c; 2017). Sequential estimation is considered in Efromovich (2012b). Müller and Van Keilegom (2012) and Müller and Schick

(2017) present more cases when a complete-case approach is optimal and show that there is no need to use imputation, work with inverse probability weights or use any other traditional remedy for missing data. Estimation of functionals is considered in Müller (2009).

4.3 Asymptotic theory of nonparametric regression estimation with MAR predictor is more complicated than its missing response counterpart. The theory and applications are presented in Efromovich (2011a, 2016c, 2017). The theory supports optimality of the proposed E-estimation methodology. See also Goldstein et al. (2014).

4.4 Conditional density estimation is a classical statistical problem, see books by Fan and Gijbels (1996) and Efromovich (1999a), as well as more recent articles by Izbicki and Lee (2016) and Bott and Kohler (2017). The asymptotic theory of E-estimation and applications may be found in Efromovich (2007g, 2010b,d).

4.5 Poisson regression for direct data is discussed in Efromovich (1999a). A number of interesting settings and applications are discussed in Ivanoff et al. (2016) where further references may be found. Discussion of a Bayesian nonparametric approach may be found in Ghosal and van der Vaart (2017).

4.6 Regression with the scale depending on the predictor is a classical topic in the regression analysis, and the regression is called heteroscedastic. Statistical analysis and efficient estimation are discussed in Efromovich (1999a, 2013a,b). Sequential estimation is discussed in Efromovich (2007d,e; 2008a). Further, not only the scale function but the distribution of regression error may be efficiently estimated by the E-estimation methodology, see Efromovich (2004h, 2005b, 2007c).

4.7 E-estimation methodology and its optimality for estimation of multivariate regression is considered in Efromovich (1999a, 2000b, 2013a,b). A further discussion of this and related topics can be found in Izenman (2008), Efromovich (2010d, 2011c), Harrell (2015) and Raghunathan (2016).

4.8 Additive regression is a natural method for avoiding the curse of multidimensionality that slows down the rate of the MISE convergence and requires a dramatic increase in the sample size to get a feasible estimation. There are no such dramatic issues with the additive regression. It also may be argued that the additive regression is the first glance on data, and it may be instrumental in finding a more appropriate model. A book-length treatment of additive models may be found in the classical book by Hastie and Tibshirani (1990), and see also Izenman (2008), Harrell (2015) and Wood (2017). The asymptotic theory of E-estimation can be found in Efromovich (2005a).

Chapter 5

Destructive Missing

This chapter is devoted to an interesting and complicated topic of dealing with the missing that may prevent us from consistent estimation based solely on available data. In other words, we are considering missing mechanisms that modify observations in such a way that no consistent estimation of an estimand, based solely on missing data (or we may say based on an M-sample defined in Section 1.3 or Chapter 4) is possible. This is the reason why we refer to this type of missing as a destructive missing. MNAR (missing not at random), when the likelihood of missing a variable depends on its value, typically implies a destructive missing. At the same time, MNAR is not necessarily destructive. For instance, we will explore a number of settings where knowledge of the availability likelihood function makes MNAR nondestructive. Another comment due is about our reference to consistent estimation. For any data, and even without data, one may propose an estimator. The point is that an estimator should be good and produce some useful information, and consistent estimation is a traditional statistical criterion for both parametric and nonparametric models. Of course, in nonparametric curve estimation we could use the criterion of the MISE vanishing instead, and this would not change the conclusions.

In general, some extra information is needed to solve the problem of consistent estimation for the case of a destructive missing, and a number of reasonable approaches will be suggested and discussed on what can be done and what can be expected.

Let us recall the used terminology. By E-estimator we understand the estimator defined in Section 2.2. Recall that to construct an E-estimator we need to propose a sample mean (or a plug-in sample mean) estimator of Fourier coefficients of an estimated function. An underlying sample of observations of random variables of interest, which in general is unknown to us due to a missing mechanism, is called an H-sample (hidden sample). Note that H-sample is a sample of direct observations of random variables of interest. A corresponding sample with missing observations is called an M-sample. Note that cases in an M-sample may be incomplete while in a corresponding H-sample all cases are complete. If an extra sample is available, then it is referred to as an E-sample (extra sample). The probability of a variable to be available (not missing) in M-sample, given all underlying observations in the corresponding H-sample, is called the availability likelihood (this function is always denoted as w). Depending on this function, the missing mechanism may be either MCAR (the availability likelihood is equal to a constant), or MAR (the availability likelihood is a function in always observed variables), or MNAR (none of the above). In what follows we will periodically remind the reader of the terminology.

The chapter begins with the topic of density estimation, which is pivotal for understanding destructive missing. This explains why each specific remedy for extracting useful information from MNAR data is discussed in a separate section. Section 5.1 serves as an introduction. It explains why MNAR may lead to destructive missing and why the problem is related to the topic of biased data. Then it is explained that if the availability likelihood is known, then MNAR does not imply a destructive missing. The latter is a pivotal conclusion because one of the main tools to "unlock" information in an MNAR dataset is to estimate the availability likelihood based on an extra sample. This is the approach discussed in Section 5.2. Namely, a sample of direct (and hence more expansive) observations is available and it is used to estimate the availability likelihood function. The main issue here is the discussion of the size of an extra sample (E-sample) that makes the approach feasible. Section 5.3 considers another remedy when there exists an auxiliary random variable which defines the missing mechanism. Under this possible scenario, using that auxiliary variable converts MNAR into MAR. Section 5.4 is devoted to regression with MNAR responses, that is the case when likelihood of missing a response depends on its value. Section 5.5 considers a problem of regression with MNAR predictors. Section 5.6 considers a regression where both predictors and responses may be missed.

5.1   Density Estimation When the Availability Likelihood is Known

The problem is to estimate the density $f^X(x)$ of a continuous random variable X supported on the unit interval $[0,1]$. As usual, it is assumed that the density is square integrable on its support, that is $\int_0^1 [f^X(x)]^2 dx < \infty$.

A classical density estimation setting is when n independent and identically distributed realizations $X_1, \ldots, X_n$ (the sample of size n) of a random variable X are available, and then the density E-estimator is suggested in Section 2.2. In some sampling procedures all n realizations may not be available due to missing. In this case the above-described sample is hidden and it is referred to as an H-sample (hidden sample), while the available sample with some realizations being missing (not available) is referred to as M-sample (sample with missing observations). We may also say that in an M-sample some cases are incomplete.

In Section 4.1 the simplest MCAR (missing completely at random) missing mechanism was considered when the probability of missing does not depend on the value of missing variable. For the MCAR, a complete-case approach, based on using only available realizations, is both feasible for small samples and asymptotically optimal. Unfortunately, as we will see shortly, this is the only case when missing allows us to consistently estimate density $f^X(x)$. In what follows it is explained why this is the case and what are the possible remedies.

We begin with discussion of the studied missing mechanism. Introduce a pair of random variables (AX, A) where X is the random variable of interest and A is a Bernoulli random variable which defines the *availability* (not missing) of the random variable X. The availability random variable A takes on only two values 0 and 1. If $A = 1$ then a realization of X is available and otherwise not. Then the available M-sample from (AX, A) is $(A_1 X_1, A_1), \ldots, (A_n X_n, A_n)$.

Remark 5.1.1 X is a continuous random variable and hence $\mathbb{P}(X = 0) = 0$. This yields that instead of considering a pair (AX, A) of variables we may consider a single variable AX. Indeed, from $\mathbb{P}(I(AX \neq 0) = A) = 1$, where $I(\cdot)$ is the indicator function, we can always restore A from the equality $A = I(AX \neq 0)$ which holds with probability 1. While it is still more convenient and transparent to work with the pair (AX, A), it is important to note that no information is lost if we observe only a sample from AX.

The joint mixed density of pair (AX, A) may be written as

$$f^{AX,A}(ax, a) = [f^X(x)w(x)]^a [1 - \int_0^1 f^X(u)w(u)du]^{1-a}, \quad x \in [0,1], \ a \in \{0,1\}, \quad (5.1.1)$$

where $\{0,1\}$ denotes a set consisting of two numbers 0 and 1, and the *availability likelihood* function $w(x)$ is defined as

$$w(x) := \mathbb{P}(A = 1|X = x), \quad x \in [0,1]. \quad (5.1.2)$$

Recall that if $w(x)$ is constant, then the missing mechanism is MCAR (missing completely at random) and otherwise it is MNAR (missing not at random). In this section (and the whole chapter) we are dealing exclusively with the MNAR.

Let us comment on (5.1.1). First of all, because X is continuous and A is discrete, we are dealing with the mixed density. Second, it is natural to think about any missing process as a two-step procedure: (i) First a realization of X occurs; (ii) Second, given the realization of X, the Bernoulli random variable A is generated with $\mathbb{P}(A = 1|X) = w(X)$. If $A = 1$ then the realization is available and otherwise not. Hence, if $A = 1$ then

$$f^{AX,A}(x, 1) = f^X(x)\mathbb{P}(A = 1|X = x) = f^X(x)w(x). \qquad (5.1.3)$$

Relation (5.1.3) explains the first factor on the right-side of formula (5.1.1) which describes the case $a = 1$. If $A = 0$, then an underlying realization of X is not available and in this case the joint density is

$$f^{AX,A}(0, 0) = \mathbb{P}(X \in [0, 1], A = 0) = \int_0^1 f^X(u)\mathbb{P}(A = 0|X = u)du$$

$$= \int_0^1 f^X(u)(1 - w(u))du = 1 - \int_0^1 f^X(u)w(u)du. \qquad (5.1.4)$$

This verifies (5.1.1) for the case $a = 0$.

It immediately follows from (5.1.1) that the joint mixed density of observations depends on the product $f^X(x)w(x)$. We know a priori that $\int_0^1 f^X(x)dx = 1$, $f^X(x) \geq 0$ and $0 \leq w(x) \leq 1$, but apart of these three relations nothing else is known. Hence, if the nuisance function $w(x)$ is unknown then consistent estimation of density $f^X(x)$, based on M-sample from (AX, A), is impossible. This is why the MNAR implies a destructive missing.

Note that the only quantity that can be estimated under the MNAR is the product $f^X(x)w(x)$, and if no extra information is available, then an E-estimate of the product may be of some interest to practitioners. Further, the only practically feasible case is when

$$w(x) \geq c_0 > 0, \quad x \in [0, 1]. \qquad (5.1.5)$$

Similarly to all previously considered settings, in what follows it is assumed that (5.1.5) holds.

To shed additional light on the MNAR problem, let us consider a subsample of complete cases, that is realizations $(A_l X_l, A_l)$ with $A_l = 1$. A random number N of complete cases in an M-sample is

$$N := \sum_{l=1}^n A_l, \qquad (5.1.6)$$

and it has the Binomial$(n, \mathbb{P}(A = 1))$ distribution with the mean $\mathbb{E}\{N\} = n\mathbb{P}(A = 1)$ and the variance $\mathbb{V}(N) = n\mathbb{P}(A = 1)(1 - \mathbb{P}(A = 1))$. Further, using (5.1.1) we can write,

$$f^{AX|A}(x|1) = \frac{f^{AX,A}(x, 1)}{\mathbb{P}(A = 1)} = \frac{f^X(x)w(x)}{\int_0^1 f^X(u)w(u)du} = f^X(x)\frac{w(x)}{E\{w(X)\}}. \qquad (5.1.7)$$

Relation (5.1.7) implies that the density $f^{AX|A}(x|1)$ of observations in complete cases is biased with respect to the estimated $f^X(x)$ and the biasing function is the availability likelihood $w(x) = \mathbb{P}(A = 1|X = x)$ (recall Section 3.1). This is an important conclusion which connects a destructive missing (and the MNAR in particular) with the topic of biased data. Let us recall a conclusion of Section 3.1 that for biased data a consistent estimation is impossible if the biasing function is unknown. As a result, it should not be a surprise that the same conclusion is valid for MNAR with unknown availability likelihood function $w(x)$.

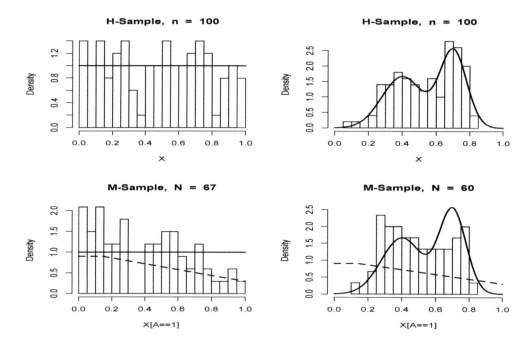

Figure 5.1 *Examples of destructive missing. Two columns of diagrams exhibit results of simulations for the Uniform and the Bimodal underlying densities shown by the solid lines. Available observations in H- and M-samples are shown by histograms. Axis label $X[A == 1]$ indicates that only available observations in M-sample are shown (the label is motivated by R operator of extracting elements of vector X corresponding to elements of vector A equal to 1). The availability likelihood function $w(x)$ is the same in the two simulations and is shown by the dashed line in the bottom diagrams. The sample size n and the number $N = \sum_{l=1}^{n} A_l$ of available observations in M-sample are shown in the corresponding titles. {The arguments are: set.corn and n control the underlying corner functions and the sample size of H-sample (recall that any custom-made corner function can be used as explained in the caption of Figure 2.3), the string w controls a function $w^*(x)$ which is bounded from above and below by constants dwU and dwL, respectively, and then the availability likelihood is $w(x) := \max(dwL, \min(dwU, w^*(x)))$.} [set.corn = c(1,3), n = 100, w = "1-0.7*x", dwL = 0.3, dwU = 0.9]*

Let us illustrate this conclusion via two simulated examples shown in Figure 5.1. The left-top diagram shows us the histogram of a hidden sample of size $n = 100$, and the solid line shows the underlying Uniform density. The sample is rather typical for the Uniform. The corresponding available observations in an MNAR sample (M-sample) are shown by the histogram in the left-bottom diagram, and the histogram is overlaid by the solid line (the underlying density) and the dashed line (the availability likelihood function). The non-constant availability likelihood yields the MNAR and the destructive missing, and the histogram of complete cases in the M-sample clearly indicates why. Indeed, the histogram is skewed to the left and its shape resembles $w(x)$ and not the underlying density, and the latter was predicted by (5.1.7). Note that the skewness becomes even more pronounced if we one more time look at the H-sample. Further, let us compare numbers of available observations in the two samples. The missing decreases the sample size $n = 100$ of the H-sample to $N = 67$ in the M-sample. This is what makes the missing so complicated even if the availability likelihood function (which is also the biasing function for the M-sample) is available, recall our discussion in Section 3.1.

A similar conclusion can be made from the second column of diagrams exhibiting an

H-sample generated by the Bimodal density and an M-sample generated by the same availability likelihood function. The H-sample clearly indicates the Bimodal density, while the missing dramatically changes the histogram of the M-sample where we also observe two modes but their magnitudes are misleading and the left one is larger and more pronounced than the right one. This is what the MNAR may do to an H-sample, and this is why this missing mechanism is destructive.

We can conclude from these two simulations that an M-sample alone does not allow us to estimate an underlying probability density. Similarly to the case of biased data, the only chance to restore the hidden density is to know the availability likelihood function $w(x)$. In this section we are considering a pivotal case when $w(x)$ is known, and other possible scenarios are explored in the following two sections.

Suppose that the availability likelihood $w(x)$ is known. This does not change the MNAR nature of an M-sample but, as we will see shortly, the missing is no longer destructive and consistent estimation of the density is possible. Let us stress one more time that MNAR does not necessarily imply a destructive missing, everything depends on the possibility to complement MNAR data by an additional information. Further, as we will see shortly, for the problem at hand the main remaining challenge is the smaller, with respect to n, number N of available observations.

First of all, let us explain why consistent estimation of the density is possible. If $w(x)$ is known, then the biasing function for complete cases in an M-sample is also known and the density E-estimator of Section 3.1 may be used to estimate f^X. Further, according to Section 3.1, this estimator does not need to know the sample size n of an underlying H-sample. This explains why consistent estimation of the density, based on the complete-case approach, is possible.

Second, recall that according to Section 2.2, to construct an E-estimator we only need to propose a sample mean (and possibly a plug-in) estimator of Fourier coefficients $\theta_j :=$ $\int_0^1 f^X(x)\varphi_j(x)dx$ where $\{\varphi_j(x)\}$ is the cosine basis on $[0,1]$. In our case, a natural sample mean estimator of θ_j is

$$\hat{\theta}_j := n^{-1}\sum_{l=1}^{n} A_l\varphi_j(A_lX_l)/w(A_lX_l). \tag{5.1.8}$$

Let us show that this estimator is indeed a sample mean estimator. Using (5.1.3) we may write,

$$\theta_j = \int_0^1 f^X(x)\varphi_j(x)$$

$$= \int_0^1 f^X(x)w(x)[\varphi_j(x)/w(x)]dx = \mathbb{E}\{A\varphi_j(AX)/w(AX)\}. \tag{5.1.9}$$

This verifies that (5.1.8) is the sample mean estimator of θ_j as well as that it is unbiased estimator of θ_j.

The Fourier estimator (5.1.8) yields the density E-estimator $\hat{f}^X(x)$, $x \in [0,1]$ of Section 2.2.

Figure 5.2 is a good tool to learn how the E-estimator performs and what can be expected when the availability likelihood function $w(x)$ is known. The structure of Figure 5.2 is similar to Figure 5.1, only here we can simultaneously observe results of 4 simulations. The missing is created by the same availability likelihood function as in Figure 5.1. The solid and dashed lines show underlying densities f^X and their E-estimates $\hat{f}^X$. Additionally, the ISEs (integrated squared errors) of E-estimates are shown in subtitles, and recall that ISE:= $\int_0^1 (f^X(x) - \hat{f}^X(x))^2 dx$. Let us look at particular simulations. The left column of diagrams explores the case of the Uniform density. We clearly see how the MNAR skews

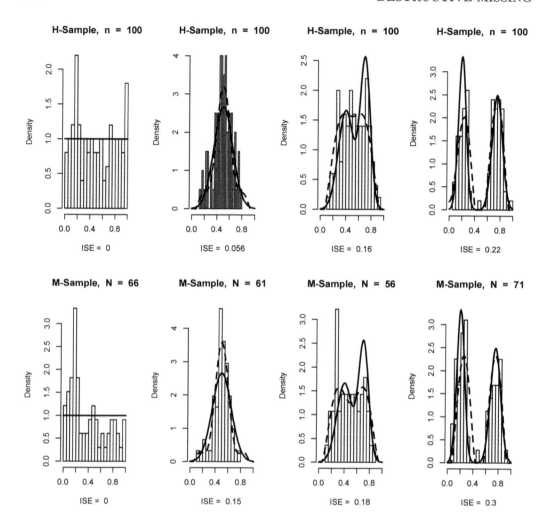

Figure 5.2 *Density estimation for MNAR data when the availability likelihood function $w(x)$ is known. Results of four simulations are shown in the four columns of diagrams whose structure is identical to those in Figure 5.1. In each diagram the histogram is overlaid by an underlying density (the solid line) and its E-estimate (the dashed line), and ISE is the integrated squared error of E-estimate. The availability likelihood is the same as in Figure 5.1. {The argument set.corn controls four corner functions and the availability likelihood $w(x)$ is defined as in Figure 5.1.} [set.corn = c(1,2,3,4), n = 100, w= "1-0.7*x", dwL = 0.3, dwU = 0.9, cJ0 = 3, cJ1 = 0.8, cTH = 4]*

available data to the left, and this is due to the decreasing $w(x)$. Nonetheless, the E-estimator correctly recovers the hidden density (note that the dashed line is hidden by the solid one). Also look at the number $N = 66$ of complete cases (available observations) in the M-sample, this is a small number for any nonparametric problem and hence it is likely that repeated simulations may produce worse estimates. The next column explores the case of the Normal density. Here again the M-sample is skewed to the left, and the E-estimate does correct the histogram and shows a symmetric bell-shaped curve around 0.5. The estimate is far from being perfect, its tails are "wrong" but note that they do describe the data. The number of available complete cases is $N = 61$, and this together with the MNAR increases the ISE almost threefold with respect to the H-sample. The number of complete cases is even smaller for the case of the Bimodal density (see the third column), it is $N = 56$. This is due to the decreasing availability likelihood and the skewed to the right shape of the Bimodal. Despite all these facts, it is fair to say that E-estimate for the M-sample indicates a bimodal shape and its ISE is just a little bit larger than its H-sample's counterpart. Of course, such an outcome is not typical but the random nature of simulations may produce atypical outcomes. For the Strata, the two E-estimates are not perfect and close to each other in terms of their shapes.

Overall, repeated simulations indicate that the missing takes a heavy toll on our ability to produce reasonable nonparametric estimates even if the availability likelihood function is known; the reduced number of available observations and their biased nature complicate the estimation. At the same time, typically the shape of an underlying density is recognizable, and this is an encouraging outcome keeping in mind that without knowing $w(x)$ we are dealing with a destructive missing which precludes us from a consistent estimation. Further, a possible practical recommendation is to consider, if possible, larger sample sizes to compensate for the MNAR.

5.2 Density Estimation with an Extra Sample

The setting is the same as in the previous Section 5.1. There is a hidden H-sample $X_1, \ldots, X_n$ from the density $f^X(x)$ supported on $[0, 1]$, this density is the estimand, and the available M-sample is $(A_1 X_1, A_1), \ldots, (A_n X_n, A_n)$. Here $A_1, \ldots, A_n$ is a sample from the availability A which is a Bernoulli random variable with the availability likelihood $\mathbb{P}(A_l = 1 | X_l = x) =: w(x) \geq c_0 > 0$, $x \in [0, 1]$. As a result, the missing is MNAR. Formula (5.1.1) describes the underlying probability model.

As we know from Section 5.1, if the availability likelihood $w(x)$ is unknown, then the MNAR is destructive and consistent estimation of the density $f^X(x)$ is impossible. To propose a remedy, some additional information should be provided.

Let us suppose that, in addition to the M-sample, we can get an extra k direct observations $X_{n+1}, \ldots, X_{n+k}$ from X, and in what follows we are referring to this extra sample as E-sample. In general these extra observations are more expensive and challenging to obtain, so $k \ll n$. The underlying idea of the remedy is to pay an extra price for E-sample to restore information contained in the M-sample. Note that in some practical situations it is possible to get an E-sample for a price. For instance: (i) An interview can be conducted (from less to more expansive options) by mail, telephone, online, in-person, using a private detective, or in a court under oath; (ii) If medical tests for pupils cannot be performed during weekends because a school is closed, a nurse can visit pupils at home; (iii) If routine measurements cannot be taken due to bad weather, then a special team can be hired for taking measurements during bad weather.

The aim is to understand how E-sample may help to recover information about f^X contained in an M-sample. Let us stress that we may use E-sample for estimation of f^X, but this is not the aim here. Further, the sample size k may be in order smaller than n. Our approach will be to use E-sample for estimation of the availability likelihood $w(x)$ and then plug it in the estimator proposed in Section 5.1.

Let us describe the proposed estimation methodology. Our first step is to use E-sample and calculate the density E-estimator $\tilde{f}^X(x)$ defined in Section 2.2. The second step is using the methodology of Section 3.7 to construct a regression E-estimator $\hat{w}(x)$ for Bernoulli regression of A on X using M-sample and the estimate $\tilde{f}(x)$. Indeed, Fourier coefficients $\eta_j := \int_0^1 w(x)\varphi_j(x)dx$ of the availability likelihood $w(x)$ may be written as $\eta_j = \mathbb{E}\{A\varphi_j(AX)/f^X(AX)\}$, and then they may be estimated by a plug-in sample mean estimator

$$\hat{\eta}_j := n^{-1} \sum_{l=1}^{n} \frac{A_l \varphi_j(A_l X_l)}{\max(\tilde{f}^X(A_l X_l), c/\ln(k))}. \tag{5.2.1}$$

Note that in (5.2.1) only observations X_l corresponding to $A_l = 1$ are used by the Fourier estimator, and hence an M-sample allows us to construct a regression E-estimate $\hat{w}(x)$ of the availability likelihood function $w(x)$. Unfortunately, we cannot further improve (5.2.1) via techniques discussed in Sections 2.3 and 2.4 and consider, for instance, the factor $A_l - \hat{\eta}_0$ in place of A_l whenever $j > 0$, because we do not observe X_l corresponding to $A_l = 0$.

The above-outlined methodology converts our setting into the one already discussed in Section 5.1. The complication here is that we use estimate $\hat{w}(x)$ in place of an unknown availability likelihood $w(x)$, and this estimate may be poor due to a relatively small size k of an extra E-sample. Further, we use the estimate in the denominator of (5.1.8) and this is always a challenge because we are forced to use a truncation. Asymptotic theory shows that the approach is feasible and $k \ll n$ can be used, but the asymptotic also tells us that both n and k should be large and the density $f^X(x)$ should be separated from zero on its support $[0, 1]$.

Is there a lesson from the asymptotic theory for small samples? Let us check this. Two figures will help us to gain experience in dealing with the destructive missing and the proposed remedy of using an extra sample of direct observations.

We begin with Figure 5.3 where four rows of diagrams show outcomes for different simulations. The sample sizes of H- and E-samples are $n = 200$ and $k = 30$; see the titles. Note that no missing occurs in E-samples and the "only" issue here is the relatively small sample size. The underlying corner densities are shown by solid lines while density E-estimates (based on E-samples) by the dashed lines; we can also see empirical ISEs. Overall, despite the very small sample, the E-estimator performs relatively well and correctly smooths the corresponding histograms. Of course, a small sample may confuse the E-estimator, and we see this, for instance, in the third (from the top) diagram. Further, look at the second (from the top) histogram for the Normal density. It is clear that the E-estimator does a remarkable job given the skewed and asymmetric data depicted by the histogram. Overall, what we see is the curse of nonparametric estimation because larger sample sizes are needed for a reliable density estimation. Nonetheless, let us continue and understand possible outcomes because our aim is to use an E-sample to recover information contained in an M-sample.

Diagrams in the middle column show us estimates $\hat{w}(X_l)$ of the availability likelihood which are nonparametric regressions of A on X. Note that in an M-sample, predictors for $A_l = 0$ are not available, and hence in general a consistent estimation of the regression is impossible. This is when our estimator of the design density of X becomes pivotal as explained in Section 3.7. Correspondingly, we use the estimate of design density, shown in a left diagram, to estimate the availability likelihood shown in a middle diagram. In a middle diagram circles show complete cases (their number $N := \sum_{l=1}^{n} A_l$ may be found in the title), triangles show underlying $w(X_l)$ and crosses show E-estimate $\hat{w}(X_l)$ for complete cases. This E-estimate is the main step of the proposed procedure, because what we see in the right column of diagrams are the density estimates based on an M-sample and $\hat{w}(x)$, and this step was discussed in Section 5.1.

Now let us look at the quality of estimates exhibited in Figure 5.3. In the left-top diagram the density f^X is estimated perfectly. The estimated availability likelihood is far from being perfect, but its shape resembles the underlying availability likelihood. Despite this imperfection, the plug-in density E-estimate is perfect. Note that we have $N = 148$ of complete cases in the M-sample. Also note how the E-sample helped us to gain information from the heavily skewed M-sample. Now let us look at results for the Normal density shown in the second from the top row. Here the density E-estimate, based on the E-sample, is relatively good and its shape is symmetric. The availability likelihood estimate correctly shows the shape of $w(x)$ but its values are smaller. This does not have a significant effect on the final density E-estimate shown in the right diagram, and also look at the ISE=0.023. The reason for this good outcome is that the density is integrated to one and the E-estimator

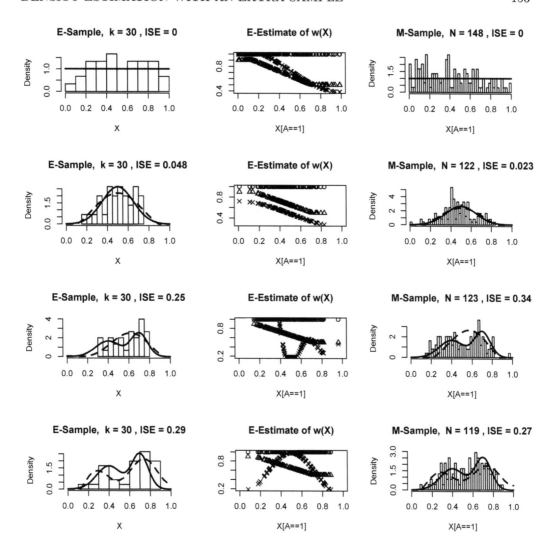

Figure 5.3 *Restoration of information in an M-sample with the help of an E-sample. This figure allows one to visualize the proposed methodology of the density E-estimation. Each row of diagrams corresponds to an underlying corner density. Underlying density and its E-estimate are shown by the solid and dashed lines, respectively. In a middle diagram pairs (X_l, A_l) with $A_l = 1$ (complete cases) are shown by circles, an underlying $w(X_l)$ and its E-estimate $\hat{w}(X_l)$ are shown by triangles and crosses, respectively. {The arguments are: set.corn controls four corner functions, k controls the sample size of E-sample, n defines the size of H-sample, w defines a function $w^*(x)$ and then the availability likelihood is $w(x) := \max(dwL, \min(dwU, w^*(x)))$. Argument c controls the parameters c used in (5.2.1). All other arguments control parameters of the E-estimator.} [set.corn = c(1,2,3,3), k = 30, n = 200, w = "1-0.7*x", dwU = 0.9, dwL = 0.5, c = 0.3, cJ0 = 3, cJ1 = 0.8, cTH = 4]*

uses this fact. Look at how nicely the E-estimate restores the bell-shaped Normal density from the skewed histogram of the M-sample.

The outcome is much worse for the third experiment with the Bimodal density, see the third row of diagrams. Here the poor density estimate, based on the E-sample, implies an extremely poor estimate of $w(x)$ which, in its turn, yields poor density E-estimate based on M-sample. Note that here the density estimate based on the M-sample with $N = 123$, according to its ISE is worse than the E-estimate based on the E-sample of size $k = 30$. This is what may be expected from small samples. In the bottom row the simulation is the same as in the third row, meaning that we observe two realizations of the same experiment. The two density E-estimates indicate two modes but they are clearly poor. Further, the estimate of $w(x)$ is also bad.

Overall, apart of the Uniform case and to some degree the Normal case, the density estimates are poor (just compare ISEs of the density estimates based on very small E-samples and much larger M-samples). There are several teachable moments here. The first one is that the asymptotic theory does matter; even for these small extra samples we may get a reasonable outcome. Second, the E-estimator is robust to imperfect estimates of $w(x)$. Finally, we need to understand why estimates of $w(x)$ may be so bad and what can be done, if any, to improve them.

Let us explore the above-observed bad estimation of the availability likelihood $w(x)$. If we look at histograms of E-samples in rows 2-4 of Figure 5.3, then it is striking how small the empirical ranges of observations in the E-samples are. For instance, for the Normal case (the left diagram in the second row) there are no observations smaller than 0.15 and larger 0.75. In other words, from the E-sample nothing can be concluded for 40% of the range of X. This is what can be expected from a small sample when an underlying density is not separated from zero. Recall that the asymptotic theory assumes that in a Bernoulli regression the design density is separated from zero, and this assumption is obviously violated here. Similar poor outcomes may be observed for the two experiments with the Bimodal density (see the left diagrams in the two bottom rows). At the same time, the outcome is much better for the Uniform which is separated from zero. What can be done for densities that are not separated from zero? There are really only two options. The former one is to increase the sample size k of E-sample, and this may not be possible in some applications. The latter is to assume that $w(x)$ cannot be too wiggly and then decrease the number of estimated Fourier coefficients used by the Bernoulli regression E-estimator $\hat{w}(x)$. But overall we are dealing with an extremely complicated problem, and the reader is advised to spend some time exploring simulations created by Figure 5.3.

Figure 5.4 allows us to explore cases with densities bounded below from zero. Here the following mixture densities are used,

$$f_v^X(x) = v f_U^X(x) + (1 - v) f_i^X(x), \quad v \in [0, 1]. \tag{5.2.2}$$

In (5.2.2) $f_U^X(x)$ is the Uniform density and $f_i^X(x)$ is the ith corner density. Further, Figure 5.4 allows us to control parameters $cJ0$ and $cJ1$ of the regression E-estimator used to estimate the availability likelihood $w(x)$. The default parameters, used in Figure 5.4, imply that the E-estimator uses no more than three first Fourier coefficients. All other features are identical to Figure 5.3.

The first two experiments, shown in the top two rows of Figure 5.4, correspond to the mixture with the Normal, and the last two to the mixture with the Bimodal. As we see, the mixtures are separated from zero, and this increases the empirical range of E-samples shown by the histograms. With the exception of the first experiment (the top row of diagrams), the density E-estimates for E-samples are poor but do reflect the histograms. At the same time, recall that we need these estimates only to estimate the availability likelihood $w(x)$. The corresponding estimates $\hat{w}(x)$ are shown in the middle column. With the exception of the third experiment, the estimates of the availability likelihood are fair, and this is reflected in the final density E-estimates for M-samples. Even for the third experiment the final estimate is better than the one based on the E-sample, and in all other cases the ISEs are significantly smaller. Further, note how the density E-estimator corrects biased M-samples, and recall that without extra samples no consistent estimation of the density, based on an M-sample, is possible. Overall, we may conclude that the asymptotic theory does help us

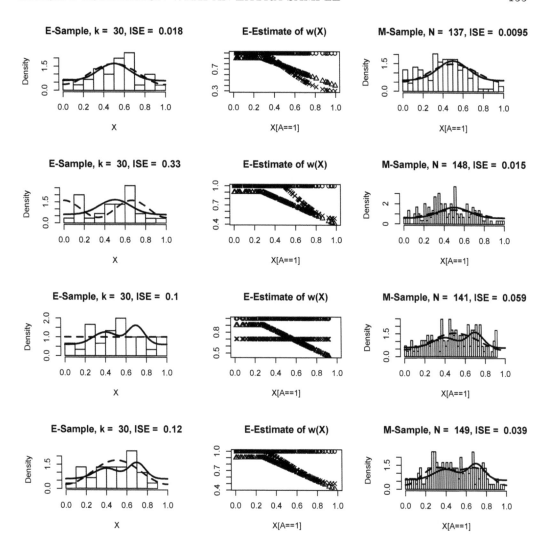

Figure 5.4 *E-estimation for a MNAR sample with extra sample. This figure is similar to Figure 5.3 but it allows one to choose underlying densities separated from zero and different parameters for the E-estimator of $w(x)$. The underlying density is the mixture of the Uniform (with weight v) and a corner density (with weight $1-v$). {New arguments are: v controls the weight in the mixture (5.2.2), setw.cJ0 and setw.cJ1 control parameters of the regression E-estimator of $w(x)$ in the 4 experiments.} [set.corn = c(2,2,3,3), n = 200, k = 30, v = 0.6, setw.cJ0 = c(2,2,2,2), setw.cJ1 = c(0,0,0,0), w = "1.1-0.7*x", dwU = 0.9, dwL = 0.3, cJ0 = 3, cJ1 = 0.8, cTH = 4]*

to understand performance of the E-estimator and why its assumptions are important. The reader is encouraged to repeat Figure 5.4, get used to the setting and the proposed solution, and try to use different parameters of the regression and density E-estimators.

Our final remark is about a possible aggregation of estimators based on E- and M-samples. Consider a mixture of two estimators of Fourier coefficients based on E- and M-samples,

$$\tilde{\theta}_j = \lambda\hat{\theta}_{Ej} + (1-\lambda)\hat{\theta}_{Mj}, \quad \lambda \in [0,1]. \tag{5.2.3}$$

Choice of the mixture coefficient λ is based on minimization of the variance of the estimator $\tilde{\theta}_j$. Then the aggregated estimator of Fourier coefficients can be used by the E-estimator. We will return to the aggregation later in Section 10.4.

5.3 Density Estimation with Auxiliary Variable

So far we have studied the case of an M-sample from pair (AX, A) where the availability A is a Bernoulli random variable with $\mathbb{P}(A = 1|X = x) = w(x)$ and $w(x)$ is not constant. This implies a destructive missing and impossibility of consistent estimation of the density $f^X(x)$ of X. Extra information is needed to restore the density, and one of the opportunities, considered in this section, is to find an always observable random variable Y (the auxiliary variable) which defines the missing mechanism.

Let us formally formulate a setting which utilizes such an opportunity. The M-sample is generated by a triplet (AX, A, Y) where Y is a continuous random variable which is referred to as the auxiliary variable. The main assumption is that, given (X, Y), the availability A depends only on Y, namely

$$\mathbb{P}(A = 1|X = x, Y = y) =: w(y) \geq c_0 > 0, \quad -\infty < y < \infty. \tag{5.3.1}$$

Note that given (5.3.1) the only interesting case is when X and Y are dependent because otherwise we are dealing with the MCAR. This is an important remark, and let us formally prove it. Suppose that X and Y are independent, (5.3.1) holds, and $f^X(x)$ is supported on $[0, 1]$. Then we can write for $x \in [0, 1]$,

$$\mathbb{P}(A = 1|X = x) = \frac{\int_{-\infty}^{\infty} f^{AX,A,Y}(x, 1, y)dy}{f^X(x)} = \frac{\int_{-\infty}^{\infty} w(y)f^{X,Y}(x, y)dy}{f^X(x)} =$$

$$= \frac{f^X(x)\int_{-\infty}^{\infty} w(y)f^Y(y)dy}{f^X(x)} = \int_{-\infty}^{\infty} w(y)f^Y(y)dy. \tag{5.3.2}$$

The integral on the right side of (5.3.2) does not depend on x and hence the missing is MCAR. We conclude that the only interesting case is when X and Y are dependent.

Now, following the E-estimation methodology, let us explain how we can estimate Fourier coefficients $\theta_j := \int_0^1 f^X(x)\varphi_j(x)dx$. First, we note that the joint mixed density of the triplet (AX, A, Y) is

$$f^{AX,A,Y}(ax, a, y) = [f^X(x)f^{Y|X}(y|x)w(y)]^a[(1 - w(y))f^Y(y)]^{1-a}, \tag{5.3.3}$$

where $x \in [0, 1]$, $y \in (-\infty, \infty)$ and $a \in \{0, 1\}$.

Second, suppose for a moment that the availability likelihood $w(y)$ is known. Then we may propose an estimator

$$\bar{\theta}_j := n^{-1} \sum_{l=1}^{n} [A_l \varphi_j(A_l X_l)/w(Y_l)]. \tag{5.3.4}$$

Let us show that (5.3.4) is a sample mean estimator. Write,

$$\mathbb{E}\{\bar{\theta}_j\} = \mathbb{E}\{A\varphi_j(AX)/w(Y)\}$$

$$= \int_0^1 f^X(x)\varphi_j(x)\left[\int_{-\infty}^{\infty} [f^{Y|X}(y|x)w(y)/w(y)]dy\right]dx = \theta_j, \tag{5.3.5}$$

where in the last equality we used $\int_{-\infty}^{\infty} f^{Y|X}(y|x)dy = 1$. This shows that (5.3.4) is the sample mean estimator and that it is unbiased.

Further, let us establish that the variance of $\bar{\theta}_j$ is

$$\mathbb{V}(\bar{\theta}_j) = n^{-1} \int_{-\infty}^{\infty} \frac{f^Y(y)}{w(y)} dy [1 + o_j(1)]. \tag{5.3.6}$$

To verify (5.3.6) we recall that $\mathbb{V}(Z) = \mathbb{E}\{Z^2\} - [\mathbb{E}\{Z\}]^2$, the variance of a sum of independent random variables is equal to the sum of variances of the variables, and that $\varphi_j^2(x) = 1 + 2^{-1/2}\varphi_{2j}(x)$. This allows us to write (recall the notation $w^{-k}(y) := 1/[w(y)]^k$),

$$\mathbb{V}(\bar{\theta}_j) = n^{-1}[\mathbb{E}\{A^2 \varphi_j^2(AX)w^{-2}(Y)\} - \theta_j^2]$$

$$= n^{-1} \int_{-\infty}^{\infty} f^Y(y)w^{-1}(y)dy + n^{-1}[2^{-1/2}E\{A^2\varphi_{2j}(AX)w^{-2}(Y)\} - \theta_j^2]$$

$$= n^{-1} \int_{-\infty}^{\infty} f^Y(y)w^{-1}(y)dy$$

$$+ n^{-1}\Big[2^{-1/2}\int_0^1 [\int_{-\infty}^{\infty} f^{Y|X}(y|x)w^{-1}(y)dy]f^X(x)\varphi_{2j}(x)dx - \theta_j^2\Big]. \tag{5.3.7}$$

Now note that due to (5.3.1) the inequality $\int_{-\infty}^{\infty} f^{Y|X}(y|x)w^{-1}(y)dy \le c_0^{-1}$ holds. Further, we always assume that $f^X(x)$ is square-integrable on $[0,1]$, and recall that according to Section 2.1 Fourier coefficients θ_j of a square-integrable $f^X(x)$ tend to zero as $j \to \infty$. These facts, together with (5.3.7), verify (5.3.6).

If the availability likelihood $w(y)$ is unknown, then it can be estimated from n available pairs $(A_1, Y_1), \ldots, (A_n, Y_n)$ using the E-estimator of Section 2.4 proposed for Bernoulli regression. Note that according to (5.3.4), we need to know an estimate of $w(y)$ only for $y = Y_l$ in complete cases (when $A_l = 1$). Another comment is about what to do if the auxiliary variable Y is not supported on $[0,1]$. Then we are using the rescaling technique explained in Chapter 2. First, observations Y_l, $l = 1, 2, \ldots, n$ are replaced by $Y_l' := (Y_l - Y_{(1)})/[Y_{(n)} - Y_{(1)}]$ where $Y_{(1)}$ and $Y_{(n)}$ are the smallest and the largest observations in the sample. Second, the Bernoulli regression E-estimator $\tilde{w}(y)$, $y \in [0,1]$ is calculated for the regression of A on Y'. Finally, the estimator is truncated from below because it is used in the denominator, that is, we set

$$\hat{w}(Y_l) := \max(c/\ln(n), \tilde{w}(Y_l')), \quad c > 0. \tag{5.3.8}$$

This is the estimator which is plugged in (5.3.4) and, as usual, c is the parameter of the estimator.

Combining the above-described steps, the proposed density E-estimator is based on the Fourier estimator,

$$\hat{\theta}_j := n^{-1} \sum_{l=1}^{n} [A_l \varphi_j(A_l X_l)/\hat{w}(Y_l)]. \tag{5.3.9}$$

Now we are in a position to test the proposed density E-estimator. We are going to consider two models for the auxiliary variable when $Y := \beta_0 + \beta_1 X + \sigma\varepsilon$ and $Y = \sigma X\varepsilon$ with ε being standard normal and independent of X. In the first case we have a classical linear relation, and in the second X defines the standard deviation of Y.

Figure 5.5 exhibits in its rows four simulations with $Y := \beta_0 + \beta_1 X + \sigma\varepsilon$, and its caption explains the diagrams. A diagram in the left column shows the scattergram of observations of (Y, A), the underlying likelihood function and its E-estimate by circles, triangles and crosses, respectively. Two top rows show independent simulations for the Normal density and two bottom ones for the Bimodal density. Due to the increasing availability likelihood function, we may expect an M-sample to be skewed to the right. And indeed, we see this

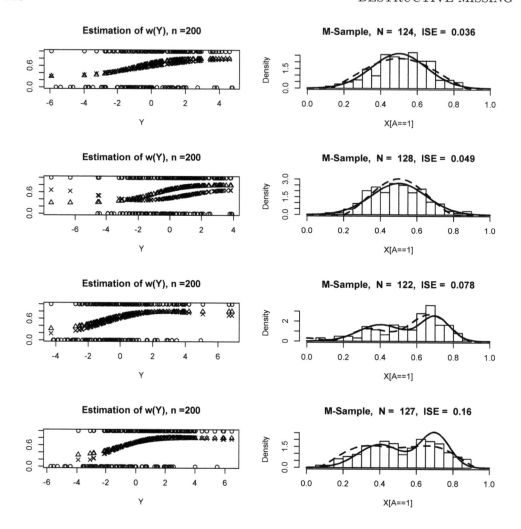

Figure 5.5 *Performance of the density E-estimator $\hat{f}^X(x)$ based on M-sample from (AX, A) and an extra sample from the auxiliary variable $Y = \beta_0 + \beta_1 X + \sigma\varepsilon$ where ε is standard normal. The variable Y defines the missing mechanism according to (5.3.1). Rows of diagrams show performance of the estimator for 4 different simulations. Circles show a scattergram of pairs (Y_l, A_l), $l = 1, \ldots, n$ while triangles and crosses show values of $w(Y_l)$ and $\hat{w}(Y_l)$ at points Y_l corresponding to $A_l = 1$. Complete cases in an M-sample are shown by the histogram while the solid and dashed lines show the underlying density f^X and its E-estimate, respectively. {Arguments set.beta = $c(\beta_0, \beta_1)$ and sigma define Y, all other arguments are the same as in Figure 5.4. [set.c = c(2,2,3,3), n = 200, c = 1, set.beta = c(0,0.3), sigma = 2, setw.cJ0 = c(3,3,3,3), setw.cJ1= c(0.3,0.3,0.3,0.3), w = "0.3+0.5*exp(1+y)/ (1+exp(1+y))", dwL = 0.3, dwU = 0.9, cJ0 = 3, cJ1 = 0.8, cTH = 4]*

in the first three simulations but not in the last one (look at the histogram in the right-bottom diagram). This is an interesting observation but not a surprise for the reader who did recommended simulations in Chapter 2. Indeed, the numbers of complete cases in the M-samples are close to 125, and we know that these sizes may produce peculiar samples that are far from expected ones.

After these remarks, let us look at particular outcomes beginning with the experiment shown in the top row. The E-estimate of the availability likelihood is pretty good. In its turn, it implies a fairly good, keeping in mind the size $N = 124$ of complete cases, E-

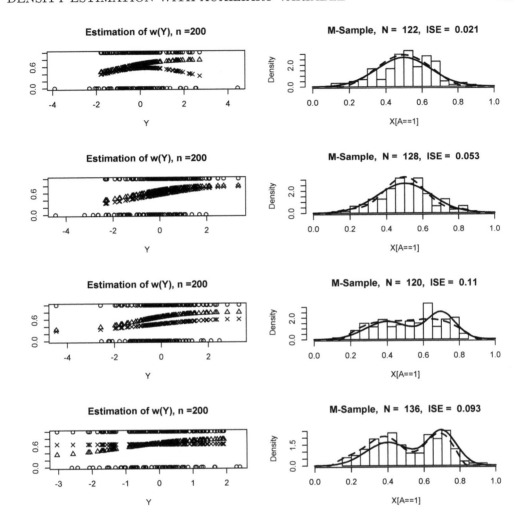

Figure 5.6 *Performance of the density E-estimator $\hat{f}^X(x)$ based on M-sample from (AX, A) and an extra sample from the auxiliary variable $Y = \sigma X \varepsilon$, where ε is standard normal. Everything else is identical to Figure 5.5. [set.c = c(2,2,3,3), n = 200, c = 1, setw.cJ0 = c(3,3,3,3), setw.cJ1 = c(0.3,0.3,0.3,0.3), sigma=2, w = "0.3+0.5*exp(1+y)/(1+exp(1+y))", dwL = 0.3, dwU = 0.9, cJ0 = 3, cJ1 = 0.8, cTH = 4]*

estimate of the Normal density. Note how the estimate corrects the biased histogram and indicates a symmetric and unimodal bell-shaped density. The next and absolutely similar simulation, shown in the second row, implies a poor estimate of $w(x)$, but is this the fault of the E-estimator? If we carefully look at the scattergram, shown by circles, we may note that the E-estimate does reflect the data at hand. Nonetheless, the poor estimate $\hat{w}(y)$ does not dramatically affect E-estimate of the Normal. Yes, it is worse than in the first simulation (compare the ISEs), but still it is symmetric and bell-shaped. The explanation of this robustness is that the particular $\hat{w}$ correctly shows the monotonicity of the underlying availability likelihood for a majority of points Y_l corresponding to $A_l = 1$.

In the last two experiments of Figure 5.5 with the Bimodal density, estimates of the availability likelihood are good, but the density estimates are different in their quality. The first one (it is the second from the bottom) is relatively good for $N = 122$ (compare with outcomes in Section 2.2 and conduct more simulations). At the same time, in the last experiment modes of the estimate have practically the same magnitudes and the estimate itself is worse (compare the ISEs). Is this the fault of the E-estimator? The answer is "no" because the estimator does exactly what it should do. Due to the biased data, the E-estimator increases the left mode and decreases the right one. In this particular M-sample, the right mode in the histogram is a bit taller than the left one, and this implies the E-estimate (the dashed line) with about the same magnitudes of the just barely pronounced modes.

Figure 5.6 allows us to explore the case $Y = 2X\varepsilon$ where ε is standard normal and independent of X, otherwise this figure and simulations are identical to Figure 5.5. The two top rows show us interesting outcomes for the Normal density. The first experiment yields a poor estimate of the availability likelihood (which is supported by the scattergram and note that the estimator knows only data), and nonetheless the density E-estimate is good for the case when only $N = 122$ observations are available in the MNAR sample. In the second experiment the situation is reversed. The availability likelihood estimate is good but the density estimate is poor (compare the ISEs). The reason for the latter is the M-sample exhibited by the histogram. We may conclude that the density E-estimator is robust toward the availability likelihood and, as we know, it always follows data in an M-sample. The same conclusion may be obtained from visualization of the two experiments for the Bimodal. We again see the importance of a "good" M-sample.

Overall, the conclusion is that in the case of a destructive missing it is prudent to search after an auxiliary variable which can explain the missing mechanism.

5.4   Regression with MNAR Responses

The underlying and hidden H-sample is $(X_1, Y_1), \ldots, (X_n, Y_n)$ from the pair (X, Y). Here X is the predictor with the design density $f^X(x)$ supported on $[0, 1]$ and $f^X(x) \geq c_* > 0$, $x \in [0, 1]$, and Y is the response which is a continuous variable that may be supported on $(-\infty, \infty)$. The available M-sample is $(X_1, A_1 Y_1, A_1), \ldots, (X_n, A_n Y_n, A_n)$ from the triplet (X, AY, A). Here A, called the availability, is a Bernoulli random variable defined by the availability likelihood

$$\mathbb{P}(A = 1 | X = x, Y = y) =: w(x, y) > c_0 > 0. \qquad (5.4.1)$$

It is assumed that $w(x, y)$ is not a constant in y, and hence we are dealing with the MNAR (missing not at random) case. The problem is to estimate the regression function

$$m(x) := E\{Y | X = x\}. \qquad (5.4.2)$$

The following example explains the setting. Suppose that we would like to predict the current salary of college graduates who graduated 5 years ago based on their GPA (grade point average). We may try to get data for n graduates, but it is likely that salaries of some graduates will not be known. What can be done in this case?

As we will see shortly, based solely on an M-sample it is impossible to suggest a consistent regression estimator, and the MNAR implies a destructive missing. Hence, an additional information is needed to restore information about regression function contained in an M-sample. Note a dramatic difference with the case of MAR responses, discussed in Section 4.2, when a complete-case approach yields optimal estimation of the regression function.

In this section a number of topics and issues are discussed, and it is convenient to consider them in corresponding subsections. We begin with the case $w(x, y) = w(y)$, explain why this implies the destructive missing, and then consider several possible sources of additional information that will allow us to consistently estimate the regression function. Then the general case of $w(x, y)$ is considered.

The case $w(\mathbf{x}, \mathbf{y}) = w(\mathbf{y})$. We are considering the setting when the likelihood of missing the response depends solely on its value. This is a classical MNAR, and let us show that no consistent estimation is possible in this case. The joint mixed density of the triplet (X, AY, A) is

$$f^{X,AY,A}(x, ay, a) = [f^{X,Y}(x,y)w(y)]^a [f^{X,A}(x,0)]^{a-1}$$

$$= [f^X(x)(f^{Y|X}(y|x)w(y))]^a \Big[\int_{-\infty}^{\infty} f^{X,Y,A}(x,y,0)dy\Big]^{a-1}$$

$$= [f^X(x)(f^{Y|X}(y|x)w(y))]^a [f^X(x)(1 - \int_{-\infty}^{\infty} f^{Y|X}(y|x)w(y)dy)]^{a-1}. \qquad (5.4.3)$$

Here $x \in [0,1]$, $y \in (-\infty, \infty)$, and $a \in \{0,1\}$. As we see, the joint density depends on the product $f^{Y|X}(y|x)w(y)$, and hence only this product or its functionals can be estimated based on an M-sample.

Inconsistency of a complete-case approach. As we know from Chapter 4, for the case of MAR responses the complete-case methodology implies a consistent and even optimal estimation. Let us check what may be expected if we use a complete-case approach for regression estimation in the MNAR case. Using (5.4.3) we can write,

$$\mathbb{E}\{AY|X, A=1\} = \int_{-\infty}^{\infty} y f^{AY|X,A}(y|x,1)dy = \frac{\int_{-\infty}^{\infty} y f^{X,AY,A}(x,y,1)dy}{f^{X,A}(x,1)}$$

$$= \frac{\int_{-\infty}^{\infty} y f^X(x)f^{Y|X}(y|x)w(y)dy}{f^X(x)\int_{-\infty}^{\infty} f^{Y|X}(y|x)w(y)dy} = \frac{\int_{-\infty}^{\infty} y f^{Y|X}(y|x)w(y)dy}{\int_{-\infty}^{\infty} f^{Y|X}(y|x)w(y)dy}. \qquad (5.4.4)$$

We conclude that the complete-case approach yields estimation of a regression function corresponding to the conditional density

$$f_*^{Y|X}(y|x) := \frac{f^{Y|X}(y|x)w(y)}{\int_{-\infty}^{\infty} f^{Y|X}(u|x)w(u)du}, \qquad (5.4.5)$$

rather than to the underlying $f^{Y|X}(y|x)$. Note that, as it could be expected, $f_*^{Y|X}$ is the biased conditional density.

Figure 5.7 complements this theoretical discussion by allowing us to look at simulations and appreciate performance of the complete-case approach. The two columns of diagrams correspond to two underlying regression functions, here the Normal and the Strata. A top diagram shows a hidden H-sample for a classical regression

$$Y_l = m(X_l) + \sigma\varepsilon_l, \quad l = 1, 2, \ldots, n \qquad (5.4.6)$$

where X_l are independent uniform random variables and ε_l are independent standard normal. The H-scattergram is shown by circles. The solid line shows the underlying regression and the dashed line shows the E-estimate. For the Normal regression the E-estimate is reasonable, it is skewed a bit and its mode is not as large as desired, but this is what the scattergram indicates and the E-estimate just follows the data. Overall the estimate is good and this conclusion is supported by the small ISE. For the Strata the E-estimate is not perfect but, as we know from previous simulations for the Strata, it is not bad either. Further, the relative magnitudes of the two modes are shown correctly.

Now let us look at the bottom diagrams in Figure 5.7 that show us MNAR samples. The MNAR is created by the availability likelihood $w(y) = \max(0.3, \min(0.9, 1-0.3y))$. Hence we are more likely to preserve smaller responses and to miss larger ones. The bottom diagrams show us complete pairs by circles and incomplete by crosses. Let us look at the left-bottom

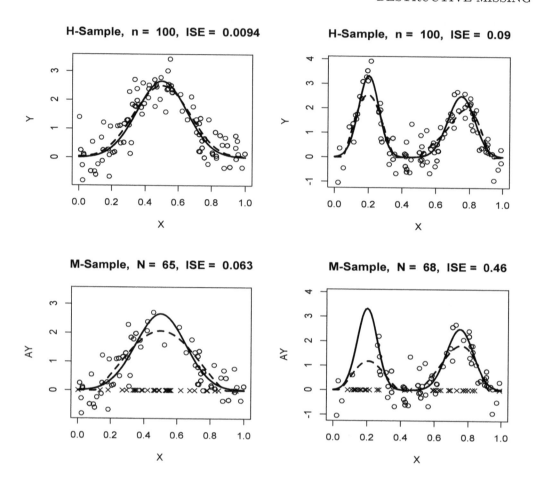

Figure 5.7 *Complete-case approach for regression with MNAR responses. The underlying regression is $Y = m(X) + \sigma\varepsilon$ where X is uniform and ε is independent standard normal. E-estimate for M-sample is based on complete cases that are shown by circles, while incomplete cases are shown by crosses. Underlying regression function and its E-estimate are shown by the solid and dashed lines, respectively. N is the number of complete cases in an M-sample. {The choice of underlying regression functions is controlled by the argument set.c. The availability likelihood is equal to* $\max(dwL, \min(dwU, w(y)))$ *and $w(y)$ is controlled by the string w.} [n = 100, set.c = c(2,4), sigma = 2, w="1-0.3*y", dwL = 0.3, dwU = 0.9, cJ0 = 3, cJ1 = 0.8, cTH = 4]*

diagram corresponding to the Normal regression. As it could be expected, a majority of missed responses are near the middle of the interval where the regression function is the largest. If we compare this diagram with the top one, then it is clear that a majority of responses near the mode are missed. As a result, the E-estimate (the dashed line), based on complete cases, shows the mode whose magnitude is dramatically smaller than magnitude of the mode in the underlying regression, and there is nothing that the complete-case approach can do about this. Further, look at how inhomogeneous the predictors in complete cases are, namely they are primarily located in the tails and there are practically no observations near the mode. This is what the MNAR does. At the same time, the symmetry is preserved, and this could be expected. Further, note that if parameter σ in (5.4.6) becomes smaller, the effect of the MNAR becomes less pronounced. It is worthwhile to think about this remark, and then repeat Figure 5.7 to get better understanding of the MNAR. Now let us look at the right-bottom diagram. Here we see a simply devastating effect of the MNAR on the left stratum where a majority of responses are missed. As a result, the E-estimate, based on complete cases, exhibits wrong magnitudes of the strata. Overall, there is nothing that can be done to improve the estimate unless extra information is available.

In conclusion, the considered MNAR is a complicated data modification, and it should not be taken lightly. It is prudent to know that no remedy exists if M-sample is the only information available. It is highly advisable to repeat Figure 5.7 with different parameters and learn about the MNAR and how destructive the missing may be.

Availability likelihood w(y) is known. Here we are continuing exploration of the MNAR with the availability likelihood $w(x, y) = w(y)$. As we know, extra information is needed for a consistent regression estimation, and in this subsection it is assumed that the availability likelihood $w(y)$ is known. To propose an E-estimator, we need to understand how to construct a sample mean estimator of Fourier coefficients θ_j of an underlying regression function (5.4.2). Let us write down a Fourier coefficient as an expectation,

$$\theta_j := \int_0^1 m(x)\varphi_j(x)dx$$

$$= \int_0^1 \mathbb{E}\{Y|X = x\}\varphi_j(x)dx = \mathbb{E}\left\{\frac{AY\varphi_j(X)}{f^X(X)w(AY)}\right\}. \tag{5.4.7}$$

Assume for a moment that the design density $f^X(x)$ of predictors is also known. Then we can suggest the following sample mean estimator,

$$\bar{\theta}_j := n^{-1}\sum_{l=1}^n \frac{A_l Y_l \varphi_j(X_l)}{f^X(X_l)w(A_l Y_l)}. \tag{5.4.8}$$

This estimator is unbiased. Let us also show that for the considered model (5.4.6) the variance of $\bar{\theta}_j$ is

$$\mathbb{V}(\bar{\theta}_j) = n^{-1}\int_{-\infty}^\infty\int_0^1 \frac{(m^2(x) + \sigma^2)f^{Y|X}(y|x)}{f^X(x)w(y)}dxdy[1 + o_j(1)]. \tag{5.4.9}$$

To verify (5.4.9) we recall that $\mathbb{V}(Z) = \mathbb{E}\{Z^2\} - [\mathbb{E}\{Z\}]^2$, the variance of a sum of independent random variables is equal to the sum of variances of the variables, and that $\varphi_j^2(x) = 1 + 2^{-1/2}\varphi_{2j}(x)$. This allows us to write,

$$\mathbb{V}(\bar{\theta}_j) = n^{-1}[\mathbb{E}\{A^2Y^2\varphi_j^2(X)[f^X(X)w(AY)]^{-2}\} - \theta_j^2]$$

$$= n^{-1}\int_{-\infty}^\infty\int_0^1 \frac{(m^2(x) + \sigma^2)f^{Y|X}(y|x)}{f^X(x)w(y)}dxdy$$

$$+ n^{-1}[2^{-1/2}E\{A^2(m^2(X) + \sigma^2)\varphi_{2j}(X)[f^X(X)w(Y)]^{-2}\} - \theta_j^2]. \tag{5.4.10}$$

Now recall that Fourier coefficients of a square-integrable function vanish as j increases, and this proves (5.4.9).

The variance (5.4.9) indicates that (5.4.8) is not an optimal Fourier estimator because it is possible to decrease the variance and remove the term $m^2(x)$. This issue was discussed in Section 2.3, and the remedy is to either subtract an estimate of $m(X_l)$ from Y_l or use a numerical integration. We skip this discussion here and will comment on it one more time in the Notes.

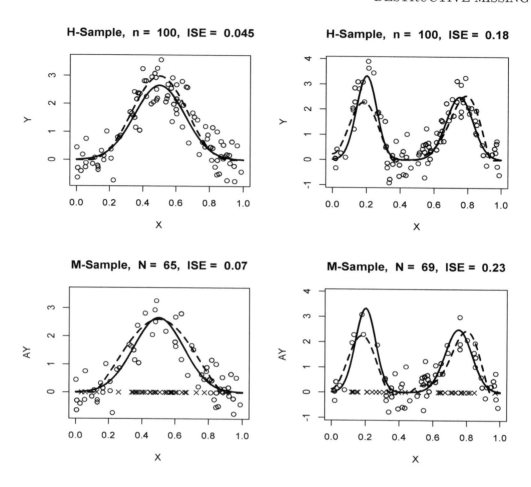

Figure 5.8 *Regression with MNAR responses when the availability likelihood w(y) is known. The underlying experiment and the structure of diagrams are the same as in Figure 5.7. [n = 100, set.c = c(2,4), sigma = 2, w = "1-0.3*y", dwL = 0.3, dwU = 0.9, c = 1, cJ0 = 3, cJ1 = 0.8, cTH = 4]*

The final remark is about the case of an unknown design density. Because all predictors are available, we can calculate the density E-estimator $\hat{f}^X(x)$ of Section 2.2, and then plug its truncated version $\max(\hat{f}^X(x), c/\ln(n))$ in (5.4.8).

Figure 5.8 shows how the proposed E-estimator performs. The simulation and the structure of diagrams is the same as in Figure 5.7. The only difference is that now we are using the E-estimator based on both M-sample and the known availability likelihood $w(y)$. As we see, despite the significantly smaller sizes N of available responses in M-samples, the E-estimates are relatively good. In the case of the Normal regression, the magnitude of the mode is shown correctly. Further, we know from Figure 5.7 that the case of the Strata regression is extremely complicated for the considered MNAR, and typically the MNAR implies that the left mode is significantly smaller than the right one. Here this is not the case, and actually the relative heights of the modes are shown better for the M-sample than for the H-sample.

The reader is advised to repeat Figure 5.8, use different parameters and realize that while the missing is MNAR, knowing the availability likelihood function allows us to restore information about a regression function contained in M-sample. Another important conclusion is that if we are able to estimate the availability likelihood function, then the MNAR is no longer destructive.

Estimation of availability likelihood w(y). As we already know, if availability likelihood $w(y)$ is unknown then no consistent regression estimation is possible based solely on an M-sample. One of the approaches to restore information about regression function from an M-sample is to use an extra sample that will allow us to estimate the availability likelihood and then convert the problem into the above-considered one. We discussed in the previous sections several possibilities that also may be used here.

One possibility is to utilize a more expansive sample from (Y, A) without missing. Suppose that we may get an extra sample (E-sample) $(Y_{n+1}, A_{n+1}), \ldots, (Y_{n+k}, A_{n+k})$ of size k which is smaller than n. Note that the E-sample alone cannot help us to estimate the regression because we do not know the predictors. On the other hand, we can use a Bernoulli regression to estimate the availability likelihood $w(y)$ and then plug the E-estimator $\hat{w}(y)$ in (5.4.8).

Another viable possibility, which is more feasible than the previous one, is to get an extra sample $Y_{n+1}, \ldots, Y_{n+k}$ from the response Y. Note that this E-sample cannot help us to estimate the regression function per se because the corresponding predictors are unknown, but it may allow us to estimate the underlying availability likelihood. Let us explain the proposed approach. First, the density of responses $f^Y(y)$ is estimated. Second, Bernoulli regression of A_l on $A_l Y_l$, based on complete cases in the M-sample, is considered. As we know from Sections 2.4 and 3.7, if the design density $f^Y(y)$ of predictors in a Bernoulli regression is known, then only "successes" may be used to estimate an underlying regression function. Finally, the estimated availability likelihood is used in (5.4.8).

We may conclude that whatever an extra opportunity exists to estimate $w(y)$, it should be explored because otherwise the M-sample is a pure loss due to the destructive MNAR.

Auxiliary variable defines the missing. Suppose that we can find an auxiliary and always observed variable Z which defines the missing, and let Z and Y be dependent. Further, the available M-sample is from the quartet (X, AY, A, Z). Then the missing model changes and it becomes MAR with the availability likelihood

$$\mathbb{P}(A = 1 | Y = y, X = x, Z = z) = \mathbb{P}(A = 1 | Z = z) =: w(z) \geq c_0 > 0. \qquad (5.4.11)$$

Let us explore regression estimation for this model assuming that Z, similar to X and Y, is a continuous random variable. The joint mixed density of the quartet for a complete case (when $A = 1$) may be written as

$$f^{X,AY,A,Z}(x, y, 1, z) = f^{X,Y,Z}(x, y, z)w(z)$$

$$= f^X(x)f^{Y|X}(y|x)f^{Z|X,Y}(z|x, y)w(z). \qquad (5.4.12)$$

This allows us to write down a Fourier coefficient of regression function (5.4.2) as an expectation. The following relations explain how this can be done. Write,

$$\theta_j = \int_0^1 m(x)\varphi_j(x)dx$$

$$= \int_0^1 \Big[\int_{-\infty}^{\infty} \int_{-\infty}^{\infty} y f^{Y|X}(y|x) \Big[\int_{-\infty}^{\infty} f^{Z|X,Y}(z|x, y)dz \Big] dy \Big] \varphi_j(x)dx$$

$$= \int_0^1 \int_{-\infty}^{\infty} \int_{-\infty}^{\infty} \frac{f^X(x)f^{Y|X}(y|x)f^{Z|X,Y}(z|x, y)w(z)y\varphi_j(x)}{f^X(x)w(z)} dz\,dy\,dx$$

$$= \mathbb{E}\Big\{ \frac{AY\varphi_j(X)}{f^X(X)w(Z)} \Big\}. \qquad (5.4.13)$$

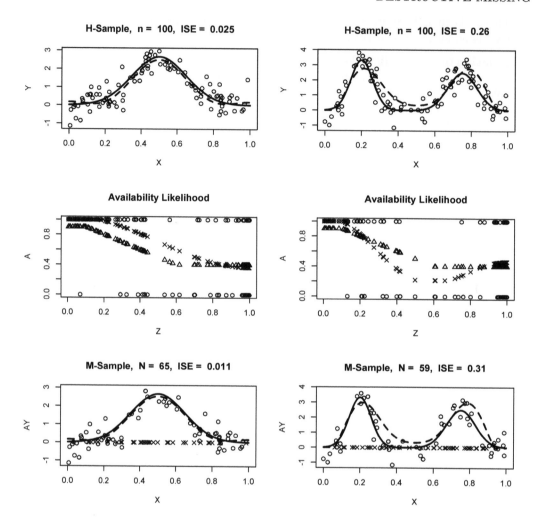

Figure 5.9 *Regression with missing responses when an auxiliary variable Z, which defines the missing, is available. The regression is $Y = m(X) + \sigma\varepsilon$ where X is uniform and ε is standard normal. The auxiliary variable is $Z = e^{1+a(Y-b)}/[1 + e^{1+a(Y-b)}]$. The availability likelihood $\mathbb{P}(A = 1|X = x, Y = y, Z = z) = \max(d_{wL}, \min(d_{wU}, w(z)))$ and $w(z) = 1 - z$. The structure of diagrams in the top and bottom rows is identical to those in Figure 5.8. In a middle diagram, circles show pairs (Z_l, A_l), $l = 1, 2, \ldots, n$ while triangles and crosses show values of the underlying availability likelihood and its E-estimate at points Z_l, $l = 1, 2, \ldots, n$. {The choice of underlying regression functions is controlled by set.c, a string w controls the choice of the availability likelihood, parameters a and b control the choice of Z, and c is used in the lower bound $c/\ln(n)$ for the density and availability likelihood E-estimates (recall that these estimates are plugged in the denominator.} [n = 100, set.c = c(2,4), sigma = 0.5, a = 3, b = 1, c = 1, w = "1-z", dwL = 0.4, dwU = 0.9, cJ0 = 3, cJ1 = 0.8, cTH = 4]*

Suppose that the design density $f^X(x)$ and the availability likelihood $w(z)$ are known, and $f^X(x) \geq c_* > 0$, $x \in [0,1]$. Then (5.4.13) implies the following sample mean Fourier estimator,

$$\bar{\theta}_j := n^{-1} \sum_{l=1}^{n} \frac{A_l Y_l \varphi_j(X_l)}{f^X(X_l) w(Z_l)}. \tag{5.4.14}$$

Because all observations of X are available, we can estimate the design density and plug its E-estimate in (5.4.14). Similarly, because all observations of the pair (Z, A) are available, the availability likelihood can be estimated via Bernoulli regression and also plugged in (5.4.14). Note that these estimates are used in the denominator, so we truncate them from below by $c/\ln(n)$. The Fourier E-estimator is constructed, and it yields the regression E-estimator $\hat{m}(x)$, $x \in [0, 1]$.

Figure 5.9 illustrates the setting. Here, similar to Figures 5.7 and 5.8, the top diagrams show regression for underlying H-samples. The middle diagrams show estimation of the availability likelihood $w(z)$ which is the regression function $\mathbb{E}\{A|Z = z\}$. Note that in both simulations the E-estimates $\hat{w}(z)$ are not good, but they do reflect the data. For instance in the right-middle diagram the E-estimate, shown by crosses, is well below the underlying availability likelihood, shown by triangles, in the middle of the unit interval. However, this behavior of the E-estimate is supported by the data. The bottom diagrams show us recovered regressions based on M-samples. Keeping in mind the small numbers N of available complete cases, the recovered regressions are good and comparable with E-estimates for H-samples.

Availability likelihood $\mathbf{w(x, y)}$. Let us consider the general case (5.4.1) when the availability likelihood depends on both X and Y. In other words, missing of the response depends on values of the predictor and the response. Of course, as we already know, in this case the missing is MNAR and M-sample alone does not allow us to suggest a consistent regression estimator. The open question is as follows. Suppose that the availability likelihood is known. Can the regression be estimated in this case?

To answer this question, consider the joint density

$$f^{X, AY, A}(x, y, 1) = f^{X,Y}(x, y)\mathbb{P}(A = 1|X = x, Y = y) = f^{X,Y}(x, y)w(x, y). \qquad (5.4.15)$$

Then a Fourier coefficient $\theta_j := \int_0^1 m(x)\varphi_j(x)dx$ of the regression function $m(x)$ may be written as

$$\theta_j = \int_0^1 \mathbb{E}\{Y|X = x\}\varphi_j(x)dx = \mathbb{E}\left\{\frac{AY\varphi_j(X)}{f^X(X)w(X, Y)}\right\}. \qquad (5.4.16)$$

Assume for a moment that the design density $f^X(x)$ is known. Then we can suggest the following sample mean estimator,

$$\bar{\theta}_j := n^{-1}\sum_{l=1}^n \frac{A_l Y_l \varphi_j(X_l)}{f^X(X_l)w(X_l, A_l Y_l)}. \qquad (5.4.17)$$

This estimator is unbiased, and for the model (5.4.6) its variance satisfies

$$\mathbb{V}(\bar{\theta}_j) = n^{-1}\int_{-\infty}^{\infty}\int_0^1 \frac{(m^2(x) + \sigma^2)f^{Y|X}(y|x)}{f^X(x)w(x, y)}dxdy(1 + o_j(1)). \qquad (5.4.18)$$

To verify (5.4.18) we recall that $\mathbb{V}(Z) = \mathbb{E}\{Z^2\} - [\mathbb{E}\{Z\}]^2$, the variance of a sum of independent random variables is equal to the sum of variances of the variables, and that $\varphi_j^2(x) = 1 + 2^{-1/2}\varphi_{2j}(x)$. This allows us to write,

$$\mathbb{V}(\bar{\theta}_j) = n^{-1}[\mathbb{E}\{A^2 Y^2 \varphi_j^2(X)[f^X(X)w(X, AY)]^{-2}\} - \theta_j^2]$$

$$= n^{-1}\int_{-\infty}^{\infty}\int_0^1 \frac{(m^2(x) + \sigma^2)f^{Y|X}(y|x)}{f^X(x)w(x, y)}dxdy$$

$$+ n^{-1}[2^{-1/2}E\{A^2(m^2(X) + \sigma^2)\varphi_{2j}(X)[f^X(X)w(X, Y)]^{-2}\} - \theta_j^2]. \qquad (5.4.19)$$

We know that Fourier coefficients of a square-integrable function vanish as j increases, and this verifies (5.4.18).

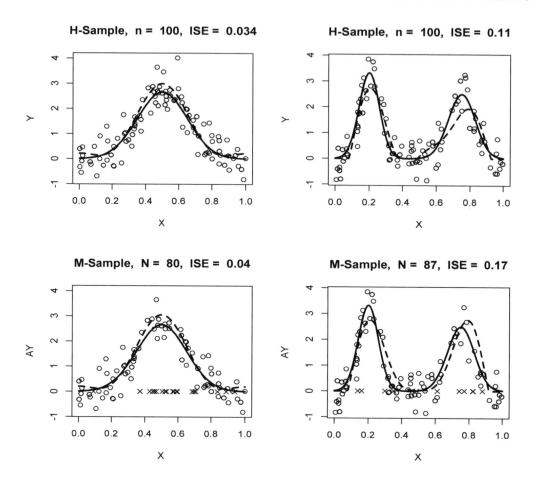

Figure 5.10 *Regression with MNAR responses when the availability likelihood function $w(x,y)$ is known. The underlying experiment and the structure of diagrams are the same as in Figure 5.7. {The availability likelihood is defined by a string w whose values are truncated from below by dwL and from above by dwU.} [n = 100, set.c = c(2,4), sigma = 2, w = "1-0.3*x*y", dwL = 0.3, dwU = 0.9, cJ0 = 3, cJ1 = 0.8, cTH = 4]*

If the design density $f^X(x)$ is unknown, it is estimated by the E-estimator and then plugged in (5.4.17).

Figure 5.10 illustrates the setting, here we use the same underlying model as in Figure 5.7 only with the availability likelihood $w(x,y) = \max(0.3, \min(0.9, 1 - 0.3xy))$. Let us look at the left column of diagrams corresponding to the Normal regression function. The underlying availability likelihood increases chances of missing responses corresponding to larger values of the product XY, and this pattern is clearly seen in the bottom diagram of the M-sample. Further, note that 20% of responses are missed. Nonetheless, the E-estimator does a good job in recovering the Normal regression function from the MNAR data. A similar pattern can be observed in the right column for the Strata regression. Here again the E-estimator does a good job in recovering the underlying regression.

The main challenge of the setting is that in general the availability likelihood $w(x,y)$ is unknown and some extra information should be provided, and this is where the main statistical issue arises. We need to estimate a bivariate function $w(x,y)$ and then use it for estimation of a univariate function $m(x)$. As we know from Section 2.5, estimation of a multivariate function is complicated by the curse of multidimensionality. In our setting this is a serious problem because the nuisance function $w(x, y)$ is bivariate while the function of interest $m(x)$ is univariate. As a result, if an extra sample of size k may be collected for estimation of $w(x, y)$, it is no longer possible to guarantee that k can be smaller in order than n.

5.5   Regression with MNAR Predictors

We observe an M-sample $(A_1 X_1, Y_1, A_1), \ldots, (A_n X_n, Y_n, A_n)$ of size n from the triplet (AX, Y, A). It is assumed that the predictor X is supported on $[0, 1]$, $f^X(x) \geq c_* > 0$ for $x \in [0, 1]$, in general the response Y is supported on $(-\infty, \infty)$, the availability A is a Bernoulli random variable, and the availability likelihood is

$$\mathbb{P}(A = 1 | Y = y, X = x) =: w(x, y) \geq c_0 > 0. \tag{5.5.1}$$

It is explicitly assumed that $w(x, y)$ depends on x and hence the missing mechanism is MNAR.

The problem is to estimate the regression function

$$m(x) := E\{Y | X = x\}, \quad x \in [0, 1]. \tag{5.5.2}$$

Of course, we can rewrite (5.5.2) in a more familiar way,

$$Y = m(X) + \sigma(X)\varepsilon, \tag{5.5.3}$$

where $E\{\varepsilon | X\} = 0$ and $E\{\varepsilon^2 | X\} = 1$ almost sure, and the regression error ε and the predictor X may be dependent.

To shed light on the problem, note that the joint mixed density of the triplet (AX, Y, A) is

$$f^{AX,Y,A}(ax, y, a) = [f^{X,Y}(x, y) w(x, y)]^a \left[\int_0^1 (1 - w(u, y)) f^{X,Y}(u, y) du \right]^{1-a}, \tag{5.5.4}$$

where $(x, y, a) \in [0, 1] \times (-\infty, \infty) \times \{0, 1\}$.

We may conclude that observations in a complete case, when $a = 1$, are biased and the biasing function is $w(x, y)$. Assuming that the availability likelihood $w(x, y)$ and the design density $f^X(x)$ are known, this remark allows us to propose an estimator of Fourier coefficients

$$\theta_j := \int_0^1 m(x) \varphi_j(x) dx \tag{5.5.5}$$

of the regression function. Indeed, using (5.5.4) we may write,

$$\theta_j = \mathbb{E}\{m(X) \varphi_j(X) [f^X(X)]^{-1}\}$$

$$= \mathbb{E}\{\mathbb{E}\{(Y | X)\} \varphi_j(X) [f^X(X)]^{-1}\}\} = \mathbb{E}\{AY \varphi_j(AX) [f^X(AX) w(AX, Y)]^{-1}\}. \tag{5.5.6}$$

Hence, the sample mean Fourier estimator is

$$\hat{\theta}_j := n^{-1} \sum_{l=1}^{n} \frac{A_l Y_l \varphi_j(A_l X_l)}{f^X(A_l X_l) w(A_l X_l, Y_l)}. \tag{5.5.7}$$

This Fourier estimator yields the regression E-estimator $\hat{m}(x)$. Note that only complete cases are used by the E-estimator.

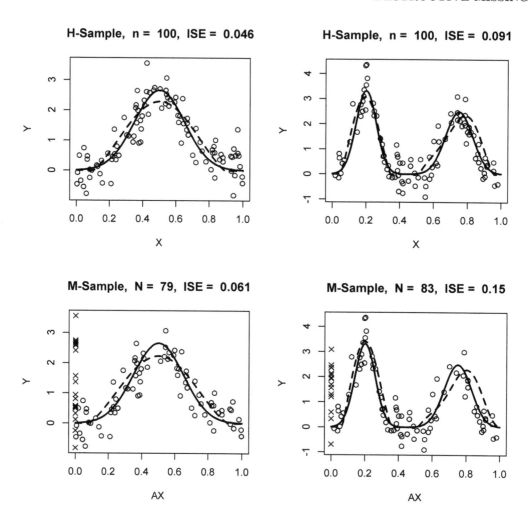

Figure 5.11 *Regression with MNAR predictors when the design density and the availability likelihood* $w(x, y)$ *are known. The underlying experiment and the structure of diagrams are the same as in Figure 5.10, only here predictors (and not responses) are missed with the same availability likelihood function.* [n = 100, set.c = c(2,4), sigma = 2, w = "1-0.3*x*y", dwL = 0.3, dwU = 0.9, cJ0 = 3, cJ1 = 0.8, cTH = 4]

Figure 5.11 illustrates the model and also shows how the E-estimator performs. Note that the underlying regression model is the same as in Figure 5.10, and the only difference is that here the predictors (and not the responses as in Figure 5.10) are missed. The latter allows us to understand and appreciate differences between the two settings. First of all, let us look at incomplete cases. It was relatively simple to recognize the pattern of $w(x, y)$ for the case of missing responses in Figure 5.10. Here, for the case of missing predictors, this is not as simple. For the case of the Normal regression function (the left column), only via comparison between the top and bottom diagrams, it is possible to recognize that the availability likelihood is decreasing in xy, and this explains the rather complicated structure of missing predictors. For the Strata (the right column of diagrams) this recognition may be simpler but again only if we compare M-sample with H-sample. In short, such an analysis is a teachable moment in recognizing the destructive nature of the MNAR. The proposed E-estimator performs relatively well.

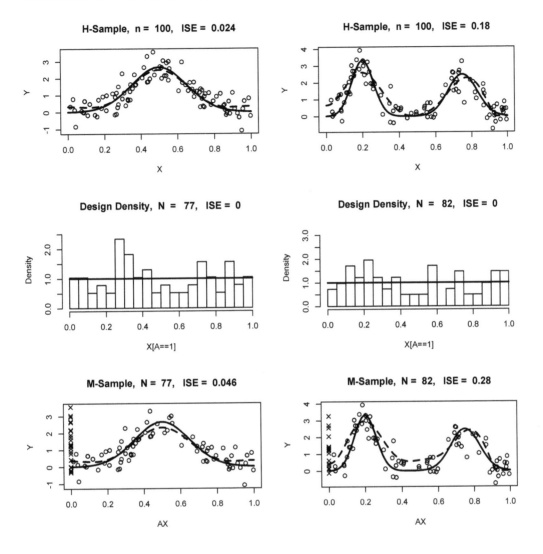

Figure 5.12 *Regression with MNAR predictors when the availability likelihood $w(x,y)$ is known. The underlying experiment and the structure of the top and bottom diagrams are the same as in Figure 5.11. The middle diagrams show histograms of available predictors and E-estimates of the design density. [n = 100, set.c = c(2,4), sigma = 2, w = " 1-0.3*x*y ", c=1, dwL = 0.3, dwU = 0.9, cJ0 = 3, cJ1 = 0.8, cTH = 4]*

Of course, the above-considered estimator used the underlying design density f^X, which in general is unknown. Is it possible to estimate it based on M-sample and an underlying availability likelihood? Let us explore this question.

To use our density E-estimator, we need to understand how to estimate Fourier coefficients

$$\kappa_j := \int_0^1 f^X(x)\varphi_j(x)dx \qquad (5.5.8)$$

of the design density $f^X(x)$, $x \in [0,1]$. Using (5.5.4) we may write that

$$\kappa_j = \int_0^1 [\int_{-\infty}^{\infty} f^{X,Y}(x,y)\varphi_j(x)dy]dx$$

$$= \int_0^1 \int_{-\infty}^{\infty} \frac{f^{AX,Y,A}(x,y,1)\varphi_j(x)}{w(x,y)} dydx = \mathbb{E}\left\{ A \frac{\varphi_j(AX)}{w(AX,Y)} \right\}. \qquad (5.5.9)$$

This allows us to define the sample mean estimator

$$\hat{\kappa}_j := n^{-1} \sum_{l=1}^{n} A_l \frac{\varphi_j(A_l X_l)}{w(A_l X_l, Y_l)}. \qquad (5.5.10)$$

Denote the corresponding density E-estimator as $\tilde{f}^X(x)$, and because it is used in denominator of (5.5.7), we plug in (5.5.7) its truncated version

$$\hat{f}^X(x) := \max(\tilde{f}^X(x), c/\ln(n)). \qquad (5.5.11)$$

As usual, it will be possible to control the choice of parameter c.

Figure 5.12 illustrates the considered setting and proposed E-estimator. As we see, while the underlying design density is uniform, the histograms of available predictors do not indicate this. The latter is explained by the fact that available predictors are biased. The particular design density E-estimates are perfect, and the corresponding regression E-estimates for the MNAR regressions are good.

It is advisable to repeat Figures 5.11 and 5.12 with different parameters and learn about the regression with MNAR predictors.

5.6 Missing Cases in Regression

So far we have considered a regression with incomplete cases where either response or predictor is missed. This type of missing is easily recognized via inspection of data and it urges for action. The situation changes if pairs (cases) are missed because then it is possible either to not realize that we are dealing with missing data or it may be tempting to ignore the missing. Further, when cases are missed, the missing cannot be MAR and it is either MCAR or MNAR. Further, if the availability likelihood is unknown, then the MNAR is destructive. The conclusion is that regression with missing cases is a challenging statistical problem that should not be taken lightly.

Let us formally describe the model and explain what can be done. We observe an M-sample $(A_1 X_1, A_1 Y_1, A_1), \ldots, (A_n X_n, A_n Y_n, A_n)$ of size n from the triplet (AX, AY, A). It is assumed that the predictor X is supported on $[0,1]$ and $f^X(x) \geq c_* > 0$ for $x \in [0,1]$, in general the response Y is supported on $(-\infty, \infty)$, the availability A is a Bernoulli random variable, and the availability likelihood is

$$\mathbb{P}(A = 1 | Y = y, X = x) =: w(x,y) \geq c_0 > 0. \qquad (5.6.1)$$

Note that unless the availability likelihood is constant, the missing mechanism is MNAR.

The problem is to estimate the regression function

$$m(x) := E\{Y | X = x\}, \quad x \in [0,1]. \qquad (5.6.2)$$

To propose a solution, we begin with analysis of the joint mixed density of the triplet (AX, AY, A). Write,

$$f^{AX,AY,A}(ax, ay, a) = [f^{X,Y}(x,y)w(x,y)]^a [1 - \mathbb{E}\{w(X,Y)\}]^{1-a}, \qquad (5.6.3)$$

where $(x,y,a) \in [0,1] \times (-\infty, \infty) \times \{0,1\}$. We may conclude that observations in a complete case, when $A = 1$, are biased and the biasing function is $w(x,y)$. As a result, while it may be tempting to simply ignore missing cases, this may lead to inconsistent estimation.

Let us assume that the availability likelihood $w(x, y)$ and the design density $f^X(x)$ are known. Following the E-estimation methodology, we need to propose a sample mean estimator of Fourier coefficients

$$\theta_j := \int_0^1 m(x)\varphi_j(x)dx \qquad (5.6.4)$$

of the regression function. Using (5.6.3) we may write,

$$\theta_j = \mathbb{E}\{m(X)\varphi_j(X)[f^X(X)]^{-1}\}$$

$$= \mathbb{E}\{\mathbb{E}\{(Y|X)\}\varphi_j(X)[f^X(X)]^{-1}\}\}$$

$$= \mathbb{E}\{AY\varphi_j(AX)[f^X(AX)w(AX, AY)]^{-1}\}. \qquad (5.6.5)$$

Hence, the sample mean Fourier estimator is

$$\hat{\theta}_j := n^{-1}\sum_{l=1}^n \frac{A_l Y_l \varphi_j(A_l X_l)}{f^X(A_l X_l)w(A_l X_l, A_l Y_l)}. \qquad (5.6.6)$$

This Fourier estimator yields the regression E-estimator $\hat{m}(x)$, $x \in [0, 1]$.

In general the design density is unknown and should be estimated. Here we again use our traditional approach of construction of a density E-estimator. Note that Fourier coefficients of $f^X(x)$ can be written as

$$\kappa_j := \int_0^1 f^X(x)\varphi_j(x)dx = \mathbb{E}\left\{A\frac{\varphi_j(AX)}{w(AX, AY)}\right\}. \qquad (5.6.7)$$

The last expression allows us to introduce the sample mean estimator

$$\hat{\kappa}_j := n^{-1}\sum_{l=1}^n A_l \frac{\varphi_j(A_l X_l)}{w(A_l X_l, A_l Y_l)}. \qquad (5.6.8)$$

Denote the corresponding density E-estimator as $\tilde{f}^X(x)$, and because it is used in denominator, we plug in (5.6.6) its truncated version

$$\hat{f}^X(x) := \max(\tilde{f}^X(x), c/\ln(n)). \qquad (5.6.9)$$

As usual, it will be possible to control the choice of the positive constant c.

Figure 5.13 illustrates the considered setting and the proposed regression E-estimator. Let us begin our analysis with the bottom diagrams that show us available M-samples. Note how easy it is to be confused and to not recognize the missing. Here only the used axis labels $X[A == 1]$ and $Y[A == 1]$ warn us about missing data. As we already know, it would be a serious mistake to ignore the missing because it inevitably yields inconsistent estimation. On the other hand, if the availability likelihood is known and the missing is taken into account, we can consistently estimate both the design density and the underlying regression function.

The particular simulations, exhibited in Figure 5.13, yield very respectful regression estimates for the M-samples. Note that even for the Strata regression, which is a complicated regression even for the case of direct observations, the outcome is good despite a small number $N = 83$ of available observations. Of course, the perfect E-estimates of design densities are helpful.

It is highly advisable to repeat Figure 5.13 with different parameters and get used to this practically important problem.

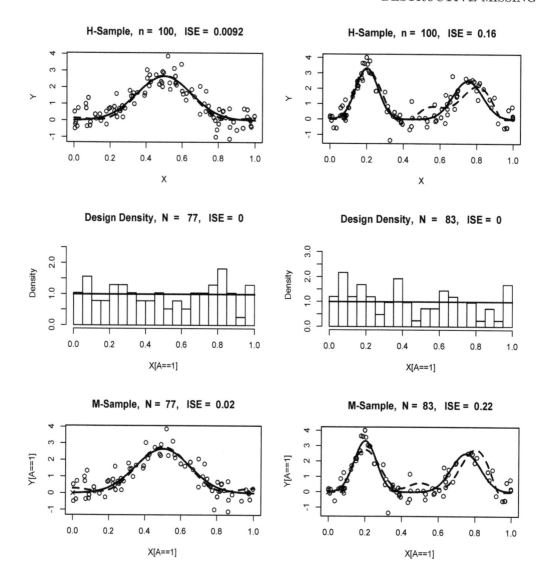

Figure 5.13 *Regression with MNAR cases when the availability likelihood $w(x,y)$ is known. The underlying regression experiment and the structure of diagrams are the same as in Figure 5.12. [n = 100, set.c = c(2,4), sigma = 2, w = "1-0.3*x*y", c = 1, dwL = 0.3, dwU = 0.9, cJ0 = 3, cJ1 = 0.8, cTH = 4]*

Our final remark is about the case when the bivariate availability likelihood $w(x,y)$ is unknown. This is a serious statistical problem with two possible approaches. The former is to conclude that a consistent estimation of the regression is impossible. The latter is to begin, if possible, collecting an extra information that will allow us to estimate the availability likelihood. This option was discussed in the previous section. The best case scenario is the presence of an auxiliary variable that defines the missing. In this case the MNAR is converted in MAR. A more complicated remedy is to obtain an extra sample that allows us to estimate $w(x,y)$. The challenge of this remedy is that we need to estimate a bivariate function, and due to the curse of multidimensionality, that estimation may require a relatively large, with respect to the sample size n, number of additional observations.

5.7   Exercises

5.1.1 Explain MCAR, MAR and MNAR missing mechanisms. Give corresponding examples.

5.1.2 Consider a continuous random variable X and a Bernoulli availability A. Explain why observing the pair (AX, A) and the single variable AX is equivalent. Also, explain the meaning of relation $A = I(AX \neq 0)$.

5.1.3 What is the availability likelihood?

5.1.4 Verify relation (5.1.3).

5.1.5 Consider the case of a not constant availability likelihood $w(x)$. Explain why knowing only an M-sample implies a destructive missing.

5.1.6* Consider the number N of available observations in an M-sample. Describe statistical characteristics of N including its mean, variance and the probability of large deviations.

5.1.7 Verify each relation in (5.1.7).

5.1.8 Explain how MNAR is related to biased data.

5.1.9 Repeat Figure 5.1 with different availability likelihoods that imply M-samples skewed to the left and to the right with respect to underlying H-samples. Explain histograms.

5.1.10 Explain the E-estimator used in Figure 5.2. Why is consistent estimation possible for the MNAR?

5.1.11* Find the variance of estimator (5.1.8).

5.1.12* Find the MISE of the density E-estimator based on Fourier estimator (5.1.8).

5.1.13 Using Figure 5.2, explore how different shapes of the availability likelihood affect density estimation.

5.1.14 Explain how $w(x)$ affects the number N of available observations.

5.1.15 Using Figure 5.2, propose better parameters of the E-estimator.

5.1.16 Explain why knowing availability likelihood makes the MNAR nondestructive.

5.1.17 What is the shape of $w(x)$ that is more damaging for estimation of the Strata density?

5.2.1 Explain why an extra sample may be needed for recovery density from MNAR data.

5.2.2 What is the idea of using an extra sample? Is it simpler to repeat sampling without missing observations? Hint: Pay attention to the fact that the size of extra sample may be in order smaller than the size n of an M-sample.

5.2.3* Why can the size of extra sample be smaller in order than n?

5.2.4 Present several examples where, for an additional price, it is possible to get an extra sample with no missing observations.

5.2.5 Explain the methodology of E-estimation for the case of an extra-sample.

5.2.6 Explain the estimator (5.2.1).

5.2.7* What is the variance of estimator (5.2.1)?

5.2.8 What is the role of parameter c in estimator (5.2.1)?

5.2.9* Evaluate the MISE of the density E-estimator based on the Fourier estimator (5.2.1).

5.2.10 Explain the simulation used in Figure 5.3.

5.2.11 E-estimates of $w(x)$, exhibited in Figure 5.3, are not satisfactory. Explain why and what can be done to improve them.

5.2.12 Using Figure 5.3, suggest better parameters of E-estimator.

5.2.13 Explain the underlying simulation used in Figure 5.4.

5.2.14 What is the difference, if any, between Figure 5.3 and Figure 5.4?

5.2.15* Explore variance of the aggregated estimator (5.2.3).

5.2.16* Suggest parameter λ which minimizes variance of estimator (5.2.3).

5.3.1 MNAR typically implies a destructive missing. Explain how an auxiliary variable can help in restoring information contained in M-sample.

5.3.2 Explain the relation (5.3.1) and its meaning.

5.3.3 Consider the case of MNAR, and suppose that (5.3.1) is valid. Can X and Y be independent?

5.3.4 Verify each relation in (5.3.2).

5.3.5 Verify (5.3.3).

5.3.6 Consider the case of a known $w(y)$ and propose a density E-estimator.

5.3.7 Explain the underlying idea of the estimator (5.3.4).

5.3.8 Show that (5.3.4) is unbiased estimator.

5.3.9* Find the variance of estimator (5.3.4).

5.3.10 Verify every equality in (5.3.5).

5.3.11 Verify every relation in (5.3.7).

5.3.12 Prove inequality $\int_{-\infty}^{\infty} f^{Y|X}(y|x)w^{-1}(y)dy < \infty$.

5.3.13* Explain how $w(y)$ may be estimated.

5.3.14* Evaluate the mean squared error of estimator $\hat{w}(y_0)$ defined in (5.3.8).

5.3.15 Explain the underlying idea of Fourier estimator (5.3.9).

5.3.16* Evaluate the mean of estimator (5.3.9).

5.3.17* Evaluate the variance of the estimator (5.3.9). Hint: Begin with the case of given $w(y)$.

5.3.18 Explain and comment on two models of the auxiliary variable Y as a function in X used in the simulations.

5.3.19 Explain the underlying simulation used in Figure 5.5.

5.3.20 How does sample size n affect estimation of the density?

5.3.21 Using Figure 5.5, explain how parameters of the linear model for Y affect the estimation.

5.3.22 Using Figure 5.5, consider all four corner functions and comment on complexity of their estimation.

5.3.23 Using Figure 5.5, propose better parameters of the E-estimator.

5.3.24 Explain the underlying simulation used in Figure 5.6.

5.3.25 Using Figure 5.6, explore how the sample size n affects the estimation.

5.3.26 Explore how parameter σ of Figure 5.6 affects estimation, and complement a numerical study by a theoretical argument. Hint: Use empirical ISE and theoretical MISE.

5.3.27 Consider all four corner functions and comment on their estimation using Figure 5.6.

5.3.28 Find better parameters of E-estimator for Figure 5.6.

5.3.29 Compare performance of E-estimator for models used in Figures 5.5 and 5.6.

5.4.1 Explain the aim of a nonparametric regression.

5.4.2 What is the definition of a nonparametric regression?

5.4.3 Formulate the model of nonparametric regression with MNAR responses.

5.4.4 What is the availability likelihood for MAR regression with missing responses?

5.4.5 What is the availability likelihood for MNAR regression with missing responses?

5.4.6* Explain why MNAR may imply destructive missing but MAR may not.

5.4.7* Describe several possible scenarios when MNAR does not imply destructive missing and a consistent estimation is possible. Explain why.

5.4.8 Explain a possibility of consistent estimation when $w(x,y) = w(y)$. Further, is the case $w(x,y) = w(x)$ of interest here?

5.4.9 Explain every equality in (5.4.3).

5.4.10 Consider the case $w(x,y) = w(y)$, and explain what functions can and cannot be consistently estimated using M-sample.

5.4.11* For the case $w(x,y) = w(y)$, propose a consistent estimator of the product $f^{Y|X}(y|x)w(y)$.

5.4.12 Explore using a complete-case approach for the MNAR with $w(x,y) = w(y)$.

5.4.13 Verify relations in (5.4.4).

5.4.14 Explain why (5.4.5) is the pivot for understanding a complete-case approach.

5.4.15 What is the underlying simulation in Figure 5.7?

5.4.16 What can be concluded from E-estimates exhibited in Figure 5.7?

5.4.17 Using Figure 5.7, explain how the availability likelihood affects M-sample, then complement your conclusion by a theoretical analysis.

5.4.18 Can an adjustment of parameters of the E-estimator improve E-estimation based on complete cases?

5.4.19 Explain all arguments of Figure 5.7.

5.4.20* Assume that the availability likelihood $w(y)$ is known. Propose density E-estimators for $f^X(x)$, $f^Y(y)$, $f^{Y|X}(y|x)$ and $f^{X,Y}(x,y)$.

5.4.21 Prove (5.4.7).

5.4.22 Explain the idea of Fourier estimator (5.4.8). Describe assumptions.

5.4.23 Show that (5.4.8) is an unbiased estimator.

5.4.24* Evaluate the variance of Fourier estimator (5.4.8). Then explain how $w(y)$ affects the mean squared error of the estimator.

5.4.25* It was explained in Section 2.3 how to decrease the variance of a sample mean estimator when the variance depends on an underlying regression function $m(x)$. Use that approach and propose a modification of (5.4.8) with a smaller variance.

5.4.26 Verify each equality in (5.4.10).

5.4.27 Explain the underlying simulation used in Figure 5.8.

5.4.28 What are the assumptions used by the E-estimator in Figure 5.8?

5.4.29 Repeat Figure 5.8 for several sample sizes and comment on the effect of the sample size on quality of estimation.

5.4.30 Using Figure 5.8, explore the effect of availability likelihood on estimation of all four corner regression functions.

5.4.31 For each corner regression function, suggest better parameters of the E-estimator used in Figure 5.8. Then compare your suggestions and comment on your findings.

5.4.32 Explain the underlying idea of using an auxiliary variable to overcome destructive missing caused by the MNAR.

5.4.33 Explain (5.4.11).

5.4.34 Under the MNAR, is it reasonable to assume that the auxiliary Z and the response Y are independent?

5.4.35 Verify (5.4.12), and explain the used assumptions.

5.4.36 Prove every equality in (5.4.13) and explain the necessity of used assumptions.

5.4.37 Explain the underlying idea of Fourier estimator (5.4.14).

5.4.38* Evaluate the mean and variance of estimator (5.4.14).

5.4.39 Explain the simulation used in Figure 5.9.

5.4.40 Does the parameter σ affect the missing mechanism?

5.4.41* Explain each step of E-estimation used in Figure 5.9.

5.4.42 Using Figure 5.9, comment on how $w(z)$ affects estimation of regression functions.

5.4.43 Explain all parameters of Figure 5.9.

5.4.44 Present several examples of MNAR missing when the availability likelihood depends on both the predictor and the response.

5.4.45 Verify (5.4.15).

5.4.46 Prove (5.4.16).

5.4.47 Explain the underlying idea of Fourier estimator (5.4.17).

5.4.48 Prove, or disprove, that (5.4.17) is an unbiased estimator. Mention used assumptions.

5.4.49* Evaluate the variance of estimator (5.4.17). Explain used assumptions.

5.4.50* Variance of estimator (5.4.17) depends on regression function. Propose a modification of the estimator which removes the dependence and makes the variance smaller. Hint: Recall Section 2.3.

5.4.51* Suppose that the assumption $f^X(x) \geq c_* > 0$, $x \in [0,1]$ is violated. Can the estimator (5.4.17) be recommended in this case?

5.4.52 Verify every equality in (5.4.19).

5.4.53 Explain how E-estimator of the design density is constructed.

5.4.54 Explain the simulation used in Figure 5.10.

5.4.55 Repeat Figure 5.10 and analyze simulated data and estimates.

5.4.56 Suggest better parameters of the E-estimator used in Figure 5.10.

5.4.57 Using bottom diagrams in Figure 5.10, try to figure out an underlying availability likelihood function $w(x,y)$.

5.4.58* Consider the case of unknown $w(x,y)$ and propose a reasonable scenario when its estimation is possible.

5.5.1 Explain models of nonparametric regression with MAR and MNAR predictors. Hint: Think about appropriate availability likelihoods.

5.5.2 Explain MNAR model (5.5.1). Give several possible examples.

5.5.3 Explain a connection between models (5.5.2) and (5.5.3).

5.5.4 Verify (5.5.4).

5.5.5 Can a complete-case approach imply a consistent estimation?

5.5.6 Prove every equality in (5.5.6).

5.5.7 What is the underlying idea of estimator (5.5.7)?

5.5.8 Is estimator (5.5.7) unbiased? Formulate assumptions needed for validity of your assertion.

5.5.9* Evaluate variance of estimator (5.5.7).

5.5.10* Variance of estimator (5.5.7) depends on an underlying regression function. Suggest a modified estimator that asymptotically has no such dependence and also has a smaller variance.

5.5.11* Explore a design density that minimizes variance of estimator (5.5.7).

5.5.12 Explain the underlying experiment used in Figure 5.11.

5.5.13 Propose better values for parameters of the E-estimator used in Figure 5.11.

5.5.14 Repeat Figure 5.11 for other corner functions and analyze diagrams and estimators.

5.5.15 Explain the motivation of estimator (5.5.10).

5.5.16* Find the mean and variance of estimator (5.5.10).

5.5.17 Explain how E-estimator of the design density is constructed.

5.5.18 What is the simulation used in Figure 5.12?

5.5.19 Explain the difference between Figures 5.11 and 5.12.

5.5.20 Using available ISEs, explore how accuracy in estimation of the design density affects the regression estimation. Hint: Use Figures 5.11 and 5.12.

5.5.21 Explain all parameters of E-estimator used in Figure 5.12.

5.6.1 Present examples of regression data where cases are missed. In other words, when either cases are complete or contain no data.

5.6.2 Describe observations in a regression model with missed cases.

5.6.3 Why is it easier to ignore complete-case missing than missing responses?

5.6.4 Can the studied missing mechanism be MAR?

5.6.5 Prove relation (5.6.3).

5.6.6 Verify (5.6.5).

5.6.7 Explain the motivation of estimator (5.6.6).

5.6.8* Find the mean and variance of estimator (5.6.6).

5.6.9* Suggest a modification of estimator (5.6.6) with smaller variance. Hint: Recall Section 2.3.

5.6.10* Explain how the design density may be estimated.

5.6.11* Find the mean and variance of estimator (5.6.8). Formulate used assumptions.

5.6.12* Propose several possible scenarios when $w(x,y)$ may be estimated.

5.6.13 What is the difference between simulations used in Figures 5.12 and 5.13?

5.6.14* Explain the difference, if any, between E-estimators used in Figures 5.12 and 5.13.

5.6.15 Repeat Figure 5.13 and, using available ISEs, explore how accuracy in estimation of the design density affects the regression estimation.

5.6.16 Suggest better parameters of the E-estimator used in Figure 5.13.

5.6.17 Explain all parameters used in Figure 5.13.

5.6.18 Explore, both theoretically and using Figure 5.13, the effect of the availability likelihood on estimation of the Normal and the Bimodal regression functions.

5.8   Notes

Missing mechanisms, implying MNAR, are considered in many books including Little and Rubin (2002), Tsiatis (2006), Molenberghs and Kenward (2007), Enders (2010), Molenberghs et al. (2014) and Raghunathan (2016).

5.1 Efficiency of the E-estimation for nonparametric density estimation is well known and the first results are due to Efromovich and Pinsker (1982) and Efromovich (1985; 2009a; 2010a,c). Sequential estimation is discussed in Efromovich (1989, 1995b, 2015). Multivariate E-estimation methodology, including the case of anisotropic densities, is discussed in Efromovich (1994b, 1999a, 2000b, 2002, 2011c).

5.2 Estimation of the density based on indirect observations and optimality of the E-estimation is discussed in Efromovich (1994c). It is known from Efromovich (2001b) that a plug-in E-estimation procedure improves the classical empirical cumulative distribution function. Further, according to Efromovich (2004c), a similar result holds for the case of biased data. These asymptotic results indicate that a similar assertion may be proved for the case of destructive missing with an extra sample.

5.3 Thresholding as an adaptive method, used by the E-estimator, is discussed in Efromovich (1995a, 1996b, 2000a). An interesting extension of the considered setting is, following Efromovich and Low (1994; 1996a,b), to consider estimation of functionals of the density. Overall, it may be expected that a plug-in procedure will yield optimal estimation. Another interesting area of research is the effect of measurement errors, see Efromovich (1997a) as well as the related Efromovich (1997b) and Efromovich and Ganzburg (1999). Is the proposed density E-estimation simultaneously optimal for the density derivatives? The result of Efromovich (1998c) points in this direction. The expansion to multivariate densities, following Efromovich (2000b, 2011c), is another interesting problem to consider.

5.4 For regression, a possibility of reducing the variance of a sample mean Fourier estimator is discussed in Efromovich (1996a, 1999a, 2005a, 2007d, 2013a), Efromovich and Pinsker (1996) and Efromovich and Samarov (2000). Two main approaches are subtracting a consistent regression estimate from the response and mimicking a numerical integration. These approaches yield efficient estimation of Fourier coefficients of a regression function. Following Efromovich (1996a), it is possible to extend the considered setting to a larger class of regression setting and prove asymptotic efficiency of the proposed E-estimation methodology.

5.5 Sequential estimation, for the considered model of MNAR predictors, is an interesting opportunity. Here results of Efromovich (2007d,e,i; 2008a,c; 2009c; 2012b) shed light on possible asymptotic results. Another interesting extension of the considered setting is, following Efromovich and Low (1994; 1996a,b), to consider estimation of functionals. Overall, it may be expected that a plug-in procedure will yield optimal estimation. It is a special and interesting new topic to explore the case of dependent observations, and then get results similar to Efromovich (1999c).

5.6 One of the interesting and new topics is developing the asymptotic theory of equivalence between the regression with missing data and the model of filtering a signal from the white Gaussian noise, see a discussion in Efromovich (1999a). Even more attractive will be results on the limits of the equivalence and how the missing affects those limits, see Efromovich and Samarov (1996), Efromovich and Low (1996a) and Efromovich (2003a). Interesting topic of uncertainty analysis is discussed in Shaw (2017). Multivariate regression with different smoothness of the regression function in covariates, called anisotropic regression, is another interesting setting to consider. While the problem is technically challenging, following Efromovich (2000b, 2002, 2005a) it is reasonable to conjecture that the E-estimation methodology will yield efficient estimation. See also an interesting discussion in Harrell (2015) where further references can be found. In a number of applications the scale function depends on auxiliary covariates, and in this case a traditional regression estimator may be improved as shown in Efromovich (2013a,b).

Chapter 6

Survival Analysis

Survival analysis traditionally focuses on the analysis of time duration until one or more events happen and, more generally, positive-valued random variables. Classical examples are the time to death in biological organisms, the time from diagnosis of a disease until death, the time between administration of a vaccine and development of an infection, the time from the start of treatment of a symptomatic disease and the suppression of symptoms, the time to failure in mechanical systems, the length of stay in a hospital, duration of a strike, the total amount paid by a health insurance, the time to getting a high school diploma. This topic may be called reliability theory or reliability analysis in engineering, duration analysis or duration modeling in economics, and event history analysis in sociology. Survival analysis attempts to answer questions such as: what is the proportion of a population which will survive past a certain time? Of those that survive, at what rate will they die or fail? Can multiple causes of death or failure be taken into account? How do particular circumstances or characteristics increase or decrease the probability of survival?

To answer such questions, it is necessary to define the notion of "lifetime". In the case of biological survival, death is unambiguous, but for mechanical reliability, failure may not be well defined, for there may well be mechanical systems in which failure is partial, a matter of degree, or not otherwise localized in time. Even in biological problems, some events (for example, heart attack or other organ failure) may have the same ambiguity. The theory outlined below assumes well-defined events at specific times; other cases may be better treated by models which explicitly account for ambiguous events.

While we are still dealing with a random variable X, that may be characterized by its cumulative distribution function $F^X(x)$ (also often referred to as the lifetime distribution function), because survival analysis is primarily interested in the time until one or more events, the random variable is assumed to be nonnegative (it is supported on $[0, \infty)$) and it is traditionally characterized by the survival function $G^X(x) := \mathbb{P}(X > x) = 1 - F^X(x)$. That is, the survival function is the probability that the time of death is later than some specified time x. The survival function is also called the survivor function or survivorship function in problems of biological survival, and the reliability function in mechanical survival problems. Usually one assumes $G(0) = 1$, although it could be less than 1 if there is the possibility of immediate death or failure. The survival function must be nonincreasing: $G^X(u) \geq G^X(t)$ if $u \geq t$. This reflects the notion that survival to a later age is only possible if all younger ages are attained. The survival function is usually assumed to approach zero as age increases without bound, i.e., $G(x) \to 0$ as $x \to \infty$, although the limit could be greater than zero if eternal life is possible. For instance, we could apply survival analysis to a mixture of stable and unstable carbon isotopes; unstable isotopes would decay sooner or later, but the stable isotopes would last indefinitely.

Typically, survival data are not fully and/or directly observed, but rather censored and the most commonly encountered form is right censoring. For instance, suppose patients are followed in a study for 12 weeks. A patient who does not experience the event of interest for the duration of the study is said to be right censored. The survival time for this person is considered to be at least as long as the duration of the study. Another example of right censoring is when a person drops out of the study before the end of the study observation time and did not experience the event. This person's survival time is said to be right censored, since we know that the event of interest did not happen while this person was under observation. Censoring is an important issue in survival analysis, representing a particular type of modified data.

Another important modification of survival data is truncation. For instance, suppose that there is an ordinary deductible D in an insurance policy. The latter means that if a loss occurs, then the amount paid by an insurance company is the loss minus deductible. Then a loss less than D may not be reported, and as a result that loss is not observable (it is truncated). Let us stress that truncation is an illusive modification because there is nothing in truncated data that manifests about the truncation, and only our experience in understanding of how data were collected may inform us about truncation. Further, to solve a statistical problem based on truncated observations, we need to know or request corresponding observations of the truncating variable.

It also should be stressed that censoring and/or truncation may preclude us from consistent estimation of the distribution of a random variable over its support, and then a feasible choice of interval of estimation becomes a pivotal part of statistical methodology. In other words, censoring or truncation may yield a destructive modification of data.

As it will be explained shortly, survival analysis of truncated and censored data requires using new statistical approaches. The main and pivotal one is to begin analysis of an underlying distribution not with the help of empirical cumulative distribution function or empirical density estimate but with estimation of a hazard rate which plays a pivotal role in survival analysis. For a continuous lifetime X, its hazard rate (also referred to as the failure rate or the force of mortality) is defined as $h^X(x) := f^X(x)/G^X(x)$. Similarly to the density or the cumulative distribution function, the hazard rate characterizes the random variable. In other words, knowing hazard rate implies knowing the distribution. As we will see shortly, using a hazard rate estimator as the first building block helps us in solving a number of complicated problems of survival analysis including dealing with censored and truncated data.

Estimation of the hazard rate has a number of its own challenges, and this explains why it is hardly ever explored for the case of direct observations. The main one is that the hazard rate is not integrated over its support (the integral is always infinity), and then there is an issue of choosing an appropriate interval of estimation. Nonetheless, estimation of the hazard rate (in place of more traditional characteristics like density or cumulative distribution function) becomes more attractive for truncated and/or censored data. Further, as it was mentioned earlier, truncation and/or censoring may preclude us from estimation of the distribution over its support, and then the problem of choosing a feasible interval of estimation becomes bona fide. Further, the hazard rate approach allows us to avoid using product-limit estimators, like a renowned Kaplan–Meier estimator, that are the more familiar alternative to the hazard rate approach. We are not using a product-limit approach because it is special in its nature, not simple for statistical analysis, and cannot be easily framed into our E-estimation methodology. At the same time, estimation of the hazard rate is based on the sample mean methodology of our E-estimation.

The above-presented comments explain why the first four sections of the chapter are devoted to estimation of the hazard rate for different types of modified data. We begin with a classical case of direct observations considered in Section 6.1. It introduces the notion of the hazard rate, explains that the hazard rate, similarly to the density or the cumulative distribution function, characterizes a random variable, and explains how to construct hazard rate E-estimator. Right censored (RC), left truncated (LT), and left truncated and right censored (LTRC) data are considered in Sections 6.2-6.4, respectively. Sections 6.5-6.7

discuss estimation of the survival function and the density. Sections 6.8 and 6.9 are devoted to nonparametric regression with censored data.

In what follows α_X and β_X denote the lower and upper bounds of the support of a random variable X.

6.1 Hazard Rate Estimation for Direct Observations

Consider a nonnegative continuous random variable X. It can be a lifetime, or the time to an event of interest (which can be the time of failure of a device, or the time of an illness relapse), or an insurance loss, or a commodity price. In all these cases it is of interest to assess the risk associated with X via the so-called *hazard rate* function

$$h^X(x) := \lim_{v \to 0} \frac{\mathbb{P}(x < X < x+v|X > x)}{v} = \frac{f^X(x)}{G^X(x)}, \quad G^X(x) > 0, \ x \geq 0, \tag{6.1.1}$$

where we use our traditional notation $f^X(x)$ for the probability density of X and $G^X(x) :=$ $\mathbb{P}(X > x) = \int_x^\infty f^X(u)du = 1 - F^X(x)$ for the survival (survivor) function, and $F^X(x)$ is the cumulative distribution function of X. If one thinks about X as a time to an event-of-interest, then $h^X(x)dx$ represents the instantaneous likelihood that the event occurs within the interval $(x, x+dx)$ given that the event has not occurred at time x. The hazard rate quantifies the trajectory of imminent risk, and it may be referred to by other names in different sciences, for instance as the failure rate in reliability theory and the force of mortality in actuarial science and sociology.

Let us consider classical properties and examples of the hazard rate. The hazard rate, similarly to the probability density or the survival function, characterizes the random variable X. Namely, if the hazard rate is known, then the corresponding probability density is

$$f^X(x) = h^X(x)e^{-\int_0^x h^X(v)dv} =: h^X(x)e^{-H^X(x)}, \tag{6.1.2}$$

where $H^X(x) := \int_0^x h^X(v)dv$ is the cumulative hazard function, and the survival function is

$$G^X(x) = e^{-\int_0^x h^X(v)dv} = e^{-H^X(x)}. \tag{6.1.3}$$

The preceding identity follows from integrating both sides of the equality

$$h^X(x) = -[dG^X(x)/dx]/G^X(x), \tag{6.1.4}$$

and then using $G^X(0) = 1$. Relation (6.1.2) follows from (6.1.1) and the verified (6.1.3).

A corollary from (6.1.3) is that for a random variable X supported on $[0, b]$, $b < \infty$ we get $\lim_{x \to b} h^X(x) = \infty$. This property of the hazard rate for a bounded lifetime plays a critical role in its estimation.

An important property of the hazard rate is that if V and U are independent lifetimes, then the hazard rate of their minimum is the sum of the hazard rates, that is, $h^{\min(U,V)}(x) = h^U(x) + h^V(x)$. Indeed, we have

$$G^{\min(U,V)}(x) = \mathbb{P}(\min(U,V) > x) = \mathbb{P}(U > x)\mathbb{P}(V > x) = G^U(x)G^V(x), \tag{6.1.5}$$

and this, together with (6.1.4), yield the assertion. This property allows us to create a wide variety of shapes for hazard rates. Another important property, following from (6.1.3) and $G^X(\infty) = 0$, is that the hazard rate is not integrable on its support, that is the hazard rate must satisfy $\int_0^\infty h^X(x)dx = \infty$. This is the reason why hazard rate estimates are constructed for a finite interval $[a, a+b] \subset [0, \infty)$ with $a = 0$ being the most popular choice. Further, similarly to the probability density, the hazard rate is nonnegative and has the same smoothness as the corresponding density because the survival function is always smoother than the density. The last but not the least remark is about scale-location transformation $Z = (X - a)/b$ of the lifetime X. This transformation allows us to study Z on the standard unit interval $[0, 1]$ instead of exploring X over $[a, a+b]$. Then the following formulae become useful,

$$G^Z(z) = G^X(a + bz), \quad f^Z(z) = bf^X(a + bz), \tag{6.1.6}$$

and

$$h^Z(z) = bh^X(a + bz), \quad h^X(x) = b^{-1}h^Z((x - a)/b). \tag{6.1.7}$$

Among examples of hazard rates for variables supported on $[0, \infty)$, the most "famous" is the constant hazard rate of an exponential random variable X with the mean $\mathbb{E}\{X\} = \lambda$. Then the hazard rate is $h^X(x) = \lambda^{-1}I(x \geq 0)$ and the cumulative hazard is $H^X(x) = (x/\lambda)I(x \geq 0)$. Indeed, the density is $f^X(x) = \lambda^{-1}e^{-x/\lambda}I(x \geq 0)$, the survival function is $G^X(x) = e^{-x/\lambda}$, and this yields the constant hazard rate. The converse is also valid and a constant hazard rate implies exponential distribution, the latter is not a major surprise keeping in mind that the hazard rate characterizes a random variable. A constant hazard rate has coined the name *memoryless* for exponential distribution. Another interesting example is the Weibull distribution whose density is $f^X(x; k, \lambda) = (k/\lambda)(x/\lambda)^{k-1}e^{-(x/\lambda)^k}I(x \geq 0)$, where $k > 0$ is the shape parameter and $\lambda > 0$ is the scale parameter. The mean is $\lambda\Gamma(1 + 1/k)$ with $\Gamma(z)$ being the Gamma function, the survivor function is $G^X(x; k, \lambda) = e^{-(x/\lambda)^k}I(x \geq 0)$, the hazard rate function is $h^X(x; k, \lambda) = (k/\lambda)(x/\lambda)^{k-1}I(x \geq 0)$, and the cumulative hazard is $H^X(x; k, \lambda) = (x/\lambda)^k I(x \geq 0)$. Note that if $k < 1$ then the hazard rate is decreasing (it is often used to model "infant mortality"), if $k > 1$ then the hazard rate is increasing (it is often used to model "aging" process), and if $k = 1$ then the Weibull distribution becomes exponential (memoryless) with a constant hazard rate.

Now we are in a position to formulate the aim of this section. Based on a sample $X_1, X_2, \ldots, X_n$ of size n from the random variable of interest (lifetime) X, we would like to estimate its hazard rate $h^X(x)$ over an interval $[a, a + b]$, $a \geq 0$, $b > 0$. Because hazard rate is the density divided by the survival function, and the survival function is always smoother than the density, the hazard rate can be estimated with the same rate as the corresponding density. Furthermore, a natural approach is to use (6.1.1) and to estimate the hazard rate by a ratio between estimates of the density and survival function. We will check the ratio-estimate shortly in Figure 6.1, and now simply note that the aim is to understand how a hazard rate may be estimated using our E-estimation methodology because for censored and truncated data, direct estimation of the density or survival function becomes a challenging problem.

To construct an E-estimator of the hazard rate, according to Section 2.2 we need to suggest a sample mean or a plug-in sample mean estimator of Fourier coefficients of the hazard rate. Remember that on $[0, 1]$ the cosine basis is $\{\varphi_0(x) = 1, \varphi_j(x) = \sqrt{2}\cos(\pi j x), j = 1, 2, \ldots\}$. Similarly, on $[a, a+b]$ the cosine basis is $\{\psi_j(x) := b^{-1/2}\varphi_j((x-a)/b), j = 0, 1, \ldots\}$. Note that the cosine basis on $[a, a + b]$ "automatically" performs the above-discussed transformation $Z := (X - a)/b$ of X. As a result, we can either work with the transformed Z and the cosine basis on $[0, 1]$ or directly with X and the corresponding cosine basis on $[a, a + b]$. Here, to master our skills in using different bases, we are using the latter approach. Suppose that $G^X(a + b) > 0$ and write for jth Fourier coefficient of $h^X(x)$, $x \in [a, a + b]$,

$$\theta_j := \int_a^{a+b} h^X(x)\psi_j(x)dx = \int_a^{a+b} \frac{f^X(x)}{G^X(x)}\psi_j(x)dx$$

$$= \mathbb{E}\{I(X \in [a, a + b])[G^X(X)]^{-1}\psi_j(X)\}. \tag{6.1.8}$$

Also, to shed light on the effect of rescaling a random variable, note that if $\kappa_j := \int_0^1 h^Z(z)\varphi_j(z)dz$ is the jth Fourier coefficient of $h^Z(z)$, $z \in [0, 1]$, then

$$\theta_j = b^{-1/2}\kappa_j. \tag{6.1.9}$$

Assume for a moment that the survival function $G^X(x)$ is known, then according to (6.1.8) we may estimate θ_j by the sample mean estimator

$$\tilde{\theta}_j := n^{-1} \sum_{l=1}^{n} \frac{\psi_j(X_l)I(X_l \in [a, a+b])}{G^X(X_l)}. \tag{6.1.10}$$

This Fourier estimator is unbiased and its variance is

$$\mathbb{V}(\tilde{\theta}_j) = n^{-1}\mathbb{V}\Big(\frac{I(X_l \in [a, a+b])\psi_j(X_l)}{G^X(X_l)}\Big). \tag{6.1.11}$$

Further, using (1.3.4) and the assumed $G^X(a+b) > 0$ we may conclude that the corresponding coefficient of difficulty is

$$d(a, a+b) := \lim_{n\to\infty} \lim_{j\to\infty} n\mathbb{V}(\tilde{\theta}_j) = b^{-1} \int_a^{a+b} h^X(x)[G^X(x)]^{-1}dx. \tag{6.1.12}$$

The coefficient of difficulty explicitly shows how the interval of estimation, the hazard rate and the survival function affect estimation of the hazard rate. As we will see shortly, the coefficient of difficulty may point upon a feasible interval of estimation.

Of course $\tilde{\theta}_j$ is an oracle-estimator which is based on an unknown survival function (note that if we know G^X, then we also know the hazard rate h^X). The purpose of introducing an oracle estimator is two-fold. First to create a benchmark to compare with, and second to be an inspiration for its mimicking by a data-driven estimator, which is typically a plug-in oracle estimator. Further, in some cases the mimicking may be so good that asymptotic variances of the estimator and oracle estimator coincide.

Let us suggest a good estimator of the survival function that may be plugged in the denominator of (6.1.10). Because X is a continuous random variable, the survival function can be written as

$$G^X(x) := \mathbb{P}(X > x) = \mathbb{P}(X \geq x) = \mathbb{E}\{I(X \geq x)\}, \tag{6.1.13}$$

and then its sample mean estimator is

$$\hat{G}^X(x) := n^{-1} \sum_{l=1}^{n} I(X_l \geq x). \tag{6.1.14}$$

Note that $\min_{k\in\{1,...,n\}} \hat{G}^X(X_k) = n^{-1} > 0$ and hence we can use the reciprocal of $\hat{G}^X(X_l)$. The sample mean estimator (6.1.14) may be referred to as an empirical survival function.

We may plug (6.1.14) in (6.1.10) and get the following data-driven estimator of Fourier coefficients θ_j of the hazard rate $h^X(x)$, $x \in [a, a+b]$,

$$\hat{\theta}_j := n^{-1} \sum_{l=1}^{n} \frac{\psi_j(X_l)I(X_l \in [a, a+b])}{\hat{G}^X(X_l)}. \tag{6.1.15}$$

This is the proposed Fourier estimator, and it is possible to show that its coefficient of difficulty is identical to (6.1.12). We may conclude that the empirical survival function is a perfect estimator for our purpose to mimic oracle-estimator (6.1.10). Further, the asymptotic theory shows that no other Fourier estimator has a smaller coefficient of difficulty, and hence the proposed Fourier estimator (6.1.15) is efficient. In its turn this result yields asymptotic efficiency of a corresponding hazard rate E-estimator $\hat{h}^X(x)$.

Let us present an example where the coefficient of difficulty is easily calculated. Consider X with exponential distribution and $\mathbb{E}\{X\} = \lambda$. Then $h^X(x) = 1/\lambda$, $G^X(x) = e^{-x/\lambda}$, and hence

$$d(a, a+b) = b^{-1} \int_a^{a+b} \lambda^{-1} e^{x/\lambda} dx = b^{-1} e^{a/\lambda} [e^{b/\lambda} - 1]. \qquad (6.1.16)$$

Note that the coefficient of difficulty increases to infinity exponentially in b. This is what makes estimation of the hazard rate and choosing a feasible interval of estimation so challenging. On the other hand, it is not difficult to suggest a plug-in sample mean estimator of the coefficient of difficulty,

$$\hat{d}(a, a+b) := n^{-1} b^{-1} \sum_{l=1}^n I(X_l \in [a, a+b]) [\hat{G}^X(X_l)]^{-2}. \qquad (6.1.17)$$

To realize that this is indeed a plug-in sample mean estimator, note that (6.1.12) can be rewritten as

$$d(a, a+b) = b^{-1} \int_a^{a+b} f^X(x) [G^X(x)]^{-2} dx = b^{-1} \mathbb{E}\{I(X \in [a, a+b]) [G^X(X)]^{-2}\}. \qquad (6.1.18)$$

This is what was wished to show. The estimator (6.1.17) may be used for choosing a feasible interval of estimation.

Figure 6.1 helps us to understand the problem of estimation of the hazard rate via analysis of a simulated sample of size $n = 400$ from the Bimodal distribution. The caption of Figure 6.1 explains its diagrams. The left-top diagram exhibits reciprocal of the survival function $G^X(x)$ by the solid line and its estimate $1/\hat{G}^X(X_l)$ by crosses. Note that x-coordinates of crosses indicate observed realizations of X, and we may note that while the support is $[0,1]$, just a few of the observations are larger than 0.85. We also observe a sharply increasing right tail of $1/\hat{G}(x)$ for $x > 0.7$, and this indicates a reasonable upper bound for intervals of estimation of the hazard rate.

Our next step is to look at two density E-estimates exhibited in the right-top diagram. The solid line is the underlying Bimodal density. The dotted line is the E-estimate constructed for interval $[a, a+b] = [0, 0.6]$ and it is based on $N = 211$ observations from this interval. The dashed line is the E-estimate based on all observations and it is constructed for the interval $[0,1]$ (the support). The subtitle shows ISEs of the two estimates. Because the sample size is relatively large, both estimates are good and have relatively small ISEs. These estimates will be used by the ratio-estimator of the hazard rate.

Diagrams in the second row show us the same estimate $\hat{d}(a, x)$ of the coefficients of difficulty for different intervals. This is done because the estimate has a sharply increasing right tail. In the left diagram we observe the logarithm of $\hat{d}(0, X_l)$, $l = 1, \ldots, n$, while the right diagram allows us to zoom in on the coefficient of difficulty by considering only $X_l \in [a1, a1+b1]$. Similarly to the left-top diagram, we conclude that interval $[0, 0.6]$ may be a good choice for estimation of the hazard rate, and we may try $[0, 0.7]$ as a more challenging one.

Diagrams in the third (from the top) row exhibit performance of the ratio-estimator

$$\check{h}^X(x) = \frac{\hat{f}^X(x)}{\hat{G}^X(x)}, \qquad (6.1.19)$$

where $\hat{f}^X$ is an E-estimate of the density. The ratio-estimate is a natural plug-in estimate that should perform well as long as the empirical survival function is not too small. The left diagram shows us the ratio-estimate based on all observations, that is, it is the ratio of the dashed line in the right-top diagram over the estimated survival function shown in the

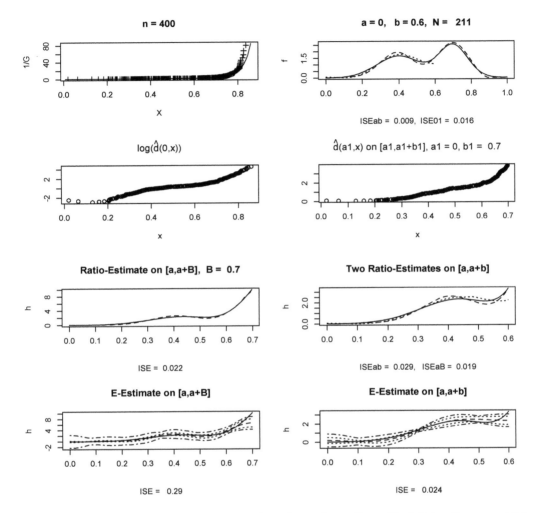

Figure 6.1 *Estimation of the hazard rate based on direct observations. The left-top diagram exhibits reciprocals of the underlying survival function (the solid line) and the empirical one (the crosses at n observed values). The right-top diagram shows the underlying density (the solid line), the E-estimate over $[a, a+b]$ based on observations from this interval (the dotted line), and the E-estimate over the support based on all observations (the dashed line). The second row of diagrams exhibits estimates $\hat{d}(a, X_l)$ over different intervals. In the third row of diagrams, the left diagram shows the hazard rate (the solid line) and its ratio-estimate on $[a, a + B]$, based on all observations, by the dashed line, while the right diagram shows ratio-estimates based on the density estimates shown in the right-top diagram. The left-bottom diagram shows the underlying hazard rate (the solid line) and the E-estimate on $[a, a + B]$ (the dashed line); it also shows the pointwise (the dotted lines) and simultaneous (the dot-dashed lines) confidence bands with confidence level $1 - \alpha$, $\alpha = 0.05$. The right-bottom diagram is similar to the left one, only here estimation over interval $[a, a + b]$ is considered. $[n = 400, corn = 3, a = 0, b = 0.6, B = 0.7, a1 = 0, b1 = 0.7, alpha = 0.05, cJ0 = 4, cJ1 = 0.5, cTH = 4]$*

left-top diagram. The estimate is shown only over the interval $[a, a + B] = [0, 0.7]$ because the reciprocal of the estimated survival function is too large beyond this interval. The solid line shows the underlying hazard rate (it increases extremely fast beyond the point 0.7). For this particular simulation the ISE=0.022 is truly impressive. The right diagram shows us two ratio-estimates for the smaller interval $[a, a + b] = [0.0.6]$; the estimates correspond to the two density estimates shown in the top-right diagram. Note that the ratio-estimate based on the density estimate for the interval $[a, a + b]$ (the dotted line) is worse than the ratio-estimate based on the density which uses all observations (the dashed line); also compare the corresponding ISEs in the subtitle. This is clearly due to the boundary effect; on the other hand, the dotted curve better fits the solid line on the inner interval $[0, 0.45]$. It will be explained in Notes at the end of the chapter how to deal with severe boundary effects.

The bottom row shows E-estimates of the underlying hazard rate for different intervals of estimation. The estimates are complemented by pointwise and simultaneous confidence variance-bands introduced in Section 2.6. The E-estimate over the larger interval $[a, a+B] = [0, 0.7]$ is bad. We have predicted the possibility of such outcome based on the analysis of the reciprocal of the survival function and the coefficient of difficulty. Estimation over the smaller interval $[a, a + b] = [0, 0.6]$, shown in the right-bottom diagram, is much better. Further, the E-estimate is better than the corresponding ratio-estimate (the dotted line in the right diagram of the third row) based on the same $N = 211$ observations from this interval. On the other hand, due to the boundary effect, the E-estimate performs worse than the ratio-estimate based on all observations. Further, note that confidence bands present another possibility to choose a feasible interval of estimation.

It is highly advisable to repeat Figure 6.1 with different corner distributions, sample sizes and intervals to get used to the notion of hazard rate and its estimation. Hazard rate is rarely studied in standard probability and statistical courses, but it is a pivotal characteristic in understanding nonparametric E-estimation in survival analysis.

6.2   Censored Data and Hazard Rate Estimation

Censoring is a typical data modification that affects survival data. We begin with the model of right censoring, which more often occurs in applications, and then explain the model of left censoring.

Right censoring (RC) occurs when a case may be removed from a study at some stage of its "natural" lifetime X and the censoring time C of removal is known. It is a standard assumption that both X and C are nonnegative random variables (lifetimes). Under a right censoring model, the smallest among the lifetime of interest X and the censoring time C is observed, and it is also known whether X or C is observed. In other words, instead of a direct sample from X, under the right censoring we observe a sample from a pair of random variables

$$(V, \Delta) := (\min(X, C), I(X \leq C)). \qquad (6.2.1)$$

Note that the indicator of censoring Δ tells us when $V = X$ or $V = C$, and because the indicator is always present in data, it is not difficult to realize that we are dealing with data modified by censoring. (The latter is a nice feature of censoring, and in Section 6.3 we will explore another modification, called truncation, when survival data do not manifest the modification.) In what follows we may use the term censoring in place of right censoring whenever no confusion with left censoring occurs.

Let us present several examples that shed light on the right censoring. Suppose that we study a time X from origination of a mortgage loan until the default on the payment; the study is based on data obtained from a large mortgage company. For each originated loan, we get either the time X of the default or the censoring time C when the loan is paid in full due to refinancing, selling a property, final payment, etc. Hence, in this study we observe the default time X only if it is not larger than the censoring time C and otherwise we observe C. Note that we also know when we observe X or C, and hence the example fits model (6.2.1). Another example is the lifetime of a light bulb that may accidentally break at time C before it burns out. But in no way the notion of censoring is restricted to event times. For instance, if an insurance policy has an upper limit C on the payment and there is a claim for a loss with X being the actual loss, then the payment will be the smaller of the loss and the limit. Insurance data typically contain information about payments and policy limits, and hence we are dealing with right censored losses. The latter is a classical example of right-censored data in actuarial science. In a clinical study, typical reasons of right censoring are end of study and withdrawal from study. Note that in all these examples true survival times are equal or greater than observed survival times. As a result, observed data are biased and skewed to the left. Further, estimation of the right tail of distribution may be difficult if not impossible. This is a specific of right censoring that we are dealing with.

Now we are in a position to explain how a hazard rate E-estimator, based on censored data, can be constructed. As usual, the key element is to propose a sample mean estimator for Fourier coefficients. From now on, we are assuming that X and C are independent and continuous random variables. We begin with presenting formulas for the distributions of interest. Because V is the minimum of two independent random variables X and C, there exists a nice formula for its survival function,

$$G^V(v) := \mathbb{P}(\min(X, C) > v) = \mathbb{P}(X > v, C > v) = G^X(v)G^C(v). \qquad (6.2.2)$$

Further, for the observed pair (V, Δ) we can write,

$$\mathbb{P}(V \leq v, \Delta = \delta) = \mathbb{P}(\Delta = \delta) - \mathbb{P}(V > v, \Delta = \delta)$$

$$= \mathbb{P}(\Delta = \delta) - \int_v^\infty [f^X(x)G^C(x)]^\delta [f^C(x)G^X(x)]^{1-\delta} dx, \quad v \geq 0, \, \delta \in \{0, 1\}. \qquad (6.2.3)$$

Differentiation of (6.2.3) with respect to v yields the following formula for the joint mixed density of the observed pair,

$$f^{V,\Delta}(v, \delta) = [f^X(v)G^C(v)]^\delta [f^C(v)G^X(v)]^{1-\delta} I(v \geq 0, \delta \in \{0, 1\}). \qquad (6.2.4)$$

The formula exhibits a remarkable symmetry with respect to δ which reflects the fact that while C censors X on the right, we may also say that the random variable X also censors C on the right whenever C is the lifetime of interest. In other words, the problem of right censoring is symmetric with respect to the two underlying random variables X and C. This is an important observation because if a data-driven estimator for distribution of X is proposed, it can be also used for estimation of the distribution of C by changing Δ on $1 - \Delta$.

Formula (6.2.4) implies that available observations of X (when $\Delta = 1$) are biased with the biasing function being the survival function $G^C(x)$ of the censored random variable (recall definitions of biased data and a biasing function in Section 3.1). Note that the biasing function is decreasing in v because larger values of X are more likely to be censored. As we know, in general the biasing function should be known for a consistent estimation of an underlying distribution. As we will see shortly, because here we observe a sample from a pair (V, Δ) of random variables, we can estimate the biasing function $G^C(x)$ and hence to estimate the distribution of X. Recall that knowing a distribution means knowing any characteristic of a random variable like its cumulative distribution function, density, hazard rate, survival function, etc. We will see shortly in Section 6.5 that estimation of the density of X is a two-step procedure. The reason for that is that censored data are biased and hence we first estimate the biasing function G^C, which is a complicated problem on its own, and only then may proceed to estimation of the density of X.

Surprisingly, there is no need to estimate the biasing function if the hazard rate is the function of interest (the estimand). This is a pivotal statistical fact to know about survival data where the cumulative distribution function and/or probability density are no longer natural characteristics of a distribution to begin estimation with. The latter is an interesting consequence of data modification caused by censoring.

Let us explain why hazard rate $h^X(x)$ is a natural estimand in survival analysis. Using (6.2.2) and (6.2.4) we may write,

$$h^X(v) := \frac{f^X(v)}{G^X(v)} = \frac{f^{V,\Delta}(v,1)}{G^C(v)G^X(v)} = \frac{f^{V,\Delta}(v,1)}{G^V(v)}. \tag{6.2.5}$$

This is a pivotal formula because (6.2.5) expresses the hazard rate of right censored X via the density and survival function of directly observed variables.

Suppose that we observe a sample of size n from (V,Δ). Denote by $\theta_j := \int_a^{a+b} h^X(x)\psi_j(x)dx$ the jth Fourier coefficient of the hazard rate $h^X(x)$, $x \in [a, a+b]$. Here and in what follows, similarly to Section 6.1, the cosine basis $\{\psi_j(v)\}$ on an interval $[a, a+b]$ is used and recall our discussion of why a hazard rate is estimated over an interval. Assume that $a + b < \beta_V$ (recall that β_V denotes the upper bound of the support of V), and note that using (6.2.5) we can write down a Fourier coefficient as an expectation of a function in V and Δ,

$$\theta_j = \mathbb{E}\left\{\frac{\Delta I(V \in [a, a+b])\psi_j(V)}{G^V(V)}\right\}. \tag{6.2.6}$$

This immediately yields the following plug-in sample mean Fourier estimator,

$$\hat{\theta}_j := n^{-1}\sum_{l=1}^{n} \frac{\Delta_l \psi_j(V_l)I(V_l \in [a, a+b])}{\hat{G}^V(V_l)}, \tag{6.2.7}$$

where

$$\hat{G}^V(v) := n^{-1}\sum_{l=1}^{n} I(V_l \geq v) \tag{6.2.8}$$

is the empirical survival function of V.

In its turn, the Fourier estimator implies the corresponding hazard rate E-estimator $\hat{h}^X(x)$. Furthermore, the coefficient of difficulty is

$$d(a, a+b) := \lim_{n\to\infty}\lim_{j\to\infty} n\mathbb{V}(\hat{\theta}_j)$$

$$= b^{-1}\int_a^{a+b} \frac{f^{V,\Delta=1}(v)}{[G^V(v)]^2}dv = b^{-1}\mathbb{E}\left\{\frac{\Delta I(V \in [a, a+b])}{[G^V(V)]^2}\right\} \tag{6.2.9}$$

$$= b^{-1}\int_a^{a+b} \frac{h^X(v)}{G^V(v)}dv = b^{-1}\int_a^{a+b} \frac{h^X(v)}{G^X(v)G^C(v)}dv. \tag{6.2.10}$$

Formula (6.2.10) clearly shows how an underlying hazard rate and a censoring variable affect accuracy of estimation. Note that the new here, with respect to formula (6.1.12) for the case of direct observations, is an extra survival function $G^C(v)$ in the denominator. This is a mathematical description of the negative effect of censoring on estimation. Because $G^C(v) \leq 1$ and the survival function decreases in v, that effect may be dramatic. Further, if $\beta_C < \beta_X$ then the censoring implies a destructive modification of data when no consistent estimation of the distribution of X is possible.

Formula (6.2.9) implies that the coefficient of difficulty may be estimated by a plug-in sample mean estimator

$$\hat{d}(a, a+b) := n^{-1}b^{-1}\sum_{l=1}^{n} \Delta_l I(V_l \in [a, a+b])[\hat{G}^V(V_l)]^{-2}. \tag{6.2.11}$$

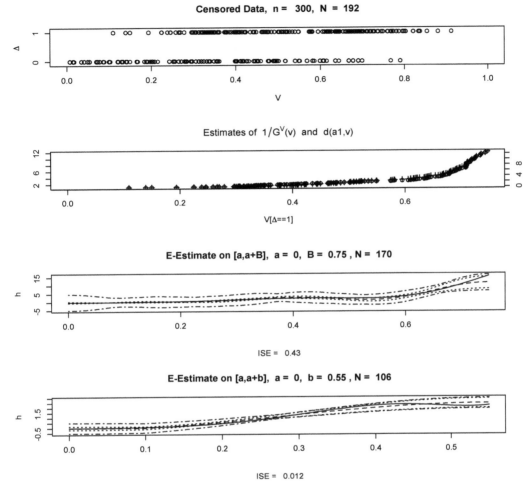

Figure 6.2 *Estimation of the hazard rate based on right censored observations. The top diagram shows $n = 300$ observations of (V, Δ), among those $N := \sum_{l=1}^{n} \Delta_l = 192$ uncensored ones. Second from the top diagram shows us by crosses values of $1/\hat{G}(V_l)$ and by circles values of $\hat{d}(a1, V_l)$ for uncensored $V_l \in [a1, a1 + b1]$; the corresponding scales are on the left and right vertical axes. Note how fast the right tails increase. The third from the top diagram shows E-estimate of the hazard rate based on observations $V_l \in [a, a + B]$, while in the bottom diagram the E-estimate is based on $V_l \in [a, a + b]$. A corresponding N shows the number of uncensored observations within an interval. The underlying hazard rate and its E-estimate are shown by the solid and dashed lines, the pointwise and simultaneous $1 - \alpha$ confidence bands are shown by dotted and dot-dashed lines. {Censoring distribution is either the default Uniform$(0, u_C)$ with $u_C = 1.5$, or Exponential(λ_C) with the default $\lambda_C = 1.5$ where λ_C is the mean. To choose an exponential censoring, set cens $= ''Expon''$. Notation for arguments controlling the above-mentioned parameters is evident.} [n = 300, corn = 3, cens = ''Unif'', uC = 1.5, lambdaC = 1.5, a = 0, b = 0.55, B = 0.75, a1 = 0, b1 = 0.75, alpha = 0.05, cJ0 = 4, cJ1 = 0.5, cTH = 4]*

Note that the new in the estimator, with respect to the case of direct observations of Section 6.1, is that $\hat{G}^X$ is replaced by $\hat{G}^V$. Because $G^V = G^X$ for the case of direct observations, this change is understandable.

The proposed estimator of Fourier coefficients allows us to construct the E-estimator of the hazard rate.

Figure 6.2 sheds light on right censoring, performance of the proposed E-estimator, and the methodology of choosing a feasible interval of estimation explained in Section 6.1. In the particular simulation X is the Bimodal and C is the Uniform(0,1.5). The top diagram shows the sample from $(V, \Delta) = (\min(X, C), I(X \leq C))$. The sample size $n = 300$ and $N := \sum_{l=1}^{n} \Delta_l = 192$ is the number of uncensored realizations of X. The latter indicates a severe censoring. The second from the top diagram allows us to evaluate complexity of the problem at hand and to choose a reasonable interval of estimation. Crosses (and the corresponding scale is shown on the left vertical axis) show values of $1/\hat{G}^V(V_l)$ for $N = 192$ uncensored observations over interval $[a1, a1 + b1] = [0, 0.75]$, while circles show values of the estimated coefficient of difficulty (its scale is on the right-vertical axis). Here we have an interesting outcome that the two functions look similar, and this is due to using different scales and rapidly increasing right tails. The interested reader may decrease $b1$ and get a better visualization of the functions for moderate values of v. Analysis of the two estimates indicates that it is better to avoid estimation of the hazard rate beyond $v = 0.75$, and probably a conservative approach is to consider intervals with the upper bound smaller than 0.6. Let us check this conclusion with the help of corresponding hazard rate E-estimates and confidence bands. The two bottom diagrams allow us to do this, here the chosen intervals are $[a, a + B] = [0, 0.75]$ and $[a, a + b] = [0, 0.55]$, respectively. At first glance, the E-estimate for the larger interval (the third from the top diagram) and its confidence bands may look attractive, but this conclusion is misleading. Indeed, please look at the scale of the bands and realize that the bands are huge! Further, the large bands "hide" the large deviation of the estimate (the dashed line) from the underlying hazard rate (the solid line). Further, the ISE = 0.43, shown in the subtitle, tells us that the estimate is far from the underlying hazard rate. The reason for this is large right tails of $1/\hat{G}^V(v)$ and $\hat{d}(0, v)$ that may be observed in the second from the top diagram. The outcome is much better for the smaller interval of estimation considered in the bottom diagram, and this is despite the smaller number $N = 106$ of available uncensored observations. Also note that the ISE is dramatically smaller.

It is fair to conclude that: (i) The proposed in Section 6.1, for the case of direct observations, methodology of choosing a feasible interval of estimation is robust and can be also recommended for censored data; (ii) The E-estimator performs well even for the case of severe censoring.

Let us finish this section by a remark about left censoring. Under a left censoring, the variable of interest X is censored on the left by a censoring variable C if available observations are from the pair $(V, \Delta) := (\max(X, C), I(X \geq C))$. For instance, when a physician asks a patient about the onset of a particular disease, the answer may be either a specific date or that the onset occurred prior to some specific date. In this case the variable of interest is left censored. Left censoring may be "translated" into a right censoring. To do this, choose a value A that is not less than all available left-censored observations and then consider new observations that are A minus left-censored observations. Then the new observations become right-censored. The latter is the reason why it is sufficient to learn about estimation for right-censored data.

6.3   Truncated Data and Hazard Rate Estimation

It is import to begin this section with a warning. While censoring manifests itself via indicators of censoring and missing manifests itself via empty cells in data, truncation is a "silent" modification of data with no indicators of truncation. Further, truncation is generated by a hidden sequential missing mechanism and we do not even know how many observations were missed. Further, an underlying sequential missing mechanism is MNAR (Missing Not At Random) and it creates biased data. The aim of this section is to explain how to recognize that observations are truncated and then, if possible, to propose an E-estimator of an underlying hazard rate. Left truncation is the most frequently occurred type of truncation, and in what follows we are concentrating on its discussion.

Let us present several examples that shed light on truncation mechanism. Our first example is a classical one in actuarial science. Suppose that we are interested in the distribution of losses incurred by policyholders of an insurance company. Also, suppose that there is no limit on a payment. An insurance company can provide us with data for policies where payments for insurable losses were made. On first glance, this may look like a clear case of direct observations of losses incurred by policyholders, but this is not necessarily the case because a typical insurance policy includes a deductible. The role of a deductible is to decrease the payment (which is the incurred loss minus the deductible) and also to discourage a policyholder from reporting small losses (note that there is no payment on a loss smaller than the deductible, plus there is a possibility of increasing the insurance premium for being a risky policyholder). As a result, smaller (with respect to deductible) incurred losses are missed in the insurance database, and the recorded losses are biased (skewed to the right) with respect to the distribution of underlying losses. Another interesting feature of the MNAR truncation mechanism is that we do not know how many insurable losses occurred. We may say that the underlying incurred loss (the random variable of interest) is left truncated by the deductible (truncating random variable). Let us stress that per se there is nothing in reported losses that manifests their truncated nature. It is up to us to realize that available observations are truncated and that the left truncation causes data to be right-skewed with respect to underlying data because larger realizations of the variable of interest are observed (included in data) with larger likelihood. Further, it is up to us to ask about corresponding observations of the truncating variable. From now on, when the actuarial example is referred to, we are assuming that available data contains reported losses and corresponding deductibles.

Our second example is more challenging. Suppose that we would like to know how long a startup technology company survives until it files for bankruptcy. To study this problem, we can look at all startups that exist at time T, learn about times of their start, and then follow them until their bankruptcy. This gives us data about lifetimes of the startups. Assuming that the process is stationary, it looks like we have desired direct observations of the lifetime of interest. Unfortunately, this is not the case because lifetimes are left truncated by time T at which we search for functioning startup companies. Indeed, if a startup company is bankrupt before time T, it is not included in the study, and furthermore we do not even know that it existed. The example looks more confusing than the actuarial one, but these two examples are similar. Indeed, let us translate the startup example into the actuarial example. We may say that an underlying lifetime of a company is an incurred loss, the time from its start to time T is the deductible, and the observed lifetime of a company is the reported loss. If we use this approach for understanding the startup data, then we may realize that the two examples are similar.

Our third example is from biostatistics where truncation is a typical phenomenon in clinical trials. Suppose that we got data from a study, conducted over the period of last 5 years, about the longevity of patients after a surgery. The data are left truncated because only patients who were alive at the baseline (which is the time of beginning of the study and in our case this was 5 years ago) can participate in the study. Please note how similar this example to the second example with startups, and you may use the above-presented methodology to "translate" it into the actuarial example.

Now, after presenting examples of left truncation, let us describe how truncation modifies underlying data. There are a nonnegative random variable of interest (lifetime) X^* and a truncation random variable T^*. Realizations of the pair (T^*, X^*) are not available directly and are hidden. If the first hidden realization (T_1^*, X_1^*) of (T^*, X^*) satisfies the inequality $X_1^* \geq T_1^*$, that is the random variable of interest is not smaller than the truncation variable, then the pair $(T_1, X_1) := (T_1^*, X_1^*)$ is observed, otherwise this hidden realization is skipped and we even do not know that it occurred. Then the underlying hidden sampling is continued until n realizations of (T, X) are available. Let us look at the truncation one more time via the actuarial example. In the actuarial example X^* is the incurred loss and T^* is the deductible. The incurred loss X^* is reported only if $X^* \geq T^*$ (strictly speaking if $X^* > T^*$ but $\mathbb{P}(X^* = T^*) = 0$ for a continuous X^* and independent T^*), and only then we may observe the reported loss $X = X^*$ and the corresponding deductible $T = T^*$. On the other hand, if $X^* < T^*$, then we do not even know that the loss occurred and also have no information about a corresponding deductible (recall that we have only information about policies with payments, and payments are triggered by reported losses).

The hidden mechanism of collecting data can be described via a negative binomial experiment such that: the experiment stops as soon as nth "success" occurs, data is collected only when a "success" occurs, there is no information on how many "failures" occurred between "successes." Assuming that X^* and T^* are independent, continuous and nonnegative random variables, the probability of "success" can be defined as

$$p := \mathbb{P}(T^* \leq X^*) = \int_0^\infty f^{T^*}(t)G^{X^*}(t)dt. \qquad (6.3.1)$$

Here $f^{T^*}(t)$ and $G^{X^*}(t)$ are the density of T^* and the survival function of X^*, respectively. Further, while we do not know the total number N of hidden "failures", the negative binomial distribution sheds some light on it. In particular, the mean and variance of N are calculated as

$$\mathbb{E}\{N\} = n(1-p)p^{-1}, \quad \mathbb{V}(N) = n(1-p)p^{-2}. \qquad (6.3.2)$$

Let us make an important remark. If $\mathbb{P}(T^* < c) = 0$ for some positive constant c then all hidden realizations of X^* smaller than c are truncated. As a result, we cannot restore the distribution of X^* for these small values. Hence in general the hazard rate may be estimated only over an interval $[a, a + b]$ with some $a > 0$, and the theory supports this conclusion. Further, as we already know from Section 6.1, in general we cannot estimate the right tail of a hazard rate, and now the left truncation may preclude us from estimating its left tail. This phenomenon will be quantified shortly.

Now we are in a position to present useful probability formulas. For the joint cumulative distribution function of the observed pair (T, X) we may write,

$$F^{T,X}(t, x) := \mathbb{P}(T \leq t, X \leq x) = \mathbb{P}(T^* \leq t, X^* \leq x | X^* \geq T^*) =$$

$$= p^{-1}\mathbb{P}(T^* \leq t, T^* \leq X^* \leq x)$$

$$= p^{-1}\int_0^t f^{T^*}(v)[\int_v^x f^{X^*}(u)du]dv, \quad 0 \leq t \leq x. \qquad (6.3.3)$$

Here p is defined in (6.3.1). Then, taking partial derivatives with respect to x and t we get the following expression for the bivariate density,

$$f^{T,X}(t, x) = p^{-1}f^{T^*}(t)f^{X^*}(x)I(0 \leq t \leq x < \infty). \qquad (6.3.4)$$

This allows us to obtain, via integration, a formula for the marginal density of X,

$$f^X(x) = f^{X^*}(x)[p^{-1}F^{T^*}(x)]. \qquad (6.3.5)$$

In its turn, for values of x such that $F^{T^*}(x) > 0$, (6.3.5) yields a formula for the density of the random variable of interest X^*,

$$f^{X^*}(x) = \frac{f^X(x)}{p^{-1}F^{T^*}(x)} \quad \text{whenever } F^{T^*}(x) > 0. \qquad (6.3.6)$$

Note that for values x such that $F^{T^*}(x) = 0$ we cannot restore the density $f^{X^*}(x)$ because all observations of X^* with such values are truncated. In other words, using our notation α_Z for a lower bound of the support of a continuous random variable Z, for consistent estimation of the distribution of X^* we need to assume that

$$\alpha_{X^*} \geq \alpha_{T^*}. \qquad (6.3.7)$$

Another useful remark is that formula (6.3.5) mathematically describes the biasing mechanism caused by the left truncation, and according to (3.1.2), the biasing function is $F^{T^*}(x)$. Note that the biasing function is increasing in x.

Let us also introduce a function, which is a probability, that plays a pivotal role in the analysis of truncated data,

$$g(x) := \mathbb{P}(T \leq x \leq X) = \mathbb{P}(T^* \leq x \leq X^*|T^* \leq X^*) = p^{-1}F^{T^*}(x)G^{X^*}(x). \qquad (6.3.8)$$

In the last equality we used the assumed independence between T^* and X^*. Note that $g(x)$ is a functional of the distributions of available observations and hence can be estimated. The latter will be used shortly.

Now we have all necessary formulas and notations to explain the method of constructing an E-estimator of the hazard rate of X^*.

Using (6.3.6) and (6.3.8), we conclude that the hazard rate of X^* can be written as

$$h^{X^*}(x) := \frac{f^{X^*}(x)}{G^{X^*}(x)} = \frac{f^X(x)}{\mathbb{P}(T \leq x \leq X)} = \frac{f^X(x)}{g(x)} \text{ whenever } F^{T^*}(x)G^{X^*}(x) > 0. \qquad (6.3.9)$$

Note that $h^{X^*}(x)$ is expressed via distributions of available random variables (X, T) and hence can be estimated. Further, note that the restriction $F^{T^*}(x)G^{X^*}(x) > 0$ is equivalent to $g(x) > 0$.

Formula (6.3.9) for the hazard rate is the key for its estimation. Indeed, consider estimation of the hazard rate over an interval $[a, a + b]$ such that $g(x) > 0$ over this interval. Similarly to the previous sections, $\{\psi_j(x)\}$ is the cosine basis on $[a, a + b]$. The proposed sample mean estimator of Fourier coefficients $\theta_j := \int_a^{a+b} \psi_j(x)h^{X^*}(x)dx$ is

$$\hat{\theta}_j := n^{-1} \sum_{l=1}^{n} \frac{\psi_j(X_l)I(X_l \in [a, a+b])}{\hat{g}(X_l)}, \qquad (6.3.10)$$

where

$$\hat{g}(x) := n^{-1} \sum_{l=1}^{n} I(T_l \leq x \leq X_l) \qquad (6.3.11)$$

is the sample mean estimator of function $g(x)$ defined in (6.3.8). Note that $\hat{g}(X_l) \geq n^{-1}$ and hence this estimator can be used in the denominator of (6.3.10).

The Fourier estimator (6.3.10) yields the corresponding E-estimator of the hazard rate. Further, the corresponding coefficient of difficulty is

$$d(a, a + b) := \lim_{n,j \to \infty} n\mathbb{V}(\hat{\theta}_j)$$

$$= b^{-1} \int_a^{a+b} \frac{h^{X^*}(x)}{g(x)}dx = b^{-1}\mathbb{E}\{I(a \leq X \leq a + b)/g^2(X)\}. \qquad (6.3.12)$$

Further, the plug-in sample mean estimator of the coefficient of difficulty is

$$\hat{d}(a, b) := n^{-1}b^{-1} \sum_{l=1}^{n} I(a \leq X_l \leq a + b)/\hat{g}^2(X_l). \qquad (6.3.13)$$

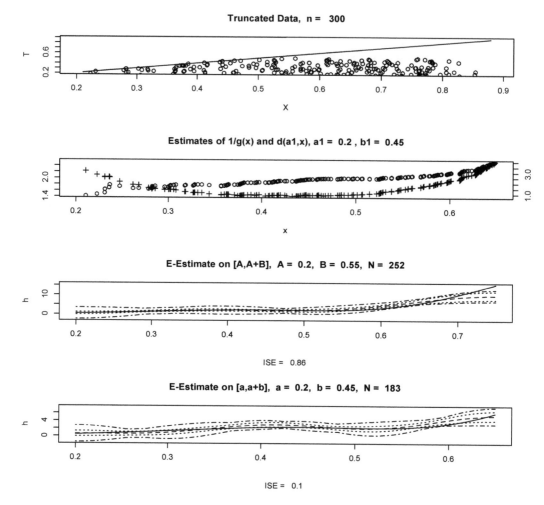

Figure 6.3 *Estimation of the hazard rate based on left truncated observations. The top diagram shows a sample of left truncated observations, the sample size $n = 300$. In the simulation X^* is the Bimodal and T^* is Uniform$(0, 0.5)$. Second from the top diagram shows by crosses the estimate $1/\hat{g}(X_l)$ and by circles the estimate $\hat{d}(a1, X_l)$, $X_l \in [a1, a1 + b1]$. Note the different scales used for these two estimates that are shown correspondingly on the left and right vertical axes. The third from the top diagram shows E-estimate of the hazard rate on interval $[A, A + B]$, while in the bottom diagram the E-estimate is for interval $[a, a + b]$. N shows the number of observations within a considered interval. The underlying hazard rate and its E-estimate are shown by the solid and dashed lines, the pointwise and simultaneous $1 - \alpha$ confidence bands are shown by dotted and dot-dashed lines, respectively. {Distribution of T is either the default Uniform$(0, u_T)$ with $u_T = 0.5$, or Exponential(λ_T) with the default $\lambda_T = 0.3$ where λ_T is the mean. Set trunc = "Expon" to choose the exponential truncation. Parameter α is controlled by alpha.} [n = 300, corn = 3, trunc = "Unif", uT = 0.5, lambdaT = 0.3, a = 0, b = 0.55, A = 0.2, B = 0.75, a1 = 0.2, b1 = 0.45, alpha = 0.05, cJ0 = 4, cJ1 = 0.5, cTH = 4]*

It is of interest to compare the proposed E-estimator with E-estimators of the previous sections. For the case of direct observations, we have $\mathbb{P}(T^* > 0) = 0$, this yields $g(x) = G^{X^*}(x)$ and hence the E-estimators coincide. For the case of censored observations, the survival function $G^V(x)$ is used in place of $g(x)$, and otherwise the E-estimators are identical. On the other hand, the important new feature of truncated data is that truncation complicates estimation of the left tail of a hazard rate.

Let us look at a particular simulation and analysis of truncated data presented in Figure 6.3. Here X^* is the Bimodal and T^* is the Uniform(0,0.5). The top diagram shows by circles the scattergram of $n = 300$ left-truncated observations of (X, T). Note that all observations are below the solid line $T = X$, this is what we must see in left-truncated observations. Another useful observation is that there are no realizations of X smaller than 0.2. This indicates that the left bound of reasonable intervals of estimation should be larger than 0.2. The second from the top diagram allows us to evaluate complexity of the problem at hand. Crosses (and the corresponding scale is shown on the left vertical axis) show values of the function $1/\hat{g}(X_l)$ for $X_l \in [a1, a1 + b1]$. Note the increasing left and right tails of the function. The coefficient of difficulty $d(a, a + b)$ is now a function of two variables, and the circles show us $\hat{d}(a1, X_l)$, $X_l \in [a1, a1 + b1]$. To see estimates for different intervals, Figure 6.3 allows to change $a1$ and $b1$. Based on the analysis of this diagram, the right side of a feasible interval of estimation should be smaller than 0.6. To test this conclusion, let us choose a relatively large $[A, A + B] = [0.2, 0.75]$ and a smaller $[a, a + b] = [0.2, 0.65]$ interval of estimation. Corresponding hazard rate E-estimates are shown in the two bottom diagrams together with confidence bands. The quality of estimation over the smaller interval is dramatically better, and the confidence bands support our preliminary conclusion about a feasible interval of estimation.

The reader might notice that confidence bands take on negative values while a hazard rate is nonnegative. It is not difficult to make them bona fide (nonnegative), but this will complicate visualization of the left tail of the estimate. Furthermore, these "not bona fide bands" help us to choose and/or justify a feasible interval of estimation.

It is highly recommended to repeat Figure 6.3 with different parameters and underlying variables. Hazard rate is rarely studied in standard probability and statistical classes, and the same can be said about truncation. Analysis of diagrams, generated by Figure 6.3, helps to gain experience in dealing with hazard rate and truncated data.

6.4   LTRC Data and Hazard Rate Estimation

In the previous sections we considered data modified by two mechanisms: right censoring and left truncation. Recall that both these mechanisms imply biased data, with right censoring favoring smaller observations and left truncation favoring larger ones. In many applications these two mechanisms act together. Let us present two examples. We begin with an actuarial example. Suppose that we have data containing payments to policyholders, and the aim is to estimate the distribution of losses incurred by policyholders. If the policies have deductibles and limits on payments, then incurred losses are left truncated by deductibles and right censored by limits on payments. In short, we may say that the available data are left truncated and right censored (LTRC). Now let us consider a clinical trial example where participants, who had a cancer surgery, are divided at the baseline into intervention group (a new medication is given to these participants) and control group (a placebo is given) and then observations of participants continue for a period of five years. The random variable of interest is the lifetime from surgery to death. As we know from the previous sections, the lifetimes in both groups are left truncated by the time from surgery to the baseline, and they may be also right censored by a number of events like the time from surgery to the end of the study, a possible moving to another area, etc. We again are dealing with the LTRC modification.

Let us formally describe a LTRC mechanism of generating a sample of size n of left truncated and right censored (LTRC) observations. There is a hidden sequential sampling from a triplet of nonnegative random variables (T^*, X^*, C^*) whose joint distribution is unknown. T^* is the truncation random variable, X^* is the random variable (lifetime) of interest, and C^* is the censoring random variable. Suppose that (T_k^*, X_k^*, C_k^*) is the kth realization of the hidden triplet and that at this moment there already exists a sample of size $l-1$, satisfying $l-1 \leq \min(k-1, n-1)$, of LTRC observations. If $T_k^* > \min(X_k^*, C_k^*)$ then the kth realization is left truncated meaning that: (i) The triplet (T_k^*, X_k^*, C_k^*) is not observed; (ii) The fact that the kth realization occurred is unknown; (iii) Next realization of the triplet occurs. On the other hand, if $T_k^* \leq \min(X_k^*, C_k^*)$ then the LTRC observation $(T_l, V_l, \Delta_l) := (T_k^*, \min(X_k^*, C_k^*), I(X_k^* \leq C_k^*))$ is added to the LTRC sample whose size becomes equal to l. The hidden sequential sampling from the triplet (T^*, X^*, C^*) stops as soon as $l = n$.

Let us stress two important facts about the model. The former is that the model includes the case when the censoring variable may be smaller than the truncation one, that is it includes the case $\mathbb{P}(C^* < T^*) > 0$. Of course, it also includes the case $\mathbb{P}(C^* \geq T^*) = 1$ with an important example being $C^* := T^* + U^*$ where $\mathbb{P}(U^* \geq 0) = 1$. The latter is that the LTRC mechanism of collecting data can be described via a negative binomial experiment such that the "success" is the event $T^* \leq \min(X^*, C^*)$, the experiment stops as soon as nth "success" occurs, data are collected only when a "success" occurs, there is no information on how many "failures" occurred between "successes", and the probability of "success" is

$$p := \mathbb{P}(T^* \leq \min(X^*, C^*)). \qquad (6.4.1)$$

Further, while we do not know the total number of hidden "failures", the negative binomial distribution sheds some light on the hidden number N_f of "failures", and in particular the mean and variance of N_f are calculated as $\mathbb{E}(N_f) = n(1-p)p^{-1}$ and $\mathrm{Var}(N_f) = n(1-p)p^{-2}$.

Now our aim is to understand how the distribution of LTRC observations is related to the distribution of the hidden realizations of the triplet (T^*, X^*, C^*). Set $V^* := \min(X^*, C^*)$. Suppose that a hidden realization of the triplet is observed, meaning that it is given that $T^* \leq V^*$ and we observe (T, V, Δ) where $T = T^*$, $V = V^*$ and $\Delta := \Delta^* := I(X^* \leq C^*)$. Recall that T^* is the underlying truncation random variable, X^* is the lifetime (random variable) of interest, C^* is the underlying censoring random variable, and p, defined in (6.4.1), is the probability of observing the hidden triplet (T^*, V^*, Δ^*). For the joint mixed distribution function of the observed triplet of random variables we can write,

$$F^{T,V,\Delta}(t, v, \delta) := \mathbb{P}(T \leq t, V \leq v, \Delta \leq \delta) = F^{T^*, V^*, \Delta^* | T^* \leq V^*}(t, v, \delta)$$

$$= p^{-1}\mathbb{P}(T^* \leq t, T^* \leq V^* \leq v, \Delta^* \leq \delta), \quad t \geq 0, v \geq 0, \delta \in \{0, 1\}. \qquad (6.4.2)$$

Let us additionally assume that random variables T^*, X^* and C^* are independent and continuous (this assumption will be relaxed later). Then, according to (6.4.1) we get $p = \int_0^\infty f^{T^*}(t)G^{X^*}(t)G^{C^*}(t)dt$, and using (6.4.2) we can write for any $0 \leq t \leq v < \infty$ and $\delta \in \{0, 1\}$,

$$\mathbb{P}(T \leq t, V \leq v, \Delta = \delta)$$

$$= p^{-1} \int_0^t f^{T^*}(\tau) \left[\int_\tau^v f^{X^*}(x)G^{C^*}(x)dx \right]^\delta \left[\int_\tau^v f^{C^*}(x)G^{X^*}(x)dx \right]^{1-\delta} d\tau. \qquad (6.4.3)$$

Taking partial derivatives of both sides in (6.4.3) with respect to t and v yields the following mixed joint probability density,

$$f^{T,V,\Delta}(t, v, \delta) = p^{-1}f^{T^*}(t)I(t \leq v)\left[f^{X^*}(v)G^{C^*}(v) \right]^\delta \left[f^{C^*}(v)G^{X^*}(v) \right]^{1-\delta}. \qquad (6.4.4)$$

Note that the density is "symmetric" with respect to X^* and C^* whenever δ is replaced on $1 - \delta$; we have already observed this fact for censored data.

Formula (6.4.4) yields the following marginal joint density

$$f^{V,\Delta}(v, 1) = p^{-1}f^{X^*}(v)G^{C^*}(v)F^{T^*}(v) = h^{X^*}(v)[p^{-1}G^{C^*}(v)F^{T^*}(v)G^{X^*}(v)]. \qquad (6.4.5)$$

In the last equality we used definition of the hazard rate $h^{X^*}(x) := f^{X^*}(x)/G^{X^*}(x)$.

The first equality in (6.4.5) yields a nice formula for the density of the random variable of interest,

$$f^{X^*}(x) = \frac{f^{V,\Delta}(x,1)}{p^{-1}G^{C^*}(x)F^{T^*}(x)} \quad \text{whenever } G^{C^*}(x)F^{T^*}(x) > 0. \qquad (6.4.6)$$

This formula quantifies the bias in available observations.

Now we are in a position to introduce the probability of an event for available variables which plays a central role in statistical analysis of LTRC data,

$$g(x) := \mathbb{P}(T \le x \le V)$$

$$= \mathbb{P}(T^* \le x \le V^* | T^* \le V^*)$$

$$= [p^{-1}G^{C^*}(x)F^{T^*}(x)]G^{X^*}(x), \quad x \in [0,\infty). \qquad (6.4.7)$$

Note that the right side of (6.4.7) contains in the square brackets the denominator of the ratio in (6.4.6). This fact, together with (6.4.6), yields two important formulae. The first one is that the underlying density of X^* may be written as

$$f^{X^*}(x) = \frac{f^{V,\Delta}(x,1)G^{X^*}(x)}{\mathbb{P}(T \le x \le V)} \quad \text{whenever } G^{C^*}(x)F^{T^*}(x) > 0. \qquad (6.4.8)$$

Next, if we divide both sides of the last equality by the survival function $G^{X^*}(x)$, then we get the following expression for the hazard rate,

$$h^{X^*}(x) = \frac{f^{V,\Delta}(x,1)}{\mathbb{P}(T \le x \le V)} \quad \text{whenever } G^{C^*}(x)F^{T^*}(x)G^{X^*}(x) > 0, \qquad (6.4.9)$$

or equivalently the formula holds whenever $g(x) > 0$ where $g(x)$ is defined in (6.4.7).

The right side of equality (6.4.9) includes characteristics of observed (and not hidden) variables that may be estimated, and this is why we can estimate the hazard rate for values of x satisfying the inequality in (6.4.9). On the other hand, in (6.4.8) the right side of the equality depends on survival function G^{X^*} of an underlying lifetime, and this is why the problem of density estimation for the LTRC is more involved and will be considered later in Section 6.7.

Now we are ready to propose an E-estimator of the hazard rate based on LTRC data. Let $[a, a+b]$ be an interval of estimation where a and b are positive and finite constants. As usual, we begin with a sample mean estimator for Fourier coefficients $\theta_j := \int_a^{a+b} h^{X^*}(x)\psi_j(x)dx$ of the hazard rate $h^{X^*}(x)$, $x \in [a, a+b]$. Recall that $\{\psi_j(x)\}$ is the cosine basis on $[a, a+b]$ introduced in Section 6.1. Using (6.4.9), together with notation (6.4.7), we may propose a plug-in sample mean estimator of the Fourier coefficients,

$$\hat{\theta}_j := n^{-1} \sum_{l=1}^{n} \Delta_l \frac{\psi_j(V_l)}{\hat{g}(V_l)} I(V_l \in [a, a+b]), \qquad (6.4.10)$$

where

$$\hat{g}(v) := n^{-1} \sum_{l=1}^{n} I(T_l \le v \le V_l). \qquad (6.4.11)$$

Statistic $\hat{g}(v)$ is the sample mean estimator of $g(v)$ (see (6.4.7)) and $\hat{g}(V_l) \ge 1/n$.

Fourier estimator (6.4.10) allows us to construct a hazard rate E-estimator $\hat{h}^{X^*}(x)$, and recall that its construction is based on the assumption that hidden random variables T^*, X^*

and C^* are continuous and mutually independent. The corresponding coefficient of difficulty is

$$d(a,b) := b^{-1} \int_a^{a+b} h^{X^*}(x) g^{-1}(x) dx, \qquad (6.4.12)$$

and its plug-in sample mean estimator is

$$\hat{d}(a,b) := n^{-1} b^{-1} \sum_{l=1}^{n} [\hat{g}(V_l)]^{-2} \Delta_l I(V_l \in [a, a+b]). \qquad (6.4.13)$$

Figure 6.4 allows us to gain experience in understanding LTRC data, choosing a feasible interval of estimation, and E-estimation. The top diagram shows by circles and triangle the scattergram of LTRC realizations of (T,V). Circles show uncensored observations, corresponding to $\Delta = 1$, and triangles show censored observations corresponding to $\Delta = 0$. Note that all observations are below the solid line $T = V$; this is what we expect from left-truncated observations. The second from the top diagram allows us to evaluate complexity of the problem at hand. Crosses (and the corresponding scale is shown on the left vertical axis) show values of $1/\hat{g}(V_l)$ for uncensored ($\Delta_l = 1$) observations that are used in estimation of Fourier coefficients; see (4.6.10). The estimated coefficient of difficulty is exhibited by circles and its scale is shown on the right vertical axis. Both estimates are shown for $V_l \in [a1, a1 + b1]$, and Figure 6.4 allows one to change the interval. For the data at hand, function $1/\hat{g}(v)$ has sharply increasing tails, and the same can be said about right tail of the coefficient of difficulty. These two estimates can be used for choosing a feasible interval of estimation for the E-estimator of the hazard rate. Two bottom diagrams show us hazard rate E-estimates for different intervals of estimation. Note that the larger interval includes areas with very large values of $1/g(v)$ and this dramatically increases the coefficient of difficulty, the confidence bands, and the ISE. The "effective" number N of uncensored V_l fallen within an interval of estimation is indicated in a corresponding title. In the bottom diagram N is almost twice smaller than in the third from the top diagram, and nonetheless the estimation is dramatically better. This sheds another light on the effect of interval of estimation and the complexity of estimating tails.

So far we have considered the case of continuous and independent underlying hidden random variables. This may be not the case in some applications. For instance, in a clinical trial we may have $C^* := T^* + \min(u, U^*)$ where $\mathbb{P}(U^* \geq 0) = 1$ and u is a positive constant that defines length of the trial. Let us consider a LTRC where a continuous lifetime X^* is independent of (T^*, C^*) while T^* and C^* may be dependent and have a mixed (continuous and discrete) joint distribution.

We begin with formulas for involved distributions. Write,

$$\mathbb{P}(V \leq v, \Delta = 1) = \mathbb{P}(X^* \leq v, X^* \leq C^* | T^* \leq \min(X^*, C^*))$$

$$= p^{-1} \mathbb{P}(X^* \leq v, X^* \leq C^*, T^* \leq X^*) = p^{-1} \int_0^v f^{X^*}(x) \mathbb{P}(T^* \leq x \leq C^*) dx. \qquad (6.4.14)$$

Here (compare with identical (6.4.1))

$$p := \mathbb{P}(T^* \leq \min(X^*, C^*)). \qquad (6.4.15)$$

Differentiation of (6.4.14) with respect to v yields a formula for the mixed density,

$$f^{V,\Delta}(v,1) = p^{-1} f^{X^*}(v) \mathbb{P}(T^* \leq v \leq C^*) = h^{X^*}(v) [p^{-1} G^{X^*}(v) \mathbb{P}(T^* \leq v \leq C^*)]. \qquad (6.4.16)$$

Further, we can write (compare with (6.4.7))

$$\mathbb{P}(T \leq x \leq V) = \mathbb{P}(T^* \leq x \leq V^* | T^* \leq V^*)$$

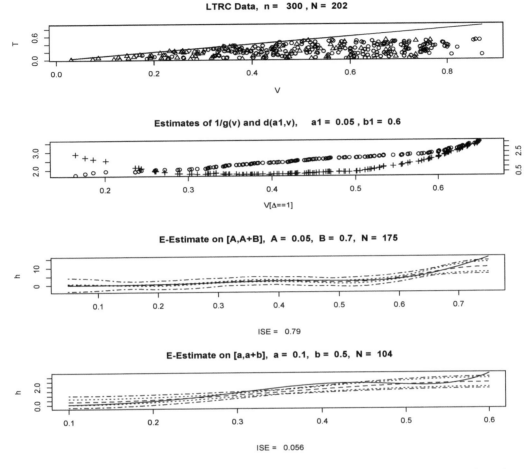

Figure 6.4 *Estimation of the hazard rate based on LTRC observations generated by independent and continuous hidden variables. The top diagram shows a sample of size $n = 300$ from (T, V, Δ). Uncensored (their number is $N := \sum_{l=1}^{n} \Delta_l = 202$) and censored observations are shown by circles and triangles, respectively. The solid line is $T = V$. In the underlying hidden model, the random variable of interest X^* is the Bimodal, the truncating variable T^* is Uniform$(0, 0.5)$, and the censoring variable C^* is Uniform$(0, 1.5)$. Second from the top diagram shows us by crosses and circles estimates $1/\hat{g}(V_l)$ and $\hat{d}(a1, V_l)$ for uncensored $V_l \in [a1, a1 + b1]$. Note the different scales for the two estimates shown on the left and right vertical axes, respectively. Two bottom diagrams show E-estimate (the dashed line), underlying hazard rate (the solid line) and pointwise and simultaneous $1 - \alpha = 0.95$ confidence bands by dotted and dot-dashed lines. The interval of estimation and the number N of observations fallen within the interval are shown in the title, the subtitle shows the ISE of E-estimate over the interval. {Distribution of T^* is either the default Uniform$(0, u_T)$ with $u_T = 0.5$, or Exponential(λ_T) with the default $\lambda_T = 0.3$ where λ_T is the mean. Censoring distribution is either the default Uniform$(0, u_C)$ with $u_C = 1.5$ or Exponential(λ_C) with the default $\lambda_C = 1.5$. For instance, to choose exponential truncation and censoring, set trunc = "Expon" and cens = "Expon".} [n = 300, corn = 3, trunc = "Unif", uT = 0.5, lambdaT = 0.3, cens = "Unif", uC = 1.5, lambdaC = 1.5, a = 0.1, b = 0.5, A = 0.05, B = 0.7, a1 = 0.05, b1 = 0.6, alpha = 0.05, cJ0 = 4, cJ1 = 0.5, cTH = 4]*

$$= p^{-1}\mathbb{P}(T^* \leq x, V^* \geq x, T^* \leq V^*) = p^{-1}\mathbb{P}(T^* \leq x, C^* \geq x, X^* \geq x)$$
$$= [p^{-1}\mathbb{P}(T^* \leq x \leq C^*)]G^{X^*}(x), \quad x \in [0, \infty). \qquad (6.4.17)$$

LTRC Data, n = 300 , N = 184

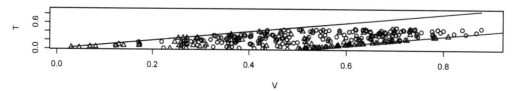

Estimates of 1/g(v) and d(a1,v),   a1 = 0.05 , b1 = 0.6

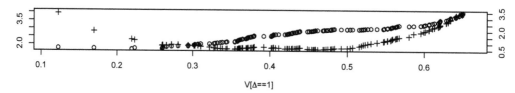

E-Estimate on [A,A+B], A = 0.05, B = 0.7, N = 171

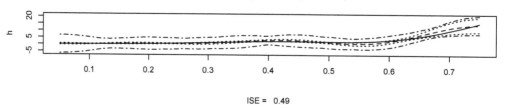

ISE = 0.49

E-Estimate on [a,a+b], a = 0.1, b = 0.5, N = 111

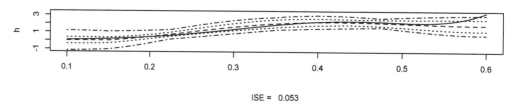

ISE = 0.053

Figure 6.5 *Estimation of the hazard rate for LTRC data with $C^* = T^* + \min(u, U^*)$ and X^* being independent of (T^*, C^*). The default U^* is Uniform$(0, u_C)$ and independent of T^*, and $u = 0.5$. The structure of diagrams is identical to Figure 6.4, only in the top diagram the second solid line is added to indicate largest possible censored observations. {Distribution of T^* is either the default Uniform$(0, u_T)$ with $u_T = 0.5$, or Exponential(λ_T) with the default $\lambda_T = 0.3$ where λ_T is the mean. Distribution of U^* is either the default Uniform$(0, u_C)$ with $u_C = 1.5$ or Exponential(λ_C) with the default $\lambda_C = 1.5$. To choose the exponential truncation and censoring, set trunc = "Expon" and cens = "Expon".} [n = 300, corn = 3, trunc = "Unif", uT = 0.5, lambdaT = 0.3, cens = "Unif", u = 0.5, uC = 1.5, lambdaC = 1.5, a = 0, b = 0.55, B = 0.75, a1 = 0, b1 = 0.75, alpha = 0.05, cJ0 = 4, cJ1 = 0.5, cTH = 4]*

Combining the last two results we conclude that

$$h^{X^*}(x) = \frac{f^{V,\Delta}(x, 1)}{\mathbb{P}(T \le x \le V)} \quad \text{whenever } \mathbb{P}(T^* \le x \le C^*) G^{X^*}(x) > 0. \tag{6.4.18}$$

Note that the inequality in (6.4.18) is equivalent to $\mathbb{P}(T \le x \le V) > 0$. As a result, we may conclude that (6.4.18) is the same formula as (6.4.9) that was established for the case of independent and continuous hidden random variables T^* and C^*. Hence the proposed E-estimator, based on this formula, is robust toward the distribution of (T^*, C^*).

Figure 6.5 allows us to look at LTRC data generated for the case $C^* = T^* + \min(u, U^*)$ and test performance of the hazard rate E-estimator. Please pay attention to the specific shape of the LTRC data in the top diagram. To shed light on the shape, recall the clinical trial example and note that all available observations in a scattergram should be between two parallel lines $T = V$ and $T = V - u$ defined by the baseline and end of the trial, respectively. The second diagram allows us to choose a feasible interval of estimation, and then we have a good chance to obtain a fair hazard rate estimate. Overall, repeated simulations show that the proposed E-estimator performs respectively well and it is robust.

It is highly recommended to repeat Figures 6.4 and 6.5 with different underlying corner distributions and parameters to gain firsthand experience in understanding of and dealing with LTRC data.

6.5   Estimation of Distributions for RC Data

In survival analysis right censoring is the most typical modification of data. For instance, if we are interested in lifetime X of a light bulb that may accidentally break at time C before it burns out, then the lifetime of interest X is right censored by the time of the accident C. Note that RC may be looked at as a missing mechanism when a missed realization of X is substituted by a corresponding realization of C. The missing mechanism is MNAR, because the missing is defined by the value of X, and hence a destructive missing may be expected. As we will see shortly, in some cases RC indeed precludes us from consistent estimation of the distribution of X.

Our aim is to understand when, based on RC observations, a consistent estimation of the distribution of random variables X and C is possible or impossible, and in the former case propose estimators for the survival functions and probability densities.

Statistical model considered in this section is as follows. The available data is a sample from a pair of random variables (V, Δ) where $V := \min(X, C)$ and $\Delta := I(X \leq C)$. It is assumed that X and C are continuous and independent random variables. Introduce a function

$$g(v) := \mathbb{P}(V \geq v) = G^V(v) = \mathbb{E}\{I(V > v)\}$$
$$= \mathbb{P}(V > v) = \mathbb{P}(X > v, C > v) = G^X(v)G^C(v), \qquad (6.5.1)$$

and note that the upper bound of the support of V is

$$\beta_V = \min(\beta_X, \beta_C). \qquad (6.5.2)$$

We can make two important conclusions from (6.5.1) and (6.5.2). The first one is that random variables X and C are absolutely symmetric in the sense that if C is the random variable of interest, then it is right censored by X and the corresponding indicator of censoring is $1 - \Delta$. As a result, if we know how to estimate G^X and f^X then the same estimators can be used for the censoring random variable C via using $1 - \Delta$ in place of Δ. Second, the distributions can be estimated only over the interval $[0, \beta_V]$. As a result, if $\beta_C < \beta_X$ then no consistent estimation of the distribution of X is possible, and if $\beta_C > \beta_X$ then no consistent estimation of the distribution of C is possible. This is why RC may be a destructive modification.

Before proceeding to exploring E-estimation for RC data, let us present a canonical in survival analysis Kaplan–Meier estimator of the survival function,

$$\tilde{G}^X(x) \; := \; 1, \; x < V_{(1)}; \quad \tilde{G}^X(x) := 0, \; x > V_{(n)};$$

$$\tilde{G}^X(x) \; := \; \prod_{i=1}^{l-1}[(n-i)/(n-i+1)]^{\Delta_{(i)}}, \quad V_{(l-1)} < x \leq V_{(l)}, \qquad (6.5.3)$$

where $(V_{(l)}, \Delta_{(l)})$ are ordered V_l's with their corresponding Δ_l, $l = 1, \ldots, n$. Kaplan–Meier estimator for G^C is obtained by replacing Δ on $1 - \Delta$.

This is not a simple task to explain the underlying idea of Kaplan–Meier estimator, and it is even more difficult to infer about its statistical properties. Let us present one possible explanation via a specific example. Consider a medical study on longevity of n individuals. Denote by $\tilde{y}_{l_1} < \tilde{y}_{l_2} < \ldots$ observations of V_{l_s} such that corresponding $\Delta_{l_s} = 1$. In other words, $\tilde{y}_{l_s}$ is the sth time of death during the study, and note that because V is a continuous random variable, the probability of simultaneous deaths is zero. To shed additional light on the notation, between times V_{l_1} and V_{l_2}, $l_2 - l_1 + 1$ individuals left the study (their survival times were censored), and after time V_{l_s} only $n - l_s$ individuals are left in the study. Suppose that we are interested in the survival function $G^X(x)$ at a time $x \in [V_{l_2}, V_{l_3})$. Using a probability formula $\mathbb{P}(A \cap B \cap C) = \mathbb{P}(A)\mathbb{P}(B|A)\mathbb{P}(C|A \cap B)$, which expresses the probability of intersection of events via conditional probabilities, we may write,

$$G^X(x) := \mathbb{P}(X > x) = \mathbb{P}\big(\text{survive in } [0, V_{l_1})\big)\mathbb{P}\big(\text{survive in } [V_{l_1}, V_{l_2})|\text{survive in } [0, V_{l_1})\big)$$

$$\times \mathbb{P}\big(\text{survive in } [V_{l_2}, x]|\text{survive in } [0, V_{l_2})\big). \qquad (6.5.4)$$

For the first probability in the right side of (6.5.4) a natural estimate is 1 because no deaths have been recorded prior to the moment V_{l_1}. For the second probability a natural estimate is $(n - l_1)/(n - l_1 + 1)$ because $n - l_1 + 1$ is the number of individuals remaining in the study before time V_{l_1} and then one individual died prior to moment V_{l_2}. Absolutely similarly, for the third probability a natural estimate is $(n - l_2)/(n - l_2 + 1)$. If we plug these estimators in (6.5.4), then we get the Kaplan–Meier estimator

$$\tilde{G}^X(x) := [(n - l_1)/(n - l_1 + 1)] \times [(n - l_2)/(n - l_2 + 1)]. \qquad (6.5.5)$$

This explanation also sheds light on the notion of product-limit estimation often applied to Kaplan–Meier estimator.

Kaplan–Meir estimator is the most popular estimator of survival function for RC data. At the same time, it is not as simple as the classical empirical (sample mean) cumulative distribution estimator $\hat{F}^X(x) := n^{-1} \sum_{l=1}^n I(X_l \le x)$ used for the case of direct observations. Can a sample mean method be used for estimation of the survival function and RC data? The answer is "yes" and below it is explained how this estimator can be constructed.

Let us explain the proposed method of estimation of the survival function and density of X based on RC data. As usual, we begin with formulas for the joint density of observed variables,

$$f^{V,\Delta}(x, 1) = f^X(x)G^C(x) = [f^X(x)/G^X(x)]g(x) = h^X(x)g(x), \qquad (6.5.6)$$

and

$$f^{V,\Delta}(x, 0) = f^C(x)G^X(x) = [f^C(x)/G^C(x)]g(x) = h^C(x)g(x), \qquad (6.5.7)$$

where $g(x)$ is defined in (6.5.1), $h^X(x)$ and $h^C(x)$ are the hazard rates of X and C. Note that (6.5.6) and (6.5.7) contain several useful expressions for the joint density that shed extra light on RC. Further, please look again at the formulas and note the symmetry of the RC with respect to the lifetime of interest and censoring variable.

We begin with the explanation of how the sample mean methodology can be used for estimation of the survival function $G^C(x)$ of the censoring random variable. Recall that it can be estimated only for $x \in [0, \beta]$ where $\beta \le \beta_V := \min(\beta_X, \beta_C)$. The idea of estimation is based on a formula which expresses the survival function via the corresponding cumulative hazard $H^C(x)$,

$$G^C(x) = \exp\{-H^C(x)\}. \qquad (6.5.8)$$

Using (6.5.1) and (6.5.7), the cumulative hazard may be written as

$$H^C(x) := \int_0^x h^C(u)du = \int_0^x [f^C(u)/G^C(u)]du$$

$$= \int_0^x [f^{V,\Delta}(u,0)/g(u)]du = \mathbb{E}\{(1-\Delta)I(V \in [0,x])/g(V)\}. \qquad (6.5.9)$$

As a result, we can use a plug-in sample mean estimator for the cumulative hazard, then plug it in (6.5.8) and get the following estimator of the survival function,

$$\hat{G}^C(x) := \exp\{-n^{-1}\sum_{l=1}^n (1-\Delta_l)I(V_l \leq x)/\hat{g}(V_l)\}. \qquad (6.5.10)$$

Here

$$\hat{g}(x) := n^{-1}\sum_{l=1}^n I(V_l \geq x) \qquad (6.5.11)$$

is the sample mean estimate of $g(x)$ defined in (6.5.1). Note that $\hat{g}(V_l) \geq n^{-1}$ and hence the estimator (6.5.10) is well defined.

The appealing feature of estimator (6.5.10) is its simple interpretation because we estimate the logarithm of the survival function by a sample mean estimator. This is why we may refer to the estimator as a sample mean estimator (it is also explained in the Notes that this estimator, written as a product-limit, becomes a Nelson–Aalen–Breslow estimator which is another canonical estimator in the survival analysis). Another important remark is that $G^X(x)$ also may be estimated by (6.5.10) with $1 - \Delta_l$ being replaced by Δ_l.

Now let us consider estimation of the probability density $f^X(x)$, $x \in [0, \beta]$. We are using notation $\{\psi_j(x)\}$ for the cosine basis on $[0, \beta]$. Fourier coefficients of $f^X(x)$ can be written, using (6.5.6), as

$$\theta_j := \int_0^\beta f^X(x)\psi_j(x)dx = \int_0^\beta [f^{V,\Delta}(x,1)\psi_j(x)/G^C(x)]dx$$

$$= \mathbb{E}\{\Delta I(V \leq \beta)\psi_j(V)/G^C(V)\}. \qquad (6.5.12)$$

This implies a plug-in sample mean estimator

$$\hat{\theta}_j := n^{-1}\sum_{l=1}^n \Delta_l I(V_l \leq \beta)\psi_j(V_l)/\hat{G}^C(V_l). \qquad (6.5.13)$$

In its turn, the Fourier estimator yields a corresponding density E-estimator $\hat{f}^X(x)$, $x \in [0, \beta]$. Further, if $\beta < \beta_C$ then the corresponding coefficient of difficulty is

$$d(0,\beta) := \beta^{-1}\mathbb{E}\{I(V \leq \beta)\Delta[G^C(V)]^{-2}\} = \beta^{-1}\int_0^\beta \frac{f^X(x)}{G^C(x)}dx, \qquad (6.5.14)$$

and it can be estimated by a sample mean estimator

$$\hat{d}(0,\beta) := n^{-1}\beta^{-1}\sum_{l=1}^n \Delta_l I(V_l \leq \beta_V)[\hat{G}^C(V_l)]^{-2}. \qquad (6.5.15)$$

Let us check how the estimator performs. Figure 6.6 exhibits a particular RC sample and the suggested estimates. In the simulation $\beta_C = 1.5 > \beta_X = 1$, and hence the distribution of X can be consistently estimated. Keeping this remark in mind, let us look at the data

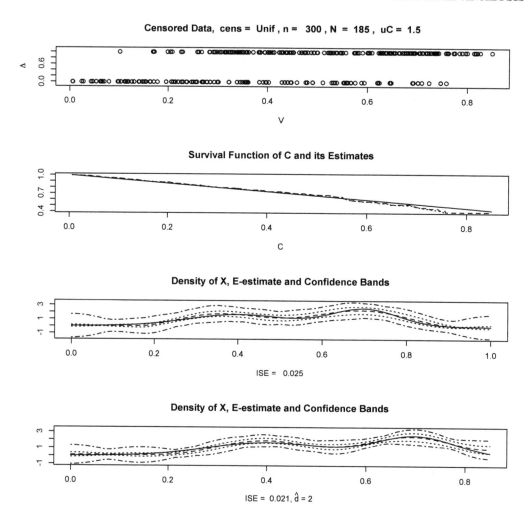

Figure 6.6 *Right-censored data with $\beta_C > \beta_X$ and estimation of distributions. The top diagram shows simulated data. The distribution of the censoring random variable C is uniform on $[0, u_C]$ and the lifetime of interest X is the Bimodal. The second from the top diagram shows by the solid line the underlying G^C, dashed and dotted lines show the sample mean and Kaplan–Meier estimates, respectively. The third from the top diagram shows the underlying density (the solid line), the E-estimate (the dashed line), and $(1 - \alpha)$ pointwise (the dotted lines) and simultaneous (the dot-dashed lines) confidence bands. The estimate is for the interval [0,1]. The bottom diagram is similar, only here the estimate is for the interval $[0, V_{(n)}]$ where $V_{(n)} := \max(V_1, \ldots, V_n)$. {For exponential censoring set cens = "Expon", and the mean of the exponential censoring variable is controlled by argument lambdaC}. [n = 300, corn = 3, cens = "Unif", uC = 1.5, lambdaC = 1.5, alpha = 0.05, cJ0 = 4, cJ1 = 0.5, cTH = 4]*

and estimates. From the title for the top diagram we note that despite the relatively large sample size $n = 300$, only $N := \sum_{l=1}^{n} \Delta_l = 185$ observations of X are not censored; similarly only $n - N = 115$ observations of C are available. Note that observations of X are shown by the horizontal coordinates of circles corresponding to $\Delta = 1$ and observations of C are similarly shown via $\Delta = 0$. Interestingly, the largest observations of X and C are far from $\beta_V = 1$ despite the large sample size. Further, despite the fact that $\beta_C > \beta_X$, the largest observed lifetime is larger than the largest observed censoring variable. The second from

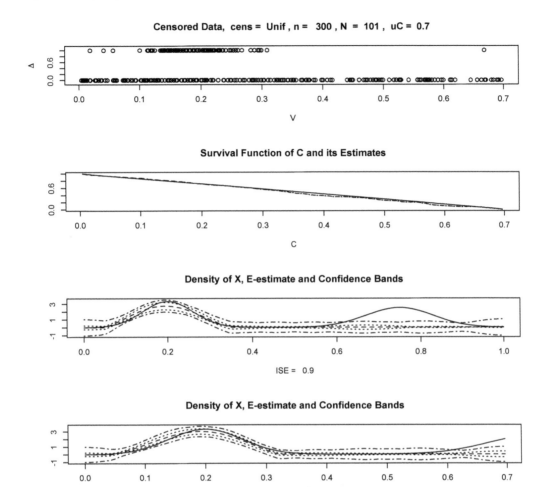

Figure 6.7 *Right-censored data with $\beta_C < \beta_X$ and estimation of distributions. This figure is created by Figure 6.6 using arguments uC = 0.7 and corn = 4. [n = 300, corn = 4, cens = "Unif", uC = 0.7, lambdaC = 1.5, alpha = 0.05, cJ0 = 4, cJ1 = 0.5, cTH = 4]*

the top diagram shows the underlying G^C, the proposed sample-mean and Kaplan–Meier estimates (consult the caption about the corresponding curves). The estimates are close to each other. Let us also stress that there is no chance to consistently estimate the right tail of the distribution of C due to the fact that $\beta_X < \beta_C$. The two bottom diagrams show how the density E-estimator, constructed for intervals $[0, 1]$ and $[0, V_{(n)}]$, performs. The first interval is of a special interest because $[0, 1]$ is the true support of the density f^X, the second one is a data-driven interval of estimation. Keeping in mind that only $N = 185$ uncensored observations are available and the Bimodal is a difficult density for estimation even for the case of direct observations, the particular density E-estimates are very good. Also look at and compare the corresponding confidence bands and ISEs.

What will be if $\beta_C < \beta_X$? Theoretically this precludes us from consistent estimation of the right tail of $f^X(x)$, but still can we estimate the density over an interval $[0, \beta]$ with $\beta < \beta_V$? The above-presented theory indicates that this is possible, and Figure 6.7 illustrates this case. This figure is generated by Figure 6.6 using $u_C = 0.7$, also note that the underlying density is the Strata. The title for the top diagram shows that only $N = 101$ uncensored observations of the lifetime X are available, and hence we are dealing with a severe censoring. Further, note that while the largest observation of C is close to $\beta_V = \beta_C = 0.7$, all but one observations of X are smaller than 0.32. The second from the top diagram exhibits very good estimates of the survival function $G^C(x)$ over the interval $[0, 0.7]$, and this is not a surprise due to the large number $n - N = 199$ of observed realizations of the censored variable. Now let us look at the two bottom diagrams. As it could be predicted, the density estimate in the second from the bottom diagram is not satisfactory because there are simply no observations to estimate the right part of the density. The E-estimate for the interval $[0, V_{(n)}]$, shown in the bottom diagram, is better but still its right tail is poor because we have just one observation of X which is larger than 0.32 and hence there is no way an estimator may indicate the large underlying right stratum.

It is recommended to repeat Figures 6.6 and 6.7 with different corner functions, censoring distributions and parameters to gain an experience in dealing with RC data.

6.6   Estimation of Distributions for LT Data

The aim is to understand how survival functions and probability densities for random variables, defining a left truncation (LT), may be estimated. Let us briefly recall the LT mechanism and involved random variables (lifetimes), more can be found in Section 6.3. A hidden random variable of interest X^* is observed only if X^* is not smaller than a hidden truncation random variable T^*, otherwise it is not even known that the event $X^* < T^*$ occurred. In what follows it is assumed that X^* and T^* are independent, continuous and nonnegative random variables (lifetimes). Recall notation α_Z and β_Z for the lower and upper bounds of the support of a random variable Z. It is assumed that supports of the hidden variables satisfy the following two relations,

$$\alpha_{X^*} \geq \alpha_{T^*}, \tag{6.6.1}$$

and

$$\beta_{X^*} \geq \beta_{T^*}. \tag{6.6.2}$$

Let us comment on these two assumptions. If $\alpha_{X^*} < \alpha_{T^*}$ then all observations of X^* less than α_{T^*} are truncated and not available. This precludes us from estimation of $f^{X^*}(x)$ for $x < \alpha_{T^*}$. If $\beta_{X^*} < \beta_{T^*}$ then the right tail of $f^{T^*}(t)$ cannot be consistently estimated. Let us also note that in what follows we are interested only in the case $\beta_{T^*} > \alpha_{X^*}$ because otherwise no truncation occurs. We will comment at the end of the section on estimation when the assumptions are not valid.

Now let us present several useful probability formulas for the LT model. The joint distribution function of the observed random variables (T, Y) is

$$F^{T,X}(t,x) = F^{T^*,X^*|T^* \leq X^*}(t,x) = p^{-1}\mathbb{P}(T^* \leq t, X^* \leq x, 0 \leq T^* \leq X^* \leq x)$$

$$= p^{-1}\int_0^t f^{T^*}(\tau)[\int_\tau^{\max(x,\tau)} f^{X^*}(u)du]d\tau, \tag{6.6.3}$$

where

$$p := \mathbb{P}(T^* \leq X^*) = \int_0^\infty f^{T^*}(t)G^{X^*}(t)dt \tag{6.6.4}$$

is the probability of X^* to be observed (not truncated by T^*). Then, by taking partial derivatives, we get the joint density of (T, X),

$$f^{T,X}(t,x) = p^{-1}f^{T^*}(t)f^{X^*}(x)I(0 \leq t \leq x < \infty). \tag{6.6.5}$$

In its turn, (6.6.5) yields two marginal densities for T and X,

$$f^T(t) = p^{-1} f^{T^*}(t) G^{X^*}(t) = \frac{f^{T^*}(t) g(t)}{F^{T^*}(t)}, \tag{6.6.6}$$

and

$$f^X(x) = p^{-1} f^{X^*}(x) F^{T^*}(x) = \frac{f^{X^*}(x) g(x)}{G^{X^*}(x)}, \tag{6.6.7}$$

where

$$g(u) := \mathbb{P}(T \le u \le X) = p^{-1} F^{T^*}(u) G^{X^*}(u). \tag{6.6.8}$$

Now we are in a position to explain proposed estimators. We begin with estimators for the boundary points of the supports,

$$\hat{\alpha}_{X^*} := X_{(1)} := \min(X_1, \dots, X_n), \quad \hat{\beta}_{T^*} := T_{(n)} := \max(T_1, \dots, T_n),$$

$$\hat{\beta}_{X^*} := X_{(n)} := \max(X_1, \dots, X_n). \tag{6.6.9}$$

Next we estimate the cumulative hazard $H^{X^*}(x) := \int_0^x h^{X^*}(u) du$ for $x < \beta_{X^*}$. Using (6.6.7) and (6.6.8) we may write,

$$H^{X^*}(x) = I(x > \alpha_{X^*}) \int_0^x h^{X^*}(u) du = I(x > \alpha_{X^*}) \int_0^x [f^{X^*}(u)/G^{X^*}(u)] du$$

$$= I(x > \alpha_{X^*}) \int_0^x [f^X(u)/g(u)] du = I(x > \alpha_{X^*}) \mathbb{E}\{I(X \le x) g^{-1}(X)\}. \tag{6.6.10}$$

Note that the cumulative hazard is written as the expectation of a function of observed variables. Hence we can estimate it by a plug-in sample mean estimator,

$$\hat{H}^{X^*}(x) = n^{-1} \sum_{l=1}^n \frac{I(X_l \le x)}{\hat{g}(X_l)}, \tag{6.6.11}$$

where

$$\hat{g}(x) := n^{-1} \sum_{l=1}^n I(T_l \le x \le X_l), \tag{6.6.12}$$

and this is a sample mean estimator of $g(x) = \mathbb{E}\{I(T \le x \le X)\}$. Note that estimator (6.6.11) is zero for $x \le X_{(1)}$ and it is equal to $\hat{H}^{X^*}(X_{(n)})$ for all $x \ge X_{(n)}$. In what follows we may refer to estimator (6.6.11) as an empirical cumulative hazard.

Now we can explain how to estimate the survival function

$$G^{X^*}(x) := \mathbb{P}(X^* > x) = e^{-\int_0^x h^{X^*}(u) du} = e^{-H^{X^*}(x)}. \tag{6.6.13}$$

We plug the empirical cumulative hazard (6.6.11) in the right side of (6.6.13) and get

$$\hat{G}^{X^*}(x) := e^{-\hat{H}^{X^*}(x)}. \tag{6.6.14}$$

The attractive feature of the estimator (6.6.14) is that its construction is simple and it is easy for statistical analysis. For instance, suppose that its asymptotic (as the sample size increases) variance is of interest. Then the asymptotic variance of the cumulative hazard is calculated straightforwardly because it is a sample mean estimator, and then the variance of the survival function is evaluated by the delta method. Let us follow these two steps and calculate the asymptotic variance. First, under the above-made assumptions the asymptotic variance of the empirical cumulative hazard is

$$\lim_{n\to\infty}[n\mathbb{V}(\hat{H}^{X^*}(x))] = \mathbb{E}\{I(X \le x)/g^2(X)\} - [H^{X^*}(x)]^2. \tag{6.6.15}$$

Second, the delta method yields

$$\lim_{n\to\infty}[n\mathbb{V}(\hat{G}^{X^*}(x))] = [G^{X^*}(x)]^2(\mathbb{E}\{I(X \le x)/g^2(X)\} - [H^{X^*}(x))]^2)$$

$$= [G^{X^*}(x)]^2\Big(p\int_0^x \frac{f^{X^*}(u)}{F^{T^*}(u)[G^{X^*}(u)]^2}du - [H^{X^*}(x))]^2\Big). \tag{6.6.16}$$

There are two lines in (6.6.16) and both are of interest to us. The top line explains how the variance may be estimated via a combination of plug-in and sample mean techniques. Furthermore, for a sample mean estimator, under a mild assumption, the central limit theorem yields asymptotic normality, then the delta method tells us that the asymptotic normality is preserved by the exponential transformation (6.6.14), and hence we can use this approach for obtaining confidence bands for the estimator (6.6.14). The bottom line in (6.6.16) is important for our understanding conditions implying consistent estimation of the survival function $G^{X^*}(x)$. For instance, if $\alpha_{T^*} < \alpha_{X^*}$, then the integral in (6.6.16) is finite. To see this, note that $0/0 = 0$ and hence the integral over $u \in [0, x]$, $x > \alpha_{X^*}$ is the same as the integral over $u \in [\alpha_{X^*}, x]$ where $F^{T^*}(u) \ge F^{T^*}(\alpha_{X^*}) > 0$ due to the assumed $\alpha_{T^*} < \alpha_{X^*}$. The situation changes if $\alpha_{T^*} = \alpha_{X^*}$, and this is a case in many applications. Depending on the ratio $f^{X^*}(u)/F^{T^*}(u)$ for u near α_{T^*}, the integral in (6.6.16) is either finite or infinity. This is what makes the effect of the LT on estimation so unpredictable because it may preclude us from consistent estimation of G^{X^*} even if $\alpha_{T^*} = \alpha_{X^*}$. Recall that we made a similar conclusion in Section 6.3 for estimation of the hazard rate. Furthermore, if $\alpha_{T^*} > \alpha_{X^*}$ then no consistent estimation of the distribution of X^* is possible, and it will be explained shortly what may be estimated in this case.

Now we are developing an E-estimator of the probability density $f^{X^*}(x)$. According to (6.6.7) we have the following relation,

$$f^{X^*}(x) = f^X(x)G^{X^*}(x)/g(x) \quad \text{whenever} \quad g(x) > 0. \tag{6.6.17}$$

Suppose that we are interested in estimation of the density over an interval $[a, a + b]$ such that $g(x)$ is positive on the interval. Denote by $\{\psi_j(x)\}$ the cosine basis on $[a, a + b]$. Then (6.6.17) allows us to write for a Fourier coefficient,

$$\theta_j := \int_a^{a+b} f^{X^*}(x)\psi_j(x)dx = \mathbb{E}\{I(X \in [a, a+b])\psi_j(X_l)G^{X^*}(X)/g(X)\}. \tag{6.6.18}$$

The expectation in (6.6.18) implies a plug-in sample mean Fourier estimator

$$\hat{\theta}_j := n^{-1}\sum_{l=1}^n I(X_l \in [a, a+b])\psi_j(X_l)\hat{G}^{X^*}(X_l)/\hat{g}(X_l), \tag{6.6.19}$$

where $\hat{G}^{X^*}$ and $\hat{g}$ are defined in (6.6.14) and (6.6.12), respectively.

In its turn, the Fourier estimator yields a density E-estimator $\hat{f}^{X^*}(x)$, $x \in [a, a + b]$. Further, the corresponding coefficient of difficulty and its estimator are

$$d(a, a + b) = b^{-1}\mathbb{E}\{[G^{X^*}(X)/g(X)]^2\}, \tag{6.6.20}$$

and

$$\hat{d}(a, a + b) := b^{-1}n^{-1}\sum_{l=1}^n[\hat{G}^{X^*}(X_l)/\hat{g}(X_l)]^2. \tag{6.6.21}$$

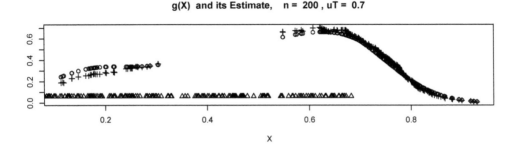

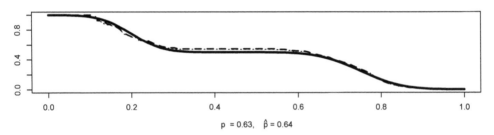

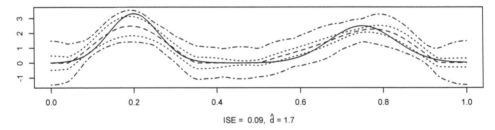

Figure 6.8 *Estimation of the distribution of the lifetime of interest X^* in an LT sample. In the hidden model X^* is distributed according to the Strata (the choice is controlled by parameter corn) and T^* is Uniform($[0, u_T]$). The top diagram shows by circles and crosses $g(X_l)$ and $\hat{g}(X_l)$, and x-coordinates of triangles show observations of the truncating variable T. In the middle diagram, the wide solid, dotted and dashed lines show the underlying survival function, Kaplan–Meyer estimate and sample mean estimate (6.6.14). The subtitle shows the underlying probability (6.6.4) and its estimate (6.6.22). The bottom diagram shows the underlying density (the solid line), the E-estimate (the dashed line), and $(1 - \alpha)$ pointwise (the dotted lines) and simultaneous (the dot-dashed lines) confidence bands. [n = 200, corn = 4, uT = 0.7, alpha = 0.05, cJ0 = 4, cJ1 = 0.5, cTH = 4]*

There is one more interesting unknown parameter that may be estimated. Recall that $p := \mathbb{P}(T^* \leq X^*)$ defines the probability of a hidden pair (T^*, X^*) to be observed, or in other words p defines the likelihood of X^* to be not truncated. Of course it is of interest to know this parameter of the LT. Equality (6.6.8), together with $F^{T^*}(\beta_{T^*}) = 1$, implies that $g(\beta_{T^*}) = p^{-1}G^{X^*}(\beta_{T^*})$. This motivates the following estimator,

$$\hat{p} := \hat{G}^{X^*}(T_{(n)})/\hat{g}(T_{(n)}). \qquad (6.6.22)$$

Note that any $t \geq T_{(n)}$ can be also used in (6.6.22) in place of $T_{(n)}$.

Figure 6.8 allows us to look at a simulated LT sample and evaluate performance of the proposed estimators, explanation of its three diagrams can be found in the caption. The top diagram allows us to visualize realizations of X and T. Here we also can compare the underlying function $g(x)$ and its sample mean estimate (6.6.12). Note how fast the right tail of $g(x)$ vanishes. Nonetheless, as we shall see shortly, this does not preclude us from estimation of the distribution of X^*. The middle diagram indicates that the sample mean and the Kaplan–Meyer estimators perform similarly. Its subtitle shows the underlying parameter p and its estimate (6.6.22). As we see, we can get a relatively fair idea about a hidden sampling which is governed by a negative binomial distribution with parameters (n, p). Finally, the bottom diagram shows us the E-estimate of the density of X^*. As we know, the Strata is a difficult density to estimate even for the case of direct observations. Here we are dealing with LT observations that are, as we know, biased. Overall the particular outcome is good because we clearly observe the two strata. The confidence bands are also reasonable, and the ISE, shown in the subtitle, is relatively small. Further, the subtitle shows us the estimated coefficient of difficulty (6.6.21).

Now let us return to the middle diagram in Figure 6.8 where the survival function and its estimates are shown. In many applications the right tail of the survival function is of a special interest. Let us explore a new idea of the tail estimation. Note that $F^{T^*}(x) = 1$ whenever $x \geq \beta_{T^*}$. This allows us to write for $x \geq \beta_{T^*}$,

$$G^X(x) = \mathbb{P}(X^* > x | T^* \leq X^*)$$

$$= p^{-1}\mathbb{P}(X^* > x, T^* \leq X^*) = p^{-1}G^{X^*}(x), \ \ x \geq \beta_{T^*}. \qquad (6.6.23)$$

We conclude that $G^{X^*}(x) = pG^X(x)$, $x \geq \beta_{T^*}$. The latter is easy to understand because no truncation occurs whenever $X^* \geq \beta_{T^*}$. As a result, for $x \geq \beta_{T^*}$ the survival function of the hidden variable of interest is proportional to the survival function of the observed variable. Parameter p is estimated by (6.6.22), and $G^X(x)$ is estimated by the empirical survival function $\hat{G}^X(x) := n^{-1}\sum_{l=1}^{n} I(X_l > x)$. This gives us an opportunity to propose a new estimator of the tail of $G^{X^*}(x)$,

$$\tilde{G}^X(x) := \hat{p}^{-1}n^{-1}\sum_{l=1}^{n} I(X_l > x), \ \ x > T_{(n)}. \qquad (6.6.24)$$

Figure 6.9 allows us to look at the zoomed-in tails of estimates produced by estimators (6.6.14) and (6.6.24). The two diagrams show results of different simulations for the same LT model. The solid, dashed and dotted lines are the underlying survival function and estimates (6.6.14) and (6.6.24). As we see, the top diagram depicts an outcome where the new estimate is better, this is also stressed by the ratio between the empirical integrated squared error (ISE) of the estimate (6.6.14) and the integrated squared error (ISEN) of the new estimate (6.6.24). Note that there are just few large observations, and this is a rather typical situation with tails. The sample size $n = 75$ is relatively small but it is chosen for better visualization of the estimates. The bottom diagram exhibits results for another simulation, and here the estimate (6.6.14) is better. This is a tie, and we may conclude that in general two simulations are not enough to compare estimators. To resolve the issue, we repeat the simulation 300 times and then statistically analyze ratios of the ISEs. Sample mean and sample median of the ratios are shown in the subtitle, and they indicate better performance of the estimator (6.6.24). Of course, the sample size is too small for estimation of the tail, but the method of choosing between two estimators is statistically sound. The reader is advised to repeat Figure 6.9 with different parameters, compare performance of the two estimators, and gain experience in choosing between several estimators.

Tail of Survival Function and its Estimates

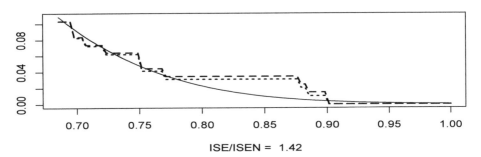

ISE/ISEN = 1.42

n = 75,   uT = 0.7,   corn = 2,   nsim = 300

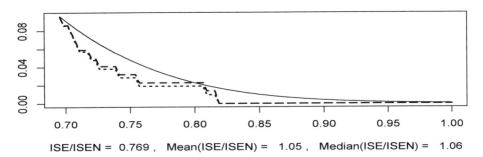

ISE/ISEN = 0.769 ,   Mean(ISE/ISEN) = 1.05 ,   Median(ISE/ISEN) = 1.06

Figure 6.9 *Estimation of the tail of survival function $G^{X^*}(x)$. The underlying simulation is the same as in Figure 6.8. The solid, dashed and dotted lines are the underlying survival function and estimates (6.6.14) and (6.6.24), respectively. {Argument nsim controls the number of simulations.} [n = 75, corn = 2, uT = 0.7, nsim = 300]*

So far we have discussed the problem of estimation of the distribution of a hidden lifetime of interest X^*. It is also of interest to estimate the distribution of a hidden truncating random variable T^*. We again begin with probability formulas. Using (6.6.6) we can write,

$$q^{T^*}(t) := \frac{f^{T^*}(t)}{F^{T^*}(t)} = \frac{f^T(t)}{g(t)} \quad \text{whenever} \ \ g(t) > 0. \tag{6.6.25}$$

Function $q^{T^*}(t)$ can be estimated using the second equality in (6.6.25), and hence we need to understand how the distribution of interest can be expressed via the function $q(t)$. This function resembles the hazard rate only now the denominator is the cumulative distribution function instead of the survival function. A straightforward algebra implies that

$$F^{T^*}(t) = \exp\left(- \int_t^{\beta_{T^*}} q^{T^*}(u)du \right) =: \exp\left(- Q^{T^*}(t) \right)$$

$$= \exp\left(- \mathbb{E}\{I(T > t)/g(T)\} \right), \quad t \in [\alpha_{T^*}, \beta_{T^*}]. \tag{6.6.26}$$

The expectation in (6.6.26) allows us to propose the following plug-in sample mean estimator of the cumulative distribution function (compare with (6.6.14))

$$\hat{F}^{T^*}(t) := \exp\left(- \hat{Q}^{T^*}(t) \right) := \exp\left(- n^{-1} \sum_{l=1}^n I(T_l > t)/\hat{g}(T_l) \right), \tag{6.6.27}$$

LT Data, n = 100, uT = 0.7

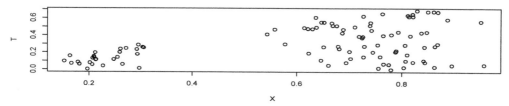

Cumulative Distribution Function of T*, its Estimate and Band

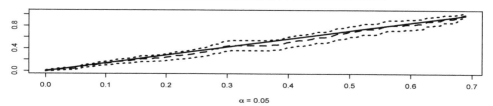

$\alpha = 0.05$

Density of T*, E-estimate and Confidence Bands

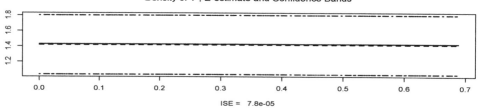

ISE = 7.8e-05

Figure 6.10 *Estimation of the distribution of T^* for LT data. Underlying simulation is the same as in Figure 6.8. Observations are shown in the top diagram. In the middle diagram the solid and dashed lines are the cumulative distribution function F^{T^*} and its estimate. The dotted lines show the $(1-\alpha)$ pointwise confidence band. Curves in the bottom diagram are the same as in the bottom diagram of Figure 6.8. [n = 100, corn = 4, uT = 0.7, alpha = 0.05, cJ0 = 4, cJ1 = 0.5, cTH = 4]*

where $\hat{g}(t)$ is defined in (6.6.12).

Now let us propose an E-estimator of the density $f^{T^*}(t)$ on an interval $[a, a+b]$ such that $g(t)$ is positive on this interval. We are using our traditional notation $\psi_j(t)$ for elements of the cosine basis on $[a, a+b]$. Formula

$$f^{T^*}(t) = f^T(t)F^{T^*}(t)/g(t) \qquad (6.6.28)$$

for the density of interest implies that its Fourier coefficients $\kappa_j := \int_a^{a+b} \psi_j(t)f^{T^*}(t)dt$ can be written as

$$\kappa_j = \int_a^{a+b} \frac{f^T(t)\psi_j(t)F^{T^*}(t)}{g(t)}dt = \mathbb{E}\Big\{ \frac{I(T \in [a, a+b])\psi_j(T)F^{T^*}(T)}{g(T)} \Big\}. \qquad (6.6.29)$$

The expectation in (6.6.29), together with (6.6.27), yields a plug-in sample mean Fourier estimator (compare with (6.6.19))

$$\hat{\kappa}_j := n^{-1}\sum_{l=1}^{n} I(T_l \in [a, a+b])\psi_j(T_l)\hat{F}^{T^*}(T_l)/\hat{g}(T_l). \qquad (6.6.30)$$

The Fourier estimator yields a density E-estimator $\hat{f}^{T^*}(t)$, $t \in [a, a+b]$ with the corresponding coefficient of difficulty

$$d(a, a+b) = b^{-1}\mathbb{E}\big\{ I(T \in [a, a+b])[F^{T^*}(T)/g(T)]^2 \big\}. \qquad (6.6.31)$$

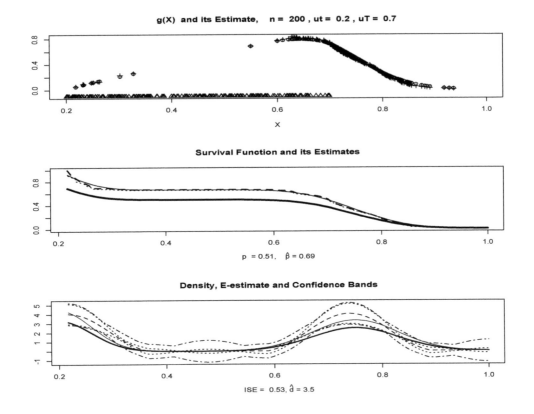

Figure 6.11 *Estimation of the distribution of* X^* *for the case* $\alpha_{X^*} < \alpha_{T^*}$. *This figure is similar to Figure 6.8 apart of two modifications. First, here* T^* *is uniform on interval* $[u_t, u_T] = [0.2, 0.7]$. *Second, the narrow solid lines in the middle and bottom diagrams show the underlying conditional survival function* $G^{X^*|X^*>u_t}(x)$ *and the underlying conditional density* $f^{X^*|X^*>u_t}(x)$, *respectively.* [n = 200, corn = 4, ut = 0.2, uT = 0.7, alpha = 0.05, cJ0 = 4, cJ1 = 0.5, cTH = 4]

Further, typically $[T_{(1)}, T_{(n)})]$ may be recommended as the interval of estimation, and note that $\mathbb{P}(T_{(n)} \le X_{(n)}) = 1$.

Figure 6.10 shows us how the proposed estimators of the cumulative distribution function and the density of the hidden truncation variable T^* perform. For the particular simulation, the estimates are good. Confidence bands for the density may look too wide, but this is only because the E-estimate is good. The reader is advised to repeat Figure 6.10 with different parameters and test performance of the estimator and confidence bands.

Note that we have developed estimators for T^* from scratch. It was explained in the previous sections that for RC data there is a symmetry between estimation of distributions of the lifetime of interest X^* and censoring variable C^*, and if an estimator for X^* is proposed, then it also can be used for C^*. Is there any type of a similar symmetry for LT data? The answer is "yes." Let γ be a positive constant such that $\gamma \ge \max(\beta_{X^*}, \beta_{T^*})$, and introduce two new random variables $X' := \gamma - T^*$ and $T' := \gamma - X^*$. Then (X', T') can be considered as underlying hidden variables for a corresponding LT data with X' being the lifetime of interest and T' being the truncating variable. This is a type of symmetry that allows us to estimate distributions of T^* using estimators developed for X^*.

Finally, let us explain what may be expected when assumptions (6.6.1) and (6.6.2) are violated. If $\alpha_{X^*} < \alpha_{T^*}$ then the LT hides left tail of the distribution of X^*, and we cannot restore it. Violation of (6.6.2) hides right tail of the distribution of T^*.

Figure 6.11 illustrates the case when $\alpha_{X^*} < \alpha_{T^*}$ (note that the diagrams are similar to the ones in Figure 6.8 whose caption explains the diagrams). First of all, let us look at the top diagram which shows realizations of X and T. If we compare them with those in Figure 6.8, then we may conclude that visualization of observations is unlikely to point upon the violation of assumption (6.6.1). The reader is advised to repeat Figures 6.8 and 6.11 and gain experience in assessing LT data. Further, it is clear from the two bottom diagrams that the estimators are inconsistent, and this is what was predicted. Nonetheless, it looks like the estimates mimic the underlying ones. Let us explore this issue and shed light on what the estimators do when $\alpha_{X^*} < \alpha_{T^*}$.

We begin with the estimator (6.6.14) of the survival function. Recall that we are considering the case $\alpha_{X^*} < \alpha_{T^*}$. Note that then $\mathbb{P}(X \leq \alpha_{T^*}) = 0$, and using (6.6.11) we can write for $x > \alpha_{T^*}$,

$$\mathbb{E}\{\hat{H}^{X^*}(x)\} = \mathbb{E}\left\{\frac{I(X < x)I(X > \alpha_{T^*})}{\hat{g}(X)}\right\}$$

$$= \mathbb{E}\left\{\frac{I(\alpha_{T^*} < X < x)}{g(X)}\right\} + \mathbb{E}\left\{I(\alpha_{T^*} < X < x)\left[\frac{1}{\hat{g}(X)} - \frac{1}{g(X)}\right]\right\}. \qquad (6.6.32)$$

The first term in (6.6.32) is what we are interested in because the second one, under a mild assumption, vanishes as n increases. Using (6.6.7) we can express the first term via the hazard rate of X^*,

$$\mathbb{E}\left\{\frac{I(\alpha_{T^*} < X < x)}{g(X)}\right\} = \int_{\alpha_{T^*}}^{x} h^{X^*}(u)du. \qquad (6.6.33)$$

In its turn, (6.6.33) implies the following relation,

$$e^{-\mathbb{E}\left\{\frac{I(\alpha_{T^*} < X < x)}{g(X)}\right\}} = e^{-\int_{\alpha_{T^*}}^{x} h^{X^*}(u)du}$$

$$= \frac{e^{-\int_0^x h^{X^*}(u)du}}{e^{-\int_0^{\alpha_{T^*}} h^{X^*}(u)du}} = \frac{G^{X^*}(x)}{G^{X^*}(\alpha_{T^*})} =: G^{X^*|X^* > \alpha_{T^*}}(x). \qquad (6.6.34)$$

On the right side of (6.6.34) we see the conditional survival function of X^* given $X^* > \alpha_{T^*}$, and this is the characteristic that estimator (6.6.14) estimates.

We may conclude that it is more accurate to say that the survival function estimator (as well as the Kaplan–Meier estimator) estimates the conditional density $G^{X^*|X^* > \alpha_{T^*}}(x)$ because $G^{X^*|X^* > \alpha_{T^*}}(x) = G^{X^*}(x)$ whenever $\alpha_{X^*} \geq \alpha_{T^*}$. This is a nice conclusion because, regardless of the assumption (6.6.1), we estimate a meaningful characteristic of the random variable of interest.

If we return to the middle diagram in Figure 6.11, then we can see that the survival function estimate is above the underlying survival function (the wide solid line), and now we know why. Further, the narrow solid line exhibits the underlying conditional survival function $G^{X^*|X^* > \alpha_{T^*}}(x)$, and we may note that the E-estimator does a good job in estimation of the conditional survival function. Further, note that the estimated probability $\hat{p} = 0.69$ of avoiding the truncation is larger than the underlying $p = 0.53$. To understand why we need to look at the estimator (6.6.22) and recall that for the considered setting the estimator $\hat{G}^{X^*}(T_{(n)})$ estimates the conditional density $G^{X^*|X^* > \alpha_{T^*}}(\beta_T) = G^{X^*}(\beta_T)/G^{X^*}(\alpha_{T^*})$. As a result, we can expect that the estimate $\hat{p}$ will be larger than the underlying p by the factor $1/G^{X^*}(\alpha_{T^*})$. And indeed, $\hat{p}/p = 1.30$ and this is fairly close to $1/G^{X^*}(\alpha_{T^*}) = 1.33$.

Now let us turn our attention to the bottom diagram in Figure 6.11 which is devoted to estimation of the density $f^{X^*}(x)$. The estimate (the dashed line) is far from the underlying density (the wide solid line), and this supports the above-made conclusion that if (6.6.1) is violated then consistent estimation of the density is impossible. Nonetheless, we may notice that the estimate mimics the density's shape. As a result, let us explore the density estimator for the case when (6.6.1) does not hold.

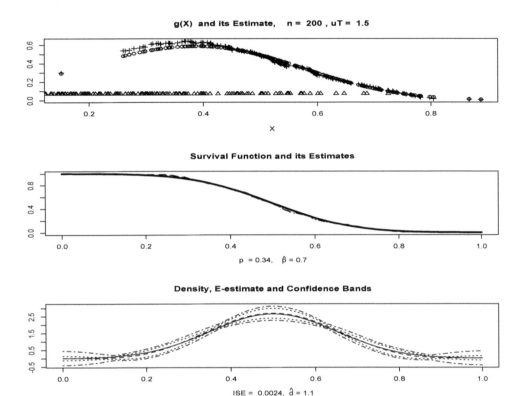

Figure 6.12 *Estimation of the distribution of X^* for the case $\beta_{T^*} > \beta_{X^*}$. The figure is created by Figure 6.8 by setting $uT = 1.5$ and corn $= 2$.*

Consider the expectation of Fourier estimator (6.6.19) given $\alpha_{X^*} < \alpha_{T^*}$. Let $[a, a+b]$ be an interval such that $g(x) > 0$ for $x \in [a, a+b]$. We may write using (6.6.7),

$$\mathbb{E}\{\hat{\theta}_j\} = \mathbb{E}\{I(X \in [a, a+b])\psi_j(X)\hat{G}^{X^*}(X)/\hat{g}(X)\}$$

$$= \mathbb{E}\{I(X \in [a, a+b])\psi_j(X)G^{X^*|X^*>\alpha_{T^*}}(X)/g(X)\}$$

$$+\mathbb{E}\left\{I(X \in [a, a+b])\psi_j(X)\left[\frac{\hat{G}^{X^*}(X)}{\hat{g}(X)} - \frac{G^{X^*|X^*>\alpha_{T^*}}(X)}{g(X)}\right]\right\}. \qquad (6.6.35)$$

The first expectation on the right side of (6.6.35) is the term of interest because the second one, under a mild assumption, vanishes as n increases. Using (6.6.17) and (6.6.34) we may write,

$$\mathbb{E}\{I(X \in [a, a+b])\psi_j(X)G^{X^*|X^*>\alpha_{T^*}}(X)/g(X)\}$$

$$= \int_a^{a+b} f^X(x)\psi_j(x)G^{X^*|X^*>\alpha_{T^*}}(x)[g(x)]^{-1}dx = \int_a^{a+b} \frac{f^{X^*}(x)\psi_j(x)}{G^{X^*}(\alpha_{T^*})}dx. \qquad (6.6.36)$$

Introduce the conditional density,

$$f^{X^*|X^*>\alpha_{T^*}}(x) := \frac{f^{X^*}(x)}{G^{X^*}(\alpha_{T^*})}I(x > \alpha_{T^*}). \qquad (6.6.37)$$

Combining (6.6.35)-(6.6.37) we conclude that the Fourier estimator (6.6.19) estimates Fourier coefficients of the conditional density (6.6.37).

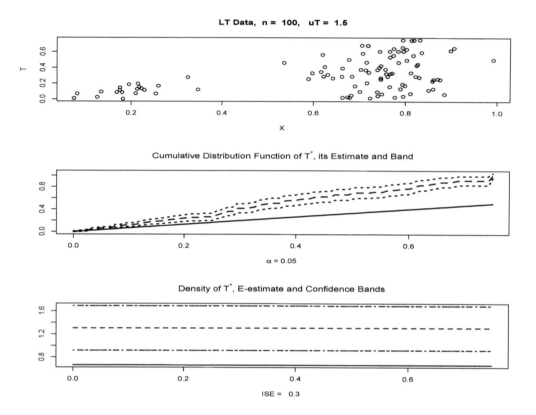

Figure 6.13 *Estimation of the distribution of T^* for the case $\beta_{T^*} > \beta_{X^*}$. The figure is created by Figure 6.10 via setting $uT = 1.5$.*

If we now return to the bottom diagram in Figure 6.11, the narrow solid line shows us the underlying conditional density. The E-estimate (the dashed line) is far from being perfect but its shape correctly indicates the two strata in the underlying conditional density.

Now let us consider the case $\beta_{T^*} > \beta_{X^*}$. Figures 6.12 and 6.13 shed light on this case. Figure 6.12 indicates that this case does not preclude us from consistent estimation of the distribution of X^*. In the top diagram x-coordinates of circles and triangles show us the observed values of X and T, respectively. Our aim is to explore the possibility to realize that $\beta_{T^*} > \beta_{X^*}$. As we discussed earlier, theoretically $T_{(n)}$ should approach $\beta_T = \beta_{X^*} = 1$. Because $X_{(n)}$ converges in probability to $\beta_X = 1$, it may be expected that $T_{(n)}$ and $X_{(n)}$ are close to each other. Unfortunately, even for the used relatively large sample size $n = 200$, $T_{(n)}$ is significantly smaller than $X_{(n)}$ and the outcome resembles what we observed in Figure 6.8. The explanation of this observation is based on formula (6.6.6) for the density $f^T(t)$. It indicates that the density is proportional to the survival function $G^{X^*}(t)$ which vanishes (and in our case relatively fast) as $t \to \beta_X = 1$. This is what significantly slows down the convergence of $T_{(n)}$ to $X_{(n)}$. The lesson learned is that data alone may not allow us to verify validity of (6.6.2). Further, we cannot consistently estimate p. At the same time, as it could be expected, consistent estimation of the distribution of X^* is possible and the exhibited results support this conclusion.

The situation clearly changes when the aim is to estimate the distribution of T^*. Figure 6.13 shows a particular outcome, and it clearly indicates our inability to estimate the distribution of T^*. At the same time, as it follows from formulas, the shape of density $f^{T^*}(t)$ over interval $[T_{(1)}, T_{(n)}]$ may be visualized, and the bottom diagram sheds light on this conclusion. The interested reader may theoretically establish what the proposed estimator estimates for the considered case $\beta_{T^*} > \beta_{X^*}$.

The reader is advised to repeat Figures 6.8–6.13 with different parameters, pay a special attention to feasible intervals of estimation, and get used to statistical analysis of LT data.

6.7 Estimation of Distributions for LTRC Data

As we already know from Section 6.4, in many applications the LT and the RC occur together, and then we are dealing with left truncated and right censored (LTRC) data. A nice example of LTRC, discussed in Section 6.4, is the actuarial example where the random variable of interest is the loss which is left truncated and right censored by the policy's deductible and limit on payment, respectively. Let us also recall several teachable moments learned in that section. First, it is important to realize that data are modified. Second, LTRC may preclude us from consistent estimation. In particular, LT may make impossible estimation of the left tail of the distribution of an underlying lifetime of interest while the RC may make impossible estimation of its right tail. Third, it is important to check assumptions about underlying distributions. In particular, it may be possible to estimate the distribution of the lifetime of interest but not the distribution of the censoring variable and vise versa. Finally, choosing a feasible interval of estimation becomes a critical part of an estimation procedure.

The aim of this section is to explore the problem of estimation of survival functions and probability densities for variables hidden by LTRC modification.

Following Section 6.4, let us briefly recall the LTRC mechanism of generating a sample of size n. There is a hidden sequential sampling from a triplet of nonnegative random variables (T^*, X^*, C^*) whose joint distribution is unknown. T^* is the truncation random variable, X^* is the random variable (lifetime) of interest, and C^* is the censoring random variable. Suppose that (T_k^*, X_k^*, C_k^*) is the kth realization of the hidden triplet and before this realization a sample of size $l-1$, $l-1 \leq \min(k-1, n-1)$ of LTRC observations is collected. Then if $T_k^* > \min(X_k^*, C_k^*)$ then left truncation of the kth realization occurs meaning that: (i) The kth triplet is not observed; (ii) The fact that the kth observation occurred is unknown; (iii) Next realization of the triplet occurs. On the other hand, if $T_k^* \leq \min(X_k^*, C_k^*)$ then the observation $(T_l, V_l, \Delta_l) := (T_k^*, \min(X_k^*, C_k^*), I(X_k^* \leq C_k^*))$ is added to the observed LTRC sample. The hidden sequential sampling from the triplet (T^*, X^*, C^*) stops as soon as $l = n$.

Now let us present basic probability formulas. The probability of observing a realization of the hidden triplet is

$$p := \mathbb{P}(T^* \leq \min(X^*, C^*)). \tag{6.7.1}$$

The joint cumulative distribution function of the observed triplet of random variables is

$$F^{T,V,\Delta}(t,v,\delta) := \mathbb{P}(T \leq t, V \leq v, \Delta \leq \delta)$$

$$= p^{-1}\mathbb{P}(T^* \leq t, T^* \leq V^* \leq v, \Delta^* \leq \delta). \tag{6.7.2}$$

If we additionally assume that hidden random variables T^*, X^* and C^* are independent and continuous, then (6.7.2) yields the following joint mixed density,

$$f^{T,V,\Delta}(t,v,\delta) = p^{-1}f^{T^*}(t)I(t \leq v)\left[f^{X^*}(v)G^{C^*}(v)\right]^\delta\left[f^{C^*}(v)G^{X^*}(v)\right]^{1-\delta}. \tag{6.7.3}$$

In its turn, (6.7.3) yields a marginal density

$$f^{V,\Delta}(v,1) = p^{-1}f^{X^*}(v)G^{C^*}(v)F^{T^*}(v) = h^{X^*}(v)[p^{-1}G^{C^*}(v)F^{T^*}(v)G^{X^*}(v)], \tag{6.7.4}$$

where $h^{X^*}(x) := f^{X^*}(x)/G^{X^*}(x)$ is the hazard rate of X^*. Further, using formula $G^{V^*}(t) = G^{X^*}(t)G^{C^*}(t)$ and (6.7.2) we get

$$f^T(t) = p^{-1}f^{T^*}(t)G^{X^*}(t)G^{C^*}(t). \tag{6.7.5}$$

Further, (6.7.4) yields

$$f^{X^*}(x) = \frac{f^{V,\Delta}(x,1)}{p^{-1}G^{C^*}(x)F^{T^*}(x)} \quad \text{whenever } G^{C^*}(x)F^{T^*}(x) > 0. \tag{6.7.6}$$

Finally, we introduce a probability that plays a key role in the analysis of LTRC data,

$$g(x) := \mathbb{P}(T \le x \le V) = p^{-1}F^{T^*}(x)G^{X^*}(x)G^{C^*}(x). \tag{6.7.7}$$

Now let us formulate assumptions motivated by previous sections. Recall notations α_Z and β_Z for the lower and upper bounds of the support of a variable Z. In what follows it is assumed, in addition to the mutual independence of continuous variables T^*, X^* and C^*, that

$$\min(\alpha_{X^*}, \alpha_{C^*}) \ge \alpha_{T^*}, \quad \beta_{X^*} \ge \beta_{T^*}, \quad \beta_{C^*} \ge \beta_{X^*}. \tag{6.7.8}$$

Now we are in a position to propose estimators for the survival function and density of the lifetime of interest X^*. We begin with estimation of the survival function $G^{X^*}(x)$. Using (6.7.6) and (6.7.7) we conclude that the density of X^* can be written as

$$f^{X^*}(x) = \frac{f^{V,\Delta}(x,1)G^{X^*}(x)}{g(x)}, \quad x \in [0, \beta_{X^*}). \tag{6.7.9}$$

Formula (6.7.9) allows us to obtain a simple formula for the cumulative hazard of X^*,

$$H^{X^*}(x) := \int_0^x [f^{X^*}(u)/G^{X^*}(u)]du$$

$$= \int_0^x [f^{V,\Delta}(u,1)/g(u)]du = \mathbb{E}\{\Delta I(V \le x)g^{-1}(V)\}, \quad x \in [0, \beta_{X^*}). \tag{6.7.10}$$

Recall that $G^{X^*}(x) = \exp\{-H^{X^*}(x)\}$, and then the expectation on the right side of (6.7.10) implies the following plug-in sample mean estimator of the survival function,

$$\hat{G}^{X^*}(x) := \exp\left\{-n^{-1}\sum_{l=1}^n \frac{\Delta_l I(V_l \le x)}{\hat{g}(V_l)}\right\}. \tag{6.7.11}$$

Here

$$\hat{g}(x) := n^{-1}\sum_{l=1}^n I(T_l \le x \le V_l) \tag{6.7.12}$$

is the sample mean estimator of the probability $g(x)$ defined in (6.7.7). Further, it is a straightforward calculation to find an asymptotic expression for the variance of empirical survival function (6.7.11),

$$\lim_{n\to\infty} n\mathbb{V}(\hat{G}^{X^*}(x)) = [G^{X^*}(x)]^2[\mathbb{E}\{\Delta I(V \le x)[g(V)]^{-2}\} - (H^{X^*}(x))^2]. \tag{6.7.13}$$

This result, together with the central limit theorem and delta method, allows us to get a pointwise confidence band for the empirical survival function.

To use our E-estimation methodology for estimation of the density $f^{X^*}(x)$ over an interval $[a, a+b] \subset [\alpha_{X^*}, \beta_{X^*})$, we need to understand how to express its Fourier coefficients as expectations. Recall our notation $\{\psi_j(x)\}$ for the cosine basis on $[a, a+b]$. We can write with the help of (6.7.6) and (6.7.7) that

$$\theta_j := \int_a^{a+b} \psi_j(x) f^{X^*}(x)dx = \int_a^{a+b} \frac{\psi_j(x) f^{V,\Delta}(x,1) G^{X^*}(x)}{g(x)}dx$$

$$= \mathbb{E}\left\{ \frac{\Delta I(V \in [a, a+b]) \psi_j(V) G^{X^*}(V)}{g(V)} \right\}. \qquad (6.7.14)$$

The expectation in (6.7.14) yields the following plug-in sample mean Fourier estimator,

$$\hat{\theta}_j := n^{-1} \sum_{l=1}^n \Delta_l I(V_l \in [a, a+b]) \psi_j(V_l) \hat{G}^{X^*}(V_l)/\hat{g}(V_l). \qquad (6.7.15)$$

Fourier estimator (6.7.15) yields a density E-estimator $\hat{f}^{X^*}(x)$ with the coefficient of difficulty

$$d(a, a+b) = b^{-1} \mathbb{E}\{\Delta I(V \in [a, a+b])[G^{X^*}(V)/g(V)]^2\}$$

$$= pb^{-1} \int_a^{a+b} \frac{f^{X^*}(x)}{F^{T^*}(x) G^{C^*}(x)}dx. \qquad (6.7.16)$$

The coefficient of difficulty is of a special interest to us because it sheds light on a feasible interval of estimation. Namely, we know that $F^{T^*}(x)$ vanishes as $x \to \alpha_{T^*}$ and $G^{C^*}(x)$ vanishes as $x \to \beta_{C^*}$, and this is what may make the integral in (6.7.16) large. The made assumption (6.7.8) allows us to avoid the case of infinite coefficient of difficulty by choosing an interval of estimation satisfying $[a, a+b] \subset [V_{(1)}, V_{(n)}]$, but still the coefficient of difficulty may be prohibitively large. At the same time, vanishing tails of an underlying density f^{X^*} may help in keeping the coefficient of difficulty reasonable, while an increasing tail makes estimation more complicated.

Now let us stress that if $\alpha_{T^*} > \alpha_{X^*}$ then, as it was explained in Section 6.6, the proposed estimators estimate the conditional survival function $G^{X^*|X^*>\alpha_{T^*}}(x)$ and the conditional density $f^{X^*|X^*>\alpha_{T^*}}(x)$. Of course, these conditional characteristics coincide with $G^{X^*}(x)$ and $f^{X^*}(x)$ whenever $\alpha_{T^*} \leq \alpha_{X^*}$, and hence we may say that in general we estimate those conditional characteristics. Statistical analysis of this setting is left as an exercise.

Figure 6.14 illustrates performance of the proposed estimators of the conditional survival function and the conditional density of the lifetime of interest. Its caption explains the simulation and diagrams. Note that the assumption (6.7.8) holds for the particular simulation, and choosing parameter $u_t > 0$ allows us to consider the case $\alpha_{T^*} > \alpha_{X^*}$ and then test estimation of the conditional characteristics.

The top diagram in Figure 6.14 shows us simulated LTRC data generated by the Normal lifetime and uniformly distributed truncating and censoring random variables. These three variables are mutually independent. The good news here is that the number $N = 230$ of available uncensored observations of the lifetime of interest is relatively large with respect to the sample size $n = 300$. If we look at the smallest observations, we may observe that α_{T^*} is close to zero, and this implies a chance for a good estimation of the left tail of the distribution of X^*. We may also conclude that it is likely that $\beta_{X^*} < \beta_{C^*}$ because a relatively large number of uncensored observations are available to the right of the largest observation of the censoring variable. The middle diagram shows us the proposed plug-in sample mean estimate (6.7.11) of $G^{X^*}(x)$, the Kaplan–Meier estimate, and the above-explained 95% confidence band for the plug-in sample mean estimate. The two estimates are practically the same. Despite a relatively large sample size $n = 300$, we see a pronounced deviation of

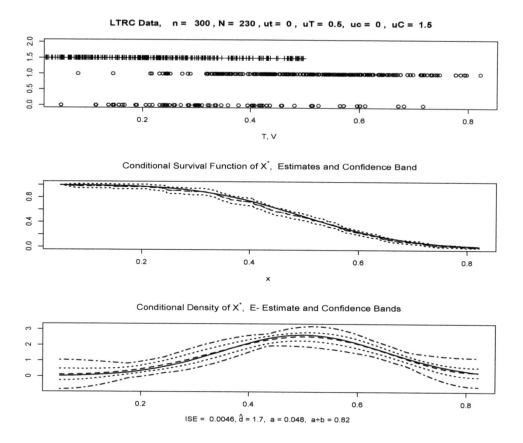

Figure 6.14 *Estimation of the conditional survival function $G^{X^*|X^*>\alpha_{T^*}}(x)$ and the conditional density $f^{X^*|X^*>\alpha_{T^*}}(x)$ for the case of LTRC observations generated by independent and continuous hidden variables. In the used simulation T^* is Uniform(u_t, u_T), X^* is the Normal and C^* is Uniform(u_c, u_C); the used parameters are shown in the main title. The top diagram shows a sample of size $n = 300$ from (T, V, Δ). Observations of (V, Δ) are shown by circles, $N := \sum_{l=1}^{n} \Delta_l = 230$ is the number of uncensored observations. Observations of the truncation variable T are shown via horizontal coordinates of crosses. In the middle diagram, the solid line is the underlying conditional survival function $G^{X^*|X^*>\alpha_{T^*}}(x)$ shown for $x \in [V_{(1)}, V_{(n)}]$, the dashed and dot-dashed lines are the sample-mean and Kaplan-Meier estimates (they are close to each other), and the dotted lines show the $(1 - \alpha)$ pointwise confidence band. The bottom diagram shows the underlying conditional density (the solid line), its E-estimate (the dashed line), and the pointwise (dotted lines) and simultaneous (dot-dashed lines) $1 - \alpha$ confidence bands. The E-estimate is for interval $[a, a + b]$ with default values $a = V_{(1)}$ and $a + b = V_{(n)}$ shown in the subtitle. {Distribution of T^* is either the Uniform(u_t, u_T) or Exponential(λ_T) where λ_T is the mean. Censoring distribution is either Uniform(u_c, u_C) or Exponential(λ_C). Parameters of underlying distributions are shown in the title of the top diagram. To choose, for instance, exponential truncation and censoring, set trunc = "Expon", cens = "Expon" and then either use default parameters or assign wished ones. To choose a manual interval $[a, a + b]$, assign wished values to arguments a and b.} [n = 300, corn = 2, trunc = "Unif", ut = 0, uT = 0.5, lambdaT = 0.3, cens = "Unif", uc = 0, uC = 1.5, lambdaC = 1.5, a = NA, b = NA, alpha = 0.05, cJ0 = 4, cJ1 = 0.5, cTH = 4]*

the estimates from the underlying survival function, and we can also see the relatively large width of the band which predicts such a possibility. In the bottom diagram, despite the skewed LTRC observations, the density E-estimate correctly shows the unimodal shape of the underlying Normal density. The estimated coefficient of difficulty, shown in the subtitle, is equal to 1.7. The latter implies that, with respect to the case of direct observations of X^*, we need 70% more LTRC observations to get the same MISE. Figure 6.14 is a good learning tool to explore LRTC data and estimators. It also allows a user to manually choose an interval of estimation $[a, a + b]$, and this will be a valuable lesson on its own.

What can be said about estimation of the distribution of C^* and T^* as well as the parameter p? Recall our discussion in Section 6.5 that X^* and C^* are "symmetric" random variables in the sense that if we consider $1 - \Delta$ instead of Δ, then formally C^* becomes the random variable of interest and X^* becomes the censoring random variable. As a result, we can use the proposed estimators for estimation of the distribution of C^*, only by doing so we need to keep in mind the assumptions and then correspondingly choose a feasible interval of estimation. The problem of estimation of the distribution of T^* and parameter p is not new for us because we can consider (T, V) as an LT realization of an underlying pair (T^*, V^*). Then Section 6.6 explains how the distribution of T^* and parameter p can be estimated.

We finish the section by considering a case where the main assumption about mutual independence and continuity of the triplet of hidden random variables is no longer valid. In some applications it is known that the censoring random variable is not smaller than the truncating variable, and the censoring random variable may have a mixed distribution. As an example, we may consider the model of a clinical trial where

$$C^* := T^* + U^* := T^* + [u_C B^* + (1 - B^*)U'], \qquad (6.7.17)$$

u_C is a positive constant, U' is a nonnegative continuous random variable with the support $[u_c, u_C]$, and B^* is a Bernoulli random variable with $\mathbb{P}(B^* = 1) = \mathbb{P}(U^* = u_C)$ being the probability that X^* is censored by the end of a clinical trial.

Assume that X^* is continuous and independent from (T^*, C^*), while truncation and censoring variables may be dependent and have a mixed (continuous and discrete) joint distribution. Can our estimators for the distribution of X^* be used in this case? In other words, are the estimators robust? To answer this question theoretically, we need to understand if formula (6.7.9), which was used to propose the estimators, still holds for the considered setting. We begin with a formula for the probability $g(x)$. Write,

$$g(x) := \mathbb{P}(T \leq x \leq V) = \mathbb{P}(T^* \leq x \leq V^* | T^* \leq V^*)$$

$$= \frac{\mathbb{P}(T^* \leq x \leq V^*, T^* \leq V^*)}{\mathbb{P}(T^* \leq V^*)} = \frac{\mathbb{P}(T^* \leq x \leq V^*)}{p}$$

$$= p^{-1}\mathbb{P}(T^* \leq x, X^* \geq x, C^* \geq x) = p^{-1}G^{X^*}(x)\mathbb{P}(T^* \leq x \leq C^*). \qquad (6.7.18)$$

In the last equality the independence of X^* and (T^*, C^*) was used. Next, we can write,

$$\mathbb{P}(V \leq x, \Delta = 1) = \mathbb{P}(X^* \leq x, X^* \leq C^* | T^* \leq V^*)$$

$$= p^{-1}\mathbb{P}(X^* \leq x, T^* \leq X^* \leq C^*) = p^{-1} \int_0^x f^{X^*}(u)\mathbb{P}(T^* \leq u \leq C^*)du. \qquad (6.7.19)$$

Differentiation of (6.7.19) with respect to x and then using (6.7.18) allows us to write,

$$f^{V,\Delta}(x, 1) = p^{-1}f^{X^*}(x)\mathbb{P}(T^* \leq x \leq C^*) = \frac{f^{X^*}(x)g(x)}{G^{X^*}(x)}. \qquad (6.7.20)$$

In its turn, (6.7.20) implies that the density of interest can be written as

$$f^{X^*}(x) = \frac{f^{V,\Delta}(x, 1)G^{X^*}(x)}{g(x)}, \quad 0 \leq x < \beta_{X^*}. \qquad (6.7.21)$$

LTRC Data,   n = 300 , N = 222 , ut = 0.2 , uT = 0.7 , uc = 0 , uC = 0.6 , censp = 0.2

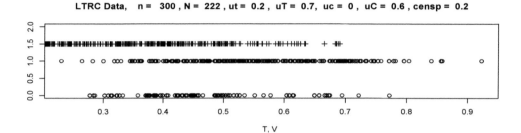

T, V

Conditional Survival Function of X˙, Estimates and Confidence Band

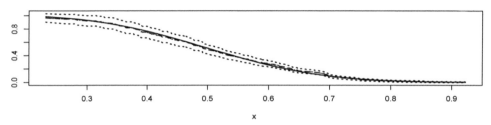

x

Conditional Density of X˙, E- Estimate and Confidence Bands

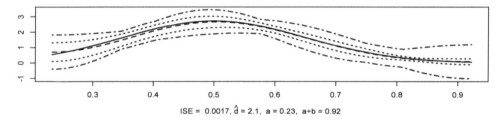

ISE = 0.0017, d̂ = 2.1, a = 0.23, a+b = 0.92

Figure 6.15 *Estimation of the conditional survival function and the conditional density of a lifetime of interest for LTRC data when $C^* := T^* + U^*$ as in (6.7.17). Variables in the triplet (X^*, T^*, U^*) are mutually independent. Variable U^* is the mixture of Uniform(u_c, u_C) random variable with the constant u_C, and the probability $\mathbb{P}(U^* = u_C)$ is controlled by the argument censp. Otherwise the underlying simulation and the structure of diagrams are as in Figure 6.14. [n = 300, corn = 2, trunc = "Unif", ut = 0.2, uT = 0.7, lambdaT = 0.3, cens = "Unif", uc = 0, uC = 0.6, lambdaC = 1.5, censp = 0.2, a = NA, b = NA, alpha = 0.05, cJ0 = 4, cJ1 = 0.5, cTH = 4]*

This is the same formula as (6.7.9) which was established for the case of independent and continuous hidden random variables. Hence an estimator, motivated by that formula, can be used for the studied case as well. Further, if $\alpha_{X^*} > \alpha_{T^*}$ then we estimate conditional characteristics $G^{X^*|X^* > \alpha_{T^*}}(x)$ and $f^{X^*|X^* > \alpha_{T^*}}(x)$.

Figure 6.15 allows us to test the made conclusion for the model (6.7.17). Its structure is similar to Figure 6.14 and the caption explains the underlying LTRC mechanism where the parameter *censp* controls the choice of $\mathbb{P}(U^* = u_C)$. In the considered simulation this parameter is equal to 0.2, meaning that in the example of a clinical trial 20% of participants are right censored by the end of the trial. The used LTRC mechanism creates challenges for the estimators because $\alpha_T^* = u_t = 0.2 > \alpha_{X^*} = 0$. As a result, theoretically we may estimate only the underlying conditional survival function and the conditional density given $X^* > u_t = 0.2$. The top diagram indicates that the truncation variable is separated from zero and it is likely that α_{T^*} is close to 0.2. Further, note that there are just few observations in the tails. The middle diagram shows that despite these challenges, the conditional survival function is estimated relatively well. The conditional density E-estimate is very good and correctly indicates the underlying conditional density $f^{X^*|X^*>\alpha_{T^*}}(x)$ despite the heavily skewed LTRC observations. The estimated coefficient of difficulty is equal to 2.1, and this, together with the large confidence bands, sheds light on the complexity of the problem and the possibility of poor estimates in other simulations. Nonetheless, it is fair to conclude that the estimators are robust to the above-discussed deviations from the basic LTRC model.

LRTC is one of the most complicated modifications of data, and it is highly recommended to use Figures 6.14 and 6.15 to learn more about this important statistical problem. Exploring different underlying distributions and parameters will help to gain necessary experience in dealing with LTRC data.

6.8 Nonparametric Regression with RC Responses

A classical regression problem, discussed in Section 2.3, is to estimate a regression function

$$m(x) = \mathbb{E}\{Y|X = x\}, \tag{6.8.1}$$

based on a sample of size n from a pair of random variables (X, Y). Here X is the predictor and Y is the response. Recall that $m(x)$ is used to predict the response given $X = x$, and also the regression is a useful tool to describe a relationship between the two variables.

We already know how regression E-estimator is constructed and how it performs for the case of direct observations from (X, Y). In survival analysis it is often the case that one of the variables is modified by censoring. In this section we are considering the case of right censored (RC) responses, and the next one explores the case of right censored predictors.

The following regression model is considered. Response Y is a lifetime (nonnegative random variable) which is right censored by a censoring random variable C. Predictor X is observed directly. As a result, we observe a sample of size n from the triplet (X, V, Δ) where $V := \min(Y, C)$ and $\Delta := I(Y \leq C)$. Note that the predictor is not censored and observed directly. In what follows we assume that the pair (X, Y) and the censoring variable C are independent, the three underlying random variables X, Y and C are continuous, and as usual we are assuming that X is supported on $[0, 1]$ according to a continuous and positive design density (the latter yields that $\min_{x \in [0,1]} f^X(x) > 0$ and hence the reciprocal of design density is finite). The problem is to propose, if possible, a consistent estimator of the regression function (6.8.1).

In previous sections a number of classical actuarial, medical and engineering examples of right censoring were presented. Let us add one more example from economics. Consider an observed purchase V which is right censored by rationing C, and Y is the underlying hidden demand that would be equal to the purchase except for the rationing. The rationing means the controlled distribution of scarce resources, goods, or services, or an artificial restriction of demand. We would like to know a relationship between the underlying demand and a predictor of interest, which may be level of inflation, income or unemployment. Note that ignoring censoring would yield a decreased demand, and it is also natural to expect that a severe censoring may preclude us from consistent estimation of the regression.

Now let us present a key probability formula. Following Section 6.5, we can write that

$$f^{X,V,\Delta}(x, y, 1) = f^X(x)f^{Y|X}(y|x)G^C(y). \tag{6.8.2}$$

We also know from Section 6.5 that the distribution of Y cannot be estimated beyond the value $\beta_V = \min(\beta_Y, \beta_C)$ (recall our notation β_Z for the upper bound of the support of a random variable Z). Hence, if

$$\beta_Y < \beta_C, \tag{6.8.3}$$

then the distribution of Y and the regression function (6.8.1) may be consistently estimated, otherwise the distribution of Y may be recovered only up to value $\beta_V = \beta_C$. As a result, let us introduce a censored (or we may say trimmed) regression function

$$m(x, \beta_V) := \mathbb{E}\{YI(Y \leq \beta_V)|X = x\}$$

$$= \int_0^{\beta_V} \frac{yf^{X,Y}(x,y)}{f^X(x)} dy = \int_0^{\beta_V} \frac{yf^{X,V,\Delta}(x,y,1)}{f^X(x)G^C(y)} dy. \qquad (6.8.4)$$

In what follows our aim is to estimate the censored regression function (6.8.4) because in general we cannot estimate the regression function (6.8.1).

Note that $m(x, \beta_V) = m(x)$ whenever (6.8.3) holds. In other words, given the assumption (6.8.3), the two characteristics of the relationship between X and Y coincide. But in general $m(x, \beta_V)$ is smaller or equal to $m(x)$, and this is a remark that bears important practical consequences. Indeed, in applications we know neither β_V nor β_Y, and instead are dealing with an empirical $\hat{\beta}_V := V_{(n)}$ which may be significantly smaller than β_Y even if (6.8.3) holds. To understand why note that $G^V(v) = G^Y(v)G^C(v)$ and hence the survival function of V decreases faster than the survival function of Y. As a result, even if assumption (6.8.3) holds, for small samples and depending on severity of censoring, a regression estimate may be significantly smaller than an underlying $m(x)$, and the latter is important to keep in mind when regression with RC response is analyzed. We will complement this discussion shortly by simulated examples.

Now let us explain how to construct E-estimator of the censored regression (6.8.4). Consider the cosine basis $\{\varphi_j(x), j = 0, 1, \ldots\}$ on $[0, 1]$. The jth Fourier coefficient of the censored regression is

$$\theta_j := \int_0^1 m(x, \beta_V)\varphi_j(x)dx = \int_0^1 \int_0^{\beta_V} \frac{vf^{X,V,\Delta}(x,v,1)\varphi_j(x)}{f^X(x)G^C(v)} dvdx$$

$$= \mathbb{E}\left\{\frac{\Delta V\varphi_j(X)}{f^X(X)G^C(V)}\right\}. \qquad (6.8.5)$$

The expectation in (6.8.5) implies that we can use a plug-in sample mean estimator. In the expectation, the design density $f^X(x)$ can be estimated by the density E-estimator $\hat{f}^X(x)$ of Section 2.2 (recall that the predictor is observed directly), and the survival function $G^C(v)$ can be estimated by the sample mean estimator (6.5.10),

$$\hat{G}^C(v) := \exp\{-n^{-1}\sum_{l=1}^n (1 - \Delta_l)I(V_l \leq v)/\hat{g}(V_l)\}, \qquad (6.8.6)$$

where

$$\hat{g}(v) := n^{-1}\sum_{l=1}^n I(V_l \geq v). \qquad (6.8.7)$$

Combining the results, the proposed Fourier estimator is

$$\hat{\theta}_j := n^{-1}\sum_{l=1}^n \frac{\Delta_l V_l \varphi_j(X_l)}{\max(\hat{f}^X(X_l), c/\ln(n))\hat{G}^C(V_l)}. \qquad (6.8.8)$$

In its turn, this Fourier estimator allows us to construct the regression E-estimator $\hat{m}(x, \beta_V)$. Further, given (6.8.3) this estimator consistently estimates $m(x)$.

Figure 6.16 allows us to understand the model, observations, and a possible estimation of the regression function. The underlying simulation and diagrams are explained in the caption. The top diagram shows us the available sample from (X, V, Δ) where uncensored cases (when $\Delta = 1$) are shown by circles and censored cases by crosses. The underlying

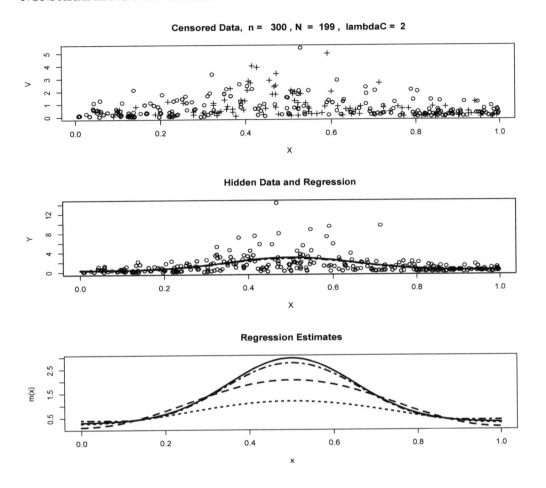

Figure 6.16 *Regression with heavily censored responses. The responses are independent exponential random variables with mean $a + f(X)$ where f is a corner function, here it is the Normal. The predictor is the Uniform. The censoring variable is exponential with the mean λ_C. The top diagram shows by circles observations with uncensored responses and by crosses observations with censored ones, $N := \sum_{l=1}^{n} \Delta_l$. Underlying scattergram of the hidden sample from (X, Y) is shown in the middle diagram, and it is overlaid by the underlying regression function (the solid line) and its E-estimate based on this sample (the dot-dashed line). The bottom diagram shows the underlying regression (the solid line), the E-estimate based on RC data shown in the top diagram (the dashed line), the E-estimate based solely on cases with uncensored responses shown by circles in the top diagram (the dotted line), and the dot-dashed line is the estimate shown in the middle diagram. {Parameter λ_C is controlled by argument lambdaC, function f is chosen by argument corn.} [n = 300, corn = 2, a = 0.3, lambdaC = 2, lambdaC = 2, cJ0 = 4, cJ1 = 0.5, cTH = 4, c = 1.]*

sample from (X, Y) is shown in the middle diagram. It is generated by the Uniform predictor X and Y being exponential variable with the mean equal to $0.3 + f(X)$ where $f(x)$ is the Normal corner function. Note that the mean is the regression function and it is shown by the solid line. Pay attention to the large volatility of the exponential regression and the range of underlying responses. The dot-dashed line is the E-estimate based on the hidden sample, and it may be better visualized in the bottom diagram. The E-estimate is good, and also the scattergram corresponds to a unimodal and symmetric regression function.

Now let us return to the top diagram which shows available data and compare them with the underlying ones shown in the middle diagram. First of all, let us compare the scales.

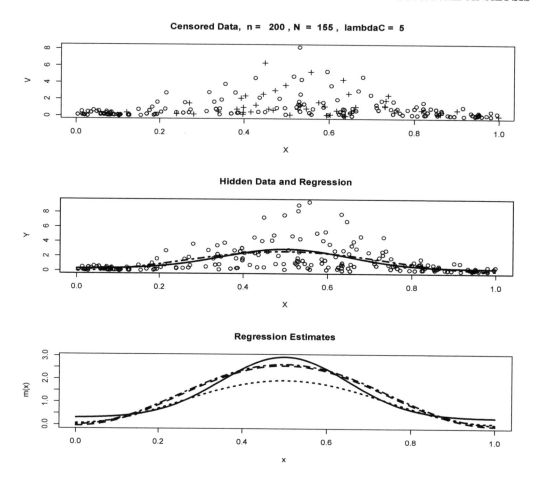

Figure 6.17 *Regression with mildly censored responses. The simulation and diagrams are the same as in Figure 6.16, only here the smaller sample size $n = 200$ and larger mean $\lambda_C = 5$ of the exponential censoring variable are used. [n = 200, corn = 2, a = 0.3, lambdaC = 5, cJ0 = 4, cJ1 = 0.5, cTH = 4, c = 1.]*

Value of the largest observed $V_{(n)}$ is close to 6 while value of the largest $Y_{(n)}$ is close to 14. This is a sign of severe censoring. Further, practically all larger underlying responses are censored, with just few uncensored responses being larger than 3. Further, among $n = 300$ underlying responses, only $N = 199$ are uncensored, that is every third response is censored. If we return to our discussion of the assumption (6.8.3) and the censored regression, we may expect here that a regression estimate, based on the censored data, may be significantly smaller than the underlying regression. The bottom diagram supports this prediction. The dashed line is the E-estimate, and it indeed significantly underestimates the underlying regression function shown by the solid line. Nonetheless, the proposed estimator does a better job than a complete-case approach which yields an estimate shown by the dotted line. Recall that a complete-case approach simply ignores cases with censored responses. What we see is that despite a relatively large sample size, the observed $V_{(n)}$ is too small and this explains the underperformance of the E-estimator.

Of course, the simulation of Figure 6.16 implies a severe censoring. Let us relax it a bit by considering a censoring variable with exponential distribution only now with a larger mean $\lambda_C = 5$. This should produce more uncensored observations and a better regression estimation. A particular outcome is shown in Figure 6.17 where also a reduced sample size $n = 200$ is chosen for better visualization of scattergrams. Otherwise, the simulation and diagrams are the same as in Figure 6.16.

Let us begin with comparison of observations in the top and middle diagrams. First of all, note that the scales are about the same, and $V_{(n)}$ is close to $Y_{(n)}$. Further, now less than a quarter of responses is censored. This is still a significant proportion but dramatically smaller than in Figure 6.16. Now let us look at the estimates. The E-estimates based on the underlying and censored data are close to each other (compare dot-dashed and dashed curves), and they are much better than the E-estimate based on complete cases (the dotted line).

It is highly recommended to repeat these two figures with different parameters and gain experience in dealing with censored responses. It is useful to manually analyze scattergrams and try to draw a reasonable regression which takes into account the nature of data. Further, use different regression functions and realize whether the E-estimator allows us to make correct conclusions about modes, namely about their number, locations and relative magnitudes.

6.9   Nonparametric Regression with RC Predictors

Consider a regression problem where the predictor X is right censored by a censoring variable C and the response Y is observed directly. In this case we observe a sample from the triplet (U, Y, Δ) where $U := \min(X, C)$ and $\Delta = I(X \leq C)$. The aim is to estimate an underlying regression

$$m(x) := \mathbb{E}\{Y|X = x\}. \qquad (6.9.1)$$

This is an interesting and possibly complicated setting because we are dealing with a modified predictor, and we know from Chapter 4 that typically this implies serious complications. And indeed, as we will see shortly, right censored predictors may preclude us from consistent estimation. At the same time, if possible a consistent estimation is surprisingly simple.

We begin our analysis of the problem with probability formulas. Assume that the pair (X, Y) and C are independent, the three variables are continuous and nonnegative, and as usual the predictor X has a continuous and positive density on the support $[0, 1]$.

Then we can write that

$$f^{U,Y,\Delta}(x, y, 1) = f^X(x)f^{Y|X}(y|x)G^C(x). \qquad (6.9.2)$$

This formula allows us to get the joint density of (U, Y) given $\Delta = 1$, that is the joint density of observations for uncensored (complete) cases. Write,

$$f^{U,Y|\Delta}(x, y|1) = \frac{f^X(x)f^{Y|X}(y|x)G^C(x)}{\mathbb{P}(\Delta = 1)}$$

$$= \left[\frac{f^X(x)G^C(x)}{\int_0^1 f^X(u)G^C(u)du} \right] f^{Y|X}(y|x) =: f^Z(x)f^{Y|X}(y|x). \qquad (6.9.3)$$

Here $f^Z(x)$ is the density of uncensored predictors, that is $f^Z(x) := f^{U|\Delta}(x|1)$.

Relation (6.9.3) is the key for understanding the data and proposed regression E-estimator. The relation tells us that, whenever $f^Z(x)$ is positive, the conditional density of Y given $U = x$ in an uncensored case, that is given $\Delta = 1$, is the same as the underlying conditional density of Y given $X = x$. As a result, for those x the uncensored-case approach is consistent. Further, even estimation of the conditional density may be based on that approach. On the other hand, if $\beta_C < \beta_X$, then $f^Z(x) = 0$ for $x \in [\beta_C, \beta_X]$ and we cannot consistently estimate a regression function over that interval. The latter is the

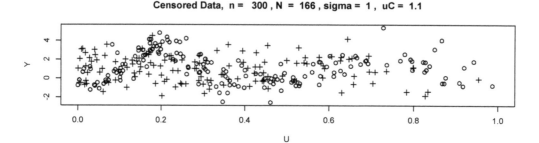

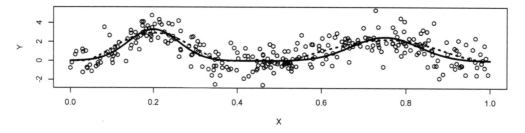

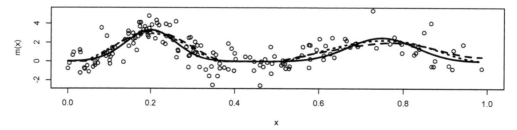

Figure 6.18 *Regression with censored predictors. The underlying predictors have the Uniform distribution, the responses are $m(X) + \sigma\varepsilon$ where $m(x)$ is the Strata and ε is standard normal, and the censoring variable is $Uniform(0, u_C)$. The top diagram shows by circles observations with uncensored predictors and by crosses observations with censored predictors, $N := \sum_{l=1}^{n} \Delta_l$ is the number of uncensored predictors. Underlying scattergram of the hidden sample from (X, Y) is shown in the middle diagram, and it is overlaid by the underlying regression function (the solid line) and its E-estimate based on this sample (the dotted line). The bottom diagram shows data and curves over an interval $[0, U_{(n)}]$. Circles show observations with uncensored predictors (they are identical to circles in the top diagram). These observations are overlaid by the underlying regression (the solid line), the E-estimate based on uncensored-case observations (the dashed line), and the E-estimate based on underlying observations (the dotted line) and it is the same as in the middle diagram. {The regression function is chosen by argument corn, parameter u_C is controlled by the argument uC.} [n = 300, corn = 4, sigma = 1, uC = 1.1, cJ0 = 4, cJ1 = 0.5, cTH = 4]*

familiar curse of severe censoring. On the other hand, we may always consistently estimate left tail of regression.

Let us compare our conclusions for regressions with MAR data (recall that the latter was discussed in Chapter 4). For the setting of MAR responses a complete-case approach is optimal, and for the setting of MAR predictors a special estimator, which uses all obser-

Censored Data, n = 300 , N = 105 , sigma = 1 , uC = 0.9

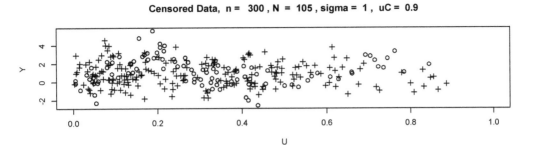

Hidden Data and Regression

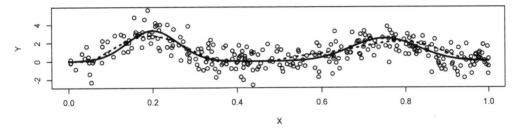

Uncensored-Case Regression

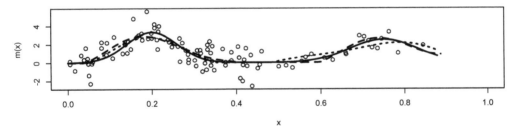

Figure 6.19 *Regression with heavily censored predictors. The simulation and the diagrams are the same as in Figure 6.18 only here $u_C = 0.9$. [n = 300, corn = 4, sigma = 1, uC = 0.9, cJ0 = 4, cJ1 = 0.5, cTH = 4]*

vations, is needed. As we now know, for regression with RC data the outcome is different. Here a complete-case approach is optimal for the setting of censored predictors and a special estimator, based on all observations, is needed for the setting of censored responses. This is a teachable moment because each modification of data has its own specifics and requires a careful statistical analysis.

Figure 6.18 allows us to understand the discussed regression with right censored predictor and the proposed solution. Its caption explains the simulation and the diagrams. The top diagram shows the scattergram of censored observations with circles indicating cases with uncensored predictors and crosses indicating cases with censored predictors. Note that the largest available predictor is near 1, and this hints that the censoring variable may take even larger values. And indeed, the censoring variable is supported on $[0, u_C]$ with $u_C = 1.1$. Further, while the underlying predictor X has the Uniform distribution, the observed U is left skewed. We see this in the diagram, and this also follows from the relation $G^U(x) = G^X(x)G^C(x)$. The latter allows us to conclude that RC predictors may create problems for estimation of a right tail of the regression. Further, note that 45% of predictors are censored, and this is a significant loss of data.

The middle diagram shows us the underlying (hidden) scattergram, the regression function and its E-estimate. The estimate is not perfect, but note the large standard deviation $\sigma=1$ of the regression error (controlled by the argument sigma) and recall that the Strata is a difficult function for a perfect estimation. The bottom diagram sheds light on the proposed uncensored-case approach. If we look at the exhibited uncensored cases, shown by circles, then we may visualize the underlying Strata regression. Of course, just few observations in the right tail complicate the visualization, but the E-estimator does a good job and the E-estimate (the dashed line) is comparable with the estimate based on all hidden observations (the dotted line).

What will be if we use a more severe censoring, for instance one with $u_C = 0.9$? A corresponding outcome is shown in Figure 6.19. The top diagram shows that the largest uncensored predictor is about 0.85, while the largest observation of the censoring variable is near 0.9. This tells us that it is likely that the support of U is defined by the censoring variable C, and this is indeed the case here. Further, note that we have just few uncensored predictors with values larger than 0.75. Further, note that only $N = 105$ predictors from underlying $n = 300$ are uncensored and they are heavily left skewed. What we see is an example of a severe censoring.

The middle diagram shows that the sample of underlying hidden observations is reasonable and, keeping in mind the large regression noise (it is possible to reduce or increase it using the argument sigma), the E-estimate is fair. The bottom diagram shows us the E-estimate based on uncensored cases (the dashed line) which can be compared with the underlying regression (the solid line) and the E-estimate of the middle diagram (the dotted line). As we see, despite the small sample size $N = 105$ of uncensored cases and the large regression noise, the uncensored-case E-estimate is relatively good. Let us stress that it is calculated for the interval $[0, U_{(n)}]$, and that we have no information about the underlying regression beyond this interval.

Repeated simulations of Figure 6.18, using different parameters and underlying regressions, may help to shed a new light on the interesting and important statistical problem of regression with censored predictor.

6.10   Exercises

6.1.1 Verify (6.1.4).

6.1.2 Prove that if the hazard rate is known, then the survival function can be calculated according to formula (6.1.3).

6.1.3 Verify (6.1.5).

6.1.4 Consider the Weibull distribution with the shape parameter k defined below line (6.1.7). Prove that if $k < 1$ then the hazard rate is decreasing and if $k > 1$ then it is increasing.

6.1.5* Suppose that X and Y are two independent lifetimes with known hazard rates and $Z := \min(X, Y)$. Find the hazard rate of Z.

6.1.6 Propose a hazard rate whose shape resembles a bathtub.

6.1.7* Prove that for any hazard rate h^Y the relation $\int_0^\infty h^Y(y)dy = \infty$ holds. Furthermore, if S is the support of Y then $\int_S h^Y(y)dy = \infty$.

6.1.8 Prove that $G^Y(y) = G^Y(a)e^{-\int_a^y h^Y(u)du}$ for any $a \in [0, y]$.

6.1.9 Verify (6.1.6) and (6.1.7). Hint: Here the scale-location transformation is considered. Begin with the cumulative distribution function and then take the derivative to get the corresponding density.

6.1.10* Propose a series estimator of the hazard rate which uses the idea of transformed $Z = (X - a)/b$.

6.1.11 Verify expression (6.1.8) for Fourier coefficients of the hazard rate.

6.1.12 Prove that the oracle-estimator $\tilde{\theta}_j^*$, defined in (6.1.10), is unbiased estimator of θ_j.

6.1.13 Find variance of the estimator (6.1.10). Hint: Note that the oracle is a classical sample mean estimate, and then use formula for the variance of the sum of independent random variables.

6.1.14 Find the mean and variance of the empirical survival function (6.1.14). Hint: Note that this estimate is the sample mean of independent and identically distributed Bernoulli random variables. Furthermore, the sum has a Binomial distribution.

6.1.15* Use Hoeffding's inequality for the analysis of large deviations of the empirical survival function (6.1.14).

6.1.16* Evaluate the mean and variance of the Fourier estimator (6.1.15). Hint: Begin with the case of known nuisance functions, consider the asymptotic as n and j increase, and then show that the plug-in methodology is valid.

6.1.17* Explain why (6.1.17) is a reasonable estimator of the coefficient of difficulty. Hint: Replace the estimated survival function by an underlying survival function G^Y, and then calculate the mean and variance of this oracle-estimator.

6.1.18* Explain why the ratio-estimator (6.1.19) may be a good estimator of the hazard rate. What are possible drawbacks of the estimator?

6.1.19 Repeat Figure 6.1 for sample sizes $n =$100, 200, 300, 400, 500 and make your conclusion about corresponding feasible intervals of estimation. Comment about your method of choosing an interval. Does a sample size affect the choice of interval?

6.1.20 Repeat Figure 6.1 twenty times, write down ISEs for the estimates, and then rank the estimates based on results of this numerical study.

6.1.21 Repeat Figure 6.1 for different underlying distributions. Choose largest feasible intervals of estimation for each distribution.

6.1.22 What are the optimal parameters of the E-estimator for the experiment considered in Figure 6.1?

6.1.23 Explain underlying simulations and all histograms in Figure 6.1.

6.1.24* Write down the hazard rate E-estimator and explain parameters and statistics used.

6.2.1 Give several examples of right censoring. Further, present an example of left censoring and explain the difference.

6.2.2 Explain the mechanism of right censoring. How are available observations in right-censored data related to underlying random variables? Hint: Check (6.2.1).

6.2.3 Are random variables V and Δ dependent, independent or is there not enough information to answer?

6.2.4 Explain formula (6.2.2). Why is its last equality valid?

6.2.5 Prove (6.2.3). Formulate assumptions needed for its validity.

6.2.6 Formulate assumptions and verify (6.2.4). Describe the support of the density. Is this a mixed density? What is the definition of a mixed density?

6.2.7 Are censored observations biased? If the answer is "yes," then what is the biasing function?

6.2.8 Explain formula (6.2.5). Does it express the hazard rate via functions that can be estimated?

6.2.9 What is the motivation behind expression (6.2.6) for Fourier coefficients?

6.2.10* Explain the Fourier estimator (6.2.7) and then describe its statistical properties. Hint: Find the mean, the variance and the probability of large deviations using the Hoeffding inequality.

6.2.11* What is the coefficient of difficulty? What is its role in nonparametric estimation?

6.2.12* Verify (6.2.9) and (6.2.10).

234 SURVIVAL ANALYSIS

6.2.13 Explain the simulation used to create Figure 6.2.

6.2.14 Repeat Figure 6.2 and write down analysis of its diagrams.

6.2.15 Describe all parameters/arguments of Figure 6.2. What do they affect?

6.2.16 Repeat Figure 6.2 about 20 times and make your own conclusion about the effect of estimator $\hat{G}^V$ on the hazard rate E-estimator.

6.2.17 Explain how the interval of estimation affects the hazard estimator. Support your conclusion using Figure 6.2.

6.2.18 Consider different distributions of the censoring variable. Then, using Figure 6.2, present a report on how those distributions affect a reasonable interval of estimation and the quality of E-estimator.

6.2.19* Explain formula (6.2.11) for the empirical coefficient of difficulty. Then calculate its mean and variance.

6.2.20* Propose a hazard rate E-estimator for left censored data.

6.3.1 Give an example of left truncated data.

6.3.2 Describe an underlying stochastic mechanism of left truncation. Is it a missing mechanism?

6.3.3 Formulate the probabilistic model of truncation.

6.3.4 Assume that the sample size of a hidden sample is n. Is the sample size of a corresponding truncated sample deterministic or stochastic? Explain your answer.

6.3.5 What are the mean and variance of the sample size of a truncated sample given that n is the size of an underlying hidden sample?

6.3.6 Explain and verify (6.3.3). What are the used assumptions?

6.3.7 Explain and verify (6.3.4). Hint: Pay attention to the support of this bivariate density.

6.3.8* Assume that T is a discrete random variable. How will (6.3.3) and (6.3.4) change?

6.3.9 Verify (6.3.5).

6.3.10* Explain and verify (6.3.6). Pay attention to the conditions when the equality holds. Will the equality be valid for points $x \leq \alpha_{T*}$?

6.3.11 Does truncation imply biasing? If the answer is "yes," then what is the biasing function?

6.3.12 Does left truncation skew a hidden sample of interest to the left or the right?

6.3.13* Explain the motivation of introducing the probability $g(x)$ in (6.3.8). Hint: Can this function be estimated based on a truncated sample? Then look at (6.3.9).

6.3.14* Verify relations in (6.3.8).

6.3.15 Verify (6.3.9).

6.3.16* How can (6.3.9) be used for estimation of the hazard rate of interest?

6.3.17 Explain the underlying idea of Fourier estimator (6.3.10). Hint: Replace the estimate $\hat{g}$ by g and show that this is a sample mean estimator.

6.3.18* What is the mean of the estimator (6.3.10)? Is it unbiased or asymptotically unbiased? Calculate the variance.

6.3.19 Explain the motivation behind the estimator (6.3.11).

6.3.20 What is the mean and variance of the estimator (6.3.11)?

6.3.21* Use Hoeffding's inequality for the analysis of the estimator (6.3.11). Then explain why this estimator may be used in the denominator of (6.3.10).

6.3.22* Verify (6.3.12).

6.3.23 Explain the motivation behind the estimator (6.3.13).

6.3.24* What are the mean and the variance of the estimator (6.3.13)?

6.3.25* Explain how an E-estimator of the hazard rate is constructed.

6.3.26 Describe the underlying simulation that creates data in Figure 6.3.

6.3.27 Repeat Figure 6.3 a number of times and recommend a feasible interval of estimation.

6.3.28 Using Figure 6.3, propose "good" values for parameters of the E-estimator.

6.3.29 Using Figure 6.3, explore the issue of how truncating variables affect quality of estimation of an underlying hazard function.

6.3.30 Can Figure 6.3 be used for statistical analysis of the used confidence bands? Test your suggestion.

6.3.31 Rank corner distributions according to difficulty in estimation of their hazard rates. Then check your conclusion using Figure 6.3.

6.3.32 Confidence bands may take on negative values. How can they be modified to take into account that a hazard rate is nonnegative?

6.3.33* Consider three presented examples of left censoring (actuarial, startups and clinical trials), and explain how each may be "translated" into another.

6.3.34* Write down the hazard rate E-estimator and explain parameters and statistics used.

6.4.1 Give several examples of LTRC data.

6.4.2 Explain an underlying stochastic mechanism of creating LTRC data.

6.4.3 Present examples when the LT is followed by the RC and vice versa.

6.4.4 Explain formula (6.4.1). Can the probability be estimated?

6.4.5 Suppose that the sample size of an underlying hidden sample of interest is n. What is the distribution of the sample size of a LTRC sample? What are its mean and variance?

6.4.6 Present examples when truncating and censoring variables are dependent and independent.

6.4.7* Explain formula (6.4.2) for the cumulative distribution function of the triplet of random variables observed in a LTRC sample. Comment on the support. What does the formula tell us about a possibility of consistent estimation of the distributions of X^*, T^* and C^*?

6.4.8 Verify formula (6.4.3) and explain assumptions under which it is correct.

6.4.9 Prove validity of (6.4.4).

6.4.10 What is the meaning of the joint density (6.4.4)? Hint: note that one of the variables is discrete.

6.4.11 Verify expression (6.4.5) for the marginal mixed density of (V, Δ).

6.4.12* Verify validity of formula (6.4.6) for the density of the variable of interest X^*. Explain the assumptions when it is valid. Can this formula be used for construction of a consistent E-estimator and what are the assumptions?

6.4.13* Why is probability (6.4.7) a pivotal step in constructing an E-estimator?

6.4.14 Explain formula (6.4.7).

6.4.15 Can the function $g(x)$ be estimated based on LTRC data?

6.4.16 Verify formulas (6.4.8) and (6.4.9). Explain the assumptions.

6.4.17* Suggest an estimator for $h^{X^*}(x)$. Hint: Use (6.4.9).

6.4.18 Is the Fourier estimator (6.4.10) a sample mean estimator?

6.4.19 Why can $\hat{g}(x)$, defined in (6.4.11), be used in the denominator of (6.4.10)?

6.4.20* Present a theoretical statistical analysis of estimators (6.4.10) and (6.4.11). Hint: Begin with the distribution and then write down the mean, the variance, and for estimator (6.4.11) use the Hoeffding inequality to describe large deviations.

6.4.21 Verify (6.4.12).

6.4.22 Explain the estimator (6.4.13). Is it asymptotically unbiased?

6.4.23 Conduct a series of simulations, using Figure 6.4, and explore the effect of estimate $\hat{g}$ on the E-estimate of the hazard rate.

6.4.24 Choose a set of sample sizes, truncated and censoring distributions, and then try to determine a feasible interval of estimation for each underlying model. Explain your choice.

6.4.25 Suggest better values of parameters of the E-estimator. Does your recommendation depend on distributions of truncating and censoring variables?

6.4.26 What is the main difference between simulations in Figures 6.4 and 6.5?

6.4.27 Present several examples where T^* and C^* are related.

6.4.28 Present several examples where $\mathbb{P}(C^* \geq T^*) = 1$ and C^* has a mixed distribution.
6.4.29* Verify each equality in (6.4.14). Explain the used assumptions.
6.4.30 Establish validity of (6.4.16). What are the used assumptions?
6.4.31 Prove (6.4.17).
6.4.32 Explain (6.4.18) and the underlying assumptions.
6.4.33 Describe the underlying simulation of Figure 6.5.
6.4.34 Explain diagrams in Figure 6.5.
6.4.35 For estimation of the hazard rate, is the model of Figure 6.5 more challenging than of Figure 6.4?
6.4.36 Using Figure 6.5, suggest better parameters of the E-estimator.
6.4.37 Using Figure 6.5, infer about performance of the confidence bands.
6.4.38* Consider a setting where the made assumptions are violated. Then propose a consistent hazard rate estimator or explain why this is impossible.

6.5.1 Present several examples of RC observations.
6.5.2 Is there something in common between RC and MNAR? If the answer is "yes," then why does MNAR typically preclude us from a consistent estimation and RC not?
6.5.3 Explain validity of (6.5.1) and the underlying assumption.
6.5.4 Prove (6.5.2).
6.5.5 Under RC, over what interval may the distribution of interest be estimated?
6.5.6 Describe assumptions that are sufficient for consistent estimation of the distribution of X.
6.5.7 Explain how the Kaplan–Meier estimator is constructed.
6.5.8* What is the underlying motivation of the Kaplan–Meier estimator? Why is it called a product limit estimator? Evaluate its mean and variance.
6.5.9 Explain formulae (6.5.6) and (6.5.7). Under what assumptions are they valid?
6.5.10 Explain how to estimate the survival function of the censoring random variable C.
6.5.11 Verify (6.5.9).
6.5.12* Find the mean and variance of the estimator (6.5.10).
6.5.13 Explain and then verify (6.5.12).
6.5.14 What is the motivation behind the Fourier estimator (6.5.13)?
6.5.15* Find the mean and variance of the estimator (6.5.13).
6.5.16 Using (6.5.14) explain how the estimator (6.5.15) is constructed.
6.5.17 Explain the simulation of Figure 6.6.
6.5.18 Explain the simulation of Figure 6.7.
6.5.19 What is the difference between simulations in Figures 6.6 and 6.7?
6.5.20 Propose better values of parameters for E-estimators used in Figures 6.6 and 6.7. Are they different? Explain your findings.
6.5.21* Consider a setting where some of the made assumptions are no longer valid. Then explore a possibility of consistent estimation.

6.6.1 Describe the model of LT.
6.6.2 Present several examples of LT observations. Based solely on LT observations, can one conclude that the observations are LT?
6.6.3 Is LT based on a missing? Is the missing MNAR? Typically MNAR precludes us from consistent estimation. Is the latter also the case for LT?
6.6.4 Explain the assumption (6.6.1) and its importance.
6.6.5 Why is the assumption (6.6.2) important?
6.6.6 Verify each equality in (6.6.3) and explain used assumptions.
6.6.7 Verify (6.6.5) and explain the used assumption.
6.6.8 Establish (6.6.6) and (6.6.7).
6.6.9* Explain the underlying idea of estimators defined in (6.6.9). Find their expectations. Can these estimators be improved?

6.6.10 Verify all relations in (6.6.10).

6.6.11* Explain construction of the estimator (6.6.11). Find its mean and variance.

6.6.12* Conduct a statistical analysis of the estimator (6.6.12). Hint: Describe the distribution and its properties.

6.6.13* Explain the motivation behind the estimator (6.6.14). Evaluate its mean and variance.

6.6.14 Verify (6.6.16).

6.6.15* Suggest an E-estimator of the density f^{X^*}.

6.6.16* Find the mean and variance of Fourier estimator (6.6.19).

6.6.17 What is the motivation behind the estimator (6.6.21)? Can you propose another feasible estimator?

6.6.18 Explain the simulation used by Figure 6.8.

6.6.19 Repeat Figure 6.8 and analyze diagrams.

6.6.20 Use Figure 6.8 and compare performance of the Kaplan–Meier estimator with the sample mean estimator.

6.6.21 Explain the underlying idea of estimator (6.6.22).

6.6.22 Use Figure 6.9 to compare performance of the two estimators.

6.6.23 Explain all relations in (6.6.23).

6.6.24 Explain diagrams in Figure 6.10. Then use it for statistical analysis of the proposed density estimator.

6.6.25 Explain the underlying simulation in Figure 6.11.

6.6.26 Repeat Figure 6.11 for different sample sizes. Write a report about your findings.

6.6.27 Suggest better values for parameters of the E-estimator used in Figure 6.11. Is your recommendation robust to changes in other arguments of the figure?

6.6.28 Explain how the cumulative distribution function of T^* can be estimated.

6.6.29 Explain how the probability density of T^* can be estimated.

6.6.30 Describe the underlying simulation in Figure 6.12.

6.6.31 Explain diagrams in Figure 6.12.

6.6.32 Describe E-estimators used in Figure 6.12.

6.6.33 Explain formula (6.6.29).

6.6.34* Evaluate the mean and variance of Fourier estimator (6.6.30).

6.6.35 Prove (6.6.31).

6.6.36 Verify (6.6.32), and then explain why we are interested in the analysis of $\mathbb{E}\{\hat{H}^{X^*}(x)\}$.

6.6.37* Show that the second expectation in the right side of (6.6.32) vanishes as n increases. Hint: Propose any needed assumptions.

6.6.38 Verify (6.6.33).

6.6.39 Prove (6.6.34). Then explain meaning of the conditional survival function.

6.6.40 Given $\alpha_{X^*} < \alpha_T^*$, explain why the distribution of X^* cannot be consistently estimated, and then explain what may be estimated.

6.6.41 Explain all curves in the middle diagram of Figure 6.11. Then repeat it and analyze the results. Is estimation of the conditional survival function robust?

6.6.42 What does $\hat{p}$ estimate in Figure 6.11?

6.6.43 Explain each relation in (6.6.35).

6.6.44* Show that the second expectation on the right side of (6.6.35) vanishes as n increases. Hint: Make a reasonable assumption.

6.6.45 Verify (6.6.36).

6.6.46 Explain the definition of conditional density (6.6.37). Then comment on what can and cannot be estimated given $\alpha_{T^*} > \alpha_{X^*}$.

6.6.47 Consider the bottom diagram in Figure 6.11 and explain the curves. Then repeat Figure 6.11 with different parameters and explore robustness of the proposed estimators.

6.6.48* Explain all steps in construction of the E-estimator $\hat{f}^{T^*}(t)$. Formulate necessary assumptions.

6.6.49 Use Figure 6.13 and analyze statistical properties of the confidence bands.
6.6.50 Is the relation $\mathbb{P}(T_{(n)} \le X_{(n)}) = 1$ valid?
6.6.51 Is there any relationship between $X_{(1)}$ and $T_{(1)}$?
6.6.52* Relax one of the used assumptions and propose a consistent estimator of $\hat{f}^{X^*}(x)$ or prove that the latter is impossible.

6.7.1 Present examples of left truncated, right censored and LTRC data.
6.7.2 What is the difference (if any) between truncated and censored data.
6.7.3 Explain how LTRC data may be generated.
6.7.4 Find a formula for the probability of an observation in a LTRC simulation.
6.7.5 Explain each equality in (6.7.2).
6.7.6* Using (6.7.3), obtain formulas for corresponding marginal densities.
6.7.7 Can the probability (6.7.7) be estimated based on LTRC data?
6.7.8* Explain assumptions (6.7.8). What will be if they do not hold?
6.7.9 Why is formula (6.7.9) critical for suggesting a density E-estimator?
6.7.10 Explain how formula (6.7.10) is obtained.
6.7.11* What is the motivation behind the estimator (6.7.11)? Find its mean and variance.
6.7.12* Present statistical analysis of the estimator $\hat{g}$. Hint: Think about its distribution.
6.7.13 Verify (6.7.13).
6.7.14 Explain all relations in (6.7.14).
6.7.15* Is (6.7.15) a sample mean Fourier estimator? Find its mean and variance.
6.7.16* Verify (6.7.16).
6.7.17 Explain how underlying distributions of the truncating and censoring random variables affect the coefficient of difficulty of the density E-estimator.
6.7.18 Explain the underlying simulation used in Figure 6.14.
6.7.19 Explain diagrams in Figure 6.14.
6.7.20 Using Figure 6.14, present statistical analysis of the E-estimator.
6.7.21 How well do the confidence bands perform? Hint: Use repeated simulations of Figure 6.14.
6.7.22 Explain every argument of Figure 6.14.
6.7.23* Write down a report about the effect of distributions of the truncated and censoring variable on quality of E-estimate. Hint: Begin with the theory based on the coefficient of difficulty and then complement your conclusion by empirical evidence created with the help of Figure 6.14.
6.7.24 What parameters of the E-estimator, used in Figure 6.14, would you recommend for sample sizes $n = 100$ and $n = 300$?
6.7.25 Use Figure 6.15 and explain how the dependence between truncated and censored variables affects the estimation.
6.7.26 Explain the motivation behind the model (6.7.17). Present several corresponding examples.
6.7.27 Explain and verify each equality in (6.7.18).
6.7.28 Verify every equality in (6.7.19).
6.7.29 Explain how formula (6.7.21) for the density of interest is obtained.
6.7.30* Using (6.7.21), suggest an E-estimator of the density. Hint: Describe all steps and assumptions.
6.7.31* Consider the case $\alpha_{X^*} < \alpha_{T^*}$ and develop the theory of estimation of the conditional survival function $G^{X^*|X^*>\alpha_{T^*}}(x)$.
6.7.32* Consider the case $\alpha_{X^*} < \alpha_{T^*}$ and develop the theory of estimation of the conditional density $f^{X^*|X^*>\alpha_{T^*}}(x)$.
6.7.33 Using Figure 6.14, explore the proposed E-estimators for the case $\alpha_{X^*} < \alpha_{T^*}$.

6.8.1 Present an example of a regression problem with direct observations. Then describe a situation when response may be censored.

6.8.2 Explain (6.8.2).

6.8.3* What is the implication, if any, of assumption (6.8.3)? What can be done, if any, if (6.8.3) does not hold?

6.8.4 What is the meaning of the censored regression function (6.8.4)? Verify each equality in (6.8.4).

6.8.5* Explain the underlying idea of consistent estimation of the regression.

6.8.6 Verify relations in (6.8.5).

6.8.7* Explain the estimator (6.8.6). Evaluate its mean and variance.

6.8.8 What is the distribution of estimator (6.8.7)?

6.8.9 Consider a RC data with Y being the random variable of interest and C being the censoring random variable. If $\hat{G}^Y(y)$ is an estimator of the survival function of Y, how can this estimator be used for estimation of $G^C(z)$?

6.8.10 Explain the estimator (6.8.8).

6.8.11* What is the mean and variance of the Fourier estimator (6.8.8)?

6.8.12* Consider a setting where the made assumptions do not hold. Then explore a possibility of consistent regression estimation.

6.8.13 Consider diagrams in Figure 6.16 and explain the underlying simulation.

6.8.14 What are the four curves in the bottom diagram of Figure 6.16? Why are they all below the underlying regression? Is this always the case?

6.8.15* Explain theoretically how the parameter λ_C affects the regression estimation, and then compare your conclusion with empirical results using Figure 6.16.

6.8.16 Suggest better parameters of the E-estimator for Figure 6.16.

6.8.17 Conduct several simulations similar to Figures 6.16 and 6.17, and then explain the results.

6.8.18 Do you believe that values of parameters of the E-estimator should be different for simulations shown in Figures 6.16 and 6.17? If the answer is "yes," then develop a general recommendation for choosing better values of the parameters.

6.9.1 Explain the mechanism of RC modification. Does this modification involve a missing mechanism? If "yes," then is it MAR or MNAR?

6.9.2 Present several examples of a regression with RC predictor.

6.9.3 What complications in regression estimation may be expected from RC predictor?

6.9.4* Write down probability formulas for all random variables involved in regression with RC predictor.

6.9.5 Can RC predictor imply a destructive modification when a consistent regression estimation is impossible?

6.9.6 For the case of RC response, the notion of a censored regression was introduced. Is there a need to use this notion for the case of RC predictor?

6.9.7 Explain a difference (if any) between regressions with censored predictor and response.

6.9.8 Is expression (6.9.2) correct? Do you need any assumptions? Prove your assertion.

6.9.9 Verify every equality in (6.9.3). Do you need any assumptions for its validity?

6.9.10* Explain why the relation (6.9.3) is the key in regression estimation.

6.9.11* Describe the random variable Z defined in (6.9.3). Propose an estimator of its density.

6.9.12* Propose an E-estimator of the conditional density $f^{Y|X}(y|x)$.

6.9.13 What are the assumptions for consistent estimation of the regression?

6.9.14 Explain the underlying simulation used in Figure 6.18.

6.9.15* Explain, step by step, how the regression E-estimator, used in Figure 6.18, is constructed.

6.9.16* Explore theoretically and empirically, using Figure 6.18, the effect of parameter u_C on estimation.

6.9.17 In your opinion, which of the corner functions are less and more difficult for estimation? Hint: Use Figure 6.18.

6.9.18 Repeat Figure 6.19. Comment on scattergrams and estimates.

6.9.19 Propose better values for parameters of the estimators used in Figures 6.18 and 6.19. Explain your recommendation. Is it robust toward different regression functions?

6.9.20* Propose E-estimator for regression of Y on C. Explain its motivation, used probability formulas and assumptions.

6.11   Notes

Survival analysis is concerned with the inference about lifetimes, that is times to an event. The corresponding problems occur in practically all applied fields ranging from medicine, biology and public health to actuarial science, engineering and economics. A common feature of available data is that observations are modified by either censoring, or truncation, or both.

There is a vast array of books devoted to this topic, ranging from those using a mathematically nonrigorous approach to mathematically rigorous books using a wide range of theories including empirical processes, martingales in continuous time and stochastic integration among others. The literature is primarily devoted to parametric and semiparametric inference as well as nonparametric estimation of the survival function, and the interested reader can find many interesting examples, ad hoc procedures, advanced theoretical results and a discussion of using different software packages in the following books: Kalbfleisch and Prentice (2002), Klein and Moeschberger (2003), Martinussen and Scheike (2006), Aalen, Borgan and Gjessing (2008), Hosmer et al. (2008), Kosorok (2008), Allison (2010, 2014), Guo (2010), Fleming and Harrington (2011), Mills (2011), Royston and Lambert (2011), van Houwelingen and Putter (2011), Wienke (2011), Chen, Sun and Peace (2012), Crowder (2012), Kleinbaum and Klein (2012), Klugman, Panjer and Willmot (2012), Liu (2012), Lee and Wang (2013), Li and Ma (2013), Allison (2014), Collett (2014), Klein et al. (2014), Harrell (2015), Zhou (2015), Moore (2016), Tutz and Schmid (2016), and Ghosal and van der Vaart (2017).

6.1 Nonparametric estimation of the hazard rate is a familiar topic in the literature. Different type of estimators, including kernel, spline, classical orthogonal series and modern wavelet methods, have been proposed. A number of adaptive, to smoothness of an underlying hazard rate function, procedures motivated by known ones for the probability density, have been developed. A relevant discussion and thorough reviews may be found in a number of classical and more recent publications including Prakasa Rao (1983), Cox and Oakes (1984), Silverman (1986), Patil (1997), Wu and Wells (2003), Wang (2005), Gill (2006), Müller and Wang (2007), Fleming and Harrington (2011), Patil and Bagkavos (2012), Lu and Min (2014) and Daepp et al. (2015) where further references may be found. Interesting results, including both estimation and testing, have been obtained for the case of known restrictions on the shape of hazard rate, see a discussion in Jankowski and Wellner (2009). The plug-in estimation approach goes back to Watson and Leadbetter (1964), and see a discussion in Bickel and Doksum (2007). Boundary effect is a serious problem in nonparametric curve estimation. Complementing a trigonometric basis by polynomial functions is a standard method of dealing with boundary effects, and it is discussed in Efromovich (1999a, 2001a, 2018a,b).

The estimator $\hat{G}^X(x)$, defined in (6.1.14), is not equal to $1 - \hat{F}^X(x)$ where $\hat{F}^X(x) := n^{-1} \sum_{l=1}^{n} I(X_l \leq x)$ is the classical empirical cumulative distribution function. The reason for this is that we use reciprocal of $\hat{G}^X(X_l)$.

Efficient estimation of the hazard rate and the effect of the interval of estimation on the MISE is discussed in Efromovich (2016a, 2017). It is proved that the E-estimation methodology yields asymptotically sharp minimax estimation.

6.2–6.4 The topic of estimation of the hazard rate from indirect observations, created by left truncation and right censoring (LTRC), as well as estimation of the distribution, have received a great deal of attention in the statistical literature with the main emphasis on parametric models and the case of RC. See, for example, books by Cox and Oakes (1984), Cohen (1991), Anderson et al. (1993), Efromovich (1999a), Klein and Moeschberger (2003), Fleming and Harrington (2011), Lee and Wang (2013), Collet (2014), Harrell (2015), as well as papers by Uzunogullari and Wang (1992), Cao, Janssen and Veraverbeke (2005), Brunel and Comte (2008), Qian and Betensky (2014), Hagar and Dukic (2015), Shi, Chen and Zhou (2015), Bremhorsta and Lamberta (2016), Dai, Restaino and Wang (2016), Talamakrouni, Van Keilegom and El Ghouch (2016), and Wang et al. (2017), where further references may be found. Estimation of the change point is an interesting and related problem in survival analysis, see a review and discussion in Rabhi and Asgharian (2017). Bayesian approach is discussed in Ghosal and van der Vaart (2017) where further references may be found.

Efficiency of the proposed E-estimation methodology is established in Efromovich and Chu (2018a,b) where a numerical study and practical examples of the analysis of cancer data and the longevity in a retirement community may be found.

6.5 Nelson–Aalen estimator of the cumulative hazard is a popular choice in survival analysis. The estimator is defined as follows. Suppose that we observe right censored survival times of n patients meaning that for some patients we only know that their true survival times exceed certain censoring times. Let us denote by $X_1 < X_2 < \ldots$ the times when deaths are observed. Then the Nelson–Aalen estimator for the cumulative hazard of X is

$$\breve{H}^X(x) := \sum_{l:\, X_l \leq x} (1/R_l), \tag{6.11.1}$$

where R_l is the number of patients at risk of death (that is alive and not censored) just prior to time X_l. Note that the estimator is nonparametric.

If we plug the Nelson-Aalen estimator in the formula $G^X(x) = \exp(-H^X(x))$ for the survival function, then the obtained estimator is referred to as Nelson-Aalen-Breslow estimator. The interested reader may compare this estimator with Kaplan-Meier estimator (6.5.3) and realize their similarity, the latter is also supported by asymptotic results. Further, estimator (6.5.10) is formally identical to the Nelson–Aalen–Breslow estimator while the underlying idea of its construction is based on the sample mean methodology. This remark sheds additional light on similar performance in the simulations of the sample mean and Kaplan–Meier estimators of survival function. Of course, there is a vast variety of different ideas and methods proposed in the literature, see the above-cited books as well as Woodroofe (1985), Dabrowska (1989), Antoniadis, Gregoire and Nason (1999), Efromovich (1999a, 2001a), De Una-Álvarez (2004), Wang (2005), Brunel and Comte (2008) and Wang et al. (2017).

Estimation under shape restrictions is an important part of survival analysis and special procedures are suggested for taking the restrictions into consideration, see a discussion in Groeneboom and Jongbloed (2014). See also Srivastava and Klassen (2016). For the E-estimation, there is no need to make any adjustment, instead, after calculating an E-estimate, it is sufficient to take a projection on the class of assumed functions and make the estimate bona fide. A theoretical justification of this approach can be found in Efromovich (2001a).

6.6 Numerical simulation is a popular statistical tool for a simultaneous analysis of several estimators, see a discussion in Efromovich (2001a) and Efromovich and Chu (2018a,b). A discussion of dependent data and further references may be found in El Ghouch and Van Keilegom (2008, 2009), Liang and de Una-Álvarez (2011) and De Una-Álvarez and Veraverbeke (2017).

6.7 An interesting extension of the discussed topic is to develop oracle inequalities for estimators under different loss functions. Here the approaches of Efromovich (2004e,f; 2007a,b) may be instrumental. Sequential estimation is another interesting topic, see Efromovich (2004d) where the case of direct observations is considered. See also Su and Wang (2012).

It is an interesting and open problem to develop a second-order efficient estimator of the survival function. Here the approach of Efromovich (2001b, 2004c) may be instrumental. Another area of developing is the efficient multivariate estimation, see a discussion in Harrell (2015) as well as corresponding results for direct observations in Efromovich (1999a, 2000b, 2010c). Specifically, an interesting approach would be to exploit the possibility of different smoothness of the density in variables. The latter is referred to as the case of an anisotropic distribution. Further, some of the variables may be discrete (the case of a mixed distribution), and this also may attenuate the curse of multidimensionality, see a discussion of the corresponding E-estmation methodology in Efromovich (2011c).

Bayesian approach may be also useful in the analysis of multivariate distributions, see Ghosal and van der Vaart (2017).

6.8–6.9 There is a number of interesting and practically important extensions of the considered problem. The first and natural one is to consider LTRC modifications. Sequential estimation with assigned risk is another natural setting, where similarly to Efromovich (2007d,e; 2008a,c) it is possible to consider estimation with assigned risk. Specific of the problem is that now censoring affects the risk, and the latter should be taken into account. Estimation of the conditional density is another important topic, here results of Efromovich (2007g, 2010b) can be helpful. See also Wang (1996), Delecroix, Lopez and Patilea (2008), Zhang and Zhou (2013), and Wang and Chan (2017).

Wavelet estimation is a popular choice for the nonparametric estimation, see a discussion in Wang (1995), Härdle et al. (1998), Mallat (1998), Efromovich (1999a), Vidakovic (1999), Nason (2008), Addison (2017), as well as an introductory text on wavelets by Nickolas (2017). The E-estimation methodology for wavelets and multiwavelets is justified in Efromovich (1999b, 2000a, 2001c, 2004e, 2007b), Tymes, Pereyra and Efromovich (2000), and Efromovich and Smirnova (2014a). Practical applications of the wavelet E-estimation are discussed in Efromovich et al. (2004), Efromovich et al. (2008), Efromovich (2009b), Efromovich and Smirnova (2014b).

Quantile regression is another traditionally studied statistical problem, see a discussion in Efromovich (1999a). Frumento and Bottai (2017) consider the problem of quantile regression under the LTRC where further references may be found. A tutorial on regression models for the analysis of multilevel survival data can be found in Austin (2017). Another related topic is the regression models for the restricted residual mean life, see a discussion in Cortese, Holmboe and Scheike (2017).

Chapter 7

Missing Data in Survival Analysis

This chapter continues exploring topics in survival analysis when, in addition to data modification caused by censoring and truncation, some observations may be missed. We already know that censoring, truncation and missing may imply biased data, and hence it will be of interest to understand how these modifications act together and how a consistent estimator should take into account the effect of these modifications. Further, we should be aware that some modifications may be destructive. Further, missing always decreases the size of available observations. All these and other related issues are discussed in this chapter. Sections 7.1-7.2 are devoted to estimation of distributions, while remaining sections explore regression problems.

7.1 MAR Indicator of Censoring in Estimation of Distribution

The main aim of this section is to explore estimation of the distribution of a right censored (RC) lifetime when the indicator of censoring may be missed according to a MAR (missing at random) mechanism. In particular, we are interested in estimation of the cumulative hazard, hazard rate, survival function and probability density.

For instance, suppose that we study the lifetime X of a light bulb that may accidentally break at time C before it burns out. In this case the lifetime of interest X is right censored by the time of the accident C, and then an observed sample is generated by a pair $(V, \Delta) := (\min(X, C), I(X \leq C))$. The case when all observations from (V, Δ) are available was considered in the previous chapter. Here we explore a more complicated case when some observations of the indicator of censoring Δ may be missed (not available). Let us stress that we always observe V but if Δ is missed then we do not know whether V is equal to X or C.

Under the above-presented scenario, we are dealing with two layers of hidden data. The first one is a hidden sample $X_1, \ldots, X_n$ from X. The second one is a hidden RC sample $(V_1, \Delta_1), \ldots, (V_n, \Delta_n)$ from (V, Δ). The observed sample is $(V_1, A_1\Delta_1, A_1), \ldots, (V_n, A_n\Delta_n, A_n)$ from $(V, A\Delta, A)$ where A is a Bernoulli random variable called the availability. It is assumed that the missing is MAR, and hence the likelihood of missing is defined by the always observed variable V, namely it is assumed that

$$\mathbb{P}(A = 1|V = v, \Delta = \delta) = \mathbb{P}(A = 1|V = v) =: w(v) \geq c_0 > 0, \qquad (7.1.1)$$

where here and in what follows $v \in [0, \infty)$ and $\delta \in \{0, 1\}$.

Is the distribution of X estimable in this case? Are the observations biased? Can the missing be ignored and a complete-case approach be used? Can the distribution of the censoring variable C be estimated? This is a type of question that we would like to answer.

We begin our exploration of the problem with probability formulas for available variables. Let us assume that X and C are continuous and independent random variables. Then, using (7.1.1), we can write for the joint mixed density of observations

$$f^{V, A\Delta, A}(v, \delta, 1) = \mathbb{P}(A = 1|V = v, \Delta = \delta)f^{V, \Delta}(v, \delta)$$

$$= w(v)[f^X(v)G^C(v)]^\delta [f^C(v)G^X(v)]^{1-\delta} I(v \in [0,\infty), \delta \in \{0,1\}). \qquad (7.1.2)$$

The formula is symmetric with respect to X and C in the sense that if Δ is replaced by $1-\Delta$ then formally C is censored by X. This tells us that it is sufficient to explore estimation of the distribution of X and then use the symmetry for estimation of the distribution of C.

Now let us recall useful notations. The first one is a probability which has several useful representations,

$$g(v) := \mathbb{P}(V \geq v) = \mathbb{E}\{I(V \geq v)\} = G^V(v) = G^X(v)G^C(v). \qquad (7.1.3)$$

Note that continuity of the random variable V was used in establishing (7.1.3). The second one is our notation β_Z for the upper bound of the support of a random variable Z. Using this notation we can write that

$$\beta_V = \min(\beta_X, \beta_C). \qquad (7.1.4)$$

Using (7.1.2)-(7.1.4) we may get the following formulas for the density $f^X(x)$ of the lifetime of interest,

$$f^X(x) = \frac{f^{V,A\Delta,A}(x,1,1)}{w(x)G^C(x)} = G^X(x)\frac{f^{V,A\Delta,A}(x,1,1)}{g(x)w(x)}, \quad x \in [0,\beta_V). \qquad (7.1.5)$$

There are two important conclusions from this relation. The former is that we cannot consistently estimate density $f^X(x)$ for points $x > \beta_V$. This conclusion is also supported by the fact that we do not have observations beyond β_V. On the other hand, if $\beta_X \leq \beta_C$ then consistent estimation of the distribution of X is possible. Of course, the opposite is true for consistent estimation of the distribution of C. This yields that unless $\beta_X = \beta_C$ the distribution of one of the two variables cannot be consistently estimated. Because we are primarily interested in estimation of the distribution of X, we assume that

$$\beta_X \leq \beta_C. \qquad (7.1.6)$$

We begin our estimation of different characteristics of the distribution of X via the cumulative hazard,

$$H^X(x) := \int_0^x [f^X(u)/G^X(u)]du, \quad x \in [0,\beta_X), \qquad (7.1.7)$$

and recall that the survival function $G^X(x) := \mathbb{P}(X > x)$ can be expressed as

$$G^X(x) = \exp(-H^X(x)). \qquad (7.1.8)$$

Estimation of the cumulative hazard involves several steps. First, using (7.1.5) we can write down the cumulative hazard as the expectation of a function of observed variables,

$$H^X(x) = \mathbb{E}\Big\{\frac{A\Delta I(V \leq x)}{g(V)w(V)}\Big\}, \quad x \in [0,\beta_V). \qquad (7.1.9)$$

To use (7.1.9) for constructing a plug-in sample mean estimator, we need to propose estimators for functions $g(v)$ and $w(v)$. As a result, our second step is to estimate function $g(v)$. According to (7.1.3), $g(v)$ may be written as the expectation of indicator $I(V \geq v)$, and this yields the sample mean estimator

$$\hat{g}(v) := n^{-1}\sum_{l=1}^n I(V_l \geq v). \qquad (7.1.10)$$

Note that $\hat{g}(V_l) \geq n^{-1}$ for $l = 1, \ldots, n$ and hence $\hat{g}(V_l)$ may be used in a denominator.

Our third step is to propose an estimator of $w(v)$. According to (7.1.1), the availability likelihood may be written as

$$w(v) = \mathbb{E}\{A|V = v\}. \tag{7.1.11}$$

We conclude that $w(v)$ is a nonparametric Bernoulli regression of A on V. Using the available sample $(V_1, A_1), \ldots, (V_n, A_n)$, we construct a regression E-estimator $\hat{w}(v)$ proposed in Section 2.4.

Now we are ready to define a plug-in sample mean estimator of the cumulative hazard,

$$\hat{H}^X(x) := n^{-1} \sum_{l=1}^{n} \frac{A_l \Delta_l I(V \leq x)}{\hat{g}(V_l) \max(\hat{w}(V_l), c/\ln(n))}, \quad x \in [0, V_{(n)}]. \tag{7.1.12}$$

Here $c > 0$ is the parameter of the estimator and recall that $V_{(n)}$ denotes the largest among n observations $V_1, \ldots, V_n$, it is used as an estimator of β_V, and note that $\mathbb{P}(V_{(n)} < \beta_V) = 1$.

In its turn, the estimator (7.1.12) and formula (7.1.8) yield a plug-in sample mean estimator of the survival function (the standard notation $\exp(z) := e^z$ is used),

$$\hat{G}^X(x) := \exp\left(-n^{-1} \sum_{l=1}^{n} \frac{A_l \Delta_l I(V \leq x)}{\hat{g}(V_l) \max(\hat{w}(V_l), c/\ln(n))}\right), \quad x \in [0, V_{(n)}]. \tag{7.1.13}$$

It is of a special interest to evaluate asymptotic variances of the empirical cumulative hazard and the empirical survival function. First, we calculate the asymptotic variance of estimator (7.1.12) of the cumulative hazard,

$$\lim_{n \to \infty} n\mathbb{V}(\hat{H}^X(x)) = \mathbb{E}\{A\Delta I(V \leq x)[g(V)w(V)]^{-2}\} - (H^X(x))^2$$

$$= \int_0^x f^X(u)[(G^X(u))^2 G^C(u)w(u)]^{-1} du - (H^X(x))^2], \quad x \in [0, \beta_V). \tag{7.1.14}$$

Then, using the delta method (or straightforwardly using a Taylor expansion of function $\exp(-x)$) we evaluate variance of the empirical survival function,

$$\lim_{n \to \infty} n\mathbb{V}(\hat{G}^X(x)) = [G^X(x)]^2 \mathbb{V}(\hat{H}^X(x))$$

$$= [G^X(x)]^2 \left[\int_0^x f^X(u)[(G^X(u))^2 G^C(u)w(u)]^{-1} du - (H^X(x))^2\right], \quad x \in [0, \beta_V). \tag{7.1.15}$$

Formulas (7.1.14) and (7.1.15) show us how distributions of X and C, together with the missing mechanism, affect accuracy of estimation. What we see is an interesting compounded effect of the two data modifications.

The survival function $G^C(x)$ of the censoring variable C may be also estimated by (7.1.13) with Δ_l being replaced by $1 - \Delta_l$, recall the above-discussed symmetry between X and C. In what follows we denote this estimator as $\hat{G}^C(x)$ and it will be used only for $x \leq V_{(n)}$.

Now let us explain how the hazard rate of X may be estimated. According to (7.1.5), the hazard rate may be written as follows,

$$h^X(x) := \frac{f^X(x)}{G^X(x)} = \frac{f^{V,A\Delta,A}(x,1,1)}{g(x)w(x)}, \quad x \in [0, \beta_V). \tag{7.1.16}$$

Recall our discussion of estimation of the hazard rate in Section 6.1 and that it may be estimated only over a finite subinterval of the support of X. Suppose that we are interested in estimation of $h^X(x)$ over an interval $[a, a+b] \subset [0, \beta_V)$. Then, using our traditional notation

$\psi_j(x)$ for elements of the cosine basis on $[a, a + b]$, we can write for Fourier coefficients of the hazard rate,

$$\kappa_j := \int_a^{a+b} h^X(x)\psi_j(x)dx = \mathbb{E}\Big\{\frac{A\Delta I(V \in [a, a + b])\psi_j(V)}{g(V)w(V)}\Big\}. \qquad (7.1.17)$$

The expectation in the right side of (7.1.17) yields the following plug-in sample mean Fourier estimator,

$$\hat{\kappa}_j := n^{-1}\sum_{l=1}^{n}\frac{A_l\Delta_l I(V_l \in [a, a + b])\psi_j(V_l)}{g(V_l)w(V_l)}. \qquad (7.1.18)$$

In its turn, the Fourier estimator implies the hazard rate E-estimator $\hat{h}^X(x)$.

Our last estimand to consider is the probability density $f^X(x)$. Similarly to the hazard rate, we estimate it over an interval $[a, a + b] \subset [0, \beta_V)$. Using (7.1.5), a Fourier coefficient of the density $f^X(x)$ can be written as the following expectation,

$$\theta_j := \int_a^{a+b} f^X(x)\psi_j(x)dx$$

$$= \int_a^{a+b} [f^{V,A\Delta,A}(x, 1, 1)\psi_j(x)/(w(x)G^C(x))]dx$$

$$= \mathbb{E}\Big\{\frac{A\Delta I(V \in [a, a + b])\psi_j(V)}{w(V)G^C(V)}\Big\}. \qquad (7.1.19)$$

This expression immediately yields a plug-in sample-mean Fourier estimator,

$$\hat{\theta}_j := n^{-1}\sum_{l=1}^{n}\frac{A_l\Delta_l I(V_l \in [a, a + b])\psi_j(V_l)}{\max(\hat{w}(V_l)\hat{G}^C(V_l), c/\ln(n))}. \qquad (7.1.20)$$

In (7.1.17), similarly to (7.1.12), we are bounding from below the denominator and the constant c is a parameter that may be manually chosen in a corresponding figure. The latter allows us to have a larger control of the E-estimator.

Fourier estimator (7.1.20) yields a corresponding density E-estimator $\hat{f}^X(x)$ defined in Section 2.2.

The coefficient of difficulty of the proposed Fourier estimator is

$$d := \lim_{n,j\to\infty} n\mathbb{V}(\hat{\theta}_j) = b^{-1}\mathbb{E}\{I(V \in [a, a + b])A\Delta[w(V)G^C(V)]^{-2}\}$$

$$= \int_a^{a+b}\frac{f^X(x)}{bw(x)G^C(x)}dx, \qquad (7.1.21)$$

and it can be estimated by the following plug-in sample mean estimator,

$$\hat{d} := n^{-1}b^{-1}\sum_{l=1}^{n}A_l\Delta_l I(V_l \in [a, a + b])[\max(\hat{w}(V_l)\hat{G}^C(V_l), c/\ln(n))]^{-2}. \qquad (7.1.22)$$

Formula (7.1.21) allows us to conclude that integral $\int_a^{a+b} f^X(x)[bw(x)G^C(x)]^{-1}dx$ should be finite for the E-estimator to be consistent. Of course, this integral is finite whenever $a + b < \beta_V$. Further, recall that the E-estimation inference, developed in Section 2.6, allows us to calculate confidence bands for the density E-estimator.

Let us check how the estimators, proposed for estimation of the survival function and the density of a lifetime of interest, perform in simulated examples. We begin with Figure 7.1.

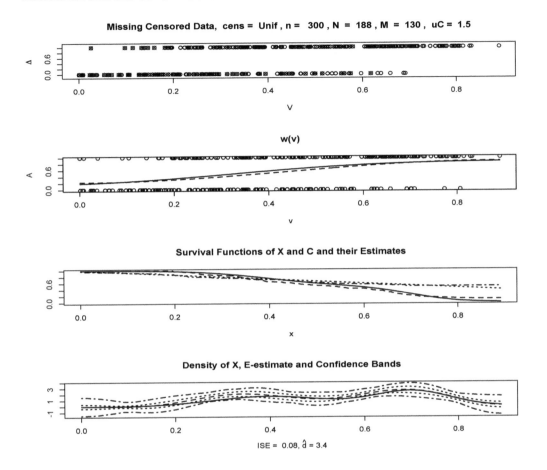

Figure 7.1 *Right censored data with MAR indicators of censoring, the case $\beta_X < \beta_C$. The RC mechanism is identical to the one described in Figure 6.6. The top diagram shows available data. Circles show available (not missed) realizations of (V, Δ) while crossed circles show hidden realizations whose values of Δ are missed. The latter allows us to visualize the hidden RC sample shown by circles. The title shows $N := \sum_{l=1}^{n} \Delta_l$, which is the number of uncensored lifetimes in the hidden RC sample, and $M := \sum_{l=1}^{n} A_l \Delta_l$, which is the number of available uncensored lifetimes in complete cases. The second from the top diagram shows by circles the Bernoulli scattergram of n realizations of (V, A) as well as the underlying availability likelihood (the solid line) and its regression E-estimate (the dashed line). The third from the top diagram shows the survival function G^X and its estimate $\hat{G}^X$ by the solid and dashed lines, respectively. It also exhibits the survival function of G^C and its estimate $\hat{G}^C$ by the dotted and dot-dashed lines, respectively. The bottom diagram shows the underlying density (the solid line) and its E-estimate (the dashed line) over interval $[0, V_{(n)}]$. It also exhibits $(1 - \alpha)$ pointwise (the dotted lines) and simultaneous (the dot-dashed lines) confidence bands. {Use cens= "Expon" for exponential censoring variable whose mean is controlled by argument lambdaC. Parameter u_C is controlled by the argument uC. Availability likelihood is defined by the string w.} [n = 300, corn = 3, w = "0.1+0.8*exp(1+6*(v-0.5))/(1+exp(1+6*(v-0.5)))", cens= "Unif", uC = 1.5, lambdaC = 1.5, alpha = 0.05, c = 1, cJ0 = 4, cJ1 = 0.5, cTH = 4]*

Because the missing mechanism acts after right censoring, it is convenient to use a familiar RC simulation of Figure 6.6. Recall that in Figure 6.6 the lifetime of interest X is generated by a corner density and C is either uniform or exponential. Then we introduce a missing mechanism for the indicator of censoring with a logistic availability likelihood function defined in the caption of Figure 7.1. In Figure 7.1 the used censoring is Uniform$(0, u_C)$, and

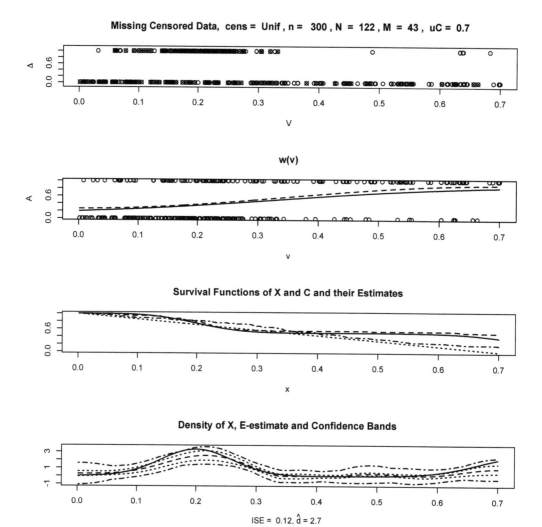

Figure 7.2 *Right censored data with MAR indicators of censoring, the case $\beta_C < \beta_X$. This figure is created by Figure 7.1 using uC = 0.7 and corn = 4. [n = 300, corn = 4, w = "0.1+0.8*exp(1+6*(v-0.5))/(1+exp(1+6*(v-0.5)))", cens = "Unif", uC = 0.7, lambdaC = 1.5, alpha = 0.05, c = 1, cJ0 = 4, cJ1 = 0.5, cTH = 4]*

more information about the simulation and diagrams may be found in the caption of Figure 7.1.

The top diagram shows us the underlying RC data by circles. In the simulation $\beta_C = 1.5 > \beta_X = 1$, and hence the distribution of X can be consistently estimated based on the underlying RC data, while the distribution of C may be estimated up to the point $\beta_V = 1$.

As we can see from the main title, in the particular simulation only $N := \sum_{l=1}^{n} \Delta_l = 188$ of underlying $n = 300$ observations of X are uncensored by the Uniform$(0,1.5)$ censoring random variable C. Further, among those 188 hidden uncensored observations of the lifetime of interest, only $M := \sum_{l=1}^{n} A_l \Delta_l = 130$ belong to complete cases. This is a dramatic decrease in the available information due to censoring and missing. Of course, it is important to stress that the information about N, the supports, and the underlying distributions are known only because the data are simulated. Otherwise we would know only the MAR RC data, n and M.

The second from the top diagram explains estimation of the availability likelihood function $w(v)$. This is a straightforward Bernoulli regression because all $n = 300$ observations of the pair (V, A) are available, and the E-estimate is good.

The third from the top diagram exhibits underlying survival functions $G^X(x)$, $G^C(x)$ and their estimates. Note that $G^C(x)$ is estimated only for $x \leq V_{(n)}$. Keeping in mind complexity of the setting and the relatively small size of available observations, the estimates are good.

The bottom diagram allows us to visualize the underlying Bimodal density (the solid line), its E-estimate (the dashed line) and the confidence bands. The density E-estimator uses the interval $[a, a+b] = [0, V_{(n)}]$. The subtitle of the diagram shows us the estimated coefficient of difficulty $\hat{d} = 3.4$. This coefficient of difficulty informs us that for the developed density E-estimator to get the same MISE as for the case of 100 direct observations of X, the sample size n should be 3.4 times larger, that is $n = 340$. Of course, here we are dealing with the estimated coefficient of difficulty, but the example explains the purpose of estimating the coefficient of difficulty. Further, recall that for Figure 6.6, with the same RC mechanism and no missing, the estimated coefficient of difficulty is 2. These numbers shed a new light on complexity of the considered model with MAR indicators of censoring. Overall, the particular density E-estimate is good, and the reader is advised to repeat Figure 7.1 and learn more about the model and proposed estimators.

The simulation in Figure 7.1 favors estimation of the distribution of X because it uses $\beta_C > \beta_X$. What will be if this condition is violated? Figure 7.2 allows us to explore such a situation. Here the underlying simulation is the same as in Figure 7.1 only the censoring variable is Uniform$(0, 0.7)$, and hence we have $\beta_C < \beta_X$. We also use the Strata as the density of X. Diagrams of Figure 7.2 are similar to diagrams of Figure 7.1.

First of all, let us look at available data and statistics presented in the top diagram. As it could be expected, all available observations of V are smaller than and close to 0.7. Further, note that there are just few uncensored observations with values larger than 0.3. Further, the censoring is so severe that in the hidden RC sample there are only $N = 122$ uncensored lifetimes, and in the available sample with missing indicators there are only 43 uncensored lifetimes that we are aware of. This is an absolutely devastating loss of information contained in 300 underlying realizations of the lifetime X.

The second from the top diagram shows us the Bernoulli regression and estimation of the availability likelihood. The third from the top diagram shows us underlying survival functions and their estimates. Under the above-explained circumstances, the estimates are reasonable. Note that no longer we can consistently estimate the distribution of X, but we can estimate its survival function up to $V_{(n)}$. Finally, the bottom diagram shows us the underlying Strata density, its E-estimate and confidence bands for $x \in [0, V_{(n)}]$. The E-estimate is far from being good, but at the same time note that it correctly indicates the main features of the Strata.

It is fair to conclude that the settings considered in Figures 7.1 and 7.2 are complicated. Repeated simulations with different parameters will be helpful in understanding MAR RC data.

7.2   MNAR Indicator of Censoring in Estimation of Distribution

We are considering the same underlying RC model as in the previous section. Namely, there is a lifetime of interest X which is right censored by a censoring variable C, and this creates a hidden sample from $(V, \Delta) := (\min(X, C), I(X \leq C))$. Additionally, the indicator of censoring Δ may be missed, and an observed sample is from a triplet $(V, A\Delta, A)$. Here A is a Bernoulli random variable, and if $A = 1$ then the indicator Δ is observed and hence we know that the lifetime is censored or not, and if $A = 0$ then we observe V but do not know if V is equal to X or C. What separates the considered setting from the one explored in Section 7.1 is that now the availability likelihood is defined not by the always observed V but by the hidden lifetime of interest X, namely it is assumed that

$$\mathbb{P}(A = 1|V = v, \Delta = \delta, X = x) = \mathbb{P}(A = 1|X = x) =: w(x) \geq c_0 > 0. \tag{7.2.1}$$

In (7.2.1) it is understood that the equality holds for all nonnegative $v = x$ if $\delta = 1$ and for all positive x and $0 \leq v < x$ if $\delta = 0$. Further, let us recall that the lifetimes X and C are nonnegative continuous random variables while A and Δ are Bernoulli.

Let us comment on this complicated missing mechanism which involves a hidden variable. The missing is not MAR because $\mathbb{P}(A = 1|V = v, \Delta = 0) \neq \mathbb{P}(A = 1|V = v)$. Hence, here we are dealing with the MNAR. We know from Chapter 5 that typically MNAR implies a destructive missing when consistent estimation is impossible. Is this the case here or are we dealing with a case where MNAR is nondestructive? We know from Chapter 5 that consistent estimation, based on MNAR data, may be possible if the MNAR is converted into MAR by an always observed auxiliary variable that defines the missing. Here we do not have an auxiliary variable. On the other hand, we know that $V = X$ given $\Delta = 1$. The latter is our only chance to convert the MNAR into MAR via considering a corresponding subsample of uncensored lifetimes, and in what follows we are exploring this opportunity to the fullest.

We begin with several probability formulas. First, let us note that events $\{V = x, \Delta = 1\}$ and $\{V = x, \Delta = 1, X = x\}$ coincide due to definition of the indicator of censoring Δ. This, together with (7.2.1), implies that

$$\mathbb{P}(A = 1|V = x, \Delta = 1) = \mathbb{P}(A = 1|V = x, \Delta = 1, X = x) = w(x). \tag{7.2.2}$$

In its turn, (7.2.2) implies that

$$\mathbb{E}\{A|V = x, \Delta = 1\} = w(x). \tag{7.2.3}$$

Second, similarly to Section 7.1, let us assume that X and C are continuous and independent random variables. Then, using (7.2.1) we can write for the joint mixed density of observations,

$$f^{V,A\Delta,A}(x, 1, 1) = \mathbb{P}(A = 1|V = x, \Delta = 1)f^{V,\Delta}(x, 1)$$

$$= \mathbb{P}(A = 1|X = x, \Delta = 1)f^X(x)G^C(x) = w(x)f^X(x)G^C(x). \tag{7.2.4}$$

Similarly,

$$f^{V,A\Delta,A}(x, 0, 1) = \mathbb{P}(A = 1|V = x, \Delta = 0)f^{V,\Delta}(x, 0)$$

$$= \mathbb{P}(A = 1|C = x, X > x)f^C(x)G^X(x)$$

$$= \frac{\mathbb{P}(A = 1, X > x|C = x)f^C(x)G^X(x)}{\mathbb{P}(X > x|C = x)}$$

$$= [\int_x^\infty w(u)f^X(u)du]f^C(x). \tag{7.2.5}$$

In the last line we used independence of X and C.

Formulas (7.2.4) and (7.2.5) indicate that, unless $w(x)$ is constant and the missing is MCAR, the symmetry between X and C, known for directly observed censored data, no longer present.

Now we are in a position to explain how we can estimate characteristics of the distribution of X, and here we are considering estimation of the cumulative hazard, survival function and the density.

Our first step is to estimate the availability likelihood $w(x)$. According to (7.2.3), for the MNAR missing $w(x)$ is the Bernoulli regression of A on V in a subsample of uncensored lifetimes, that is when $\Delta = 1$. The Bernoulli regression E-estimator $\hat{w}(x)$, introduced in Section 2.4, can be used to estimate the availability likelihood.

Our second step is to estimate the cumulative hazard of X. Using (7.2.4) we write,

$$H^X(x) := \int_0^x [f^X(u)/G^X(u)]du = \mathbb{E}\left\{\frac{A\Delta I(V \le x)}{g(V)w(V)}\right\}, \quad x \in [0, \beta_V). \tag{7.2.6}$$

Here

$$g(v) := \mathbb{P}(V \ge v) = \mathbb{E}\{I(V \ge v)\} = G^V(v) = G^X(v)G^C(v), \tag{7.2.7}$$

where we used the assumption that X and C are continuous and independent random variables.

The probability $g(v)$ may be estimated by a sample mean estimator,

$$\hat{g}(v) := n^{-1} \sum_{l=1}^n I(V_l \ge v). \tag{7.2.8}$$

Combining the obtained results, we get a plug-in sample mean estimator of the cumulative hazard,

$$\hat{H}^X(x) := n^{-1} \sum_{l=1}^n \frac{A_l \Delta_l I(V_l \le x)}{\hat{g}(V_l)\max(\hat{w}(V_l), c/\ln(n))}, \quad x \in [0, V_{(n)}]. \tag{7.2.9}$$

Here, as usual, $c > 0$ is a parameter that may be chosen for a corresponding figure.

In its turn, (7.2.9) allows us to estimate the survival function $G^X(x)$. Recall that $G^X(x) = \exp(-H^X(x))$, and this, together with (7.2.9), yields the plug-in sample mean estimator of the survival function,

$$\hat{G}^X(x) := e^{-\hat{H}^X(x)}, \quad x \in [0, V_{(n)}]. \tag{7.2.10}$$

Estimation of the density $f^X(x)$ over an interval $[a, a+b] \subset [0, \beta_V)$ is based on using formula (7.2.4). Denote by $\psi_j(x)$ elements of the cosine basis on $[a, a+b]$ and write a Fourier coefficient of the density $f^X(x)$ as

$$\theta_j := \int_a^{a+b} f^X(x)\psi_j(x)dx = \mathbb{E}\left\{\frac{A\Delta I(V \in [a, a+b])\psi_j(V)}{w(V)G^C(V)}\right\}. \tag{7.2.11}$$

To use the sample mean approach for estimation of the expectation in the right side of (7.2.11) we need to know functions $w(v)$ and $G^C(v)$. The former is estimated by the the above-defined Bernoulli regression E-estimator $\hat{w}(x)$. There are several possibilities to estimate $G^C(v)$, and here we are using a plug-in estimator motivated by (7.2.7), namely we set

$$\hat{G}^C(v) := \hat{g}(v)/\hat{G}^X(v), \tag{7.2.12}$$

where $\hat{g}(v)$ is defined in (7.2.8) and $\hat{G}^X(x)$ in (7.2.10). These estimators allow us to define a plug-in sample mean Fourier estimator,

$$\hat{\theta}_j := n^{-1} \sum_{l=1}^n \frac{A_l \Delta_l I(V_l \in [a, a+b])\psi_j(V_l)}{\max(\hat{w}(V_l)\hat{G}^C(V_l), c/\ln(n))}. \tag{7.2.13}$$

Note that both censoring and missing creates bias in observed data, and then the random factor $A\Delta[\hat{w}(V)\hat{G}^C(V)]^{-1}$, used in (7.2.13), corrects the compounded bias.

Missing Censored Data, cens = Unif , n = 300 , N = 174 , M = 129 , uC = 1.3

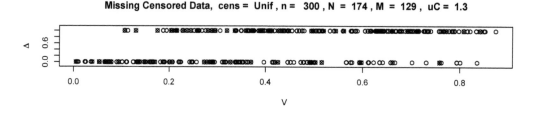

V

w(x) and its Estimate

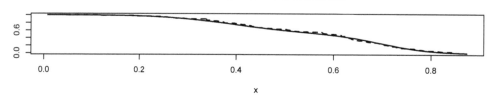

X

Survival Function of X and its Estimate

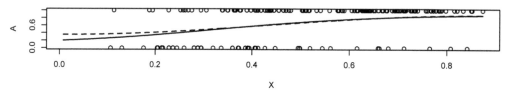

x

Density of X, E-estimate and Confidence Bands

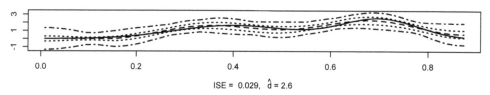

ISE = 0.029, $\hat{d}$ = 2.6

Figure 7.3 *Right censored data with MNAR indicators of censoring. The MNAR is defined by the lifetime of interest X according to the availability likelihood w indicated below. All other components of the simulation and the diagrams are the same as in Figure 7.1. [n = 300, corn = 3, w = "0.1+0.8*exp(1+6*(x-0.5))/(1+exp(1+6*(x-0.5)))", cens = "Unif", uC = 1.5, c = 1, alpha = 0.05, cJ0 = 4, cJ1 = 0.5, cTH = 4]*

The Fourier estimator yields the density E-estimator $\hat{f}(x)$, $x \in [a, a + b]$.

Figure 7.3 illustrates the setting and the proposed solution, and its diagrams are similar to those in Figure 7.1. The top diagram shows us the available data. Note that the hidden sample from X contains $n = 300$ observations, only $N := \sum_{l=1}^{n} \Delta_l = 174$ of them are uncensored, and of those only $M := \sum_{l=1}^{n} A_l \Delta_n = 129$ contain indicators of censoring. In other words, we have only 129 observations of X to work with, and they are biased by both censoring and MNAR. These observations are shown by horizontal coordinates of circles with y-coordinates equal to 1. They are also shown by horizontal coordinates of circles, corresponding to $A = 1$, in the second diagram.

The second from the top diagram shows us the scatterplot of pairs (X_l, A_l) with corresponding indicators $\Delta_l = 1$. Note that if $\Delta_l = 1$, then $V_l = X_l$ and hence we know X_l. The underlying regression function (the solid line) is the availability likelihood. Its regression E-estimate is shown by the dashed line, and it is based on $M = 129$ observations. Note that among 129 observations of X only a few have values less than 0.2 and larger than 0.8. This is a serious complication, and for this particular simulation the E-estimator does a good job. At the same time, let us recall that according to (7.2.9) and (7.2.13), we need to know values of $w(v)$ only for points V_l corresponding to cases with $A_l = \Delta_l = 1$.

The third diagram shows us the underlying survival function of the Bimodal distribution and the sample mean estimate (7.2.10). The estimate is good. Finally, the density estimate is shown in the bottom diagram together with its confidence bands. It is worthwhile to pay attention to a relatively large coefficient of difficulty shown in the subtitle. It raises a red flag about complexity of the considered estimation problem. The reader is advised to repeat the simulation, use different parameters, and appreciate complexity of the problem of estimation of the density based on censored observations with MNAR indicators of censoring.

7.3   MAR Censored Responses

We are considering a regression problem with two sequentially created layers of hidden observations. The first layer of hidden observations is created by a sample $(X_1, Y_1), \ldots, (X_n, Y_n)$ of size n from a pair of continuous random variables (X, Y). Here X is the predictor, Y is the response, and the aim is to estimate the regression function

$$m(x) := \mathbb{E}\{Y|X = x\}. \tag{7.3.1}$$

This is a classical regression problem, and if the sample would be known, then we could use the regression E-estimator of Section 2.3.

The second layer of hidden observations is created by right censoring of the response Y by a censoring variable C. The hidden sample is $(X_1, V_1, \Delta_1), \ldots, (X_n, V_n, \Delta_n)$ where $V_l := \min(Y_l, C_l)$ and $\Delta_l := I(Y_l \leq C_l)$. Note that this sample is from (X, V, Δ) where $V := \min(Y, C)$ and $\Delta := I(Y \leq C)$. If this sample was known, we could use the regression E-estimator of Section 6.8.

Finally, censored responses are subject to MAR (missing at random) with the likelihood of missing defined by the always observed predictor. Namely, what we observe is a sample $(X_1, A_1V_1, \Delta_1, A_1), \ldots, (X_n, A_nV_n, \Delta_n, A_n)$ from (X, AV, Δ, A) where the availability A is a Bernoulli random variable and

$$\mathbb{P}(A = 1|X = x, Y = y, V = v, \Delta = \delta, C = u)$$

$$= \mathbb{P}(A = 1|X = x) =: w(x) \geq c_0 > 0. \tag{7.3.2}$$

Based on this twice modified sample, first by RC of the response and then by MAR of the censored response, the aim is to estimate the underlying regression function (7.3.1). In what follows it is assumed that X, Y and C are continuous variables, and the pair (X, Y) is independent of C. As usual, the predictor X is supported on $[0, 1]$ according to the design density $f^X(x) \geq c_* > 0$, $x \in [0, 1]$, and because we are considering survival data, both Y and C are nonnegative. Also recall our notation β_Z for the right boundary point of the support of a continuous random variable Z.

To propose a solution and understand when a consistent estimation of the regression function is possible, we begin with probability formulas for observed variables. The joint mixed density of (X, AV, Δ, A) may be written as

$$f^{X,AV,\Delta,A}(x, v, \delta, 1) = w(x)f^X(x)[f^{Y|X}(v|x)G^C(v)]^\delta[f^C(v)G^{Y|X}(v|x)]^{1-\delta}$$

$$\times I(x \in [0, 1], v \in [0, \beta_V], \delta \in \{0, 1\}). \tag{7.3.3}$$

It is stressed in (7.3.3) that the observed random variable V cannot be larger than $\beta_V = \min(\beta_Y, \beta_C)$, and note that this is the only variable that may give us information about a hidden response Y. As a result, as we know from previous sections and as it follows from (7.3.3), we may estimate any characteristic of the distribution of Y only up to point β_V. This is bad news because it implies that we may consistently estimate the regression function only if $\beta_Y \leq \beta_V$, or equivalently if

$$\beta_Y \leq \beta_C. \tag{7.3.4}$$

Hence, as it was explained in Section 6.8, a reasonable estimand is a censored (trimmed) regression function

$$m(x, \beta_V) := \mathbb{E}\{YI(Y \leq \beta_V)|X = x\}. \tag{7.3.5}$$

If (7.3.4) holds, then the censored regression function is equal to the regression of interest $m(x)$ defined in (7.3.1). Further, in general $m(x, \beta_V) \leq m(x)$, and more discussion can be found in Section 6.8.

To use our methodology of E-estimation, we need to understand how to construct a sample mean estimator of Fourier coefficients of the function $m(x, \beta_V)$ assuming for now that β_V is given. This is not a simple procedure and it involves a number of steps. In what follows we use our traditional notation $\varphi_j(x)$ for elements of the cosine basis on $[0, 1]$. Using (7.3.3) we can write for a Fourier coefficient of $m(x, \beta_V)$,

$$
\begin{aligned}
\theta_j &:= \int_0^1 m(x, \beta_V)\varphi_j(x)dx \\[2mm]
&= \int_0^1 \Big[\int_0^{\beta_V} y f^{Y|X}(y|x)dy\Big]\varphi_j(x)dx \\[2mm]
&= \int_0^1 \Big[\int_0^{\beta_V} \frac{v f^{X,AV,\Delta,A}(x,v,1,1)\varphi_j(x)}{w(x)f^X(x)G^C(v)}dv\Big]dx \\[2mm]
&= \mathbb{E}\Big\{\frac{A\Delta V\varphi_j(X)}{[w(X)f^X(X)]G^C(V)}\Big\}.
\end{aligned}
\tag{7.3.6}
$$

Now let us make the following remark. For some previous settings we made a useful observation that there is no need to estimate the availability likelihood $w(x)$ because $w(x)f^X(x)/\mathbb{P}(A = 1)$ is the density of predictors in a complete-case subsample. This would be a welcome outcome here because to estimate (7.3.6) we need to know this product, and note that the product is highlighted by the square brackets in the denominator of (7.3.6). Unfortunately, this is not the case for the subsample satisfying $A_l\Delta_l = 1$ which is considered in (7.3.6), because even given (7.3.4) the integral $\int_0^{\beta_V} f^{Y|X}(v|x)G^C(v)dv$ depends on x. At the same time, if we restrict our attention to complete-case pairs (X_l, V_l) given $A_l = 1$, then the density of predictors in that subsample is

$$p(x) := f^{X|A}(x|1) = \frac{w(x)f^X(x)}{\mathbb{P}(A = 1)} = \frac{w(x)f^X(x)}{\int_0^1 w(u)f^X(u)du}. \tag{7.3.7}$$

This allows us to estimate the product $w(x)f^X(x)$ by $\hat{p}(x)[n^{-1}\sum_{l=1}^n A_l]$ where $\hat{p}(x)$ is the density E-estimator of Section 2.2 based on observations of X in complete cases.

Another possible approach is to estimate $f^X(x)$ by the density E-estimator $\hat{f}^X(x)$ based on n available observations of the predictor. Further, note that n observations of (X, A) are available, and hence we can estimate $w(x)$ by our Bernoulli regression E-estimator $\hat{w}(x)$. As we see, we have possibilities to either estimate these two functions separately or estimate their product.

Now let us return to (7.3.6). To use this expectation as a guide for a sample mean estimation of Fourier coefficients, in addition to knowing the product $w(x)f^X(x)$ we need to know the survival function $G^C(v)$ of the censoring variable. Its estimation is an interesting and not a simple problem due to the missing of censored responses. We begin explanation of a proposed procedure of estimation of $G^C(v)$ with writing down the joint mixed density of (X, AV) and the events $\Delta = 0$ and $A = 1$,

$$f^{X,AV,\Delta,A}(x, v, 0, 1) = w(x)f^X(x)f^C(v)G^{Y|X}(v|x)I(x \in [0,1], v \in [0, \beta_V]). \qquad (7.3.8)$$

This formula implies allows us to write (recall that (X, Y) and C are independent)

$$\frac{f^C(v)}{G^C(v)} = \frac{f^{X,AV,\Delta,A}(x, v, 0, 1)}{w(x)f^X(x)G^C(v)G^{Y|X}(v|x)}$$

$$= \frac{f^{X,AV,\Delta,A}(x, v, 0, 1)}{w(x)f^X(x)G^{V|X}(v|x)}, \quad x \in [0,1],\ v \in [0, \beta_V). \qquad (7.3.9)$$

In its turn, formula (7.3.9) allows us to get a useful formula for the cumulative hazard of C,

$$H^C(v) := \int_0^v [f^C(y)/G^C(y)]dy = \int_0^v \int_0^1 [f^C(y)/G^C(y)]dxdy$$

$$= \int_0^v \int_0^1 \frac{f^{X,AV,\Delta,A}(x, y, 0, 1)}{w(x)f^X(x)G^{V|X}(y|x)}dxdy$$

$$= \mathbb{E}\Big\{ \frac{(1-\Delta)AI(AV \le v)}{[w(X)f^X(X)]G^{V|X}(AV|X)} \Big\}, \quad v \in [0, \beta_V). \qquad (7.3.10)$$

This result shows that we can express the cumulative hazard of C as an expectation of observed variables. In that expectation we know how to estimate the product $w(x)f^X(x)$, and we need to explain how to estimate the conditional survival function $G^{V|X}(v|x) := \mathbb{P}(V > v|X = x)$. This is a new problem for us and it is of interest on its own.

Using the assumed continuity of the random variable V we may write,

$$G^{V|X}(v|x) := \mathbb{P}(V > v|X = x) = \mathbb{P}(V \ge v|X = x) = \mathbb{E}\{I(V \ge v)|X = x\}. \qquad (7.3.11)$$

The right side of (7.3.11) reveals that for a given v the conditional survival function $G^{V|X}(v|x)$ is a Bernoulli regression of the response $I(V \ge v)$ on the predictor X. The only complication here is that some realizations of the response $I(V \ge v)$ are missed. The good news is that the missing is MAR and it is defined by the predictor. Hence, according to Section 4.2, we may use a complete-case approach together with our regression E-estimator. This gives us the regression E-estimator $\hat{G}^{V|X}(v|x)$.

Now we are in a position to finish our explanation of how to estimate the survival function $G^C(v)$. Recall that this function can be expressed via the cumulative hazard as $G^C(v) = \exp(-H^C(v))$, and hence, according to (7.3.7) and (7.3.10), the plug-in sample mean estimator of the survival function is

$$\hat{G}^C(v) := \exp\Big\{ -\sum_{l=1}^n \frac{(1-\Delta_l)A_l I(A_l V_l \le v)}{\hat{p}^X(X_l)[\sum_{k=1}^n A_k]\hat{G}^{V|X}(A_l V_l|X_l)} \Big\}, \quad v \in [0, (AV)_{(n)}]. \qquad (7.3.12)$$

All estimators of nuisance functions, needed for mimicking (7.3.6), are constructed, and this allows us to propose the following plug-in sample mean Fourier estimator,

$$\hat{\theta}_j := \sum_{l=1}^n \frac{\Delta_l A_l V_l \varphi_j(X_l)}{[\sum_{k=1}^n A_k]\max(\hat{p}^X(X_l)\hat{G}^C(A_l V_l), c/\ln(n))}. \qquad (7.3.13)$$

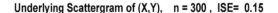

Underlying Scattergram of (X,Y),   n = 300 , ISE= 0.15

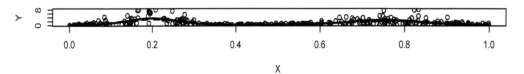

RC Data, N = 225 , lambdaC = 3 , ISE = 0.39

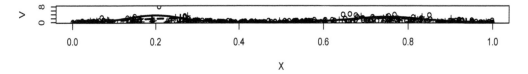

MAR RC Data, M = 147

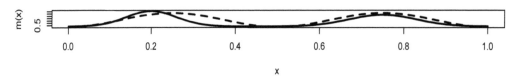

Survival Function of C and its Estimate

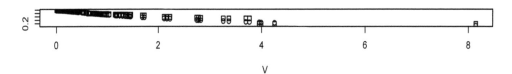

Regression Function and its Estimate, ISE= 0.5

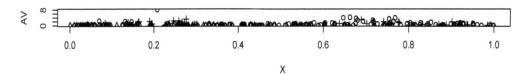

Figure 7.4 *Regression with MAR right censored response. The simulation and the diagrams are explained in the text. [n = 300, corn = 4, a = 0.3, w =" 0.1+0.8*exp(1+6*(x-0.5))/(1+exp(1+6*(x-0.5)))", lambdaC = 3, c = 1, cJ0 = 4, cJ1 = 0.5, cTH = 4]*

In its turn, the Fourier estimator yields the regression E-estimator $\hat{m}(x, \beta_V)$. Note that while we do not explicitly estimate parameter β_V because this is not needed, using $(AV)_{(n)}$ would be a natural choice.

Figure 7.4 sheds additional light on the problem and the proposed solution. We begin with explanation of the diagrams and the simulation. The top diagram shows an underlying hidden scattergram created by the Uniform predictor X and the response Y having exponential distribution with the mean $a + f(X)$ where a is a positive constant and f is a corner function, here $a = 0.3$ and f is the Strata. As usual, the solid and dashed lines show us the underlying regression and its E-estimate. The sample size $n = 300$ and the integrated squared error (ISE = 0.15) of the E-estimate are shown in the title. The second (from the top) diagram shows us the scattergram where the responses are right censored by an exponential censoring variable C with the mean λ_C. Circles show observations with uncensored responses and crosses show censored ones. Again, the solid and dashed lines are the underlying regression and its E-estimate of Section 6.8 based on the censored data. The title indicates that the number of uncensored responses is $N := \sum_{l=1}^{n} \Delta_l = 225$ and the mean of the censoring variable is 3. The third diagram shows us the available data when the censored responses are MAR according to the availability likelihood $w(x)$ defined in the caption. Incomplete cases are shown by triangles and in the title $M := \sum_{l=1}^{n} A_l \Delta_l = 147$ shows the number of available uncensored responses. The fourth diagram shows us by circles and squares the underlying survival function $G^C(V_l)$ and its estimate $\hat{G}^C(V_l)$ at points corresponding to $A_l \Delta_l = 1$. The bottom diagram shows the underlying regression and its E-estimate based on data exhibited in the third diagram.

Now we are ready to analyze the simulated data and estimates presented in Figure 7.4. The top two diagrams show us the sequentially simulated layers of hidden data. The underlying scattergram of direct observations from (X, Y) is shown in the top diagram by circles. Note that we are dealing with the response that is exponentially distributed and the underlying regression is the vertically shifted Strata. Visual analysis of the scattergram supports the E-estimate. The second diagram exhibits a scattergram where responses, shown in the top diagram, are right censored by an exponential random variable. The E-estimate (the dashed line), introduced in Section 6.8, is both visually and in terms of the ISE is worse than the one in the top diagram. The latter should be expected and note that the size of uncensored responses reduced from 300 to 225. Further, note that larger responses have a larger likelihood to be censored, and this dramatically affects visualization of the left stratum. The latter is the primary reason of the poor estimation. Nonetheless, even these censored observations are not available to us. Instead, we get observations shown in the third diagram where some responses are missed according to a missing at random (MAR) mechanism with the availability likelihood $w(x)$ defined in the caption. As a result, in the available dataset we have only 147 uncensored responses. Further, the availability function is increasing in x, and this additionally complicates estimation of the left tail of the underlying regression. The two bottom diagrams are devoted to the proposed E-estimation. The fourth diagram shows us by circles values of the underlying survival function $G^C(v)$ and the squares show us its estimate (7.3.12). Note that we need to know the survival function only at points V_l corresponding to $A_l \Delta_l = 1$, and these are the points at which the diagram shows us values of the survival function and its estimate. This diagram highlights estimation of a more challenging function among nuisance ones, and it also sheds light on the hidden censoring mechanism. Finally, the bottom diagram exhibits the underlying regression and its E-estimate based on the MAR RC data shown in the third diagram. As it could be expected, the estimate is worse than the two E-estimates based on the hidden samples, but keeping in mind complexity of the problem and that only 147 uncensored and complete cases are available, the outcome is very respectful. Indeed, we do see the two strata, the right one is exhibited almost perfectly, the left one is shifted to the right but its magnitude is shown well. It is a good idea now to look one more time at the scattergram in the third diagram and try to recognize the underlying regression based on the data. The latter is definitely a complicated task, and it may be instructive to look again at how the E-estimator solved it.

It is highly recommended to repeat Figure 7.4 with different parameters and gain experience in analyzing the scattergrams and understanding how the E-estimator performs.

7.4   Censored Responses and MAR Predictors

We are considering a regression problem when first responses are right censored and then predictors may be missed. For this setting we are dealing with two sequentially created layers of hidden observations. The first layer of hidden observations is created by a sample $(X_1, Y_1), \ldots, (X_n, Y_n)$ of size n from a pair of continuous random variables (X, Y). Here X is the predictor, Y is the response, and the aim is to estimate the regression function

$$m(x) = \mathbb{E}\{Y|X = x\}. \tag{7.4.1}$$

The second layer of hidden observations is created by right censoring of the response Y by a censoring variable C. This creates a sample $(X_1, V_1, \Delta_1), \ldots, (X_n, V_n, \Delta_n)$ where $V_l := \min(Y_l, C_l)$ and $\Delta_l := I(Y_l \leq C_l)$. This sample is from (X, V, Δ) where $V := \min(Y, C)$ and $\Delta := I(Y \leq C)$. Finally, predictors are subject to MAR (missing at random) with the likelihood of missing defined by the always observed variable V. Namely, we observe a sample $(A_1 X_1, V_1, \Delta_1, A_1), \ldots, (A_n X_n, V_n, \Delta_n, A_n)$ from (AX, V, Δ, A) where the availability A is Bernoulli and

$$\mathbb{P}(A = 1|X = x, Y = y, V = v, \Delta = \delta, C = u) = \mathbb{P}(A = 1|V = v) =: w(v) \geq c_0 > 0. \tag{7.4.2}$$

Based on this twice modified original sample from (X, Y), first by RC of the response and then by MAR of the predictor, the aim is to estimate the underlying regression function (7.4.1). In what follows it is assumed that X, Y and C are continuous variables, the pair (X, Y) is independent of C, the predictor X has a continuous and positive design density supported on $[0, 1]$, and both Y and C are nonnegative. Also recall our notation β_Z for the right boundary point of the support of a continuous random variable Z.

To explain the proposed solution, we begin with a formula for the joint mixed density of (AX, V, Δ, A),

$$f^{AX, V, \Delta, A}(x, v, \delta, 1) = w(v) f^X(x) [f^{Y|X}(v|x) G^C(v)]^\delta [f^C(v) G^{Y|X}(v|x)]^{1-\delta}$$

$$\times I(x \in [0, 1], v \in [0, \beta_V], \delta \in \{0, 1\}). \tag{7.4.3}$$

This formula, together with the discussion in Sections 6.8 and 7.3, implies that a consistent estimation of the regression function is possible only if

$$\beta_Y \leq \beta_C. \tag{7.4.4}$$

Hence, as it was explained in Sections 6.8 and 7.3, a reasonable estimand is a censored regression function

$$m(x, \beta_V) := \mathbb{E}\{Y I(Y \leq \beta_V)|X = x\}, \tag{7.4.5}$$

which is equal to $m(x)$ whenever (7.4.4) holds.

To use our methodology of E-estimation, we need to understand how to construct a sample mean estimator of Fourier coefficients of $m(x, \beta_V)$. Using our traditional notation $\varphi_j(x)$ for elements of the cosine basis on $[0, 1]$ and with the help of (7.4.3) we can write,

$$\theta_j := \int_0^1 m(x, \beta_V) \varphi_j(x) dx$$

$$= \int_0^1 \left[\int_0^{\beta_V} v f^{Y|X}(v|x) dv \right] \varphi_j(x) dx$$

$$= \int_0^1 \left[\int_0^{\beta_V} \frac{v f^{AX, V, \Delta, A}(x, v, 1, 1) \varphi_j(x)}{w(v) f^X(x) G^C(v)} dv \right] dx$$

$$= \mathbb{E}\Big\{ \frac{A\Delta V \varphi_j(AX)}{w(V)f^X(AX)G^C(V)} \Big\}. \tag{7.4.6}$$

To use (7.4.6) for constructing a plug-in sample mean estimator, we need to estimate functions $w(v)$, $f^X(x)$ and $G^C(v)$. We begin with estimation of the availability likelihood $w(v)$. Using (7.4.2) we note that

$$w(v) = \mathbb{E}\{A|V=v\}, \; v \in [0, \beta_V]. \tag{7.4.7}$$

This implies that $w(v)$ is a regression function. Because all n realizations of pair (V, A) are available, we can use our Bernoulli regression E-estimator $\hat{w}(v)$ of Section 2.4 for estimation of the availability likelihood.

Our second step is to estimate the design density $f^X(x)$. To apply our density E-estimator, we need to understand how to express Fourier coefficients of the density as expectations. Recall that observations of the predictor are MAR, and we can write that

$$f^{AX,V,A}(x,v,1) = w(v)f^X(x)f^{V|X}(v|x)I(x \in [0,1], v \in [0, \beta_V]). \tag{7.4.8}$$

This formula allows us to write down a Fourier coefficient of the design density as follows,

$$\kappa_j := \int_0^1 f^X(x)\varphi_j(x)dx = \int_0^1 \int_0^{\beta_V} f^X(x)\varphi_j(x)f^{V|X}(v|x)dvdx$$

$$= \int_0^1 \int_0^{\beta_V} [f^{AX,V,A}(x,v,1)\varphi_j(x)/w(v)]dvdx = \mathbb{E}\Big\{ \frac{A\varphi_j(AX)}{w(V)} \Big\}. \tag{7.4.9}$$

The expectation in (7.4.9) implies the following plug-in sample mean estimator of Fourier coefficients of the design density,

$$\hat{\kappa}_j := n^{-1} \sum_{l=1}^n \frac{A_l\varphi_j(A_lX_l)}{\max(\hat{w}(V_l), c/\ln(n))}. \tag{7.4.10}$$

This Fourier estimator yields the density E-estimator $\hat{f}^X(x)$ defined in Section 2.2.

Our third step is to estimate the survival function $G^C(v)$ of the censoring variable C. The estimation is based on the relation $G^C(v) = \exp\{-H^C(v)\}$ between the survival function and the cumulative hazard $H^C(v)$. Let us explain how $H^C(v)$ may be estimated. The joint mixed density of (AX, V) and the events $\Delta = 0$ and $A = 1$ can be written as

$$f^{AX,V,\Delta,A}(x,v,0,1) = w(v)f^X(x)f^C(v)G^{Y|X}(v|x)I(x \in [0,1], v \in [0, \beta_V]). \tag{7.4.11}$$

This, together with formula $G^{V|X}(v|x) = G^C(v)G^{Y|X}(v)$, yields that

$$\frac{f^C(v)}{G^C(v)} = \frac{f^{AX,V,\Delta,A}(x,v,0,1)}{w(v)f^X(x)G^C(v)G^{Y|X}(v|x)}$$

$$= \frac{f^{AX,V,\Delta,A}(x,v,0,1)}{w(v)f^X(x)G^{V|X}(v|x)}, \quad x \in [0,1], \; v \in [0, \beta_V). \tag{7.4.12}$$

In its turn, (7.4.12) allows us to write for the cumulative hazard of C,

$$H^C(v) := \int_0^v [f^C(y)/G^C(y)]dy = \int_0^v \int_0^1 [f^C(y)/G^C(y)]dxdy$$

$$= \int_0^v \int_0^1 \frac{f^{AX,V,\Delta,A}(x,y,0,1)}{w(y)f^X(x)G^{V|X}(y|x)}dxdy$$

$$= \mathbb{E}\Big\{ \frac{(1-\Delta)AI(V \le v)}{w(V)f^X(AX)G^{V|X}(V|AX)} \Big\}, \quad v \in [0, \beta_V). \tag{7.4.13}$$

To mimic (7.4.13) by a plug-in sample mean estimator, we need to make an extra step and estimate the conditional cumulative hazard $G^{V|X}(v|x)$. Recall that V is a continuous random variable and write,

$$G^{V|X}(v|x) := \mathbb{P}(V > v|X = x) = \mathbb{P}(V \ge v|X = x) = \mathbb{E}\{I(V \ge v)|X = x\}. \tag{7.4.14}$$

Note that for a fixed v we converted the problem of estimation of the conditional survival function $G^{V|X}(v|x)$ into a regression problem based on n observations from the triplet $(AX, I(V \ge v), A)$. This is a regression problem with MAR predictor considered in Section 4.3. Let us recall its solution for the data at hand. For a fixed v, using (7.4.8) allows us to write down a Fourier coefficient of $G^{V|X}(v|x)$ as

$$\nu_j := \int_0^1 G^{V|X}(v|x)\varphi_j(x)dx = \int_0^1 \int_v^\infty f^{V|X}(u|x)\varphi_j(x)dudx$$

$$= \mathbb{E}\Big\{ \frac{AI(V \ge v)\varphi_j(AX)}{w(V)f^X(AX)} \Big\}. \tag{7.4.15}$$

Using our E-estimators $\hat{w}(v)$ and $\hat{f}^X(x)$ we may propose the following plug-in sample mean Fourier estimator,

$$\hat{\nu}_j := n^{-1} \sum_{l=1}^n \frac{A_l I(V_l \ge v)\varphi_j(A_l X_l)}{\max(\hat{w}(V_l)\hat{f}^X(A_l X_l), c/\ln(n))}. \tag{7.4.16}$$

This allows us to construct the regression E-estimator $\hat{G}^{V|X}(v|x)$ for a given v, and note that we need to calculate $\hat{G}^{V|X}(V_l|A_l X_l)$ only for cases with $A_l(1-\Delta_l) = 1$.

We have estimated all nuisance functions used in (7.4.13). These estimators, together with formula $G^C(v) = \exp(-H^C(v))$ and (7.4.13), yield the following plug-in sample mean estimator of the survival function of censoring variable C,

$$\hat{G}^C(v) := \exp\Big\{ -n^{-1} \sum_{l=1}^n \frac{(1-\Delta_l)A_l I(V_l \le v)}{\hat{w}(V_l)\hat{f}^X(A_l X_l)\hat{G}^{V|X}(V_l|A_l X_l)} \Big\}, \quad v \in [0, V_{(n)}]. \tag{7.4.17}$$

This finishes the third step of the regression estimation.

Our last step is to calculate a plug-in sample mean Fourier estimator of the censored regression function $m(x, \beta_V)$. Using (7.4.6) and the above-introduced estimators of nuisance functions we define the following Fourier estimator,

$$\hat{\theta}_j := n^{-1} \sum_{l=1}^n \frac{A_l \Delta_l V_l \varphi_j(A_l X_l)}{\max(\hat{w}(V_l)\hat{f}^X(A_l X_l)\hat{G}^C(V_l), c/\ln(n))}. \tag{7.4.18}$$

The Fourier estimator yields the regression E-estimator $\hat{m}(x, \beta_V)$.

The considered regression problem with missing and censored data, as well as the proposed E-estimation procedure, are complicated. As a result, it is of a special interest to look at a simulation and estimates exhibited in Figure 7.5. The underlying simulation of X, Y and C is the same as in Figure 7.4. The difference here is that now the predictor is MAR. The top diagram shows a scattergram generated from (X, Y). The sample size $n = 300$ is large but the E-estimate is not very good. Its shape can be justified by the data, but in any case the outcome was much better in Figure 7.4 where the same simulation and the same E-estimator are utilized. This tells us that even sample sizes of several hundreds may not be enough for a reliable estimation of a curve like the Strata. Another issue here is

Underlying Scattergram of (X,Y),   n = 300 , ISE= 0.22

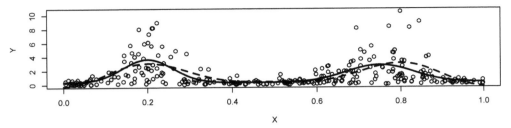

RC Data, N = 219 , lambdaC = 3 , ISE = 0.34

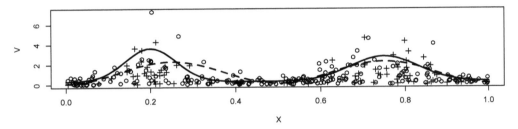

MAR RC Data and Regression, M = 76 , ISE = 0.41

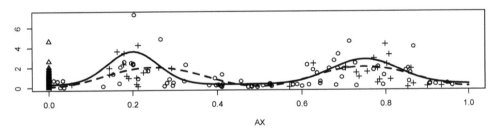

Figure 7.5 *Regression with right censored responses and MAR predictors. The simulation of* (X, Y, C) *is the same as in Figure 7.4, and the availability likelihood* $w(v)$ *is defined below. The solid line and the dashed line show the underlying regression and its estimate based on data shown in a diagram. In the top diagram circles show the hidden simulation from* (X, Y). *In the middle diagram circles show uncensored cases and crosses show censored cases.* $N := \sum_{l=1}^{n} \Delta_l$ *is shown in the title. The bottom diagram shows by circles uncensored and complete cases, by crosses censored and complete cases, and by triangles incomplete cases.* $M := \sum_{l=1}^{n} A_l \Delta_l$ *is shown in the title.* $[n =$ *300, corn = 4, a = 0.3, lambdaC = 3, w =* $"0.3+0.5*exp(1+2*(v-2))/(1+exp(1+2*(v-2)))"$, *c = 1, cJ0 = 4, cJ1 = 0.5, cTH = 4]*

that the exponential regression, considered here, has a large variance which complicates the estimation.

The middle diagram shows data with right censored responses. It is a valuable learning moment to compare circles in this diagram with circles in the top diagram, and then to realize how damaging the censoring is for the regression. Note that we have only $N = 219$ uncensored responses. As a result, the poor visual appeal of the E-estimate is well understood. On the other hand, in terms of the ISE, this estimate is on par with the estimate in Figure 7.4. Let us stress that so far we have being dealing with just another simulation of Figure 7.4. The difference between the underlying simulations is in the missing mechanism.

Recall that data shown in the two top diagrams are not available and hidden. We may visualize them only due to the underlying simulation, and the diagrams allow us to appreciate complexity of the underlying simulation. Available observations are shown in the bottom diagram. Here the underlying hidden predictors, shown in the middle diagram, are missing at random. The underlying availability likelihood is indicated in the caption, it is increasing in v, and the latter favors not missing X corresponding to a larger V. Keeping this remark in mind, it is a valuable lesson to look at data and realize how the missing modified data. Further, note that we are left only with $M = 76$ complete and uncensored pairs, and this is just a quarter of the initial $n = 300$ pairs. The E-estimate (the dashed line) is clearly far from being good, but it is close to the one in the middle diagram and does indicate two strata.

The considered setting is one of the most complicated ones considered so far. It is highly advisable to repeat Figure 7.5 with different parameters and get used to the setting and the proposed solution.

7.5   Censored Predictors and MAR Responses

We are considering a regression problem when first the predictor is right censored and then the response is missing at random. The underlying model is as follows. The first layer of hidden observations is created by a sample $(X_1, Y_1), \ldots, (X_n, Y_n)$ of size n from a pair of continuous random variables (X, Y). Here X is the predictor, Y is the response, and the the regression function of interest is

$$m(x) = \mathbb{E}\{Y | X = x\}. \tag{7.5.1}$$

The second layer of hidden observations is created by right censoring of the predictor X by a censoring variable C. The hidden observations are $(U_1, Y_1, \Delta_1), \ldots, (U_n, Y_n, \Delta_n)$ where $U_l := \min(X_l, C_l)$ and $\Delta_l := I(X_l \leq C_l)$. This sample is from (U, Y, Δ) where $U := \min(X, C)$ and $\Delta := I(X \leq C)$. Finally, the responses are missing at random and we observe a sample $(U_1, A_1 Y_1, \Delta_1, A_1), \ldots, (U_n, A_n Y_n, \Delta_n, A_n)$ from (U, AY, Δ, A). Here A is the availability which is a Bernoulli random variable and

$$\mathbb{P}(A = 1 | U = u, X = x, Y = y, \Delta = \delta, C = z) = \mathbb{P}(A = 1 | U = u) =: w(u) \geq c_0 > 0. \tag{7.5.2}$$

Based on this twofold modified sample, first by RC of the predictor and then by MAR of the response, the aim is to estimate the underlying regression function (7.5.1). In what follows it is assumed that X, Y and C are continuous variables, the pair (X, Y) is independent of nonnegative C, the predictor X is supported on $[0, 1]$ according to a continuous and positive design density, and (recall notation β_Z for the right boundary point of the support of a continuous random variable Z)

$$\beta_X \leq \beta_C. \tag{7.5.3}$$

We begin the explanation of how to construct a regression estimator with a formula for the joint mixed density of (U, AY, Δ, A) for the case $\Delta = 1$ and $A = 1$ (or equivalently $A\Delta = 1$), that is when no censoring of the predictor and nor missing of the response occurs. Write,

$$f^{U, AY, \Delta, A}(u, y, 1, 1) = [w(u) f^X(u) G^C(u)] f^{Y|X}(y|u) I(u \in [0, 1], y \in (-\infty, \infty)). \tag{7.5.4}$$

Now let us make a remark that sheds light on the formula. Consider the probability density of U given $A\Delta = 1$. Using (7.5.4) that conditional density may be written as

$$f^{U|A\Delta}(u|1) = \frac{w(u) f^X(u) G^C(u)}{\mathbb{P}(A\Delta = 1)}, \quad u \in [0, 1]. \tag{7.5.5}$$

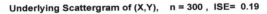

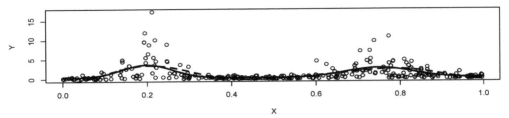

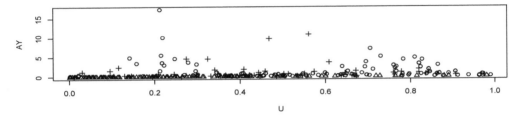

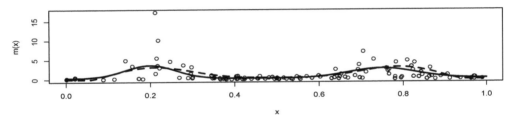

Figure 7.6 *Regression with right censored predictors and missing at random responses. The top diagram shows an underlying hidden scattergram created by the Uniform predictor X and the response Y having exponential distribution with the mean $a + f(X)$ where f is a corner function, here it is the Strata. The solid and dashed lines show the underlying regression and its E-estimate. The sample size n and the integrated squared error (ISE) of the E-estimate are shown in the title. The middle diagram shows the available data where predictors are censored by a Uniform$(0, u_C)$ variable C, and then responses are missed according to the availability function $w(u)$ defined below. Uncensored and complete cases are shown by circles, censored and complete cases are shown by crosses, incomplete cases are shown by triangles. The bottom diagram shows us by circles the scattergram of uncensored and complete cases, and the underlying regression and its E-estimate, based on these observations, are shown by the solid and dashed lines, respectively. In the title $M := \sum_{l=1}^{n} A_l \Delta_l$ is the number of uncensored and complete cases. [n = 300, corn = 4, a = 0.3, u_C = 1.5, w = "0.1+0.8*exp(1+6*(u-0.5))/(1+exp(1+6*(u-0.5)))", c = 1, cJ0 = 4, cJ1 = 0.5, cTH = 4]*

Note that this density, up to the denominator $\mathbb{P}(A\Delta = 1)$, is the factor in the square brackets in (7.5.4). This remark sheds a new light on the joint distribution of (U, AY), and it also highlights a possibility of using an uncensored-complete-case approach for the regression E-estimation. Let us check that this approach is applicable here.

To construct a regression E-estimator, we need to propose a sample mean estimator of Fourier coefficients of regression $m(x)$,

$$\theta_j := \int_0^1 m(x)\varphi_j(x)dx = \int_0^1 \int_{-\infty}^{\infty} yf^{Y|X}(y|x)\varphi_j(x)dydx. \tag{7.5.6}$$

Here $\varphi_j(x)$, $j = 0, 1, \ldots$ are elements of the cosine basis on $[0, 1]$.

Using (7.5.4) and (7.5.5) we can continue (7.5.6) and write,

$$\theta_j = \mathbb{E}\Big\{ \frac{A\Delta Y \varphi_j(U)}{\mathbb{P}(A\Delta = 1) f^{U|A\Delta}(U|1)} \Big\}. \tag{7.5.7}$$

Set $M := \sum_{l=1}^{n} A_l \Delta_l$, note that the sample mean estimator of $\mathbb{P}(A\Delta = 1)$ is M/n, and denote by $\hat{f}^{U|A\Delta}(u|1)$ the density E-estimator of Section 2.2 based on uncensored and complete cases. Then the expectation in the right side of (7.5.7) yields the plug-in sample mean Fourier estimator,

$$\hat{\theta}_j := M^{-1} \sum_{l=1}^{n} \frac{A_l \Delta_l \varphi_j(U_l)}{\max(\hat{f}^{U|A\Delta}(U_l|1), c/\ln(n))}. \tag{7.5.8}$$

Note that this estimator is based solely on uncensored and complete cases.

In its turn, this Fourier estimator allows us to construct the regression E-estimator $\hat{m}(x)$. Further, this is the same estimator that one would get by using E-estimator of Section 2.3 based on uncensored and complete cases. The latter proves the above-made conjecture about the applicability here of the uncensored-complete-case approach. This is a welcome news for the considered regression setting dealing with twice modified data.

Another way to look at the proposed solution is as follows. Formula (7.5.4) implies that

$$f^{Y|X}(y|x) = f^{AY|U,A\Delta}(y|x, 1). \tag{7.5.9}$$

In words, the underlying conditional distribution of Y given X is the same as the conditional distribution of AY given U in the subsample with uncensored and complete cases. This is an interesting and important conclusion on its own because it simplifies estimation of the conditional distribution. It also immediately implies that, as we already know, the corresponding regressions also coincide.

Figure 7.6 illustrates the setting and E-estimation based on uncensored and complete cases. The top diagram exhibits the underlying scattergram, the regression function and its E-estimate. Note the high volatility of data. The middle diagram shows us the available data. The bottom diagram shows us the scattergram of uncensored and complete cases where $A\Delta = 1$, and the proposed E-estimate. We may see the Strata pattern in the scattergram, and this supports the theoretical conclusion about feasibility of the uncensored-complete-case approach. The main statistical complication here is the dramatic reduction in the size of available cases from $n = 300$ to $M = 113$. Keeping this in mind, the E-estimate does a good job in recovering the underlying regression.

7.6 Truncated Predictors and MAR Responses

So far we have considered data modified by censoring. In this section we are considering a left truncation (LT) which is another classical data modification considered in survival analysis.

The following example will help us to understand a regression setting with left truncated predictor and missing response. Consider an actuarial example dealing with payments on a home insurance where X^* is the (monetary) loss incurred by a policyholder, Y^* is the age of roof, and T^* is the deductible in the insurance policy. The aim is to predict the age of roof given a reported loss, and it is natural to solve the prediction problem using nonparametric regression. As we know from Section 6.3, we may get information about a triplet $(X, Y, T) := (X^*, Y^*, T^*)$ only if the loss X^* is not smaller than the deductible T^*, and this creates the LT modification of underlying insurable losses. Further, the data may not contain age of the roof for some policies, and hence we are dealing with missing responses. We may conclude that the available sample is from a quartet (X, AY, T, A) where A is a Bernoulli random variable, and to avoid MNAR let us assume that the likelihood of missing is defined by the value of insurable loss,

$$\mathbb{P}(A = 1|X = x, Y = y, T = t) = \mathbb{P}(A = 1|X = x) = w(x) \geq c_0 > 0. \qquad (7.6.1)$$

Keeping this example in mind, let us formulate the considered regression problem. There is a hidden sequential sampling $(X_1^*, Y_1^*, T_1^*), (X_2^*, Y_2^*, T_2^*), \ldots$ from a triplet (X^*, Y^*, T^*). Here X^* is the predictor, Y^* is the response, and T^* is the truncation variable. The problem is to estimate the regression function

$$m(x) = \mathbb{E}\{Y^*|X^* = x\}. \qquad (7.6.2)$$

Hidden realizations from (X^*, Y^*, T^*) create a second layer of hidden observations $(X_1, Y_1, T_1), \ldots, (X_n, Y_n, T_n)$ according to the left truncation modification. Namely, if $X_1^* \geq T_1^*$, that is the predictor is not smaller than the truncation variable, then the triplet $(X_1, Y_1, T_1) := (X_1^*, Y_1^*, T_1^*)$ is observed, otherwise this hidden realization is skipped and we even do not know that it occurred. Then the second realization (X_2^*, Y_2^*, T_2^*) is considered, and the process is stopped as soon as nth observation (X_n, Y_n, T_n) is obtained. After the second layer of hidden truncated observations is created, responses are MAR and we observe a sample $(X_1, A_1Y_1, T_1, A_1), \ldots, (X_n, A_nY_n, T_n, A_n)$ where A is the availability and it is a Bernoulli variable with the availability likelihood (7.6.1).

In what follows it is assumed that X^*, Y^* and T^* are continuous lifetimes (nonnegative variables), the pair (X^*, Y^*) is independent of T^*, and X^* has a continuous and positive design density on the support $[0, 1]$.

To explain a proposed solution of the regression problem, we begin with probability formulas for joint mixed densities. Write,

$$f^{X,AY,T,A}(x, y, t, 1)$$

$$= w(x)f^{T^*}(t)f^{X^*}(x)f^{Y^*|X^*}(y|x)[\mathbb{P}(T^* \leq X^*)]^{-1}I(0 \leq t \leq x \leq 1)I(y \geq 0). \qquad (7.6.3)$$

We can integrate both sides of (7.6.3) with respect to t and get

$$f^{X,AY,A}(x, y, 1)$$

$$= w(x)f^{X^*}(x)f^{Y^*|X^*}(y|x)F^{T^*}(x)[\mathbb{P}(T^* \leq X^*)]^{-1}I(0 \leq x \leq 1, y \geq 0). \qquad (7.6.4)$$

Further, we integrate both sides of (7.6.4) with respect to y and get

$$f^{X,A}(x, 1) = w(x)f^{X^*}(x)F^{T^*}(x)[\mathbb{P}(T^* \leq X^*)]^{-1}I(0 \leq x \leq 1). \qquad (7.6.5)$$

Now we are ready to explain a proposed solution. Recall our notation α_Z for the left boundary point of the support of a continuous variable Z. It is clear from (7.6.3), as well as from the definition of the left truncation, that a consistent estimation of the regression is possible only if

$$\alpha_{T^*} \leq \alpha_{X^*}. \qquad (7.6.6)$$

Otherwise, we may estimate $m(x)$ only for $x \geq \max(\alpha_{T^*}, \alpha_{X^*})$.

Assume that (7.6.6) holds. To construct a regression E-estimator, we need to understand how to express Fourier coefficients of $m(x)$ via an expectation. Recall our traditional notation $\varphi_j(x)$ for elements of the cosine basis on $[0, 1]$ and write,

$$\theta_j := \int_0^1 m(x)\varphi_j(x)dx = \int_0^1 [\int_0^\infty yf^{Y^*|X^*}(y|x)dy]\varphi_j(x)dx. \qquad (7.6.7)$$

Using (7.6.4) we continue,

$$\theta_j = \int_0^1 \int_0^\infty \frac{y f^{X,AY,A}(x,y,1)\varphi_j(x)}{w(x) f^{X^*}(x) F^{T^*}(x)[\mathbb{P}(X^* \geq T^*)]^{-1}} dy \, dx. \tag{7.6.8}$$

Now note that, according to (7.6.5), the denominator in (7.6.8) is equal to $f^{X,A}(x,1)$. This allows us to continue (7.6.8) and write,

$$\theta_j = \mathbb{E}\left\{\frac{AY\varphi_j(X)}{f^{X,A}(X,1)}\right\}. \tag{7.6.9}$$

This is the key formula because Fourier coefficients are expressed as expectations. The mixed density $f^{X,A}(x,1) = \mathbb{P}(A=1) f^{X|A}(x|1)$ can be estimated by $(N/n)\hat{f}^{X|A}(x|1)$ where $N := \sum_{l=1}^n A_l$ is the number of complete cases and $\hat{f}^{X|A}(x|1)$ is the density E-estimator of Section 2.2 based on predictors in complete cases. This and (7.6.9) yield the plug-in sample mean Fourier estimator

$$\hat{\theta}_j := N^{-1} \sum_{l=1}^n \frac{A_l Y_l \varphi_j(X_l)}{\max(\hat{f}^{X|A}(X_l|1), c/\ln(n))}. \tag{7.6.10}$$

In its turn, the Fourier estimator yields the regression E-estimator $\hat{m}(x)$.

If we look at formula (7.6.10) one more time, then it is not difficult to realize that the regression E-estimator is based on complete cases. In other words, a complete-case approach is consistent for the considered regression with truncated predictors and MAR responses. This is a welcome conclusion for the otherwise complicated statistical setting.

Figure 7.7 illustrates the setting and the proposed estimation, and its caption explains the simulation and the diagrams. The hidden left truncated data is shown in the top diagram. Note that the underlying predictor X^* is uniform on $[0,1]$, and despite this and the large sample size of 300 observations, only a few LT predictors are less than 0.2. Note that without LT there should be on average about 60 predictors smaller than 0.2. This tells us about very strong truncation which makes a reasonable estimation of left tail of a regression function practically impossible. And indeed, the E-estimator does a very good job in restoring the underlying regression apart of the left tail. The middle diagram shows us the same data only with MAR responses. Cases with missed responses are highlighted by crossed circles. As we see, the missing dramatically reduced the number of complete cases from 300 to 213. Further, we may observe that now just a few complete cases have predictors smaller than 0.3. To understand why, it is worthwhile to look at the availability likelihood whose E-estimate and the corresponding scattergram are shown in the bottom diagram. As we see, the availability likelihood is an increasing function with a small left tail. This availability likelihood confounds the already complicated left-tail issue.

Note that estimates of nuisance functions, here the availability likelihood, may be a useful statistical tool in analysis of modified data. Confidence bands are another useful statistical tool, they are not shown here to avoid overcrowding the diagrams.

Now let us look at the top diagram in Figure 7.8 where the case $\alpha_{T^*} > \alpha_{X^*}$ is considered. Here we observe a simulation which is similar to the one in Figure 7.7, again the left tail of the E-estimate is bad, and we know the root of the problem. The middle diagram exhibits the availability likelihood (the solid line) and its E-estimate (the dashed line). The used availability likelihood again compounds difficulties in estimation of the left tail. Nonetheless, while the diagrams in Figures 7.7. and 7.8 look similar, in Figure 7.8 we are dealing with destructive left truncation because T is uniform on $[0.1, 0.8]$, and hence even if $n \to \infty$ we cannot recover the regression $m(x)$ for $x \in [0, 0.1)$. Again, compare the diagrams in Figures 7.7 and 7.8 and think about a possibility to recognize the fact that in the former simulation the LT is nondestructive, while in the latter it is destructive.

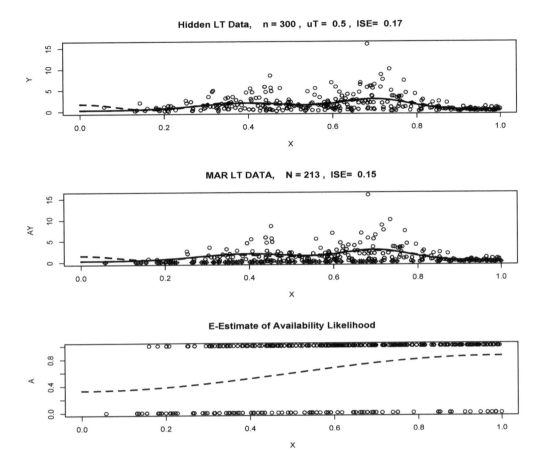

Figure 7.7 *Regression with left truncated predictors and then missing at random responses. The top diagram shows the hidden left truncated scattergram created by the Uniform predictor X^*, the response Y^* having exponential distribution with the mean $a + f(X^*)$ where f is a corner function, here it is the Bimodal, and the truncation variable T^* which is Uniform$(0, u_T)$. The solid and dashed lines show the underlying regression and its E-estimate. The middle diagram shows us the available data when the responses are missing according to the availability function $w(x)$ defined below. The incomplete cases are highlighted by crossed circles, $N := \sum_{l=1}^{n} A_l$ is the number of complete cases. The solid and dashed lines show the underlying regression and its E-estimate. The bottom diagram shows the regression E-estimate of the availability likelihood based on the scattergram of pairs (X_l, A_l), $l = 1, \ldots, n$. [$n = 300$, corn = 3, $a = 0.3$, $u_T = 0.5$, $w = "0.1+0.8*exp(1+6*(x-0.5))/(1+exp(1+6*(x-0.5)))"$, $c = 1$, cJ0 = 4, cJ1 = 0.5, cTH = 4]*

We may conclude that in general it may be prudent to estimate regression $m(x)$ over an interval based on data. For instance, it may be an interval between smallest and largest available predictors. In our particular case, only the choice of the left boundary of the interval of estimation is of interest, and Figure 7.8 allows us to choose the interval of estimation $[a_{manual}, 1]$ manually. Namely, after the two top diagrams are drawn, the program stops and we can visualize data and the availability likelihood. Then, at the R prompt >, enter a wished value for a_{manual}, then press "enter", after next R prompt > press "enter" and the bottom diagram will appear. For the particular simulation in Figure 7.8, the value $a_{manual} = 0.2$ was chosen. The bottom diagram clearly shows the improvement in the left tail of the E-estimate. Note that $M = 232$ is the number of complete cases with predictors

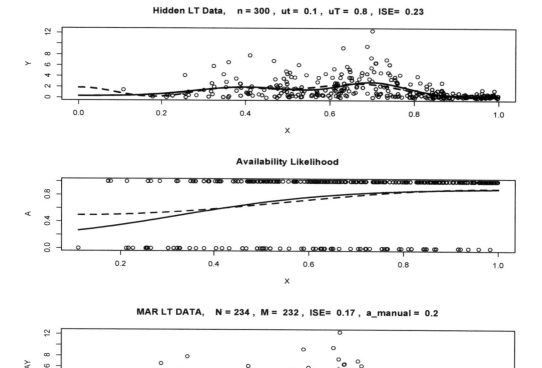

Figure 7.8 *Left truncated predictors and MAR responses, the case $\alpha_{T*} > \alpha_{X*}$. The simulation and histograms, apart of their order, are the same as in Figure 7.7 with only two differences. The former is that the truncation variable is now Uniform(u_t, u_T). The latter is that after exhibiting two top diagrams the program stops and asks about input of a_{manual} that is used by the E-estimator to calculate E-estimate over the interval $[a_{manual}, 1]$. The title shows the number N of complete cases and the number M of predictors in complete cases belonging to the interval of estimation. {To enter a wished value for a_{manual}, which must be between 0 and 1, after R prompt > type a wished number, press "return", and after next R prompt > press "return".} [n = 300, corn = 3, a = 0.3, ut = 0.1, uT = 0.8, w = "0.1+0.8*exp(1+6*(x-0.5))/(1+exp(1+6*(x-0.5)))", c = 1, cJ0 = 4, cJ1 = 0.5, cTH = 4]*

not smaller than a_{manual}, in other words this is the number of complete cases used by the E-estimator.

Figure 7.8 is a useful tool to learn about effects of the LT predictor and the MAR response on regression estimation and how to choose a feasible interval of estimation. The reader is advised to repeat it with different parameters and gain a necessary experience in dealing with this complex problem.

Finally, let us again stress that considered in this and previous sections exponential regression is a complicated regression model due to possibly large regression errors. Indeed, an exponential variable with mean λ has variance equal to λ^2. For exponential regression with regression function $m(x)$ this implies that

$$Y = m(X) + m(X)\varepsilon, \quad \text{where} \quad \mathbb{V}(\varepsilon) = 1. \qquad (7.6.11)$$

Because considered regression functions take large values, this implies a large volatility in responses. It may be an important teachable moment to compare the top diagram in Figure 7.8 with the diagrams in Figure 2.7 where regressions with additive normal errors are exhibited.

7.7 LTRC Predictors and MAR Responses

We are considering a regression problem with modified data where first the predictor is left truncated and right censored (LTRC), and then the response is missing at random with the probability of missing defined by the modified predictor.

Let us formulate a studied model of generating data. There is a hidden sequential sampling $(X_1^*, Y_1^*, T_1^*, C_1^*), (X_2^*, Y_2^*, T_2^*, C_2^*), \ldots$ from a quartet (X^*, Y^*, T^*, C^*). The sequential sampling is defined by a left truncation that will be explained shortly. In that quartet of random variables X^* is the predictor, Y^* is the response, T^* is the truncation variable, and C^* is the censoring variable. Realizations of the sequential sampling create the first layer of hidden observations. Then observations of the predictor are right censored by the censoring variable, and this modifies (X^*, C^*) into a pair $(U^*, \Delta^*) := (\min(X^*, C^*), I(X^* \leq C^*))$. This creates a second layer of hidden sequential observations from the censored quartet $(U^*, \Delta^*, Y^*, T^*)$. The third layer of hidden observations is a sample of size n created by the truncation variable T^*. Namely, the first hidden realization $(U_1^*, \Delta_1^*, Y_1^*, T_1^*)$ of the censored quartet either becomes the first element $(U_1, \Delta_1, Y_1, T_1)$ of that sample if $U_1^* \geq T_1^*$ or it is skipped and we even do not know that it occurred. Then the second hidden realization $(U_2^*, \Delta_2^*, Y_2^*, T_2^*)$ is similarly analyzed, and the process is sequentially continued until the sample of size n from (U, Δ, Y, T) is collected. The final modification is the MAR of the response, and the available sample of size n is from the quintet (U, Δ, AY, T, A). Here A is the availability which is a Bernoulli random variable, and the availability likelihood is

$$\mathbb{P}(A = 1 | U = u, \Delta = \delta, Y = y, T = t) = \mathbb{P}(A = 1 | U = u) = w(u) \geq c_0 > 0. \qquad (7.7.1)$$

Based on a sample $(U_1, \Delta_1, A_1 Y_1, T_1, A_1), \ldots, (U_n, \Delta_n, A_n Y_n, T_n, A_n)$ the aim is to estimate the regression function for the first layer of hidden observations,

$$m(x) = \mathbb{E}\{Y^* | X^* = x\}. \qquad (7.7.2)$$

Note that here we are dealing with a rather intricate modification of the underlying sample from (X^*, Y^*).

To understand a possible solution of the regression problem, we begin with probability formulas that shed light on the problem. Write,

$$\mathbb{P}(U \leq u, \Delta = 1, AY \leq y, A = 1) = \mathbb{P}(U^* \leq u, \Delta^* = 1, Y^* \leq y, A = 1 | U^* \geq T^*)$$

$$= \frac{\mathbb{P}(X^* \leq u, C^* \geq X^*, Y^* \leq y, A = 1, X^* \geq T^*)}{\mathbb{P}(U^* \geq T^*)}. \qquad (7.7.3)$$

Suppose that pairs (X^*, Y^*) and (C^*, T^*) are independent, predictor X^* and response Y^* are continuous random variables, and X^* has a continuous and positive density on its support $[0, 1]$. Then we may continue (7.7.3),

$$\mathbb{P}(U \leq u, \Delta = 1, AY \leq y, A = 1)$$

$$= \frac{\int_{-\infty}^{y} [\int_0^u w(x) f^{X^*}(x) f^{Y^*|X^*}(v|x) \mathbb{P}(T^* \leq x \leq C^*) dx] dv}{\mathbb{P}(U^* \geq T^*)}. \qquad (7.7.4)$$

By taking partial derivatives with respect to u and y, we get the corresponding joint mixed density,

$$f^{U,\Delta,AY,A}(u,1,y,1) = \left[\frac{w(u)f^{X^*}(u)\mathbb{P}(T^* \leq u \leq C^*)}{\mathbb{P}(U^* \geq T^*)}\right]f^{Y^*|X^*}(y|u). \qquad (7.7.5)$$

In its turn, via integration with respect to y, formula (7.7.5) yields the marginal mixed density

$$f^{U,\Delta,A}(u,1,1) = \frac{w(u)f^{X^*}(u)\mathbb{P}(T^* \leq u \leq C^*)}{\mathbb{P}(U^* \geq T^*)}. \qquad (7.7.6)$$

Using (7.7.5) and (7.7.6) we conclude that for values of u such that

$$w(u)f^{X^*}(u)\mathbb{P}(T^* \leq u \leq C^*) > 0, \qquad (7.7.7)$$

we have the following pivotal relation,

$$f^{AY|U,\Delta,A}(y|u,1,1) = f^{Y^*|X^*}(y|u). \qquad (7.7.8)$$

We conclude that given (7.7.7) and assumed independence between pairs (X^*, Y^*) and (C^*, T^*), the conditional distribution of the underlying response Y^* given the underlying predictor X^* is the same as the conditional distribution of the observed AY given the observed U in uncensored and complete cases. This is a fantastic news for considered regression problem because the complicated problem of estimating a regression in the bottom layer of hidden observations, based on threefold modified data, is converted into a standard regression for the above-described subsample of available data. As a result, the main complication here is the smaller sample size of the subsample.

Let us check validity of the proposed solution via simulations. We begin with the case of independent and continuous truncation and censoring variables. Figure 7.9 illustrates the setting and the proposed estimation, and its caption explains the simulation and the diagrams. The top diagram illustrates hidden LTRC data where only cases with uncensored predictors are shown. Note that these cases may be used for consistent regression E-estimation. The censoring reduced the original sample size $n = 300$ to $N = 204$. Further, the underlying predictor X^* is uniform, and note how light the tails of the observed predictors are after the LTRC modification. This is a glum reality of dealing with LTRC predictors.

Because there may be no observations in the left and right tail that point upon an underlying relationship between the predictor and the response, it is prudent to estimate the regression only over the range of available uncensored predictors. The corresponding E-estimate is shown by the dashed line, while the underlying regression is shown by the solid line over the whole interval $[0,1]$. The E-estimate is good, and this is also reflected by its ISE shown in the title. Of course, we know the hidden observations only due to the simulation, on the other hand we look at an important regression on its own whenever no missing of responses occurs.

The middle diagram sheds light on the MAR. It exhibits Bernoulli regression for the problem of estimation of the underlying availability likelihood. The estimate is not used by the regression E-estimator but it is an important tool to understand an underlying missing mechanism. Note that the E-estimate is far from being perfect but it correctly describes the data at hand. In other words, while we do know the underlying Bernoulli regression (the solid line) thanks to the simulation, the E-estimator knows only data and describes the data. It is important to note that here we are dealing with $n = 300$ pairs of observations. Further, the diagram helps us to visualize LTRC predictors in complete cases.

Hidden LTRC Data, n = 300 , N = 204 , ut = 0 , uT = 0.5, uc = 0 , uC = 1.5 , ISE= 0.076

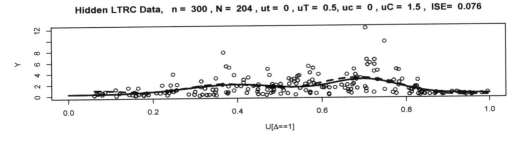

Availability Likelihood

Data with MAR Response and LTRC Predictor, N = 202 , M = 140 , ISE= 0.12

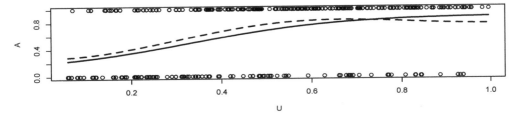

Figure 7.9 *Regression with left truncated and right censored predictors and MAR responses. Solid and dashed lines in diagrams are an underlying curve and its E-estimate, respectively. The top diagram shows by circles a subsample with uncensored predictors whose size $N := \sum_{l=1}^{n} \Delta_l$ is shown in the title. In the simulation T^* is Uniform(u_t, u_T), X^* is the Uniform, C^* is Uniform(u_c, u_C), and Y^* is generated as in Figure 7.7; see the parameters in the title. The E-estimate is calculated over the range of uncensored predictors, while the underlying regression (the solid line) is shown over $[0, 1]$. The middle diagram shows the scattergram of (U, A) overlaid by the underlying availability likelihood, controlled by the argument w, and its E-estimate. The bottom diagram shows the scattergram from $(U, AY, \Delta = 1)$, the regression function and its E-estimate. In the title $N := \sum_{l=1}^{n} A_l = 202$ is the total number of available responses and $M := \sum_{l=1}^{n} A_l \Delta_l = 140$ is the number of complete pairs with uncensored predictor. Note that only 140 pairs are used to calculate the uncensored-complete-case E-estimate, and these pairs are shown by circles. The E-estimate is calculated over an interval defined by the range of uncensored predictors in complete cases. {Distribution of T^* is either the Uniform(u_t, u_T) or Exponential(λ_T) where λ_T is the mean. Censoring distribution is either Uniform(u_c, u_C) or Exponential(λ_C). For instance, to choose exponential truncation and censoring, set trunc = "Expon", cens = "Expon" and then choose wished parameters. The parameters will be shown in the title and point upon underlying distributions.} [n = 300, a = 0.3, corn = 3, trunc = "Unif", ut = 0, uT = 0.5, lambdaT = 0.3, uc = 0, uC = 1.5, cens = "Unif", lambdaC = 1.5, w = "0.1+0.8*exp(1+6*(u-0.5))/(1+exp(1+6*(u-0.5)))", c=1, cJ0 = 4, cJ1 = 0.5, cTH = 4]*

The exhibited Bernoulli regression E-estimate indicates that the MAR of responses may dramatically aggravate complexity of estimation of the left tail because the probability of an incomplete pair increases for smaller values of the LTRC predictor. And indeed, let us look at available MAR data exhibited in the bottom diagram. The E-estimator is based solely on pairs shown by circles, and note that there are only 8 predictors with values less than 0.2. Further, the title shows that there is a total of $N = 202$ complete pairs from underlying $n = 300$, and among those only $M = 140$ with uncensored predictors.

Recall that the above-presented theory asserts that the conditional distribution of AY given U in complete pairs with uncensored predictors coincides with the underlying conditional distribution of Y^* given X^*. Hence, the regression functions are also the same. This phenomenon is used by the proposed regression E-estimator based on the uncensored-complete-case approach. Of course, the design densities of the underlying uniform X^* and observed U, given $A\Delta = 1$, are different, and we can clearly see this in the bottom diagram where we have just a few observations in the left tail.

The regression E-estimator is constructed over the range of complete and uncensored pairs (see the circles), it is fairly good and nicely indicates the two modes. Of course, it is a challenge to estimate the left tail of the regression function due to the LTRC of the predictor and the special shape of the availability likelihood.

The reader is advised to repeat this simulation with different parameters to gain experience in dealing with this complicated modification of regression data.

So far we have considered the case of independent and continuous censoring and truncation variables. The above-presented theory asserts that the proposed uncensored-complete-case approach is valid also for a general case where these variables may be dependent and have a mixed distribution. This setting is examined in Figure 7.10 where the model $C^* := T^* + W^*$ is considered. Here $W^* := (1 - B)Z^* + Bu_C$ where Z^* is Uniform(u_c, u_C) and B is an independent Bernoulli(p_c). Note that parameter p_c controls the frequency with which W^* is equal to the largest possible value of Z^*. A typical example is when censoring may occur only after truncation, T^* is the baseline for a study of a lifetime X^*, and u_C is the length of the study when the lifetime is necessarily censored.

Let us look at the diagrams. The top diagram shows us the hidden simulated LTRC data. Here we use T^* with exponential distribution, and this fact is highlighted by showing its mean $\lambda_T = 0.3$ in the title of the top diagram. Note that in Figure 7.9, where uniform truncation is used, its parameters u_t and u_T are shown. With respect to Figure 7.9, here the exponential truncation may affect larger values of the predictor, and note that in Figure 7.9 no values of X^* larger than 0.5 are truncated. Further, here the censoring and truncation variables are dependent, and the chosen joint distribution dramatically affects the number of censored predictors. The title shows that only $N = 150$ from underlying $n = 300$ predictors are uncensored. This is indeed a dramatic decrease from what we have seen in Figure 7.9. Also, please look at the tails with just a few predictors.

All these factors definitely affect the shown regression E-estimate. At the same time, the E-estimate correctly indicates two modes of the underlying Bimodal regression, and even their magnitudes are exhibited relatively well.

The middle diagram shows us the Bernoulli regression E-estimate of the availability likelihood $w(u)$. Note that here we are dealing with direct observations and a large sample size $n = 300$. Nonetheless, the estimate looks bad. Why does it look bad? Because we compare it with the underlying availability likelihood which generated the data. On the other hand, the estimate clearly follows the data, and its left tail is well justified. Here we again are dealing with a small number of predictors in the tails despite the Uniform distribution of X^*. Why is this the case? Because U is the LTRC version of X^*, and as a result U has lighter tails of its density. We may conclude that even the sample size $n = 300$ does not preclude us from anomalies that we see in the data.

The proposed E-estimator of the underlying regression (7.7.2) does not use the availability likelihood and is based on the uncensored-complete-case approach. The bottom diagram shows us a relatively fair estimate which indicates a regression with two modes. Note that

Hidden LTRC Data, n = 300 , N = 150 , lambdaT = 0.3, uc = 0 , uC = 0.6 , censp = 0.2 , ISE= 0.15

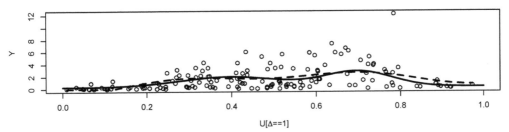

Availability Likelihood

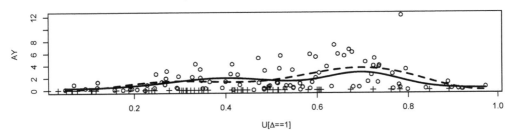

Data with MAR Response and LTRC Predictor, N = 189 , M = 98 , ISE= 0.35

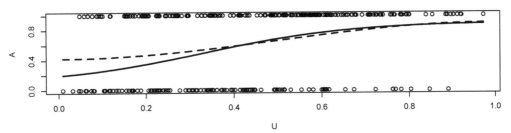

Figure 7.10 *Regression with LTRC predictors and MAR responses when censoring and truncating variables are dependent and $C^* = T^* + W^*$. In the simulation variables X^*, T^* and W^* are mutually independent. Variable W^* is a mixture of Uniform(u_c, u_C) with a constant u_C, namely $\mathbb{P}(W^* = u_C) = p_c$ and otherwise W^* has the uniform distribution. Parameter p_c is controlled by the argument censp. The used truncation is exponential. Otherwise the simulation and the structure of diagrams are identical to Figure 7.9. {Distribution of T^* is either Uniform(u_t, u_T) or Exponential(λ_T) where the parameter λ_T is the mean. To choose uniform truncation set trunc = "Unif ".} [n = 300, a = 0.3, corn = 3, trunc = "Expon", ut = 0, uT = 0.5, lambdaT = 0.3, uc = 0, uC = 0.6, censp = 0.2, w = "0.1+0.8*exp(1+6*(u-0.5))/(1+exp(1+6*(u-0.5)))", c=1, cJ0 = 4, cJ1 = 0.5, cTH = 4]*

there are only $M = 98$ pairs that are neither censored nor incomplete, and this is a relatively small sample size for estimation of the Bimodal regression function. On the top of the small sample, we are dealing with the small number of available predictors in the tails, and the latter is caused by the LTRC predictors and MAR responses.

It is highly advisable to repeat Figure 7.10 with different parameters and gain first-hand experience in dealing with this complicated modification of regression data.

7.8   Exercises

7.1.1 Explain a right censoring mechanism and present several examples.

7.1.2 Under the considered RC of a lifetime of interest X, is the censored data biased? Prove your assertion analytically and define, if data is biased, the biasing function.

7.1.3 Explain the assumption (7.1.1). Hint: Think about possible MAR and MNAR mechanisms.

7.1.4 What will be if in (7.1.1) the probability in the left side depends on δ? Can a consistent estimator be constructed in this case?

7.1.5 Explain each equality in (7.1.2). Then prove them.

7.1.6 There is some type of symmetry in (7.1.2) with respect to the lifetime of interest X and the censoring variable C. Explain it. Further, explain how the symmetry may be used.

7.1.7 Is the function $g(v)$, defined in (7.1.3), equal to the survival function of V? If the answer is "no" then explain the motivation behind its definition.

7.1.8 Explain each equality in (7.1.3). What are the used assumptions?

7.1.9 Verify equality (7.1.4). Is it valid if X and C are dependent?

7.1.10 Explain each equality in (7.1.5). Pay attention to assumptions.

7.1.11 Why do we have a restriction on x in (7.1.5)? Explain both mathematically and via the underlying probability model.

7.1.12 Why do we assume (7.1.6)?

7.1.13 Is it possible to relax (7.1.6) if we want to estimate the distribution of X?

7.1.14 Explain the definition of the cumulative hazard. Hint: Use (7.1.7).

7.1.15 Write down (7.1.7) using the hazard rate function. What is the relationship between the cumulative hazard and the hazard rate?

7.1.16 Can a hazard rate function take on negative values? Is it, similarly to a probability density, integrated to 1?

7.1.17 Prove formula (7.1.8) which expresses the survival function via the cumulative hazard. Is an assumption required?

7.1.18 Explain why formula (7.1.9) is valid and formulate assumptions.

7.1.19* Is the estimator (7.1.10) unbiased? What is it variance? What is the probability of a large deviation?

7.1.20 Why is estimator (7.1.10) convenient for plugging in a denominator?

7.1.21 Explain how the availability likelihood affects distribution estimation.

7.1.22 Explain validity of formula (7.1.11). Why is this representation of the availability likelihood important?

7.1.23 What is the motivation of the estimator (7.1.12)? What are the used assumptions?

7.1.24* Find asymptotic expressions for the mean and the variance of estimator (7.1.12). Hint: Begin with known plug-in functions. Use assumptions.

7.1.25* Find asymptotic expressions for the mean and the variance of estimator (7.1.13). Hint: Recall the delta method and begin with known plug-in functions.

7.1.26 Verify (7.1.14).

7.1.27 Verify (7.1.15).

7.1.28 Explain how the survival function G^C may be estimated by a given estimator of G^X.

7.1.29* Verify relations in (7.1.16) and (7.1.17).

7.1.30* Calculate the mean and the variance of Fourier estimator (7.1.18).

7.1.31 Verify relations in (7.1.19). What are the used assumptions?

7.1.32* Explain how the Fourier estimator (7.1.20) is constructed. Evaluate its mean and variance.

7.1.33 Is there a necessity to bound from below the denominator in (7.1.20)? Can a different approach be proposed?

EXERCISES

7.1.34 Explain definition of the coefficient of difficulty (7.1.21), its role in nonparametric estimation, and then verify relations in (7.1.21).

7.1.35* Explain the underlying idea of the estimator (7.1.22). Then evaluate its mean and variance.

7.1.36 Formulate a sufficient condition for a consistent density estimation.

7.1.37 Explain the simulation used in Figure 7.1.

7.1.38 Repeat Figure 7.1 and conduct an analysis of the diagrams.

7.1.39 Explain how censoring affects estimation of the survival function. Use both the theory and Figure 7.1 to justify your answer.

7.1.40* Propose better parameters of the E-estimator used in Figure 7.1. Does your answer depend on an underlying density, censoring and sample size? Comment on your answer and support it by simulations.

7.1.41 What is the difference between cases considered in Figures 7.1 and 7.2? Is a consistent estimation possible in both cases?

7.1.42 Repeat Figure 7.2 with different parameters and present analysis of your observations. Do you have any suggestions on how to improve the estimators?

7.1.43 Explain limitations of estimators used in Figure 7.2.

7.2.1 What is the difference, if any, between MAR and MNAR?

7.2.2 Explain a right censoring model with the MNAR satisfying (7.2.1).

7.2.3 Present an example of a model considered in Section 7.2. Explain all variables and the missing mechanism.

7.2.4 Does the availability likelihood (7.2.1) imply MAR or MNAR?

7.2.5* Explain why the considered MNAR is nondestructive. Hint: Recall a discussion in Chapter 5.

7.2.6 Verify all equalities in (7.2.4).

7.2.7 Verify (7.2.5).

7.2.8 Compare (7.2.4) with (7.2.5) and explain why the expressions are different and there is no traditional symmetry.

7.2.9 Explain why the cumulative hazard may be written as (7.2.6). Then comment on the inequality.

7.2.10 Verify (7.2.7). Is any assumption needed?

7.2.11* Find the mean and the variance of estimator (7.2.8). Explore the probability of large deviations.

7.2.12* Explain the underlying idea of the estimator (7.2.9) of the cumulative hazard. Then evaluate its mean and variance.

7.2.13 Is it possible to increase the interval for which estimator (7.2.9) is proposed?

7.2.14 Explain how the survival function of X may be estimated.

7.2.15* Find the mean and variance of the estimator (7.2.10).

7.2.16 Explain the expression (7.2.11) for the Fourier coefficient of the density $f^X(x)$.

7.2.17 What is the underlying idea of Fourier estimator (7.2.13)?

7.2.18* Is it possible to avoid bounding from below the denominator in (7.2.13)?

7.2.19* Find the mean and variance of the estimator (7.2.13).

7.2.20 Explain the simulation used in Figure 7.3.

7.2.21* Repeat Figure 7.3 using different shapes of the availability likelihood function. Then comment on shapes that benefit or make worse estimation of the density. Is your conclusion robust toward an underlying density?

7.2.22 Repeat Figure 7.3 with different distributions of the censoring variable. Comment on your observations.

7.2.23* Consider the case when the MNAR is defined by the hidden censoring variable C. Propose estimators for the survival function and the density of the lifetime of interest. Hint:

Begin with writing down the assumption about the MNAR,

$$\mathbb{P}(A = 1|V = v, \Delta = \delta, C = u) = \mathbb{P}(A = 1|C = u) =: w(u) \geq c_0 > 0. \qquad (7.8.1)$$

Then follow the approach outlined in Section 7.2.

7.2.24* Propose an estimator of the hazard rate. Hint: Look at the hazard rate estimator proposed in Section 7.1.

7.3.1 Explain why and in what sense the regression (7.3.1) is the best predictor of Y given $X = x$. Hint: Think about the mean and the variance.

7.3.2 Explain the considered model of MAR censored responses. How many layers of hidden observations does this model involve?

7.3.3 Explain the assumption (7.3.2) and present several plausible examples.

7.3.4 Suppose that the support of X is an interval $[a, a + b]$. How can this setting be transformed into the considered one?

7.3.5 Verify (7.3.3).

7.3.6 Does the formula (7.3.3) imply any type of symmetry between variables?

7.3.7 Why do we see the conditional survival function in (7.3.3)? What is the definition of a conditional survival function?

7.3.8 What is β_V? How can β_V be expressed via β_Y and β_C?

7.3.9* Explain when a consistent regression estimation is possible.

7.3.10 Why is the censored regression (7.3.5) introduced?

7.3.11 When do the censored regression (7.3.5) and the classical regression (7.3.1) coincide? What is a general relationship between them?

7.3.12 Explain each equality in establishing the relation (7.3.6). Why do we want to express the Fourier coefficient as an expectation?

7.3.13* Replace $m(x, \beta_V)$ by $m(x)$ in the left side of (7.3.6), and then try to express it as an expectation. Point upon a place where a complication may arise.

7.3.14 What is the assumption used in establishing (7.3.6)?

7.3.15 Find the density of predictors in the subsample of complete-case pairs.

7.3.16 Verify (7.3.7). Then explain its statistical consequences.

7.3.17* Propose an estimator for the product $w(x)f^X(x)$. Then calculate its mean and variance.

7.3.18* Propose an estimator of the survival function G^C of the censoring variable.

7.3.19 Prove the validity of (7.3.8).

7.3.20 Verify (7.3.9) and explain used assumptions.

7.3.21* Write down a formula for the cumulative hazard of the censoring variable and then express it as an expectation. Then compare your result with (7.3.10).

7.3.22 Verify and explain formula (7.3.11). Why do we need to know the conditional survival function?

7.3.23 How can a Bernoulli regression be used to estimate the conditional survival function $G^{V|X}(v|x)$?

7.3.24 Explain the underlying motivation of the estimator (7.3.12).

7.3.25* Evaluate the mean and the variance of the estimator (7.3.12).

7.3.26 Explain the underlying motivation of Fourier estimator (7.3.13).

7.3.27* Find the coefficient of difficulty of the Fourier estimator (7.3.13). Hint: Coefficient of difficulty is the asymptotic variance of a Fourier estimator as the sample size and the frequency tend to infinity.

7.3.28* Explain how a nonparametric regression E-estimator may be constructed for a regression with MAR RC response. Hint: Describe the model, present all necessary probability formulas, then explain all steps in construction of a plug-in sample mean Fourier estimator.

7.3.29 Explain the simulation used in Figure 7.4.

7.3.30 Repeat Figure 7.4 and present a statistical analysis of results. Hint: Use ISEs.

7.3.31* Explain, both theoretically and using Figure 7.4, how censoring and MAR affect the regression estimation. Is your conclusion robust toward different underlying regression function?

7.3.32 Suggest better parameters of the E-estimator used in Figure 7.4.

7.4.1 Explain a regression model with censored response and MAR predictor.

7.4.2 What are the hidden layers of observations in the model?

7.4.3 Explain the assumption (7.4.2).

7.4.4 What may be expected if the availability likelihood depends on both V and Y?

7.4.5* Relax the assumption that C is a continuous and independent variable.

7.4.6 Verify (7.4.3). Do we have any type of symmetry that may be used in estimation?

7.4.7 Explain the assumption (7.4.4).

7.4.8* Suppose that (7.4.4) does not hold. Explain a consequence and propose a possible solution.

7.4.9 Explain the underlying idea of introducing the censored regression function (7.4.5).

7.4.10 Explain each equality in (7.4.6).

7.4.11 What nuisance functions are needed to be estimated for construction of a plug-in sample mean Fourier estimator?

7.4.12 Explain how expression (7.4.7) may be used for estimation of the availability likelihood. Hint: Think about Bernoulli regression.

7.4.13 Propose a method of estimation of the design density f^X.

7.4.14 Explain (7.4.8). What are the assumptions?

7.4.15 Verify all equalities in (7.4.9). Write down and explain used assumptions.

7.4.16* Find the coefficient of difficulty of Fourier estimator (7.4.10). Write down used assumptions.

7.4.17 Explain the underlying idea (steps) in estimation of the survival function of the censoring variable. What are the assumptions?

7.4.18* Can the survival function $G^C(v)$ be consistently estimated under the made assumptions? Is its consistent estimation necessary for solving the regression problem?

7.4.19* Suppose that the task is to consistently estimate the distribution of the censoring variable. Propose an estimator of the survival function and the density. Hint: Think about assumptions.

7.4.20 Verify (7.4.11). What are the used assumptions?

7.4.21 Prove validity of (7.4.12). Explain the assumptions.

7.4.22 Explain how the cumulative hazard of the censoring variable may be written as an expectation.

7.4.23 Verify (7.4.13). Explain used assumptions.

7.4.24 What is the definition of a conditional survival function? Is the notion of a conditional survival function feasible only for continuous variables? If the answer is "no," then does (7.4.14) hold?

7.4.25* Explain the method of estimation of the conditional survival function $G^{V|X}(v|x)$. Hint: Note that this is a bivariate function, and nonetheless the estimation procedure is univariate.

7.4.26* Calculate the coefficient of difficulty of the Fourier estimator (7.4.16).

7.4.27 Explain how the estimator (7.4.17) is constructed.

7.4.28* Evaluate the mean and variance of the empirical survival function (7.4.17).

7.4.29 Is the empirical survival function (7.4.17) unbiased? Is it asymptotically unbiased? Prove your assertion.

7.4.30 Explain the underlying idea of Fourier estimator (7.4.18). Hint: Is it a plug-in sample mean estimator?

7.4.31* Calculate the coefficient of difficulty of the Fourier estimator (7.4.18).

7.4.32 Is there a need for truncation in the denominator of (7.4.18)?

7.4.33* Write down all steps of construction of the proposed regression E-estimator.

7.4.34 Explain the underlying simulation used by Figure 7.5.

7.4.35 Repeat Figure 7.5 with different censoring and availability likelihood. Explain how these nuisance factors affect estimation of the regression. Is your conclusion robust toward an underlying regression function?

7.4.36 Repeat Figure 7.5 with different sample sizes, make hard copies and then write a report with statistical analysis of obtained results. Hint: Use ISEs.

7.4.37 Explain all diagrams in Figure 7.5. Repeat simulation and compare outcomes. Write a report about your findings.

7.5.1 Explain the considered model of censored predictor and MAR response.

7.5.2 Describe a simulation that generates the considered regression with RC predictor and MAR response. Hint: Think about several layers of hidden observations.

7.5.3 Explain the motivation behind assumption (7.5.2).

7.5.4* Suppose that (7.5.2) does not hold and the availability likelihood depends on the response. What can be done in this case? Hint: The missing is MNAR and destructive. Recall our methods of dealing with such settings.

7.5.5* Formulate all made assumptions. Then, one by one, relax them and propose a feasible solution.

7.5.6 Verify (7.5.4).

7.5.7* Find a formula for the joint mixed density $f^{U,AY,\Delta,A}(u, y, \delta, a)$. Hint: Follow (7.5.4).

7.5.8 Explain and verify formula (7.5.5). Hint: Recall the notion of conditional density.

7.5.9 Verify (7.5.6). Can its right side be written as an expectation?

7.5.10 In (7.5.7), what can be said about the denominator?

7.5.11 Explain how the Fourier estimator (7.5.8) is proposed..

7.5.12* Find the coefficient of difficulty of Fourier estimator (7.5.8).

7.5.13* Explain why the proposed uncensored-complete-case approach is consistent.

7.5.14* Prove (7.5.9). Then propose a procedure for estimation of the conditional density $f^{Y|X}(y|x)$.

7.5.15 Explain the underlying experiment in Figure 7.6.

7.5.16 Repeat Figure 7.6 a number of times, make hard copies, analyze diagrams, comment on obtained estimates.

7.5.17 Using Figure 7.6, explain how shape of the availability likelihood affects estimation of different underlying regression functions.

7.5.18 Find better values of parameters of the E-estimator used in Figure 7.6.

7.5.19 Repeat Figure 7.6 with different censoring distributions and explain outcomes.

7.5.20 Comment on how censoring and missing affect regression estimation. Hint: Think about main issues in a traditional regression estimation.

7.6.1 Describe a regression problem with truncated predictor and MAR response.

7.6.2 Explain a simulation that implies a truncated predictor and MAR response. Hint: Think about several layers of hidden observations, and one of them is a sequential sampling.

7.6.3 Explain the assumption (7.6.1).

7.6.4 The aim is to estimate the underlying regression function (7.6.2). Explain why there is a hope to construct a consistent estimator based on the considered model where both the predictor and the response are modified.

7.6.5* Formulate the used assumptions. Then relax the assumed independence between (X^*, Y^*) and T^*. What can be done in this case?

7.6.6 Verify (7.6.3). Hint: Use assumptions.

7.6.7 Prove (7.6.4). May we refer to this density as marginal?

7.6.8 Verify (7.6.5). Then explore the issue of biased predictors and find the biasing function. Use Figure 7.7 to check the conclusion.

7.6.9 Explain the assumption (7.6.6).

7.6.10* Suppose that (7.6.6) is not valid. What and how can be estimated in this case?
7.6.11 Explain (7.6.7).
7.6.12 Prove (7.6.8).
7.6.13 Prove (7.6.9). Why do we want to express a Fourier coefficient as an expectation? Is any expectation useful?
7.6.14* Find the coefficient of difficulty of the Fourier estimator (7.6.10).
7.6.15 Explain how and why Fourier estimator (7.6.10) is proposed.
7.6.16* Explain why the complete-case approach is feasible for the considered model.
7.6.17 Describe the simulation used in Figure 7.7.
7.6.18 The regression E-estimator does not use the availability likelihood. Why is it of interest to visualize it?
7.6.19 Repeat Figure 7.7 several times and write a report about analysis of scattergrams and estimates.
7.6.20 Using Figure 7.7, conduct a statistical experiment that explains how the considered data modification affects the ISE.
7.6.21 Using both theory and Figure 7.7, explain how the data modification decreases the effective sample size.
7.6.22 What is the difference between simulations in Figures 7.7. and 7.8?
7.6.23 What are the roots of the problem with tail estimation? What can be done to resolve it?
7.6.24 Repeat Figure 7.8 a number of times and suggest a better interval of estimation.
7.6.25 What is the difference, if any, between the classical additive regression and the considered exponential one?
7.6.26 Explain the model (7.6.11).
7.6.27* According to (7.6.11), function $m(x)$ is both the mean and the scale. Can the scale E-estimator of Section 3.6 be used here? Is it a better estimator than the proposed one? Is it a good idea to aggregate the two estimators?
7.6.28* The considered missing of the response Y is MAR because according to (7.6.1) it is defined by the always observed predictor X. On the other hand, in general X and Y are dependent and then $\mathbb{P}(A = 1|Y = y)$ is a function in y. Is this a contradiction? Explain your answer.
7.6.29* Propose an E-estimator for the conditional density $f^{Y^*|X^*}(y|x)$. Hint: Think about feasibility of a complete-case approach.

7.7.1 Explain how LTRC modification of a random variable works.
7.7.2 Describe an example of a regression with LTRC predictor and MAR response.
7.7.3 Suppose that you would like to simulate the studied data. Explain your simulation. Hint: Think about several layers of hidden observations.
7.7.4 Explain the assumption (7.7.1).
7.7.5* Suppose that (7.7.1) does not hold and the availability likelihood depends on the value of the response. What can be done in this case?
7.7.6 Verify each equality in (7.7.3). What are the used assumptions?
7.7.7 Is there a name for the probability in left side of (7.7.3)?
7.7.8* Prove (7.7.4) and comment on the assumption.
7.7.9 Prove (7.7.5). What are the assumptions?
7.7.10* Similarly to (7.7.5), find a formula for $f^{U,\Delta,Y,A}(u,\delta,y,a)$.
7.7.11 Verify (7.7.6). Formulate the used assumptions.
7.7.12 Prove (7.7.8). What are the used assumptions?
7.7.13* What is an important statistical conclusion from (7.7.8)? Hint: Think about a simple method of estimation of the conditional density and regression. Then explain why this formula holds.
7.7.14 Explain the simulation and diagrams in Figure 7.9.

7.7.15* Using both theory and simulations, explain how the considered modification of data affects estimation of tails.

7.7.16* Propose a feasible interval of estimation. Justify your choice.

7.7.17 Suggest an availability likelihood that simplifies estimation of tails.

7.7.18 Explain the simulation used in Figure 7.10.

7.7.19 Present an example where the censoring mechanism of Figure 7.10 is a reasonable model.

7.7.20 Compare simulations used in Figures 7.9 and 7.10, and then explain how they affect estimation of the underlying regression.

7.7.21 Repeat Figure 7.10 with different parameters, and write a report about your analysis of data and estimates.

7.7.22 Is the proposed regression E-estimator affected by performance of the availability likelihood E-estimator?

7.9 Notes

It is possible to consider censoring as a special example of missing when a logical NA (not available), which indicates a missed value, is replaced by the value of a censoring variable and by an indicator of this replacement. Truncation is also related to missing via a hidden underlying sequential sampling with missing observations. Further, all these modifications imply biased data, and this sheds light on similarity in the estimation methodology. Nonetheless, it is a long standing tradition to consider survival analysis and missing data as separate branches of statistical science.

A review of the literature devoted to analysis of truncated and censored data by methods of the theory of missing data can be found in the book by van Buuren (2012) where a number of imputation procedures and softwares are discussed. The prevention of missing data in clinical trials is discussed in Little et al. (2012). An example of the treatment of missing data in survival analysis of a large clinical study can be found in Little R. et al. (2016). Klein et al. (2014), Allison (2014), Harrell (2015), and Little T.D. et al. (2016) cover a number of interesting settings and practical examples.

The literature on optimal (efficient) nonparametric estimation for missing LTRC data is practically next to none, and this is an interesting and new area of research, see Efromovich (2017). It is reasonable to conjecture that the E-estimation approach yields efficient estimation for considered problems.

7.1-7.2 For the literature, Chen and Cai (2017) and Zou and Liang (2017) are among more recent publications where further references may be found. Applications and Bayesian approach are discussed in Allen (2017). Functional data analysis is an interesting extension, see Kokoszka and Reimherr (2017).

7.3-7.7 Developing the asymptotic theory of efficient nonparametric regression estimation for missing survival data is an open problem. It is possible to conjecture that the E-estimation methodology is still efficient and implies sharp minimax results. Here the asymptotic theory developed in Efromovich (1996a; 2000b; 2011a,b,d; 2012a; 2013a; 2014c,f; 2016a,b; 2017; 2018a,b) will be instrumental. A book-length treatment of interval-censored failure time data can be found in Sun and Zhao (2013).

Sequential estimation is a natural approach for the considered settings. Following Efromovich (2007d,e; 2008a,c; 2009c) it is possible to consider estimation with assigned risk. The problem becomes complicated because now both the missing and the underlying LTRC affect the choice of optimal sequential estimator. Multivariate regression is another attractive topic of research. See also Efromovich (1980a,b; 1989; 2004b,d,g; 2007f,g).

Chapter 8

Time Series

So far we have considered cases of samples where observations are independent. In many statistical applications the observations are dependent. Time series, stochastic process, Markov chain, Brownian motion, mixing, weak dependence and long-memory are just a few examples of the terminology used to describe dependent observations. Dependency may create dramatic complications in statistical analysis that should be understood and taken into account. It is worthwhile to note that, in a number of practically interesting cases, dependent observations may be considered as a modification of independent ones, and this is exactly how many classical stochastic processes are defined and/or generated. Further, the dependency will allow us to test the limits of the proposed E-estimation methodology.

Dependent observations are considered in this and next chapters. This chapter is devoted to stationary time series with the main emphasis on estimation of the spectral density and missing data, while the next chapter is primarily devoted to nonstationary dependent observations.

Sections 8.1 and 8.2 serve as the introduction to dependent observations and the spectral density, respectively. Sections 8.3 and 8.4 consider estimation of the spectral density for time series with missing observations. In particular, Section 8.4 presents a case of destructive missing. Estimation of the spectral density for censored time series is discussed in Section 8.5. Probability density estimation for dependent observations is explored in Section 8.6. Finally, Section 8.7 is devoted to the problem of nonparametric autoregression.

8.1 Discrete-Time Series

A discrete-time series (which also may be referred to as a discrete-time process, or a process, or a time series) is a set of pairs of observations (t_1, X_1), (t_2, X_2), ..., (t_n, X_n) where each response X_s has been recorded at a specific time t_s, and traditionally $t_1 < t_2 < \cdots < t_n$. The setting resembles regression, only now the main issue is that responses X_s, $s = 1, 2, \ldots$ may be dependent and the main task is to explore the relationship between the responses and understand how to solve practical problems that involve dependent observations. (In general t_i may not necessarily be a time but a location like in spatial geostatistical data or the order of insurance claim; while there are some specifics in those cases, the general methodology remains the same.) A typical feature of a time series is that predictors t_l are equidistant integers. Then, without loss of generality, we may set $t_l = l$, and the corresponding time series is completely described by the responses $\{X_l,\ l = 1, 2, \ldots\}$, which may be treated as a sequence of regular observations in time, and this explains why such a sequence is called a *time series*. Many practical examples are indeed sequences in time, but there are plenty of other examples; for instance, data may be collected in space. In the latter case the data are often referred to as *spatial data*, and there is even a special branch in statistics, known as geostatistics, that is primarily concerned with the analysis of such data. In this chapter, for the sake of clarity, we shall use only time series terminology and assume that data are collected sequentially in time.

A time series is called *strictly stationary* (or *stationary*) if the joint distribution of $(X_{s_1}, X_{s_2}, \ldots, X_{s_m})$ and $(X_{s_1+k}, X_{s_2+k}, \ldots, X_{s_m+k})$ are the same for all sets $(s_1, \ldots, s_m)$ and all integers k, m. In other words, a shift in time does not change the joint distribution and thus the time series is stationary in time. Note that no assumption about moments is made, and for instance a time series of independent realizations of a Cauchy random variable is a strictly stationary time series.

A time series $\{X_t\}$ is called *zero-mean* if $\mathbb{E}\{X_t\} = 0$ for all t. Note that a zero-mean time series assumes existence of the first moment, but no other assumptions about moments or the distribution is made.

A time series $\{X_t\} := \{\ldots, X_{-1}, X_0, X_1, \ldots\}$ is called *second-order stationary* time series if: (i) $\mathbb{E}\{X_t^2\} < \infty$ for all t, that is, the second moment is finite; (ii) $\mathbb{E}\{X_t\} = \mu$ for all t, that is, the expectation is constant; (iii) the *autocovariance function* $\gamma^X(l, s) := \mathbb{E}\{(X_l - \mu)(X_s - \mu)\}$ satisfies the relation $\gamma^X(l, s) = \gamma^X(l + k, s + k)$ for all integers l, s, and k, that is, a translation in time does not affect the autocovariance function. The property (iii) implies that $\gamma^X(l, s) =: \gamma^X(l-s) = \gamma^X(s-l)$. To see this just set $k = -s$ and $k = -l$. Thus a zero-mean and second-order stationary time series is characterized by its autocovariance function $\gamma^X(k)$ at the *lag* k, and further there is a nice relation $\gamma^X(0) = \mathbb{E}\{X_t^2\} = \mathbb{V}(X_t)$ which holds for all t. Also note that no assumptions about higher moments is made for a second-order stationary time series.

Now let us comment about estimation of the mean of a second-order stationary time series $\{Y_t\} := \{\mu + X_t\}$ where $\{X_t\}$ is a zero-mean and second-order stationary time series with the autocovariance function $\gamma^X(t) := \mathbb{E}\{(X_t - \mathbb{E}\{X_t\})(X_0 - \mathbb{E}\{X_0\})\} = \mathbb{E}\{X_t X_0\}$. Suppose that we observe a realization $Y_1, \ldots, Y_n$ of $\{Y_t\}$. First of all we note that

$$\mathbb{E}\{Y_t\} = \mathbb{E}\{\mu + X_t\} = \mu + \mathbb{E}\{X_t\} = \mu. \tag{8.1.1}$$

This formula immediately yields the following sample mean estimator of μ,

$$\bar{\mu} := n^{-1} \sum_{l=1}^{n} Y_l. \tag{8.1.2}$$

Expectation of the sum of random variables is always (regardless of dependence between that variables) the sum of expectations of the random variables, and hence the sample mean estimator (8.1.2) is unbiased. Let us also note that $\gamma^X(j) = \gamma^Y(j)$ for all j.

What can be said about variance of the sample mean estimator (8.1.2)? So far we have been dealing only with independent observations and it is of interest to understand how to deal with dependent variables. There is a special technique for dealing with sums of dependent variables and we should learn it. Write,

$$\mathbb{V}(\bar{\mu}) = \mathbb{E}\{[n^{-1} \sum_{l=1}^{n} Y_j - \mu]^2\}$$

$$= \mathbb{E}\{[n^{-1} \sum_{l=1}^{n} (Y_l - \mu)]^2\} = \mathbb{E}\{[n^{-1} \sum_{l=1}^{n} X_l]^2\}. \tag{8.1.3}$$

The squared sum in the right side of (8.1.3) may be written as a double sum, and we continue (8.1.3),

$$\mathbb{V}(\bar{\mu}) = \mathbb{E}\{[n^{-1} \sum_{l=1}^{n} X_l]^2\} = \mathbb{E}\{n^{-2} \sum_{l,t=1}^{n} X_l X_t\}$$

$$= n^{-2} \sum_{l,t=1}^{n} \gamma^X(l, t) = n^{-1}[n^{-1} \sum_{l=1}^{n} \{\sum_{t=1}^{n} \gamma^X(l - t)\}]. \tag{8.1.4}$$

Note that only in the last equality we used the second-order stationarity of $\{X_t\}$.

Relation (8.1.4) is the result that we need. Let us consider several possible scenarios. If the observations are independent, then $\gamma^X(l) = 0$ for any $l \neq 0$, and (8.1.4) together with $\gamma^X(0) = \mathbb{V}(X_1)$ imply the familiar formula $\mathbb{V}(\bar{\mu}) = n^{-1}\mathbb{V}(X_1)$. Similarly, if $\sum_{l=0}^{\infty} |\gamma^X(l)| < c < \infty$ then $\mathbb{V}(\bar{\mu}) < cn^{-1}$ and we again have the parametric rate of the variance convergence. In the latter case the stochastic process may be referred to as a *short-memory* time series. However, if the sum $\sum_{l=0}^{n} \gamma^X(l)$ diverges, then we lose the rate n^{-1}. For instance, consider the case of a *long-memory* time series when $\gamma^X(t)$ is proportional to $|t|^{-\alpha}$, $0 < \alpha < 1$. Then the sum in the curly brackets on the right side of (8.1.4) is proportional to $n^{1-\alpha}$ and the variance is proportional to $n^{-\alpha}$. This is a dramatic slowing down of the rate of convergence caused by dependence between observations. The above-explained phenomenon sheds light on complexity of dealing with stochastic processes.

The simplest zero-mean and second-order stationary time series is a process in which the random variables $\{X_t\}$ are uncorrelated (that is, $\gamma^X(t) = 0$ for $t \neq 0$) and have zero mean and unit variance. Let us denote this time series as $\{W_t\}$ and call it a *standard (discrete time) white noise*. A classical example is a time series of independent standard Gaussian random variables, which is the white noise that will be used in the following simulations, and we call it a *standard Gaussian white noise*.

In its turn, a white noise allows us to define a wide variety of dependent second-order stationary and zero-mean processes via a set of linear difference equations. This leads us to the notion of an *autoregressive moving average process of orders p and q*, an ARMA(p, q) process for short. By definition, the process $\{X_t, t = \ldots, -1, 0, 1, \ldots\}$ is said to be an ARMA(p, q) process if $\{X_t\}$ is zero-mean and second-order stationary, and for every t

$$X_t - a_1 X_{t-1} - \cdots - a_p X_{t-p} = \sigma(W_t + b_1 W_{t-1} + \cdots + b_q W_{t-q}), \qquad (8.1.5)$$

where $\{W_t\}$ is a standard white noise, $\sigma > 0$, the orders p and q are nonnegative integers, and $a_1, \ldots, a_p, b_1, \ldots, b_p$ are real numbers. For the case of a Gaussian white noise we shall refer to the corresponding ARMA process as a *Gaussian ARMA process*.

Two particular classical examples of an ARMA process are a *moving average* MA(q) process, which is a moving average of $q + 1$ consecutive realizations of a white noise,

$$X_t = \sigma(W_t + b_1 W_{t-1} + \cdots + b_q W_{t-q}), \qquad (8.1.6)$$

and an *autoregressive* AR(p) process satisfying the difference equation

$$X_t - a_1 X_{t-1} - \cdots - a_p X_{t-p} = \sigma W_t. \qquad (8.1.7)$$

The MA and AR processes play an important role in the analysis of time series. For instance, prediction of values $\{X_t, t \geq n+1\}$ in terms of $\{X_1, \ldots, X_n\}$ is relatively simple and well understood for an autoregressive process because $\mathbb{E}\{X_t | X_{t-1}, X_{t-2}, \ldots\} = a_1 X_{t-1} + \ldots + a_p X_{t-p}$. Also, for a given autocovariance function it is simpler to find an AR process with a similar autocovariance function. More precisely, if an autocovariance function $\gamma^X(j)$ vanishes as $j \to \infty$, then for any integer k one can find an AR(k) process with the autocovariance function equal to $\gamma^X(j)$ for $|j| \leq k$. The "negative" side of an AR process is that it is not a simple issue to find a stationary solution for (8.1.7), and moreover, it may not exist. For instance, the difference equation $X_t - X_{t-1} = \sigma W_t$ has no stationary solution, and consequently there is no AR(1) process with $a_1 = 1$. A thorough discussion of this issue is beyond this short introduction, and in what follows a range for the coefficients that "keeps us out of trouble" will be always specified.

The advantages of a moving average process are its simple simulation, the given expression for a second-order stationary solution, and that it is very close by its nature to white noise, namely, while realizations of a white noise are uncorrelated, realizations of an MA(q)

process are also uncorrelated whenever the lag is larger than q. The disadvantages, with respect to AR processes, are more complicated procedures for prediction and estimation of parameters. Thus, among the two, typically AR processes are used for modeling and prediction. Also, AR processes are often used to approximate an ARMA process.

For a time series, and specifically ARMA processes, the notion of causality (future independence) plays an important role. The idea is that for a causal ARMA process $\{X_t\}$ (or more specifically, a causal process with respect to an underlying white noise $\{W_t\}$) it is quite natural to expect that an ARMA time series $\{X_t\}$ depends only on current and previous (but not future!) realizations of the white noise. Thus, we say that an ARMA process $\{X_t\}$ generated by a white noise $\{W_t\}$ is *causal* if $X_t = \sum_{j=0}^{\infty} c_j W_{t-j}$, where the coefficients c_j are absolutely summable. Clearly, MA(q) processes are causal, but not all AR(p) processes are; for instance, a stationary process corresponding to the difference equation $X_t - 2X_{t-1} = W_t$ is not causal. We shall not elaborate more on this issue and only note that in what follows we are considering simulations of Gaussian ARMA(1,1) processes corresponding to the difference equation $X_t - aX_{t-1} = \sigma(W_t + bW_{t-1})$ with $|a| < 1$ and $-a \neq b$. It may be directly verified that for such a this equation has a stationary and causal solution $X_t = \sigma W_t + \sigma(a+b)\sum_{j=1}^{\infty} a^{j-1} W_{t-j}$.

As we know from our discussion of (8.1.4), for a statistical inference it is important to know how fast autocovariance function $\gamma^X(k)$ decreases in k. Introduce a class of autocovariance functions

$$\mathcal{A}(Q, q, \beta, r) = \{\gamma^X : \ 0 < q < \gamma^X(0) \leq Q < \infty,$$

$$|\gamma^X(k)| \leq Q(k+1)^\beta e^{-rk}(1 + o_k(1)), \ k = 1, 2, \ldots\}. \tag{8.1.8}$$

Recall that $o_k(1)$ denote generic sequences such that $o_k(1) \to 0$ as $k \to \infty$. The importance of class (8.1.8) is explained by the fact that the autocovariance function of a causal ARMA(p, q) time series decreases exponentially (compare with the analytic class (2.1.2)). Furthermore, the parameter r in (8.1.8) is the logarithm of the minimal modulus of zeroes of the autoregressive polynomial, and an ARMA process is causal if and only if r is positive. Furthermore, for any quartet $\{Q, q, \beta, r\}$ there exists a causal ARMA(p, q) process such that in (8.1.8) we get equalities. Of course, class (8.1.8) includes not only spectral densities of ARMA time series. For instance, note that covariance of the product of two independent zero-mean time series is the product of covariances of these series. As a consequence if a zero-mean and second order stationary time series is multiplied by a time series from class (8.1.8) (in particular by an ARMA time series), then the product also belongs to class (8.1.8).

Apart of linear (with respect to an underlying white noise) processes, in many applications an underlying time series may not be linear. For instance, many interesting processes may be modeled by a nonparametric autoregressive process $X_t = q(X_{t-1}, \ldots, X_{t-p}) + \sigma(X_{t-1}, \ldots, X_{t-p})W_t$ where $q(\cdot)$ and $\sigma(\cdot)$ are p-variate functions. We will continue discussion of this process in Section 8.7.

So far we have been using a time domain (dynamic) approach to describe a process via evolution of a stochastic process in time. Another approach, considered in the next section, is to look at a second-order stationary process in the spectral (frequency) domain. And yet another approach is to explore the dependence between events as the time between them increases. The theory that uses this approach is called the *mixing* theory. Let us consider a stationary time series $\{X_t\}$ and introduce a class $\mathcal{M}_{-\infty}^k$ of all possible events of interest generated by variables $\{\ldots, X_{k-1}, X_k\}$ and a class $\mathcal{M}_k^\infty$ of all possible events of interest generated by variables $\{X_k, X_{k+1}, \ldots\}$ (a rigorous definition of these classes can be found in references mentioned in the Notes where the class is called a sigma-algebra). Consider a particular t and two events $G_t \in \mathcal{M}_{-\infty}^t$ and $G^{t+s} \in \mathcal{M}_{t+s}^\infty$. If the events are independent, then we know that

$$\mathbb{P}(G_t \cap G^{t+s}) - \mathbb{P}(G_t)\mathbb{P}(G^{t+s}) = 0. \tag{8.1.9}$$

Note that due to stationarity we can set $t = 0$, in other words if (8.1.9) holds for a particular $t = t_0$, then it holds for all t. Then the underlying idea of the mixing theory is to describe the dependence between separated in time events via deviation of the left side of (8.1.9) from zero. For instance, for a stationary time series $\{X_t\}$ we can introduce a *mixing coefficient*

$$\alpha^X(s) := \sup_{G_0 \in \mathcal{M}^0_{-\infty}, G^s \in \mathcal{M}^\infty_s} \{|\mathbb{P}(G_0 \cap G^s) - \mathbb{P}(G_0)\mathbb{P}(G^s)|\}, \ s > 0. \tag{8.1.10}$$

Let us present several general properties and examples of mixing coefficients. Mixing coefficient $\alpha^X(s)$ is either positive or equal to zero, and not increasing in s. For the case of a stationary time series $\{X_t\}$ of independent variables we have $\alpha^X(s) = 0$. Another important example is an *m-dependent* series satisfying $\alpha^X(s) = 0$ for $s > m$. The meaning of m-dependence is that variables, separated in time for more than m time-units, are independent. Another class of time series, often considered in the mixing theory, is when for $\tau > 8$

$$\sum_{s=0}^{\infty}(s+1)^{\tau/2-1}\alpha^X(s) \le Q < \infty. \tag{8.1.11}$$

Further, an important Kolmogorov-Rosanov result for a Gaussian stationary time series $\{X_t\}$ is that the covariance $\gamma^X(s)$ is of the same order as the mixing coefficient $\alpha^X(s)$, and this result bridges the classical second-order stationary time series theory with the mixing theory.

Let us also mention several classical results of the mixing theory that allow us to analyze expectations. Consider a stationary time series $\{X_t\}$, a random variable Y_0 which is a function of $\{\ldots, X_{-1}, X_0\}$, and a random variable Z_s which is a function of $\{X_s, X_{s+1}, \ldots\}$, $s > 0$. For a time series of independent variables we have $\mathbb{E}\{Y_0 Z_s\} - E\{Y_0\}E\{Z_s\} = 0$, and for an m-dependent series we have $\mathbb{E}\{Y_0 Z_s\} - E\{Y_0\}E\{Z_s\} = 0$ for $s > m$. Further, suppose that $\mathbb{E}\{|Y_0|^{2+\delta}\} < \infty$ and $\mathbb{E}\{|Z_s|^{2+\delta}\} < \infty$ for some $\delta > 0$, then there exists a finite constant c such that

$$|\mathbb{E}\{Y_0 Z_s\} - E\{Y_0\}E\{Z_s\}| \le c[\alpha^X(s)]^{\delta/(2+\delta)}. \tag{8.1.12}$$

Further, if $\mathbb{P}(|Y_0| > c_1) = 0$ and $\mathbb{P}(|Z_s| > c_2) = 0$ then

$$|\mathbb{E}\{Y_0 Z_s\} - E\{Y_0\}E\{Z_s\}| \le 4c_1 c_2 \alpha^X(s). \tag{8.1.13}$$

These are types of results that allow us to develop statistical inference in problems with stationary dependent variables, more results (including rigorous measure-theoretical formulations) may be found in the references mentioned in the Notes.

The reader may also recall that a discrete time Markov chain is another classical approach for modeling and analysis of dependent observations. It will be briefly discussed in Section 8.3.

8.2 Spectral Density and Its Estimation

Let $\{X_t\}$ be a zero-mean and second-order stationary time series with the autocovariance function $\gamma^X(j) := E\{X_t X_{t+j}\}$. The second-order properties of this time series are completely described by its autocovariance function, or, equivalently, under mild conditions (for instance, a sufficient condition is $\sum_{j=-\infty}^{\infty} |\gamma^X(j)| < \infty$), by its Fourier transform, which is called the *spectral density* function,

$$g^X(\lambda) \quad := \quad (2\pi)^{-1} \sum_{j=-\infty}^{\infty} \gamma^X(j)\cos(j\lambda) \tag{8.2.1}$$

$$= \quad (2\pi)^{-1}\gamma^X(0) + \pi^{-1}\sum_{j=1}^{\infty}\gamma^X(j)\cos(j\lambda), \quad -\pi < \lambda \le \pi. \tag{8.2.2}$$

Here the frequency λ is in units radians/time, and to establish the equality (8.2.2) we used the relation $\gamma^X(-j) = \gamma^X(j)$. The spectral density is symmetric in λ about 0, i.e., the spectral density is an even function. Thus, it is customary to consider a spectral density on the interval $[0, \pi]$. The spectral density is also a nonnegative function (like the probability density), and this explains why it is called a density.

One of the important applications of the spectral density is searching for a deterministic periodic component (often referred to as a seasonal component) in nonstationary time series. Namely, a peak in $g^X(\lambda)$ at frequency λ^* indicates a possible periodic phenomenon with period

$$T^* = \frac{2\pi}{\lambda^*}. \tag{8.2.3}$$

This formula explains why spectral domain analysis is the main tool in searching after the period of a seasonal component, and the estimator will be discussed in Section 9.3.

Let us explain how the spectral density may be estimated using our E-estimation methodology. But first let us pause for a moment and stress the following important remark. By its definition, spectral density is a *cosine series*, and this is an example where the basis is chosen not due to its convenience, as we did in the cases of E-estimation of the probability density and the regression, but due to definition of the estimand. In other words, spectral density estimation is the most appealing example of using a series approach and the cosine basis. Now let us return to estimation of the spectral density.

Denote by $X_1, \ldots, X_n$ the realization of a second-order stationary and zero-mean time series. The classical *empirical autocovariance estimator* is defined as

$$\tilde{\gamma}^X(j) := n^{-1} \sum_{l=1}^{n-j} X_l X_{l+j}, \quad j = 0, 1, \ldots, n-1, \tag{8.2.4}$$

while the sample mean estimator is

$$\hat{\gamma}^X(j) := (n-j)^{-1} \sum_{l=1}^{n-j} X_l X_{l+j}, \quad j = 0, 1, \ldots, n-1. \tag{8.2.5}$$

Note that in the empirical autocovariance the divisor n is not equal to the number $n-j$ of terms in the sum, and hence it is a biased estimator. On the other hand, this divisor ensures that an estimate corresponds to some second-order stationary series. For all our purposes there is no difference between using the two estimators, but it is always a good idea to check which one is used by a statistical software. In what follows, proposed E-estimators will be based on the sample mean autocovariance estimator (8.2.5), and the reason is to follow our methodology of sample mean estimation. On the other hand, many classical spectral estimators, like the periodogram discussed below, use the estimator (8.2.4).

Based on (8.2.2), if one wants to estimate a spectral density and is not familiar with basics of nonparametric estimation discussed in Chapter 2, it is natural to plug the empirical autocovariance (8.2.4) in place of unknown autocovariance. And sure enough, this approach is well known and the resulting estimator (up to the factor $1/2\pi$) is called a *periodogram*,

$$\mathcal{I}^X(\lambda) := \tilde{\gamma}^X(0) + 2\sum_{j=1}^{n-1} \tilde{\gamma}^X(j)\cos(j\lambda) = n^{-1}\Big|\sum_{l=1}^{n} X_l e^{-il\lambda}\Big|^2. \tag{8.2.6}$$

Here i is the imaginary unit, i.e., $i^2 := -1$, $e^{ix} = \cos(x) + i\sin(x)$, and the periodogram is defined at the so-called *Fourier frequencies* $\lambda_k := 2\pi k/n$, where k are integers satisfying $-\pi < \lambda_k \le \pi$.

Periodogram, as a tool for spectral-domain analysis, was proposed in the late nineteenth century. It has been both the glory and the curse of the spectral analysis. The glory, because many interesting practical problems were solved at a time when no computers were available. The curse, because the periodogram, which had demonstrated its value for locating periodicities (recall (8.2.3)), proved to be an erratic and inconsistent estimator. The reason for the failure of the periodogram is clear from the point of view of nonparametric curve estimation theory discussed in Chapter 2. Indeed, based on n observations, the periodogram estimates n Fourier coefficients, and this explains the erratic performance and inconsistency. Nonetheless, it is still a popular estimator.

Using the sample mean estimator (8.2.5), we may use the E-estimator of Section 2.2 for estimation of the spectral density. Of course, the theory of E-estimation was explained for the case of independent observations, but as we will see shortly, it can be extended to dependent observations. We are beginning with a simulated example which sheds light on the problem, periodogram and E-estimator, and then explore the theory.

Figure 8.1 allows us to visualize an ARMA process, its spectral density and the two above-defined estimates of the spectral density. A particular realization of the Gaussian ARMA(1, 1) time series $X_t + 0.3X_{t-1} = 0.5(W_t - 0.6W_{t-1})$ is shown in the top diagram. Note how fast observations oscillate over time. This is because here the covariance between X_t and X_{t-1} is negative. This follows from the following formula for calculating the autocovariance function of the causal ARMA(1, 1) process $X_t - aX_{t-1} = \sigma(W_t + bW_{t-1})$ with $|a| < 1$,

$$\gamma^X(0) = \frac{\sigma^2[(a+b)^2 + 1 - a^2]}{(1-a^2)}, \quad \gamma^X(1) = \frac{\sigma^2(a+b)(1+ab)}{(1-a^2)},$$

$$\gamma^X(j) = a^{j-1}\gamma(1), \quad j \ge 2. \tag{8.2.7}$$

Note that if $a > 0$ and $b > 0$, then $\gamma(1) > 0$, and a realization of the time series will "slowly" change over time. On the other hand, if $a + b < 0$ and $1 + ab > 0$ then a realization of the time series may change its sign almost every time. Thus, depending on a and b, we may see either slow or fast oscillations in a realization of an ARMA(1, 1) process. Figure 8.1 allows us to change parameters of the ARMA process and observe different interesting patterns in this pure stochastic process. In particular, to make the process slowly changing and even see interesting repeated patterns in a time series, choose positive parameters a and b.

The solid line in the bottom diagram of Figure 8.1 shows us the underlying theoretical spectral density of the ARMA(1, 1) process. As we see, because here both a and b are negative, in the spectral domain high frequencies dominate low frequencies. The formula for calculating the spectral density is $g^X(\lambda) = \sigma^2|1 + be^{i\lambda}|^2/[2\pi|1 - ae^{i\lambda}|^2]$, and it is a particular case of the following formula for a causal ARMA(p, q) process (8.1.5),

$$g^X(\lambda) = \frac{\sigma^2 \left|1 + \sum_{j=1}^{q} b_j e^{-ij\lambda}\right|^2}{2\pi \left|1 - \sum_{j=1}^{p} a_j e^{-ij\lambda}\right|^2}. \tag{8.2.8}$$

The middle diagram shows us that the periodogram has a pronounced mode at frequency $\lambda^* \approx 2.6$ which, according to (8.2.3), indicates a possibility of a deterministic periodic (seasonal) component with the period which is either 2 or 3. One may see or not see such a component in the data, but thanks to the simulation we do know that there is no periodic component in the data. The reader is advised to repeat this figure and get used to reading a periodogram because it is commonly used by statistical softwares. The bottom diagram exhibits the spectral density E-estimate which correctly shows the absence of any periodic

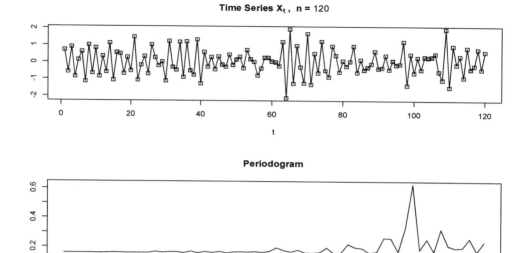

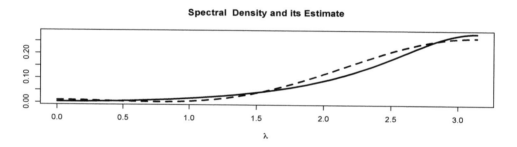

Figure 8.1 *ARMA(1,1) time series and two estimates of the spectral density. The top diagram shows a particular realization of a Gaussian ARMA(1,1) time series* $Y_t - aY_{t-1} = \sigma(W_t + bW_{t-1})$, $t = 1, 2, \ldots, n$, *where* $a = -0.3$, $b = -0.6$, $\sigma = 0.5$, *and* $n = 120$. *The middle diagram shows the periodogram. The spectral density E-estimate (the dashed line) and the underlying spectral density (the solid line) are exhibited in the bottom diagram. {The length* n *of a realization is controlled by the argument* n. *Parameters of simulated ARMA(1,1) process are controlled by the arguments* sigma, a, *and* b. *Use* $|a| < 1$. *All the other arguments control parameters of the E-estimator. Note that the string sp is added to these arguments to indicate that they control coefficients of the spectral density E-estimator.} [n = 120, sigma = 0.5, a = -0.3, b = -0.6, cJ0sp = 2, cJ1sp = 0.5, cTHsp = 4]*

(seasonal) component, and the E-estimate nicely resembles the underlying spectral density. Please pay attention to the relatively small sample size $n = 120$. Again, it is important to repeat this figure, with different sample sizes and for different ARMA processes, to get first-hand experience in spectral analysis of stationary time series.

We are finishing this section with theoretical analysis of the MISE of a series estimator. This is an interesting and technically challenging problem because a relatively simple technique of inference for a sum of independent variables no longer is applicable.

First, let us begin with an example of using the Parseval identity. We can write that

$$\int_{-\pi}^{\pi} [g^X(\lambda)]^2 d\lambda = (2\pi)^{-1}[\gamma^X(0)]^2 + \pi^{-1}\sum_{j=1}^{\infty}[\gamma^X(j)]^2. \qquad (8.2.9)$$

Similarly, for a series estimator

$$\bar{g}^X(\lambda, J) := (2\pi)^{-1}\hat{\gamma}^X(0) + \pi^{-1}\sum_{j=1}^{J}\hat{\gamma}^X(j)\cos(j\lambda), \tag{8.2.10}$$

we get the following expression for its MISE,

$$\text{MISE}(\bar{g}^X(\lambda, J), g^X(\lambda)) := \mathbb{E}\{\int_{-\pi}^{\pi}[\bar{g}^X(\lambda, J) - g^X(\lambda)]^2 d\lambda\}$$

$$= \left[(2\pi)^{-1}\mathbb{E}\{[\hat{\gamma}^X(0) - \gamma^X(0)]^2\} + \pi^{-1}\sum_{j=1}^{J}\mathbb{E}\{[\hat{\gamma}^X(j) - \gamma^X(j)]^2\}\right] + \pi^{-1}\sum_{j>J}[\gamma^X(j)]^2. \tag{8.2.11}$$

In (8.2.11) the term in the large square brackets is the integrated variance (or simply variance) of $\bar{g}^X(\lambda, J)$, and the last term is the integrated squared bias of $\bar{g}^X(\lambda, J)$. This is a classical decomposition of the MISE (recall our discussion in Chapter 2).

Now we need to learn the technique of inference for a sum of dependent variables. In what follows we are considering $j < n$ and assume that $\{X_t\}$ is a Gaussian zero-mean and second-order stationary time series. Recall that the expectation of a sum is always the sum of expectations, and hence the mean of the sample mean autocovariance (8.2.5) is

$$\mathbb{E}\{\hat{\gamma}^X(j)\} = (n - j)^{-1}\sum_{l=1}^{n-j}\mathbb{E}\{X_l X_{l+j}\} = \gamma^X(j). \tag{8.2.12}$$

We conclude that the sample mean estimator (8.3.5) is unbiased and that the dependence does not change this nice property of a sample mean estimator. Now we are considering the variance of the sample mean estimator, and this problem is more involved because we need to learn several technical steps. Write,

$$\mathbb{V}(\hat{\gamma}^X(j)) = (n - j)^{-2}\mathbb{E}\{[\sum_{l=1}^{n-j}(X_l X_{l+j} - \gamma^X(j)]^2\}. \tag{8.2.13}$$

This is the place where we may either continue by going from the squared sum to a double sum and then do a number of calculations, or use the following nice formula. Consider a zero-mean and second-order stationary time series $\{Z_t\}$. Then, using the technique of (8.1.3)-(8.1.4) we get a formula

$$\mathbb{E}\{[\sum_{l=1}^{k} Z_l]^2\} = \sum_{l=-k}^{k}(k - |l|)\mathbb{E}\{Z_0 Z_l\}. \tag{8.2.14}$$

Note how simple, nice and symmetric the formula is, and it allows us to write down the variance of a sum of dependent variables as a sum of expectations. Using (8.2.14) in (8.2.13) we get for $j < n$,

$$\mathbb{V}(\hat{\gamma}^X(j)) = (n - j)^{-2}\sum_{l=-n+j}^{n-j}(n - j - |l|)[\mathbb{E}\{X_0 X_j X_l X_{l+j}\} - (\gamma^X(j))^2]. \tag{8.2.15}$$

Our next step is to understand how to evaluate terms $\mathbb{E}\{X_0 X_j X_l X_{l+j}\}$. In Section 8.1 several possible paths, depending on the assumption about dependency, were discussed. Here it is assumed that the time series is Gaussian, and this implies the following nice formula,

$$\mathbb{E}\{X_0 X_j X_l X_{l+j}\} = (\gamma^X(j))^2 + (\gamma^X(l))^2 + \gamma^X(l+j)\gamma^X(l-j). \tag{8.2.16}$$

Using (8.2.16) in the right side of (8.2.15) yields for $j < n$,

$$\mathbb{V}(\hat{\gamma}^X(j)) = (n-j)^2 \sum_{l=-n+j}^{n-j} (n-j-|l|)[(\gamma^X(l))^2 + \gamma^X(l+j)\gamma^X(l-j)]. \qquad (8.2.17)$$

We need to add one more assumption that holds for all short-memory time series (including ARMA)

$$\sum_{l=0}^{\infty} |\gamma^X(l)| < \infty. \qquad (8.2.18)$$

Note that the Cauchy-Schwarz inequality implies that $|\gamma^X(j)| \leq \gamma^X(0)$, and this together with (8.2.18) yields another useful inequality

$$\sum_{l=0}^{\infty} [\gamma^X(l)]^2 \leq \gamma^X(0) \sum_{l=0}^{\infty} |\gamma^X(l)| < \infty. \qquad (8.2.19)$$

Using these two inequalities in (8.2.17), together with the Cauchy inequality

$$2|\gamma^X(l+j)\gamma^X(l-j)| \leq \rho[\gamma^X(l+j)]^2 + \rho^{-1}[\gamma^X(l-j)]^2, \quad \rho > 0, \qquad (8.2.20)$$

which is valid for any positive number ρ, we conclude that

$$\mathbb{V}(\hat{\gamma}^X(j)) = n^{-1}d(g^X)(1 + c_{n,j}), \qquad (8.2.21)$$

where $c_{n,j} \to 0$ as both n and j increase in such a away that $j < J_n = o_n(1)n$, and

$$d(g^X) := [\gamma^X(0)]^2 + 2\sum_{l=1}^{\infty}[\gamma^X(l)]^2 = 2\pi \int_{-\pi}^{\pi} [g^X(\lambda)]^2 d\lambda. \qquad (8.2.22)$$

In (8.2.22) the equality is valid due to (8.2.9).

Relation (8.2.21) is important because it shows us that asymptotically, given the weak dependence property (8.2.18), the variance (and the mean squared error) of the sample mean autocovariance decreases with the same classical parametric rate n^{-1} as the variance of the sample mean of independent observations. Furthermore, there is a simple expression for the coefficient of difficulty $d(g^X)$ which is proportional to the integrated squared spectral density.

Using (8.2.21) in (8.2.11) for $J = J_n$ satisfying $J_n = o_n(1)n$ and $J_n \to \infty$ as $n \to \infty$, we conclude that

$$\text{MISE}(\bar{g}^X(\lambda, J_n), g^X(\lambda)) = J_n n^{-1} 2 \int_{-\pi}^{\pi} [g^X(\lambda)]^2 d\lambda (1 + o_n(1)) + \pi^{-1} \sum_{j > J_n} [\gamma^X(j)]^2. \qquad (8.2.23)$$

This is a general expression for the MISE of a series estimator $\bar{g}^X(\lambda, J_n)$ defined in (8.2.10). To simplify it further, we need to add an assumption which will allow us to evaluate the second term (the ISB) on the right side of (8.2.23). As an example, let us additionally assume that the considered Gaussian time series is ARMA. Then the autocovariance function belongs to a class $\mathcal{A}(Q, q, \beta, r)$ defined in (8.1.8). For this class the assumption (8.2.18) holds, we can bound from above the sum in (8.2.23) and get the upper bound for the MISE,

$$\text{MISE}(\bar{g}^X(\lambda, J_n), g^X(\lambda))$$

$$\leq [2\int_{-\pi}^{\pi} [g^X(\lambda)]^2 d\lambda J_n n^{-1} + \pi^{-1} Q^2 \sum_{j > J_n} (j+1)^{2\beta} e^{-2rj}](1 + o_n(1)). \qquad (8.2.24)$$

Now we can minimize the right side of (8.2.24) with respect to cutoff J_n. We have a large choice of sequences J_n that may do this. For instance, we can choose $J_n = J'_n$ where J'_n is the largest integer smaller than $(2r)^{-1}\ln(n)(1 + 1/\sqrt{\ln(n)})$. This yields

$$\frac{\text{MISE}(\bar{g}^X(\lambda, J'_n), g^X(\lambda))}{\int_{-\pi}^{\pi}[g^X(\lambda)]^2 d\lambda} \leq r^{-1}\ln(n)n^{-1}(1 + o_n(1)). \qquad (8.2.25)$$

Note that this choice of the cutoff makes the integrated squared bias asymptotically smaller in order than the variance. This is an important property which is typical for ARMA processes. Further, this also means that in (8.2.25) we can replace the inequality on equality and get

$$\frac{\text{MISE}(\bar{g}^X(\lambda, J'_n), g^X(\lambda))}{\int_{-\pi}^{\pi}[g^X(\lambda)]^2 d\lambda} = r^{-1}\ln(n)n^{-1}(1 + o_n(1)). \qquad (8.2.26)$$

Further, note that J'_n uses only parameter r of the class (8.1.8).

Asymptotic theory shows that no other estimator can improve the right side of (8.2.26) uniformly over the class (8.1.8), namely for any (not necessarily series) estimator $\breve{g}^X(\lambda)$ the following lower bound holds,

$$\sup_{g^X \in \mathcal{A}(Q,q,\beta,r)} \left\{ \frac{\text{MISE}(\breve{g}^X(\lambda), g^X(\lambda))}{\int_{-\pi}^{\pi}[g^X(\lambda)]^2 d\lambda} \right\} \geq r^{-1}\ln(n)n^{-1}(1 + o_n(1)). \qquad (8.2.27)$$

The right sides of (8.2.26) and (8.2.27) coincide up to a factor $1 + o_n(1)$, and this allows us to say that the lower bound (8.2.27) is sharp and that the series estimator $\bar{g}^X(\lambda, J'_n)$ is asymptotically minimax (optimal, efficient). Of course, J'_n depends on parameter r of the analytic class, and this is why the E-estimator chooses a cutoff using data, or we may say that E-estimator is an adaptive estimator because it adapts to an underlying class of spectral densities.

There is another interesting conclusion from (8.2.23) that can be made about a reasonable estimator that does not require adaptation to the class (8.1.8) and may choose a cutoff J_n a priori before getting data. Let us consider $J_n = J^*_n$ which is the largest integer smaller than $(c_n/2)\ln(n)$ where $c_n \to \infty$ as slow as desired, say $c_n = \ln(\ln(\ln(n)))$. Then a direct calculation shows that

$$\frac{\text{MISE}(\bar{g}^X(\lambda, J^*_n), g^X(\lambda))}{\int_{-\pi}^{\pi}[g^X(\lambda)]^2 d\lambda} = c_n \ln(n)n^{-1}(1 + o_n(1)). \qquad (8.2.28)$$

Note that the rate of the MISE convergence is just "slightly" slower than the minimax rate $\ln(n)n^{-1}$.

As we have seen, estimation of spectral densities is an exciting topic with rich history, fascinating asymptotic theory, and numerous practical applications.

8.3 Bernoulli Missing

Similarly to the previous section, in what follows $X_1, \ldots, X_n$ is a realization of a zero-mean and second-order stationary time series. Additionally, only to simplify notation, it is also assumed that $\mathbb{P}(X_t = 0) = 0$. The new here is that some observations in the realization may be missed. The missing mechanism is described by a stationary Bernoulli time series $\{A_t\}$ which is independent of $\{X_t\}$ and such that A_t takes on either value 0 or 1. To avoid a case when all observations are missed, it is assumed that $\mathbb{P}(A_t = 1) > 0$. Recall that in Chapters 4 and 5 availability variables $A_1, A_2, \ldots$ were independent Bernoulli variables; in this section they are depended and this is why we refer to them as a Bernoulli time series. Then, similarly to our previous missing models, the available observations are

$$Y_l := A_l X_l, \quad l = 1, 2, \ldots, n. \tag{8.3.1}$$

As a result, X_l is missed whenever $A_l = 0$. The above-made assumption $\mathbb{P}(X_t = 0) = 0$ implies that $\mathbb{P}(A_l = I(Y_l \neq 0)) = 1$, and hence we do not need to keep track of the time series $\{A_t\}$.

The problem is to estimate the spectral density $g^X(\lambda)$ of a hidden underlying process $\{X_t\}$ given observations (8.3.1). In this section two models of the missing mechanism are considered, and more models will be presented in the next section.

The first missing model is called a *Markov–Bernoulli* model and it is generated by a Markov–Bernoulli time series (chain) $\{A_t\}$. Let us recall some basic facts about Markov chains. A discrete time Markov chain is a time series (stochastic process) that undergoes transitions from one state to another, between a finite or countable number of possible states. In our special case of the Bernoulli process there are just two states: 0 and 1. This random process is memoryless meaning that the probability of transition into the next state depends only on the current state and not on the sequence of events that preceded it. This specific kind of "memorylessness" is called the Markov property. The changes of state of the Markov chain are called transitions, and the probabilities associated with various state-changes are called transition probabilities. If transition probabilities do not change in time, then the Markov chain is called stationary. An example is the dietary habits of a creature who eats either grapes or lettuce exactly once a day conforming to the following habits. If it ate grapes today then tomorrow, regardless of what it ate in previous days, it will eat grape with probability 1/3 and lettuce with probability 2/3; do you see that these two probabilities are correctly added to 1? If it ate lettuce today then tomorrow, regardless of what it ate in previous days, it will eat grape with probability 3/5 and lettuce with probability 2/5. Suppose that an experiment on the creature (say a blood test or a training session) is conducted only on the day when grapes are eaten. Then an observation is missed at the day when lettuce is eaten. Correspondingly, the transition probabilities for the availability are $\mathbb{P}(A_t = 1|A_{t-1} = 1) = 1/3$, $\mathbb{P}(A_t = 0|A_{t-1} = 1) = 2/3$, $\mathbb{P}(A_t = 1|A_{t-1} = 0) = 3/5$, and $\mathbb{P}(A_t = 0|A_{t-1} = 0) = 2/5$. This is an example of a Markov–Bernoulli missing.

For a stationary Markov–Bernoulli missing mechanism, it is of a special interest to learn about random length of the batch of missing observations. Denote this length as L and, as an example, consider a time series with $\ldots, A_t = 1, A_{t+1} = 0, A_{t+2} = 0, \ldots, A_{t+L} = 0, A_{t+L+1} = 1, \ldots$ Here we observe, after the available observation X_t, the batch of missing observations $X_{t+1}, \ldots, X_{t+L}$ of length L. Given that $L \geq 1$, the distribution of L is geometric with $\mathbb{P}(L = k) = \alpha^{k-1}(1 - \alpha)$, $k = 1, 2, 3, \ldots$ where $\alpha = \mathbb{P}(A_{t+1} = 0|A_t = 0)$. The mean length $\mathbb{E}\{L\}$ of the batch is $(1 - \alpha)\alpha^{-1}$ and the variance is $(1 - \alpha)\alpha^{-2}$. Another interesting property, which shows that the probability of larger batches of missing observations decreases exponentially, is described by the inequality

$$\mathbb{P}(L > k) \leq [|\ln(\alpha)|^{-1}(1 - \alpha)]e^{-k|\ln(\alpha)|}. \tag{8.3.2}$$

To conclude our brief introduction to Markov chains, let us note that a Markov chain of order m (or a Markov chain with memory m), where m is a finite positive integer, is a process satisfying

$$\mathbb{P}(A_t = a_t|A_{t-1} = a_{t-1}, A_{t-2} = a_{t-2}, \ldots)$$

$$= \mathbb{P}(A_t = a_t|A_{t-1} = a_{t-1}, A_{t-2} = a_{t-2}, \ldots, A_{t-m} = a_{t-m}).$$

In other words, only last m past states define the future state. Note that the classical Markov chain may be referred to as the Markov chain of order 1.

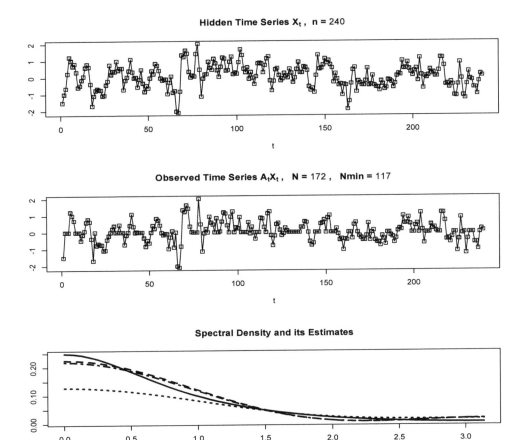

Figure 8.2 *Markov–Bernoulli missing mechanism. Markov chain is generated according to transition probabilities* $\mathbb{P}(A_{t+1} = 0 | A_t = 0) = \alpha$ *and* $\mathbb{P}(A_{t+1} = 1 | A_t = 1) = \beta$. *The top diagram shows* n *realizations of an underlying Gaussian ARMA(1,1) time series* X_t *defined in Figure 8.1 only here parameters are* $a = 0.4$, $b = 0.5$, $\sigma = 0.5$. *The middle diagram shows the observed time series* $\{A_t X_t\}$ *with missing observations. Set* $N_j := \sum_{l=1}^{n-j} A_l A_{l+j}$, *and then* $N := N_0$ *and* $Nmin := \min_{0 \leq j \leq c_{J0sp} + c_{J1sp} \ln(n)} N_j$ *are shown in the title. The bottom diagram shows by the solid line the underlying spectral density as well as the following three estimates. The proposed E-estimate is shown by the dashed line. The naïve E-estimate (the dotted line) is based on available observations of time series* $\{A_t X_t\}$ *shown in the middle diagram. Oracle's E-estimate, based on the hidden underlying realizations of* $\{X_t\}$, *is shown by the dot-dashed line. [n = 240, sigma = 0.5, a = 0.4, b = 0.5, alpha = 0.4, beta = 0.8, cJ0sp = 2, cJ1sp = 0.5, cTHsp = 4]*

Figure 8.2 helps us to understand how a Markov–Bernoulli missing mechanism performs. The top diagram shows realization of a Gaussian ARMA(1, 1) time series $\{X_t\}$. Recall that the top diagram in Figure 8.1 also shows us a realization of a Gaussian ARMA(1, 1) time series, only here we use positive values for parameters a and b (see the caption and compare with the time series in the top diagram in Figure 8.1 where negative parameters are used). The positive parameters create a slowly oscillating time series, it is even possible to imagine a deterministic periodic (seasonal) component, but here we do know that everything is purely stochastic. It is natural to expect that in the spectral domain low frequencies dominate high frequencies, and the solid line in the bottom diagram, which is the plot of the underlying spectral density, supports this conclusion. The observed time series $\{A_t X_t\}$ is shown in the middle diagram. Note that the missing data are easily visualized, and hence there is no need to plot the time series $A_1, \ldots, A_n$. Another observation, based on the middle diagram, is that some batches of missing observations are relatively long. The number $N := \sum_{l=1}^{n} A_l$ of available observations is shown in the title of the middle diagram. The particular number $N = 172$ shows that almost 30% of observations are missed. Can a data-driven spectral density estimator be proposed that performs reasonably well under these circumstances? Further, is the number N the only one to look after? These are the questions that we would like to explore.

Let us explain how a data-driven E-estimator, based on missing data (8.3.1), can be constructed. The aim is to propose a "good" estimate of the autocovariance function $\gamma^X(j)$ of the time series of interest $\{X_t\}$. Our approach is as follows. We begin with a sample mean autocovariance estimator of the observed time series $\{Y_t\} := \{A_t X_t\}$, and then decide on what changes should be done to estimate $\gamma^X(j)$. Using independence between time series $\{A_t\}$ and $\{X_t\}$, together with the assumption that $\mathbb{E}\{X_t\} = 0$, we conclude that $\mathbb{E}\{A_t X_t\} = 0$. Then the sample mean autocovariance is

$$\hat{\gamma}^Y(j) := (n-j)^{-1} \sum_{l=1}^{n-j} Y_l Y_{l+j} = (n-j)^{-1} \sum_{l=1}^{n-j} (A_l X_l)(A_{l+j} X_{l+j}). \qquad (8.3.3)$$

Using the assumed second-order stationarity of $\{X_t\}$ and stationarity of $\{A_t\}$, we can evaluate the expectation of the sample mean autocovariance. Write,

$$\mathbb{E}\{\hat{\gamma}^Y(j)\} = \mathbb{E}\{(n-j)^{-1} \sum_{l=1}^{n-j} (A_l X_l)(A_{l+j} X_{l+j})\}$$

$$= (n-j)^{-1} \sum_{l=1}^{n-j} \mathbb{E}\{A_l A_{l+j}\} \mathbb{E}\{X_l X_{l+j}\}$$

$$= \gamma^X(j)[(n-j)^{-1} \sum_{l=1}^{n-j} \mathbb{E}\{A_l A_{l+j}\}] = \gamma^X(j) \mathbb{E}\{A_1 A_{1+j}\}. \qquad (8.3.4)$$

Note that the expectation $\mathbb{E}\{A_1 A_{1+j}\}$, which we see in (8.3.4), looks like the autocovariance but it is not because the expectation of A_t is not zero.

We conclude that the expectation of the estimator $\hat{\gamma}^Y(j)$, based on the observed process Y_t, is the product of the underlying autocovariance of interest $\gamma^X(j)$ and the function $\mathbb{E}\{A_1 A_{1+j}\} = \mathbb{P}(A_1 A_{1+j} = 1)$. The function $\mathbb{E}\{A_1 A_{1+j}\}$ can be estimated by its sample mean, and this yields the following plug-in sample mean estimator of the autocovariance of interest $\gamma^X(j)$,

$$\hat{\gamma}^X(j) := \frac{(n-j)^{-1} \sum_{l=1}^{n-j} Y_l Y_{l+j}}{(n-j)^{-1} \sum_{l=1}^{n-j} I(Y_l Y_{l+j} \neq 0)} = \frac{\sum_{l=1}^{n-j} Y_l Y_{l+j}}{\sum_{l=1}^{n-j} I(Y_l Y_{l+j} \neq 0)}. \qquad (8.3.5)$$

Similarly to Section 4.1, we need to comment on the term $N_j := \sum_{l=1}^{n-j} I(Y_l Y_{l+j} \neq 0)$ $= \sum_{l=1}^{n-j} A_l A_{l+j}$ which is used in the denominator of (8.3.5). Theoretically the number of available pairs of observations may be zero, and then using the assumed $0/0 := 0$ in (8.3.5) we get $\hat{\gamma}^X(j) = 0$. This is a reasonable outcome for the case when no information about $\gamma^X(j)$ is available. Another remedy is to consider only samples with $N_{\min} := \min_{\{0 \leq j \leq c_{J0sp} + c_{J1sp} \ln(n)\}} N_j > k$ for some $k \geq 0$, because $c_{J0sp} + c_{J1sp} \ln(n)$ is the largest frequency used by the E-estimator (recall (2.2.4)).

Is the plug-in estimator (8.3.5) unbiased whenever $N_j > k \geq 0$? It is often not the case when a statistic is plugged in the denominator of an unbiased estimator, and hence it is of interest to check this property. Using independence between $\{A_t\}$ and $\{X_t\}$ we can write,

$$\mathbb{E}\{\hat{\gamma}^X(j)|N_j > k\} = \mathbb{E}\left\{\mathbb{E}\left\{\frac{\sum_{l=1}^{n-j}(A_l X_l)(A_{l+j} X_{l+j})}{\sum_{l=1}^{n-j} A_l A_{l+j}}\middle| A_1, \ldots, A_n, N_j > k\right\}\right\}$$

$$= \mathbb{E}\left\{\frac{\sum_{l=1}^{n-j} A_l A_{l+j}\mathbb{E}\{X_l X_{l+j}\}}{\sum_{l=1}^{n-j} A_l A_{l+j}}\middle| N_j > k\right\} = \gamma^X(j). \qquad (8.3.6)$$

This establishes that, whenever $N_{\min} > 0$, the proposed autocovariance estimator (8.3.5) is unbiased, and it can be used to construct the spectral density E-estimator $\hat{g}^X(\lambda)$.

Now let us return to Figure 8.2. The title of the middle diagram indicates $N_{\min} = 117$, and this tells us about complexity of the problem of the spectral density estimation with missing data when the size $n = 240$ of the hidden time series is decreased to $N = 172$ available observations, and then the minimal number of available pairs for estimation of the autocovariance is decreased to 117, that is more than the twofold decrease of n. It is also of interest to present the underlying $N_0, \ldots, N_5$ that are 172, 134, 120, 117, 122 and 119, respectively. We may conclude that, in the analysis of a time series with missing observations, it is important to take into account not only n and N but also $N_{\min}$.

The bottom diagram in Figure 8.2 exhibits three estimates and the underlying spectral density of interest $g^X(\lambda)$ (the solid line). The dashed line is the proposed data-driven E-estimate. Note that it correctly exhibits the slowly decreasing shape of the underlying density (compare with the solid line which is the underlying spectral density). The dotted line shows us the E-estimate based on $\hat{\gamma}^Y(j)$, in other words, this is a naïve estimate which ignores the missing data and deals with $\{Y_t\}$ like it is the time series of interest. While overall it is a poor estimate of the underlying spectral density, note that it correctly shows smaller power of the observed time series at low frequencies. The dot-dashed line shows us oracle's E-estimate based on the hidden time series $\{X_t\}$ shown in the top diagram. The bottom diagram allows us to compare performance of the same E-estimator based on three different datasets, and based on this single simulation we may conclude that the proposed data-driven E-estimator performs relatively well, and ignoring the missing, as the naïve E-estimator does, is a mistake. The interested reader is advised to repeat this figure, possibly with different parameters, to get first-hand experience in dealing with the Markov-Bernoulli missing mechanism.

Now let us consider a different mechanism of creating missing observations. Here the number L of consecutive missing observations, that is the length of a batch of missing observations, is generated by a random variable. Correspondingly, we will generate missing observations by choosing the distribution of L and refer to this missing mechanism as batch-Bernoulli. The following example clarifies the definition. Suppose that each hour we need to conduct an experiment whose outcomes create a time series of hourly observations. However, this task is of lower priority with respect to other jobs that may arrive in batches with each job requiring one hour to be fulfilled. As a result we may have intervals of time when the experiment is not conducted. A modification of that example is when the experiment cannot be performed if the equipment malfunctions and then a random number of hours is required to fix the equipment.

A distribution of L, which is often used to model batches, is Poisson with $\mathbb{E}\{L\} = \lambda$, and definition of this distribution can be found in Section 1.3. To simulate the corresponding batch-Bernoulli missing, we generate a sample $L_1, L_2, \ldots$ of independent random variables from L. If $L_1 = 0$ then no missing occurs and if $L_1 = k > 0$ then k consecutive observations $X_1, \ldots, X_k$ are missed, etc. The following inequality sheds light on how large Poisson batches can be,

$$\mathbb{P}(L \geq k) \leq e^{-\lambda}k^{-k}(e\lambda)^k = e^{-k\ln(k/e\lambda)-\lambda}, \quad k > \lambda. \qquad (8.3.7)$$

Hidden Time Series Xₜ , n = 240

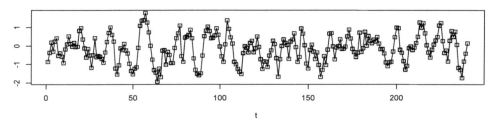

Observed Time Series AₜXₜ , N = 127 , Nmin = 61

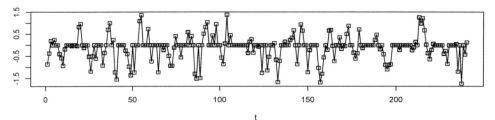

Spectral Density and its Estimates

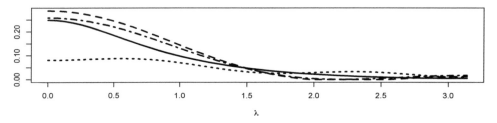

Figure 8.3 *Batch-Bernoulli missing mechanism. Lengths $L_1, L_2, \ldots$ of missing batches are indepen-dent Poisson random variables with mean $\mathbb{E}\{L\} = \lambda$. Otherwise the simulation and structure of diagrams are similar to Figure 8.2. {Parameter λ is controlled by the argument lambda.} [n = 240, sigma = 0.5, a = 0.4, b = 0.5, lambda=0.5, cJ0sp = 2, cJ1sp = 0.5, cTHsp = 4]*

For a batch-Bernoulli missing the relation (8.3.6) still holds and hence the same spectral density E-estimator may be used.

Figure 8.3 illustrates the batch-Bernoulli missing mechanism, and here $\lambda = 0.5$. Apart of the new missing mechanism, the simulation and diagrams are the same as in Figure 8.2. The top diagram shows a realization of the ARMA process. The middle diagram shows us the missing pattern, and it sheds new light on the name of the missing mechanism. We can also note from the title that in this particular simulation, the Poisson batches decreased the number $n = 240$ of hidden observations to just $N = 127$ available observations, in other words, almost a half of hidden underlying observations is missed. Further, the minimal number of available pairs for calculation autocovariance coefficients of the E-estimator is $N_{\min} = 61$, and it is almost a quarter of $n = 240$. Let us also add the information about the underlying $N_0, \ldots, N_5$ that are 127, 71, 66, 62, 61 and 64, respectively. Not surprisingly, these small numbers of pairs, available for calculation of the autocovariance, take a toll on the E-estimate (compare the dashed line of the E-estimate with the dot-dashed line of the oracle's E-estimate based on $n = 240$ hidden realizations of $\{X_t\}$). At the same time, the proposed estimate still nicely exhibits shape of the underlying spectral density and it is dramatically better than the naïve E-estimate (the dotted line) which ignores the missing and treats $\{Y_t\}$ as the time series of interest. It is a good exercise to repeat this figure with different parameters and then analyze patterns in time series and performance of estimators.

8.4   Amplitude-Modulated Missing

The considered problem is to estimate the spectral density $g^X(\lambda)$ of a time series $\{X_t\}$. Observations of the time series are hidden and instead observations of a time series $\{Y_t\} :=$ $\{U_t A_t X_t\}$ are available. Here, similarly to the previous section, $\{A_t\}$ is a stationary Bernoulli time series which defines an underlying missing mechanism, $\{U_t\}$ is a zero-mean and second-order stationary time series which performs amplitude-modulation. Let us make several additional assumptions. Similarly to the previous section we assume that $\mathbb{P}(X_t = 0) = 0$. To separate the effects of A_t and U_t, it is assumed that $\mathbb{P}(U_t = 0) = 0$ and this prevents U_t from affecting the missing and allows us to say that $\{U_t\}$ is a pure amplitude-modulating time series. (Of course, a time series $\{A_t\}$ may be also looked at as an amplitude-modulating time series, and this terminology is well accepted in the literature.) Further, it is assumed that $U_1, U_2, \ldots$ are independent and identically distributed realizations of a random variable U such that

$$\mathbb{P}(U = 0) = 0, \quad \mathbb{E}\{U\} = \mu \neq 0, \quad \mathbb{E}\{U^2\} = \mu_2. \tag{8.4.1}$$

Further, it is assumed that the times series $\{U_t\}$, $\{A_t\}$ and $\{X_t\}$ are mutually independent.

Is it possible to estimate the spectral density $g^X(\lambda)$ of an underlying time series $\{X_t\}$ based on amplitude-modulated observations $Y_t = U_t A_t X_t$, $t = 1, 2, \ldots, n$? To answer this question, let us look at the autocovariance function $\gamma^Y(j)$ of the observed time series $\{Y_t\}$. Introduce a sample mean autocovariance,

$$\hat{\gamma}^Y(j) := (n - j)^{-1} \sum_{l=1}^{n} Y_l = (n - j)^{-1} \sum_{l=1}^{n} A_l U_l X_l. \tag{8.4.2}$$

Because $\{X_t\}$ is zero-mean and it is assumed that the three time series are mutually independent, the observed time series $\{Y_t\}$ is also zero-mean. Then, using (8.4.1) we may write,

$$\mathbb{E}\{\hat{\gamma}^Y(j)\} = \mathbb{E}\{(U_1 A_1 X_1)(U_{1+j} A_{1+j} X_{1+j})\}$$

$$= \mathbb{E}\{U_1 U_{1+j}\} \mathbb{E}\{A_1 A_{1+j}\} \mathbb{E}\{X_1 X_{1+j}\}$$

$$= [\mu_2 I(j = 0) + \mu^2 I(j \neq 0)] \mathbb{E}\{A_1 A_{1+j}\} \gamma^X(j). \tag{8.4.3}$$

Equation (8.4.3) sheds light on a possibility to solve the problem of estimation of the spectral density $g^X(\lambda)$. Namely, if parameters μ_2 and μ are known, then it is possible to estimate the spectral density using the approach of the previous section. Further, if the two parameters are unknown but a sample from U is available, then the two parameters can be estimated by sample mean estimators. Note that in both cases some extra information is needed. On the other hand, (8.4.3) indicates that, based solely on amplitude-modulated data we cannot consistently estimate the spectral density $g^X(\lambda)$. In other words, the considered modification is destructive.

Despite the above-made gloom conclusion, the following approach may be a feasible remedy in some practical situations. Introduce a function $s^X(\lambda)$ which is called the spectral *shape* or the shape of spectral density,

$$s^X(\lambda) := \pi^{-1} \sum_{j=1}^{\infty} \gamma^X(j) \cos(j\lambda). \tag{8.4.4}$$

Note that $g^X(\lambda) = (2\pi)^{-1}\gamma^X(0) + s^X(\lambda)$, and hence in a graphic the spectral shape is just shifted (in vertical direction) spectral density, and the shift is such that the integral of the shape over $[0,\pi]$ is zero. Further, apart of the variance of X_t, the spectral shape provides us with all values of the autocovariance function $\gamma^X(j)$, $j > 0$. Further, in practical applications the spectral density is often of interest in terms of its modes, and then knowing either the spectral density, or the spectral shape, or the scaled spectral shape $s^X(\lambda,\mu)$ defined as

$$s^X(\lambda,\mu) = \mu^2 s^X(\lambda) =: \pi^{-1}\sum_{j=1}^{\infty}(\mu^2\gamma^X(j))\cos(j\lambda) =: \pi^{-1}\sum_{j=1}^{\infty}\eta^X(j)\cos(j\lambda) \qquad (8.4.5)$$

is equivalent. Recall that $\mu := \mathbb{E}\{U\}$ is an unknown parameter (the mean) of the amplitude-modulating distribution.

The above-made remark makes the problem of estimation of the scaled spectral shape of a practical interest. Furthermore, estimation of the scaled shape of spectral density is possible based on the available amplitude-modulated time series $\{Y_t\}$. Indeed, consider the estimator

$$\hat{\eta}^X(j) := \frac{\sum_{l=1}^{n-j}Y_lY_{l+j}}{\sum_{l=1}^{n-j}I(Y_lY_{l+j} \neq 0)}, \quad j \geq 1 \qquad (8.4.6)$$

of coefficients $\eta^X(j)$ in the cosine expansion (8.4.5). Following Section 8.3, set $N_j := \sum_{l=1}^{n-j}A_lA_{l+j}$, $N := N_0$ and $N_{\min} := \min_{\{0 \leq j \leq c_{J0sp}+c_{J1sp}\ln(n)\}} N_j$. Let us explore the expectation of the estimator (8.4.6) given $N_j > 0$. Write,

$$\mathbb{E}\{\hat{\eta}^X(j)|N_j > 0\} = \mathbb{E}\left\{\mathbb{E}\left\{\frac{\sum_{l=1}^{n-j}Y_lY_{l+j}}{\sum_{l=1}^{n-j}I(Y_lY_{l+j} \neq 0)}\Big|A_1,\ldots,A_n,N_j > 0\right\}\right\}$$

$$= \mathbb{E}\left\{\frac{\sum_{l=1}^{n-j}(A_lA_{l+j})\mathbb{E}\{U_lU_{l+j}\}\mathbb{E}\{X_lX_{l+j}\}}{\sum_{l=1}^{n-j}A_lA_{l+j}}\Big|N_j > 0\right\}$$

$$= \mu^2\gamma^X(j) = \eta_j^X, \quad j \geq 1. \qquad (8.4.7)$$

We conclude that whenever $N_{\min} > 0$ the plug-in sample mean estimator (8.4.6) is unbiased for all j used by the E-estimator. Hence, this estimator can be used for construction of the E-estimator $\hat{s}^X(\lambda,\mu)$ of the scaled shape. Note that the parameter μ is still unknown, and hence the underlying spectral shape $s^X(\lambda)$ will be estimated up to an unknown factor $\mu^2 = [\mathbb{E}\{U\}]^2$.

Figure 8.4 illustrates considered problem of amplitude-modulated time series. The top diagram shows a realization of a hidden time series of interest. The observed amplitude-modulated time series is shown in the middle diagram. The used Markov-Bernoulli process $\{A_t\}$ is the same as in Figure 8.2. The process $U_t := \mu U_t^*/\mathbb{E}\{U_t^*\}$ where $U_1^*,\ldots,U_n^*$ are independent random variables generated according to one of our corner densities; here the density is the Normal. The title shows that only $N = 180$ from the underlying $n = 240$ observations are available. This is not too bad, but $N_{\min} = 125$ informs us about a heavy missing. Let us also present information about $N_0,\ldots,N_5$ that are 180, 143, 139, 130, 128 and 125, respectively. We see that the missing implies an almost twofold decrease in the number of available pairs for calculating the autocovariance function. In the bottom diagram the underlying scaled shape $s^X(\lambda,\mu)$ is shown by the solid line, the E-estimate and the oracle's estimate, based on the underlying realization of $\{X_t\}$, are shown by the dashed and dot-dashed lines, respectively, the dotted curve shows the naïve E-estimate based on the available observations $Y_1,\ldots,Y_n$. The E-estimate is close to its oracle's benchmark while the naïve estimate, which ignores the amplitude-modulated nature of the observed time series, is far from being even close to the underlying scaled shape.

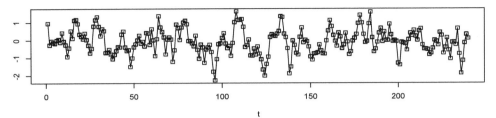

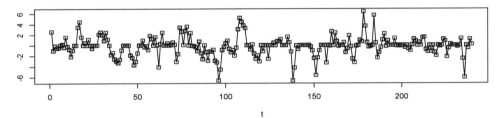

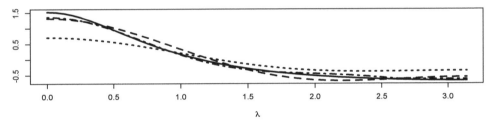

Figure 8.4 *Amplitude-modulated time series and estimation of the scaled shape of the spectral density. Amplitude-modulated observations* $Y_t := U_t A_t X_t$, $t = 1, 2, \ldots, n$ *are generated by a Markov-Bernoulli time series* $\{A_t\}$ *used in Figure 8.2, and* $U_1, U_2, \ldots, U_n$ *is a sample from a scaled corner density whose mean* $\mathbb{E}\{U_t\} = \mu$. *The structure of diagrams is identical to Figure 8.2. For the considered setting only estimation of the scaled shape* $s^X(\lambda, \mu)$ *of the spectral density is possible, and the scaled shape and its estimates are shown in the bottom diagram.* {*Parameter* μ *is controlled by the argument mu, the choice of corner function is controlled by argument corn.*} *[n = 240, sigma = 0.5, a = 0.4, b = 0.5, alpha = 0.4, beta = 0.8, corn = 2, mu = 3, cJ0sp = 2, cJ1sp = 0.5, cTHsp = 4]*

Now let us consider another interesting example of an amplitude-modulated time series. Assume that $Z_1, \ldots, Z_n$ are independent realizations of (sample from) a Poisson random variable Z with the mean λ, and define the observed time series as $\{Y_t\} := \{Z_t X_t\}$. In this example, variables Z_t create both the missing (when $Z_t = 0$) and the amplitude-modulation when $Z_t > 0$. The example is of interest because it happens in practical applications and because, as we will see shortly, the Poisson amplitude-modulation is not destructive.

Let us explain a proposed E-estimation of the spectral density $g^X(\lambda)$. Note that $\mathbb{E}\{I(Z_t = 0)\} = \mathbb{P}(Z_t = 0) = e^{-\lambda}$, and hence the parameter λ can be estimated by the plug-in sample mean estimator

$$\hat{\lambda} := -\ln(\sum_{l=1}^{n} I(Y_l = 0)/n). \tag{8.4.8}$$

Hidden Time Series X$_t$, n = 240

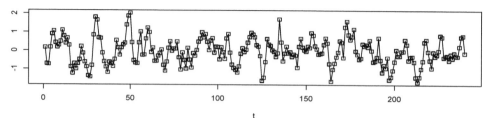

Observed Time Series Z$_t$X$_t$, N = 193, Nmin = 149

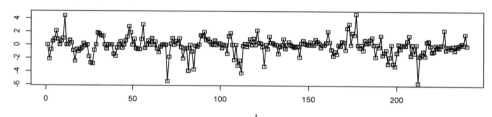

Spectral Density and its Estimates

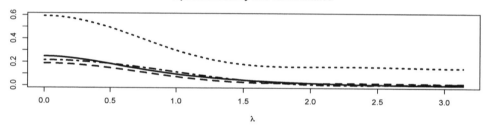

Figure 8.5 *Amplitude-modulation by Poisson variable. Observed time series is $Y_t = Z_t X_t$, $t = 1, 2, \ldots, n$ where $\{X_t\}$ is the ARMA process defined in the caption of Figure 8.2 and $Z_1, Z_2, \ldots, Z_n$ is a sample from a Poisson random variable Z with the mean $\mathbb{E}\{Z\} = \lambda$. The structure of diagrams is similar to those in Figure 8.2. {The argument lambda controls λ.} [n = 240, sigma = 0.5, a = 0.4, b = 0.5, lambda = 1.5, cJ0sp = 2, cJ1sp = 0.5, cTHsp = 4]*

Further, we can write

$$\gamma^Y(j) = [(\lambda + \lambda^2)I(j = 0) + \lambda^2 I(j \neq 0)]\gamma^X(j), \quad j = 0, 1, \ldots \qquad (8.4.9)$$

This allows us to propose an estimator of the autocovariance function

$$\hat{\gamma}^X(j) := [(\hat{\lambda} + \hat{\lambda}^2)I(j = 0) + \hat{\lambda}^2 I(j \neq 0)]^{-1}\hat{\gamma}^Y(j), \quad j = 0, 1, \ldots \qquad (8.4.10)$$

In its turn, the autocovariance estimator yields the E-estimator of the spectral density.

Figure 8.5, whose structure is similar to Figure 8.2, illustrates both the setting and the E-estimator. The top diagram shows the underlying time series of interest. The middle diagram shows the observed amplitude-modulated time series, as well as the number $N := \sum_{l=1}^n I(Z_l X_l \neq 0) = 193$ of available observations, and $N_{\min} = 149$. Also, let us present numbers $N_0, \ldots, N_5$ of available pairs 193, 151, 157, 150, 149 and 153, respectively. Returning to the middle diagram, please note how visually different are realizations of the amplitude-modulated time series and the underlying one. Further, note the difference in their scales. It is difficult to believe that it is possible to restore the underlying spectral density of the hidden time series $\{X_t\}$ after its amplitude-modulation by the Poisson variable. Nonetheless, the bottom diagram shows that the spectral density E-estimate is close to the oracle's estimate based on the hidden time series, and overall it is very good. On the other hand, the naïve spectral density E-estimate of Section 8.2, based on observations $Z_1 X_1, \ldots, Z_n X_n$, is clearly poor. Note that it indicates a dramatically larger overall power of the observed time series, especially on low frequencies.

The overall conclusion is that it is absolutely prudent to pay attention to a possible modification of an underlying time series of interest.

8.5   Censored Time Series

The aim is to explore the problem of estimation of the spectral density for a time series modified by a right censoring. An underlying zero-mean and stationary time series of interest $\{X_t\}$ is hidden and not available for a statistical inference. Instead, a bivariate time series $\{V_t, \Delta_t\}$ is observed where $V_t := \min(X_t, C_t)$, $\Delta_t := I(X_t \leq C_t)$, and $\{C_t\}$ is a stationary censoring time series. It what follows it is assumed that all random variables are continuous, their distributions are unknown, time series $\{X_t\}$ and $\{C_t\}$ are independent, and $\beta_{X_t} \leq \beta_{C_t}$ (recall that β_Z denotes the upper bound of the support of Z). Recall that the right censoring (RC) was discussed in great details in Chapter 6, only here we no longer assume that the variables are nonnegative because we are dealing with zero-mean time series.

To explore the problem of estimation of the spectral density $g^X(\lambda)$, we begin with a formula for a joint density of two pairs of observed variables,

$$f^{V_1, V_{1+j}, \Delta_1, \Delta_{1+j}}(v_1, v_{1+j}, 1, 1) = f^{X_1, X_{1+j}}(v_1, v_{1+j}) G^{C_1, C_{1+j}}(v_1, v_{1+j}), \qquad (8.5.1)$$

where $G^{C_1, C_{1+j}}(v_1, v_{1+j}) = \mathbb{P}(C_1 > v_1, C_{1+j} > v_{1+j})$ is the bivariate survival function. This formula allows us to write,

$$\mathbb{E}\{V_1 V_{1+j} \Delta_1 \Delta_{1+j}\}$$

$$= \int_{-\infty}^{\infty} \int_{-\infty}^{\infty} v_1 v_{1+j} f^{X_1, X_{1+j}}(v_1, v_{1+j}) G^{C_1, C_{1+j}}(v_1, v_{1+j}) dv_1 dv_{1+j}. \qquad (8.5.2)$$

This relation points upon a possible solution. Recall that $\{X_t\}$ is a zero-mean and stationary time series. Using this assumption, together with (8.5.1), allows us to write down the autocovariance function of interest as

$$\gamma^X(j) := \mathbb{E}\{X_1 X_{1+j}\}$$

$$= \int_{-\infty}^{\infty} \int_{-\infty}^{\infty} v_1 v_{1+j} f^{X_1, X_{1+j}}(v_1, v_{1+j}) dv_1 dv_{1+j}$$

$$= \mathbb{E}\left\{ \frac{V_1 V_{1+j} \Delta_1 \Delta_{1+j}}{G^{C_1, C_{1+j}}(V_1, V_{1+j})} \right\}. \qquad (8.5.3)$$

In its turn, (8.5.3) implies that if the bivariate survival function $G^{C_1, C_{1+j}}(v, u)$ of the censoring time series is known (for instance from previous or extra studies), then the sample mean estimator of the autocovariance function $\gamma^X(j)$ is

$$\tilde{\gamma}^X(j) := (n-j)^{-1} \sum_{l=1}^{n-j} \frac{V_l V_{j+j} \Delta_l \Delta_{l+j}}{G^{C_1, C_{1+j}}(V_l, V_{l+j})}. \qquad (8.5.4)$$

Note that the estimator is unbiased.

If the bivariate survival function is unknown, then we need to estimate it. Here we restrict our attention to the case when $C_1, C_2, \ldots, C_n$ are independent and identically distributed realizations of a random variable C. To propose an estimator, we note that

$$G^{V_t}(v) = G^{X_t}(v)G^C(v), \qquad (8.5.5)$$

where $G^Z(z) := \mathbb{P}(Z > z)$ denotes the survival function of Z. This allows us to write

$$f^{V_t, \Delta_t}(v, 0) = f^C(v)G^{X_t}(v) = \frac{f^C(v)G^{V_t}(v)}{G^C(v)}. \qquad (8.5.6)$$

Next, recall that the hazard rate function of the censoring variable is defined as

$$h^C(v) := \frac{f^C(v)}{G^C(v)}. \qquad (8.5.7)$$

Using this relation in (8.5.6) yields that

$$f^{V_t, \Delta_t}(v, 0) = h^C(v)G^{V_t}(v). \qquad (8.5.8)$$

In its turn, (8.5.8) implies that

$$h^C(v) = \frac{f^{V_t, \Delta_t}(v, 0)}{G^{V_t}(v)} \qquad (8.5.9)$$

whenever $G^{V_t}(v) > 0$.

The numerator and denominator on the right side of (8.5.9) can be consistently estimated because these are characteristics of observed random variables, then the ratio of the estimators may be used to estimate the hazard rate. Recall that the hazard rate is a characteristic of the distribution, and hence an estimator of the hazard rate can be used as a plug-in for a corresponding estimator of the survival function. There is a simple relation between a survival function and the corresponding cumulative hazard that will be used shortly.

The cumulative hazard of the censoring variable is defined as

$$H^C(v) := \int_{-\infty}^{v} h^C(y)dy. \qquad (8.5.10)$$

Using (8.5.9) allows us to continue (8.5.10),

$$H^C(v) := \int_{-\infty}^{v} \frac{f^{V_t, \Delta_t}(y, 0)}{G^{V_t}(y)}dy$$

$$= \mathbb{E}\{(1 - \Delta_t)I(V_t \le v)[G^{V_t}(V_t)]^{-1}\}. \qquad (8.5.11)$$

To use this expectation for constructing a sample mean estimator, we need to estimate the survival function of V_t. Note that the time series $\{V_t\}$ is stationary, and then the following sample mean estimator is a natural choice,

$$\hat{G}^{V_t}(v) := n^{-1} \sum_{l=1}^{n} I(V_l \ge v). \qquad (8.5.12)$$

Using this estimator in (8.5.11) implies the sample mean estimator of the cumulative hazard $H^C(v)$,

$$\hat{H}^C(v) := n^{-1} \sum_{l=1}^{n} \frac{(1 - \Delta_l)I(V_l \le v)}{\hat{G}^{V_l}(V_l)}. \qquad (8.5.13)$$

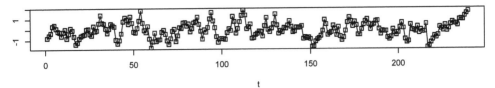

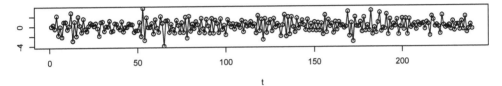

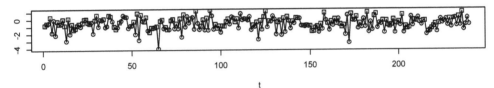

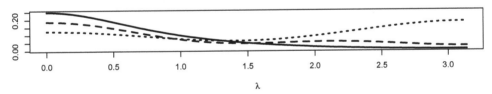

Figure 8.6 *Censored time series. The underlying time series $\{X_t\}$ is generated by the same Gaussian ARMA process as in Figure 8.2. The censoring time series $\{C_t\}$ is generated by a sample from a Gaussian variable with zero mean and standard deviation σ_C. In the third from the top diagram squares and circles show cases with uncensored and censored X_t, respectively. $N := \sum_{l=1}^{n} \Delta_l$ is the number of uncensored observations. In the bottom diagram the underlying spectral density $g^X(\lambda)$, its E-estimate and naïve E-estimate are shown by the solid, dashed and dotted lines, respectively. {Parameter σ_C of the Gaussian censoring time series is controlled by the argument sigmaC.} [n = 240, sigma = 0.5, a = 0.4, b = 0.5, sigmaC = 1, cJ0sp = 2, cJ1sp = 0.5, cTHsp = 4]*

As soon as the cumulative hazard is estimated, we use the familiar from Chapter 4 relation

$$G^C(v) = \exp(-H^C(v)) \tag{8.5.14}$$

to estimate the survival function of C by the plug-in estimator,

$$\hat{G}^C(v) = \exp(-\hat{H}^C(v)). \tag{8.5.15}$$

Note that in the considered case $G^{C_1, C_{1+j}}(v, u) = G^C(u)G^C(v)$, and hence estimator (8.5.15) may be used in (8.5.4) whenever the bivariate survival function is unknown.

Figure 8.6 sheds light on the setting and the proposed estimation. The top diagram shows us by squares a realization of an underlying time series $\{X_t\}$. It is generated by the same Gaussian ARMA process as in Figure 8.2. The second from the top diagram shows us by circles the censoring time series which is generated by a white Gaussian process with the unit standard deviation. Note that $\{X_t\}$ and $\{C_t\}$ have a different behavior in terms of oscillations and they have different shapes of the spectral densities. The censored time series is shown in the third from the top diagram. Here we see $V_t = \min(X_t, C_t)$, $t = 1, \ldots, n$, and if $\Delta_t = 1$ then the value of V_t is shown by the square and if $\Delta = 0$ then by the circle. In other words, uncensored observations of $\{X_t\}$ are shown by squares (and you can also see them in the top diagram) while the censored observations are hidden and instead you can see the corresponding values of $\{C_t\}$ shown by circles (they may be also observed in the second from the top diagram). Note that only $N = 120$ observations are uncensored, and this is a severe loss of information. The third diagram contains all available observations that can be used in statistical analysis. A visual analysis clearly indicates that the dynamic and spectral patterns of time series $\{V_t\}$ and $\{X_t\}$ are different. The bottom diagram shows us the underlying spectral density $g^X(\lambda)$ (the solid line) and the proposed estimate (the dashed line). The dotted line shows the spectral density E-estimate of the time series $\{V_t\}$, which is a naïve estimate that ignores the censoring.

Censored time series is a complicated stochastic modification when underlying observations are missing not at random and then substituted by values of a censoring time series. Repeated simulations, generated by Figure 8.6, may help in understanding the modification and how the proposed estimation performs.

8.6   Probability Density Estimation

Density estimation for the case of a sample $X_1, X_2, \ldots, X_n$ from a continuous random variable X, when the observations are independent, was considered in Section 2.2. In this section we would like to consider the density estimation problem for the case of a stationary time series $\{X_t\}$ of continuous variables.

Let us first briefly recall the idea of E-estimation presented in Section 2.2 for the case of a sample of independent observations. It is assumed that the density $f^X(x)$ is supported and square-integrable on $[0, 1]$, and we use the cosine basis $\varphi_0(x) := 1$, $\varphi_j(x) = 2^{1/2} \cos(\pi j x)$, $j = 1, 2, \ldots$ on $[0, 1]$. Then the density can be written as a Fourier expansion,

$$f^X(x) = \sum_{j=0}^{\infty} \theta_j \varphi_j(x), \tag{8.6.1}$$

where

$$\theta_j := \int_0^1 \varphi_j(x) f^X(x) dx \tag{8.6.2}$$

are Fourier coefficients of $f^X(x)$. The idea of E-estimation is based on the fact that the right side of (8.6.2) can be written as the expectation, namely

$$\theta_j = \mathbb{E}\{\varphi_j(X)\}. \tag{8.6.3}$$

The expectation immediately implies the following sample mean Fourier estimator,

$$\hat{\theta}_j := n^{-1} \sum_{l=1}^{n} \varphi_j(X_l). \tag{8.6.4}$$

The Fourier estimator is unbiased and

$$\mathbb{V}(\hat{\theta}_j) = n^{-1}(1 + o_j(1)). \tag{8.6.5}$$

Using this Fourier estimator by the E-estimator of Section 2.2 yields the wished density estimator. Further, Section 2.2 shows via simulations a good performance of the E-estimator for relatively small samples. In short, E-estimation is based on a good sample mean Fourier estimator.

Now let us relax the assumption about independence and consider a realization $X_1, X_2, \ldots, X_n$ of a stationary time series $\{X_t\}$. The problem is again to estimate the density $f^X(x)$, and note that now this density is marginal with respect to the joint density $f^{X_1, \ldots, X_n}(x_1, \ldots, x_n)$. Further, let us relax another assumption that the support of the density is known. Further, what will be if some of the dependent observations are missed? These are the issues that we would like to address.

First, let us relax the assumption about independence while still considering densities supported on $[0, 1]$. Let us carefully look at the Fourier estimator (8.6.4). For a stationary time series the estimator is still a sample mean estimator and unbiased. Indeed, we know that the expectation of a sum of random variables is always a sum of expectations, regardless of dependence between the random variables, and hence even if $X_1, X_2, \ldots, X_n$ are dependent we get

$$\mathbb{E}\{\hat{\theta}_j\} = \mathbb{E}\{n^{-1}\sum_{l=1}^{n}\varphi_j(X_l)\} = n^{-1}\sum_{l=1}^{n}\mathbb{E}\{\varphi_j(X_l)\}. \tag{8.6.6}$$

Using the assumed stationarity of the time series we can continue,

$$\mathbb{E}\{\hat{\theta}_j\} = n^{-1}n\int_0^1 f^X(x)\varphi_j(x)dx = \theta_j. \tag{8.6.7}$$

This proves that Fourier estimator (8.6.4) is unbiased. Further, the proof of (8.6.7) is based only on the property that the expectation $\mathbb{E}\{\varphi_j(X_t)$ does not depend on t, and this is a weaker assumption than the stationarity.

Next we need to evaluate the variance of the estimator (8.6.4). Write,

$$\mathbb{V}(\hat{\theta}_j) = \mathbb{E}\{[n^{-1}\sum_{l=1}^{n}\varphi_j(X_l) - \theta_j]^2\} = \mathbb{E}\{[n^{-1}\sum_{l=1}^{n}(\varphi_j(X_l) - \theta_j)]^2\}$$

$$= n^{-2}\sum_{l,t=1}^{n}\mathbb{E}\{(\varphi_j(X_l) - \theta_j)(\varphi_j(X_t) - \theta_j)\} = n^{-2}\sum_{l,t=1}^{n}\mathbb{E}\{\varphi_j(X_l)\varphi_j(X_t)\} - \theta_j^2. \tag{8.6.8}$$

In the last equality we used (8.6.7) and the stationarity.

Consider the expectation on the right side of (8.6.8). Using stationarity of $\{X_t\}$ we get

$$\mathbb{E}\{\varphi_j(X_l)\varphi_j(X_t)\} = \mathbb{E}\{\varphi_j(X_0)\varphi_j(X_{|t-l|})\}. \tag{8.6.9}$$

Using this result and (8.2.14) in (8.6.8) yields,

$$\mathbb{V}(\hat{\theta}_j) = n^{-2}\sum_{l=-n}^{n}(n - |l|)\mathbb{E}\{\varphi_j(X_0)\varphi_j(X_l)\} - \theta_j^2. \tag{8.6.10}$$

The expression (8.6.10) allows us to make the following important conclusion. If the following sum converges,

$$\sum_{l=-n}^{n}|\mathbb{E}\{\varphi_j(X_0)\varphi_j(X_l)\}| < c^* < \infty, \tag{8.6.11}$$

then the variance of $\hat{\theta}_j$ decreases with the parametric rate n^{-1} despite the dependence between the observations, namely

$$\mathbb{V}(\hat{\theta}_j) < c^* n^{-1}. \tag{8.6.12}$$

Otherwise the dependence may slow down the rate of the variance decrease and hence make the estimation less accurate.

As an example, consider the case of a stationary series $\{X_t\}$ with the mixing coefficient $\alpha^X(s)$ defined in (8.1.10). Then using (8.1.13) we get

$$|\mathbb{E}\{\varphi_j(X_0)\varphi_j(X_s)\} - \theta_j^2| \le 8\alpha^X(s). \tag{8.6.13}$$

Using this relation we conclude that if the mixing coefficients are summable, that is

$$\sum_{s=0}^{\infty} \alpha^X(s) < \infty, \tag{8.6.14}$$

then (8.6.11) holds and we get the classical rate n^{-1} for the variance. If (8.6.14) holds then we are dealing with the case of *weak* dependency (*short memory*). For instance, if $\alpha^X(s)$ is proportional to $s^{-\beta}$ with $\beta > 1$ then this is the case of weak dependency. Another classical case of weak dependency is a Gaussian ARMA time series where mixing coefficients decrease exponentially. On the other hand, if $\beta < 1$, and an example will be presented shortly, the variance convergence slows down to $n^{-\beta}$. In the latter case the dependence is called strong and we may say that we are dealing with variables having *long memory* of order β.

Now we are explaining our approach for the case when the support of X is unknown and may be a real line like in the case of a Gaussian stationary time series. Denote by $X_{(1)}$ and $X_{(n)}$ the smallest and largest observations (recall that this is our traditional notation for ordered observations), define rescaled on $[0,1]$ observations $Y_l := (X_l - X_{(1)})/(X_{(n)} - (X_{(1)}))$, construct for the rescaled observations the density E-estimator $\hat{f}^Y(y)$, and then define the the rescaled back density E-estimator

$$\hat{f}^X(x) := \frac{\hat{f}^Y((x - X_{(1)})/(X_{(n)} - X_{(1)}))}{X_{(n)} - X_{(1)}} I(x \in [X_{(1)}, X_{(n)}]). \tag{8.6.15}$$

Let us explain the underlying idea of (8.6.15). Consider a variable Y' with density $f^{Y'}(y)$ supported on $[0,1]$. Introduce a new variable $X' := a + bY'$ which is a scale-location transformation of Y'. Then the cumulative distribution function of X' is

$$F^{X'}(x) = \mathbb{P}(X' \le x) = \mathbb{P}(Y' \le (x - a)/b)$$

$$= F^{Y'}((x - a)/b) = \int_0^{(x-a)/b} f^{Y'}(y) dy. \tag{8.6.16}$$

Note that the support of X' is the interval $[a, a+b]$, and hence via differentiation of (8.6.16) we get

$$f^{X'}(x) = b^{-1} f^{Y'}((x - a)/b) I(x \in [a, a + b]). \tag{8.6.17}$$

This is a classical formula for a scale-location transformation, and it motivated (8.6.15).

Now let us get a feeling of the effect of dependency on estimation of the density via simulated examples. Figure 8.7 allows us to understand performance of the E-estimator and complexity of the problem for the case of short-memory processes. The figure allows us to consider two ARMA(1,1) processes with different parameters. In the left column we consider the case of a highly oscillated zero-mean Gaussian ARMA(1,1) process $\{X_t\}$. In the right column we are also considering the case of a zero-mean Gaussian ARMA(1,1)

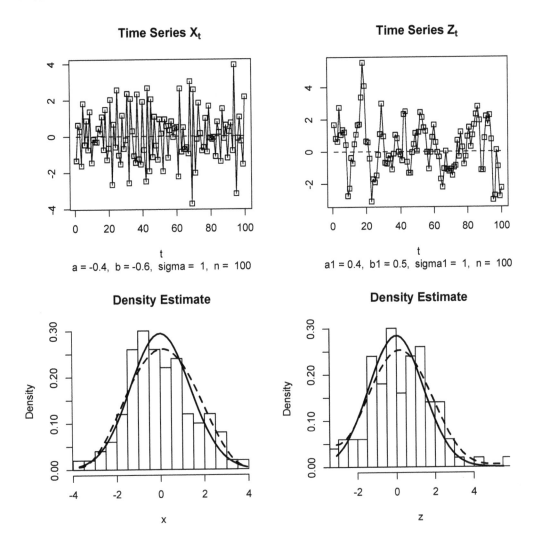

Figure 8.7 *Density estimation for short-memory stationary time series. Realizations of time series* $\{X_t\}$ *and* $\{Z_t\}$ *are generated by Gaussian ARMA(1,1) processes whose parameters are shown in the subtitles. Note that they are similar to those in Figures 8.1 and 8.2, respectively. The histograms are overlaid by the underlying density (the solid line) and the E-estimate (the dashed line) shown over the estimated support.* {*Parameters a, b and sigma control* $\{X_t\}$ *while a1, b1 and sigma1 control* $\{Z_t\}$.} [n = 100, sigma = 1, a = -0.4, b = -0.6, sigma1 = 1, a1 = 0.4, b1 = 0.5 , cJ0 = 4, cJ1 = 0.5, cTH = 4]

process $\{Z_t\}$ only with parameters implying slower oscillations. Note that the horizontal dashed line in the top diagrams helps us to visualize oscillations around zero. Both $\{X_t\}$ and $\{Z_t\}$ are weak-dependent and short-memory processes with exponentially decreasing mixing coefficients, but they have different shapes of spectral densities, see Figures 8.1 and 8.2, respectively. For each process, corresponding bottom diagrams shows us the histogram of available observations as well as the underlying Gaussian density and its E-estimate. Let us analyze and compare these diagrams. First of all, the left histogram is clearly more symmetric about zero than the right one. The right histogram is skewed to the right, and this reflects the fact that the observed series $\{Z_t\}$ spends more time above zero than below. This is because $\{Z_t\}$ has more power on lower frequencies and it requires more time (larger n) to exhibit its stationarity and zero-mean property. In terms of modes, the E-estimates indicate a pronounced single mode, but again the more frequently oscillating $\{X_t\}$ yields the better shape of the E-estimate. Finally, let us look at the effect of using the empirical support (note that this is the same support used by the famous empirical cumulative distribution function $\hat{F}^X(x) := n^{-1}\sum_{l=1}^{n} I(X_l \le x)$). Of course, a Gaussian variable is supported on a real line, and here we used a finite empirical support. This did a good job for $\{X_t\}$ and a reasonable one for $\{Z_t\}$.

Now let us consider an even more extreme case of dependence, a long-memory time series. The top diagram in Figure 8.8 exhibits a zero-mean Gaussian time series with long memory of order $\beta = 0.4$. It looks like the above-discussed time series $\{Z_t\}$ on steroids. Note that it begins above the zero and rarely goes into negative territory (the horizontal dashed line helps us to see this). Of course, eventually the series will stay negative a long time, but much larger samples are needed to see this. The reader is advised to generate more series and get used to processes with long-memory. In the second (from the top) diagram we see the histogram of available observations overlaid by the underlying density (the solid line) and its E-estimate (the dashed line) shown over the range of observations. The data is clearly skewed to the right, the estimated support is skewed to the right, and there is nothing that can be done about this. We simply need more observations to correctly estimate the density, and this is an important lesson to learn.

What will be if some observations are missed? How does missing, coupled with dependency, affect E-estimation of the density? First, let us answer these natural questions analytically. Consider a time series $\{A_t X_t\}$ where the availabilities A_t are independent Bernoulli random variables and $\mathbb{P}(A_t = 1|\{X_t\}) = w$, $w \in (0,1]$. Additionally, we are assuming that $\mathbb{P}(X_t = 0) = 0$ and hence $\mathbb{P}(A_t = I(A_t X_t \ne 0)) = 1$. As a result, even if we do not directly observe a particular A_t, we do know it from the available time series $\{A_t X_t\}$.

We are dealing with MCAR (missing completely at random) and hence it is natural to try a complete-case approach which yields the following Fourier estimator (compare with (8.6.4)),

$$\tilde{\theta}_j := \frac{\sum_{l=1}^{n} I(A_l X_l \ne 0)\varphi_j(A_l X_l)}{\sum_{l=1}^{n} I(A_l X_l \ne 0)}. \tag{8.6.18}$$

Let us check that the Fourier estimator is unbiased given $N := \sum_{l=1}^{n} A_l > 0$. Using the rule of calculation of an expectation via the expectation of a conditional expectation, together with $\mathbb{E}\{\varphi_j(X_l)\} = \theta_j$, we get

$$\mathbb{E}\{\tilde{\theta}_j|N > 0\} = \mathbb{E}\left\{\mathbb{E}\left\{\frac{\sum_{l=1}^{n} I(A_l X_l \ne 0)\varphi_j(A_l X_l)}{\sum_{l=1}^{n} I(A_l X_l \ne 0)}\Big|\{A_t\}, N > 0\right\}\Big|N > 0\right\}$$

$$= \mathbb{E}\left\{\frac{\sum_{l=1}^{n} A_l \theta_j}{\sum_{l=1}^{n} A_l}\Big|N > 0\right\} = \theta_j. \tag{8.6.19}$$

This is a pivotal result which yields unbiasedness of the sample mean estimator for the case of missing dependent observations. Hence we again may use our density E-estimator based on complete cases of time series $\{A_t X_t\}$.

Let us look at a simulation of a long-memory time series with missing observations. The third (from the top) diagram in Figure 8.8 shows us such a realization, and note that the time series of interest is shown in the top diagram. Clearly the density of available observations will be skewed to the right because there are only several negative observations. Further, note that we have dramatically less, just $N = 58$, available realizations of the long-memory time series. How will the E-estimator perform under these circumstances? The bottom diagram shows us the underlying density and the E-estimate. Yes, the estimate is skewed, and support is shown incorrectly, but overall this estimate is on par with its benchmark

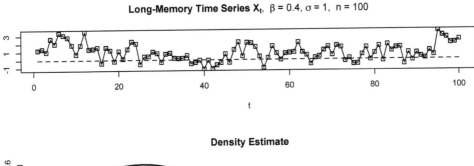

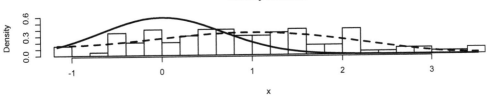

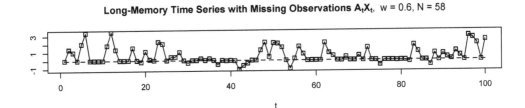

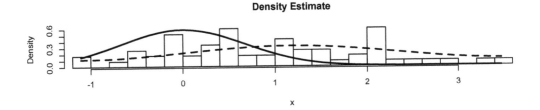

Figure 8.8 *Density estimation for long-memory of order β zero-mean Gaussian time series $\{X_t\}$ without and with missing observations. The missing is created by a time series $\{A_t\}$ of independent Bernoulli variables with $\mathbb{P}(A = 1) = w$, and then the available time series is $\{A_t X_t\}$. The title of the third diagram shows w and the number $N := \sum_{l=1}^{n} A_l$ of available observations. In the second and fourth diagrams the histograms are based on data in the first and third diagrams, respectively. The histograms are overlaid by solid and dashed lines showing the underlying density and its E-estimate over the range of available observations. {Parameter β is controlled by argument beta. Standard deviation of the time series is controlled by argument sigma.} [n = 100, sigma = 1, beta = 0.4, w = 0.6, cJ0 = 4, cJ1 = 0.5, cTH = 4]*

in the second (from the top) diagram. Despite all the complications, we do see a unimodal and surprisingly symmetric shape of the E-estimate. Note that the missing makes available observations less dependent, and then the main complication is the smaller number N of available observations.

A conclusion is that dependency in observations should not be taken lightly, and typically larger sample size is the remedy. Further, while short-memory dependence rarely produces a dramatic effect on statistical estimation or inference, a long-memory dependence, coupled with a relatively small sample size, may be destructive. Working with and repeating Figures 8.7 and 8.8 will help the reader to understand and appreciate the dependency.

8.7 Nonparametric Autoregression

Consider a stationary time series $\{X_t\}$ generated according to the model

$$X_t = m(X_{t-1}) + \sigma(X_{t-1})W_t, \tag{8.7.1}$$

where $m(x)$ and $\sigma(x) > 0$ are smooth functions, and $\{W_t\}$ is a standard white noise, that is a time series of independent and identically distributed variables with zero mean and unit variance.

The time series (8.7.1) has many applications, interpretations, and it is known under different names. If in (8.7.1) $m(X_t) = aX_t$ and $\sigma(X_t) = \sigma$ then, according to (8.1.7), the process becomes AR(1) autoregression. This explains the name *nonparametric autoregression* for (8.7.1). Further, consider the problem of prediction (forecasting) X_t given X_{t-1}. For instance, we would like to predict the temperature for tomorrow based on the temperature today, or predict a stock return for tomorrow, etc. Then the best predictor, that minimizes the mean squared error, is $m(x) := \mathbb{E}\{X_t|X_{t-1} = x\}$ and it may be referred to as a nonlinear predictor or nonparametric regression. Further, in the theory of dynamic models, the equation is called a nonlinear dynamic model, X_t is called a state of the model, $m(x)$ is called an iterative map, and $s(x)$ is called a scale map. Note that if $\sigma(x) = 0$, then $X_t = m(X_{t-1})$ and a current state of this dynamic model is defined solely by its previous state (the states are iterated). This explains the name iterative map of $m(x)$.

Let us explain how we may estimate the autoregression function $m(x)$ in model (8.7.1). Set $Z_t := X_{t-1}$ and rewrite (8.7.1) as

$$X_t := m(Z_t) + \sigma(Z_t)W_t, \ \ t = 1, 2, \ldots, n. \tag{8.7.2}$$

Let us look at (8.7.2) more closely. First, we observe $n - 1$ pairs $(X_2, Z_2), (X_3, Z_3), \ldots, (X_n, Z_n)$. Second, we can write using independence of $Z_t := X_{t-1}$ and W_t, together with the zero mean property of W_t,

$$\mathbb{E}\{X_t|Z_t = z\} = m(z) + \sigma(z)\mathbb{E}\{W_t\} = m(z). \tag{8.7.3}$$

We may conclude that (8.7.2) is a regression problem with Z_t being the predictor and X_t being the response. Hence we can use the regression E-estimator of Section 2.3 to estimate $m(z)$ and the scale E-estimator of Section 3.6 to estimate $\sigma(z)$.

Further, we can generalize model (8.7.1) and consider

$$X_t = m(X_{t-1}) + \sigma(X_{t-1})U_t, \tag{8.7.4}$$

where $\{U_t\}$ is a stationary and unit-variance time series satisfying $\mathbb{E}\{U_t|X_{t-1}\} = 0$. Then $\mathbb{E}\{X_t|X_{t-1} = x\} = m(x)$ and we again may use the regression and scale E-estimators.

Figure 8.9 presents the proposed statistical analysis of the nonparametric autoregression (8.7.4) with $\{U_t\}$ being a Gaussian ARMA(1,1) process; this process allows us to test robustness of the E-estimator to the assumption $\mathbb{E}\{U_t|X_{t-1}\} = 0$. The underlying simulation is explained in the caption. Diagram 1 shows us a particular realization. It is difficult to gain anything feasible from its visualization, and it looks like there is nothing special in this highly oscillated time series. Keeping in mind that the autoregression is often used to model a stock price over a short period of time, it is easy to understand why trading stocks is a complicated issue. Diagram 2 shows us the scattergram of pairs (X_{t-1}, X_t), it sheds light on the underlying iterative process of autoregression, and look at how inhomogeneous

1. Nonparameteric Autoregression, n = 240

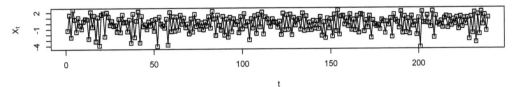

2. Scattergram of X_t Versus X_{t-1}

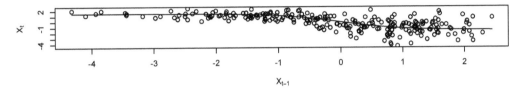

3. Estimate of m(x)

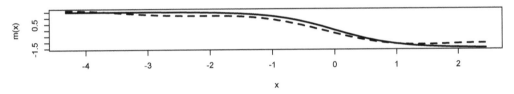

4. Estimate of Scale

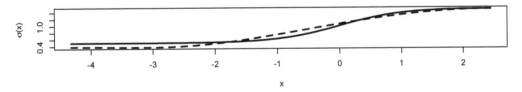

Figure 8.9 *Nonparametric autoregression. The simulation uses model (8.7.4) where* $m(x) := C \exp(\lambda_m x)/(1 + \exp(\lambda_m x))$, $\sigma(x) := a_1 + b_1 \exp(\lambda_s x)/(1 + \exp(\lambda_s x))$, *and* $\{U_t\}$ *is a Gaussian ARMA(1, 1) time series with parameters* (a, b, σ). *In second and third diagrams the solid line is the underlying function* $m(x)$, *and in Diagram 4 the solid line is the underlying scale* $\sigma(x)$. *The dashed lines show E-estimates.* {*Parameters* λ_m *and* λ_s *are controlled by arguments lambdam and lambdas, respectively.*} [n = 240, a = -0.3, b = 0.6, sigma = 1, a1 = 0.5, b1 = 1, lambdam = -2, lambdas = 2, C = 3, cJ0 = 4, cJ1= 0.5, cTH = 4]

the scattergram is. The solid line is the underlying autoregression function $m(x)$, and it definitely sheds light on the data (of course, for real data this line would not be available). The interesting and rather typical feature is a small number of observations in the tails. Another interesting feature of the data is that the variability of observations depends on the predictor X_{t-1} and it is clearly larger for positive predictors.

Now an important remark is due. In a nonparametric autoregression the role of noise $\{U_t\}$ is absolutely crucial because it forces $\{X_t\}$ to have a sufficiently large range of values which, in its turn, allows us to estimate the autoregression function. To appreciate the role

1. Nonparameteric Autoregression with Missing Data, n = 240 , N = 175

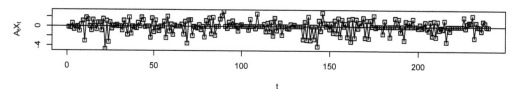

t

2. Scattergram of Available X$_t$ Versus Available X$_{t-1}$, M = 135

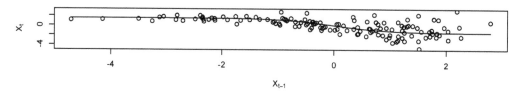

X$_{t-1}$

3. Estimate of m(x)

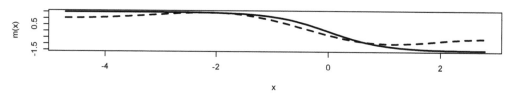

x

4. Estimate of Scale

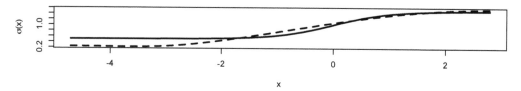

x

Figure 8.10 *Nonparametric autoregression with Markov-Bernoulli missing. Markov chain* $\{A_t\}$ *is generated according to transition probabilities* $\mathbb{P}(A_{t+1} = 0|A_t = 0) = \alpha$ *and* $\mathbb{P}(A_{t+1} = 1|A_t = 1) = \beta$. *Diagram 1 shows the observed time series* $\{A_t X_t\}$ *where* $\{X_t\}$ *is generated as in Figure 8.9.* $N := \sum_{l=1}^{n} A_l$ *is the number of available observations of* X_t *while* $M := \sum_{l=2}^{n} A_{l-1} A_l$ *is the number of available pairs of observations; these statistics are shown in the titles.* [n = 240, alpha = 0.4, beta = 0.8, a = -0.3, b = 0.6, sigma = 1, a1 = 0.5, b1 = 1, lambdam = -2, lambdas = 2, C = 3, cJ0 = 4, cJ1 = 0.5, cTH = 4]

of the noise, just set it to zero and then check the outcome theoretically and using Figure 8.9 (in the figure set $a1 = 0$ and $b1 = 0$).

Diagram 3 shows us the estimated autoregression function, and the E-estimate is good. Diagram 4 shows us the E-estimate of the scale function, and it is also good. Note that its left tail is smaller than the underlying scale but you can check diagram 2 and conclude that the data do support opinion of the E-estimate.

Now let us consider the same model only when some observations are missing according to a Markov-Bernoulli time series $\{A_t\}$ discussed in Section 8.3. This model is illustrated in Figure 8.10 whose caption explains the simulation and notation. The underlying process

$\{X_t\}$ is the same as in Figure 8.9. Diagram 1 shows us realization of the available time series $\{A_t X_t\}$ where only $N = \sum_{l=1}^{n} A_l = 175$ from $n = 240$ observations are available. Now note that E-estimation is based on pairs (X_{l-1}, X_l), $l = 2, \ldots, n$ and these pairs are available only if $A_{l-1} A_l = 1$. The available pairs are shown in diagram 2, and the number M of available pairs is 135. Note that we lost almost a half of underlying pairs (compare with Diagram 2 in Figure 8.9). This loss definitely has affected estimation of the autoregression function (Diagram 3) and the scale function (Diagram 4). At the same time, with the help of Diagram 2 we may conclude that the estimates reflect the data. Indeed, let us look at the left tail in Diagram 2. Clearly all observations are below the solid line (the underlying autoregression), and this is what the E-estimate in diagram 3 tells us. Further, note that observations of X_t in the left tail of Diagram 2 exhibit a minuscular variability, and this is correctly reflected by the E-estimate in Diagram 4. It is possible to make a similar conclusion about the right tail.

We conclude that the proposed methodology of E-estimation is robust and performs relatively well for the complicated model of nonparametric autoregression with missing observations. The reader is advised to repeat Figures 8.9 and 8.10 with different parameters and get used to this important stochastic model.

8.8   Exercises

8.1.1 Consider a not necessarily stationary time series $\{X_t\}$ with a uniformly bounded second moment, that is $\mathbb{E}\{X_t^2\} \leq c < \infty$ for any t. Is the first moment of X_t uniformly bounded?

8.1.2 Consider a stationary time series with a finite second moment. Does the first moment exist? If the answer is "yes," then can the first moment $\mathbb{E}\{X_t\}$ change in time?

8.1.3 Consider a zero-mean and second-order stationary time series $\{X_t\}$. Prove that the autocovariance function $\gamma^X(0) = \mathbb{E}\{X_t^2\} = \mathbb{V}(X_t)$ for any t.

8.1.4 Verify (8.1.3) and (8.1.4).

8.1.5* Show that for a short-memory second-order stationary time series the sample mean estimator of its mean has variance which vanishes with the rate n^{-1}.

8.1.6* Consider the case of a long-memory second-order stationary time series with $\gamma^X(t)$ proportional to $|t|^{-\alpha}$, $0 < \alpha < 1$. Evaluate the variance of the sample mean estimator $\bar{\mu} := n^{-1} \sum_{l=1}^{n} X_l$.

8.1.7 Let $\{W_t\}$ be a standard Gaussian white noise. For the following time series, verify the second-order stationarity and calculate the mean and autocovariance function:
(i) $X_t = a + b W_t$.
(ii) $X_t = a + b W_t \sin(ct)$.
(iii) $X_t = W_t W_{t-2}$.
(iv) $X_t = a W_t \cos(ct) + b W_{t-2} \sin(ct)$.
(v) $Z_t = a + b W_t + c W_t^2$.

8.1.8 Suppose that $\{X_t\}$ and $\{Z_t\}$ are two uncorrelated second-order stationary time series. What can be said about a time series $\{Y_t\} := a\{X_t\} + b\{Z_t\}$ where the sum is understood as the elementwise sum. Hint: Calculate the mean and autocovariance.

8.1.9 Prove that an autocovariance function always satisfies the inequality $\gamma^X(k) \leq \gamma^X(0)$. Hint: Use Cauchy-Schwarz inequality.

8.1.10* Consider a realization $X_1, \ldots, X_n$, $n \geq p$ of a causal AR(p) process (8.1.7). Prove that the estimator $\hat{X}_{n+1} := \sum_{k=1}^{p} a_k X_{n+1-k}$ is the best linear predictor of X_{n+1} that minimizes the mean squared error $\mathbb{E}\{(\hat{X}_{n+1} - X_{n+1})^2\}$ over all linear estimators $\tilde{X}_{n+1} = \sum_{k=1}^{n} b_k X_{n+1-k}$. Hint: Write down the mean squared error and then minimize it with respect to $b_1, \ldots, b_n$. Also, note that the mean squared error is not smaller than $\mathbb{E}\{W_{n+1}^2\}$.

8.1.11 What is the definition of a causal ARMA process? Why are these processes of interest?

8.1.12* Consider the process $X_t = X_{t-1} + W_t$. Show that it does not have a stationary solution.

8.1.13 Explain how MA(q) and AR(p) processes can be simulated.

8.1.14* Consider a Gaussian ARMA(1,1) process $\{X_t\}$ defined by $X_t - aX_{t-1} = \sigma(W_t + bW_{t-1})$ with $|a| < 1$ and $-a \neq b$. Show that the process is stationary and causal. Hint: Show that the process may be written as $X_t = \sigma W_t + \sigma(a + b)\sum_{j=1}^{\infty} a^{j-1}W_{t-j}$.

8.1.15 Consider the product of two independent second-order stationary time series one of which belongs to class (8.1.8). Does the product belong to class (8.1.8)?

8.1.16 For a second-order stationary times series $\{X_t\}$ from a class (8.1.8), what can be said about variance of a linear sum $\sum_{l=1}^{n} a_l X_l$? Hint: Introduce reasonable restrictions on numbers $a_l, l = 1, 2, \ldots, n$.

8.1.17 Consider a MA(2) process. What can be concluded about its mixing coefficient (8.1.10)?

8.1.18 Give an ARMA example of m-dependent time series.

8.1.19 Consider a stationary time series $\{X_t\}$ with a known mixing coefficient $\alpha^X(s)$. Evaluate the mean and variance of $Y := n^{-1}\sum_{l=1}^{n} \sin(X_l)$. Hint: Use (8.1.13) and any additional assumption needed.

8.1.20 Consider a stationary Gaussian ARMA process. What can be said about its mixing coefficient (8.1.10)? Hint: Use the Kolmogorov-Rosanov result.

8.1.21* Prove (8.1.12).

8.1.22* Verify (8.1.13).

8.2.1 Is the spectral density symmetric about zero (even function)? Also, calculate $\int_{-\pi}^{\pi} g^X(\lambda)d\lambda$.

8.2.2* Show that a spectral density is a nonnegative function.

8.2.3 How can Fourier coefficients of a spectral density $g^X(\lambda)$ be expressed via the autocovariance function?

8.2.4* Explain formula (8.2.3). Present a motivating example.

8.2.5 What is the difference, if any, between estimators (8.2.4) and (8.2.5)?

8.2.6 Compare biases of autocovariance estimators (8.2.4) and (8.2.5) for a zero-mean and second-order stationary time series.

8.2.7* Calculate variance and the mean squared error of the autocovariance estimator (8.2.4). Hint: Make your own assumptions.

8.2.8 Calculate variance of the estimator (8.2.5). Hint: Make your own assumptions.

8.2.9* Calculate the mean and variance of the periodogram.

8.2.10 Using your knowledge of nonparametric estimation, explain why a periodogram cannot be a consistent estimator.

8.2.11 What is the definition of a Gaussian ARMA(1,1) process?

8.2.12* Verify (8.2.7).

8.2.13 Prove that the autocovariance of an ARMA(1,1) process decreases exponentially.

8.2.14 In Figure 8.1 the ARMA process exhibits high fluctuations. Suggest parameters of an ARMA process that fluctuates slower. Use Figure 8.1 to verify your recommendation.

8.2.15 The periodogram in Figure 8.1 indicates a possible periodic (so-called seasonal) component in the underlying process. Use formula (8.2.3) to find the period of a possible periodic component. Do you believe that this component is present in the process shown in the top diagram?

8.2.16 Explain how the spectral density E-estimator is constructed.

8.2.17 Using Figure 8.1, propose better parameters of the E-estimator.

8.2.18 Using Figure 8.1, conduct simulations with different values of parameter σ, and report your findings.

8.2.19 Is it reasonable to believe that, as (8.2.8) indicates, the spectral density of an ARMA process is proportional to σ^2?

8.2.20 Explain relation (8.2.9).

8.2.21 In the right side of (8.2.11), one part is called the integrated variance (or simply variance), and another the integrated squared bias. Write down these two components and explain their names.

8.2.22 Prove (8.2.12).

8.2.23* Evaluate the variance of the sample autocovariance.

8.2.24 Verify (8.2.14). Hint: Write down the squared sum via a corresponding double sum, and then think about addends as elements of a matrix.

8.2.25 Verify (8.2.15).

8.2.26* Prove (8.2.16). Hint: Use the assumption that the time series is Gaussian.

8.2.27 Verify (8.2.17).

8.2.28 Why do we need the assumption (8.2.18)? Does it hold for ARMA processes?

8.2.29 Using Cauchy-Schwarz inequality, establish (8.2.19).

8.2.30 Prove (8.2.20).

8.2.31* Prove (8.2.21).

8.2.32 What do (8.2.21) and (8.2.22) tell us about the sample autocovariance estimator?

8.2.33* Use formula (8.2.23) and find optimal cutoff for an ARMA process. Then compare estimation of the spectral density with estimation of a single parameter.

8.2.34 Verify (8.2.25).

8.2.35 Explain how (8.2.28) is obtained and why the proposed estimator $\bar{g}^X$ presents a practical interest.

8.2.36* Suppose that (8.2.26) is correct. The aim is to propose an estimator and to choose a minimal sample size that the MISE does not exceed a fixed positive constant ε. Propose such an estimator and the sample size.

8.3.1 Explain the missing mechanism (8.3.1).

8.3.2 In the case of a time series with missing observations, we observe realizations of two time series $\{A_t X_t\}$ and $\{A_t\}$. Explain why under the assumption $\mathbb{P}(X_t = 0) = 0$ it is sufficient to know realizations of only one time series $\{A_t X_t\}$.

8.3.3 Prove that if $\mathbb{P}(X_t = 0) = 0$, then $\mathbb{P}(A_t = I(Y_t \neq 0)) = 1$ where Y_t is defyned in (8.3.1). Does this imply that $A_t = I(Y_t \neq 0)$?

8.3.4 Suppose that the chance of rain tomorrow depends only on rain or no rain today. Suppose that if it rains today, then tomorrow it will rain with probability α, and if it is no rain today, then tomorrow it will rain with probability β. Find the probability that if today is no rain, then two days from today there will be rain. Hint: Consider a two-state Markov chain.

8.3.5* For the previous problem, consider 10 consecutive days and find the expected number of rainy days.

8.3.6* Consider a Markov-Bernoulli missing mechanism with $\alpha := \mathbb{P}(A_{t+1} = 0 | A_t = 0)$. Let L be the length of a batch of missing cases. Explain why, given the batch length $L \geq 1$, the distribution of L is geometric with $\mathbb{P}(L = k) = \alpha^{k-1}(1 - \alpha)$, $k = 1, 2, 3, \ldots$

8.3.7* For the setting of Exercise 8.3.6, find the mean and the variance of the batch length L.

8.3.8 Is the autocovariance $\gamma^A(j)$ equal to $\mathbb{E}\{A_1 A_{1+j}\}$?

8.3.9 Give the definition of a Markov chain.

8.3.10 Is an ARMA(1,1) process a Markov chain? If the answer is "yes," then what is the order of the Markov chain? Hint: Think about the effect of parameters of an ARMA processes.

8.3.11 Explain the simulation used to create Figure 8.2.

8.3.12 Repeat Figure 8.2 for different Markov-Bernoulli processes. Explain how its parameters affect the missing and estimation of the spectral density.

8.3.13* Why does the naïve estimate in Figure 8.2 indicate a lower (with respect to the E-estimate) spectrum power on low frequencies?

8.3.14 Explain the three estimates shown in Figure 8.2.

8.3.15 Using repeated simulations of Figure 8.2, propose better parameters of the E-estimator.

8.3.16* Find the mean and variance of the available number $N = \sum_{l=1}^{n} A_l$ of observations.

8.3.17 Verify (8.3.3).

8.3.18 Explain every equality in (8.3.4). Do not forget to comment on used assumptions.

8.3.19 What is the underlying idea of the estimator (8.3.5)? Explain its numerator and denominator.

8.3.20 Show that the autocovariance estimator (8.3.5) is unbiased.

8.3.21* Calculate the variance of estimator (8.3.5). Hint: Propose your assumptions.

8.3.22 Explain the underlying simulation in Figure 8.3.

8.3.23 Explain how parameter λ affects the number N of available observations in the simulation of Figure 8.3.

8.3.24 Formulate basic statistical properties of a batch-Bernoulli process.

8.3.25* Explain how the three estimates, shown in Figure 8.3, are constructed.

8.3.26* Given the same number of available observations, is Markov-Bernoulli or batch-Bernoulli missing mechanism better for estimation? You may use either a theoretical approach or simulations to answer the question.

8.3.27 How many parameters are needed to define a stationary Markov chain of order 2? Hint: It may be helpful to begin with a Markov chain of order 1.

8.4.1 Explain an amplitude-modulated missing mechanism. Give several examples.

8.4.2 Suppose that $\{X_t\}$, $\{A_t\}$ and $\{U_t\}$ are (second-order) stationary time series. Is their product a (second-order) stationary time series?

8.4.3 What do we need the assumption (8.4.1) for?

8.4.4* Consider the case when $\{A_t\}$ and $\{U_t\}$ are dependent. Does this affect estimation of the spectral density g^X?

8.4.5* Consider the case when $\{A_t\}$ and $\{X_t\}$ are dependent. Does this affect estimation of the spectral density g^X?

8.4.6* Find the mean and variance of the sample mean autocovariance (8.4.2).

8.4.7 Verify each equality in (8.4.3). Explain where and how the made assumptions about the three processes are used.

8.4.8 Can the spectral density g^X be consistently estimated based on amplitude-modulated observations? In other words, is this missing destructive?

8.4.9* Give definition of the shape of a function. Then explain when and how shape of the spectral density may be estimated for amplitude-modulated data.

8.4.10* Find the mean and variance of estimator (8.4.6).

8.4.11 Verify and explain all steps in establishing (8.4.7).

8.4.12 Explain the simulation used in Figure 8.4.

8.4.13 Explain how parameters of the processes $\{A_t\}$ and $\{U_t\}$ affect the number N of available observations. Support your conclusion using Figure 8.4.

8.4.14 Find better parameters of the E-estimator used in Figure 8.4.

8.4.15 Explain why the naïve estimate in Figure 8.4 indicates a smaller spectrum power on low frequencies.

8.4.16 Explain the model of amplitude-modulation by a Poisson variable.

8.4.17 In general, an amplitude-modulation implies inconsistent estimation of the spectral density. On the other hand, the Poisson amplitude modulation does allow a consistent estimation. Why?

8.4.18* Explain the estimator (8.4.8) of the mean of a Poisson distribution. Then evaluate its mean and variance.

8.4.19* Prove (8.4.9). Explain the used assumption.

8.4.20* Find the mean and variance of the estimator (8.4.10).

8.4.21* Consider the case when $\{X_t\}$ and Poisson $\{U_t\}$ are dependent. Explore the possibility of a consistent estimation of the spectral density or its shape.

8.4.22 Explain the simulation used in Figure 8.5.

8.4.23 Find the mean and variance of the available number N of observations in Figure 8.5. Does an underlying (hidden) time series $\{X_t\}$ affect N?

8.4.24 Consider the bottom diagram in Figure 8.5. The naïve estimate exhibits a larger spectrum power at all frequencies. Why?

8.4.25 Use Figure 8.5 to answer the following question. How do parameters of the underlying ARMA process affect estimation of its spectral density?

8.5.1 Explain the model of right censored time series. Present examples.

8.5.2 What are the available observations when time series is censored?

8.5.3 Explain formula (8.5.1).

8.5.4 What is the definition of a bivariate survival function? What are its properties?

8.5.5 Prove (8.5.3).

8.5.6* Find the mean and variance of estimator (8.5.4). Is it unbiased?

8.5.7* Explain the method of estimation of the survival function $G^C(v)$.

8.5.8 Verify (8.5.5).

8.5.9 Explain why the formula (8.5.6) is of interest.

8.5.10 What is the definition of a hazard rate? What are its properties?

8.5.11 Assume that the hazard rate is known. Suggest a formula for the corresponding probability density.

8.5.12 Verify (8.5.9).

8.5.13* Explain how the numerator and denominator in (8.5.9) may be estimated.

8.5.14 Prove validity of (8.5.11).

8.5.15* Find the mean and variance of the estimator (8.5.12).

8.5.16* Use an exponential inequality to infer about estimator (8.5.12).

8.5.17* Evaluate the mean and variance of estimator (8.5.13). Is it unbiased? Is it asymptotically unbiased?

8.5.18 Prove (8.5.14).

8.5.19 Explain why (8.5.15) is a reasonable estimator of the survival function.

8.5.20 Explain the simulation used to create Figure 8.6.

8.5.21 Explain all diagrams in Figure 8.6.

8.5.22 Consider Figure 8.6 and answer the following question. Why is the censored time series highly oscillated, while the underlying $\{X_t\}$ is not?

8.5.23 Explain how the naïve spectral density estimate is constructed.

8.5.24 The considered problem is complicated. Repeat Figure 8.6 a number of times and make your own conclusion about the E-estimator.

8.5.25 Use Figure 8.6 and then explain your observations about the size N of uncensored observations.

8.5.26 Suggest better parameters for the E-estimator. Hint: Use Figure 8.6 with different sample sizes and parameters.

8.5.27* Consider the case of a stationary time series $\{C_t\}$. Propose a consistent estimator of the bivariate survival function $G^{C_1,C_{1+j}}(v,u)$.

8.5.28* Consider the same setting only for time series of lifetimes. Propose a spectral density estimator and justify its choice. Hint: Estimate the mean, subtract it, and then check how this step affects statistical properties of the E-estimator.

8.6.1 Give definition of the probability density $f^X(x)$ of a continuous random variable X.

8.6.2 Suppose that $[0,1]$ is the support for a random variable X (or we may say the support of the probability density f^X). What is the meaning of this phrase?

8.6.3 Find the mean and variance of the sample mean estimator (8.6.4) for the case of independent observations of (a sample from) X.

8.6.4 Consider a time series $\{X_t\}$. What is the assumption that allows us to define the density f^{X_t}? What is the assumption that makes feasible the problem of estimation of the density f^{X_t}?

8.6.5 Explain why (8.6.6) is still valid for the case of dependent observations.

8.6.6 Is the sample mean estimator $\hat{\theta}_j$ unbiased? Is it robust toward dependence between observations?

8.6.7* Find the variance of the sample mean estimate $\hat{\theta}_j$ for the case of dependent observations. Explain all steps and made assumptions.

8.6.8 Explain all steps in establishing (8.6.8). Write down all used assumptions.

8.6.9 What is the assumption needed for validity of (8.6.9)?

8.6.10 Explain how the equality (8.6.10) was obtained.

8.6.11 Why is the assumption (8.6.11) important?

8.6.12 Explain importance of conclusion (8.6.12) for estimation of the probability density.

8.6.13 Give a definition of a weak dependence. Compare with the case of processes with long memory.

8.6.14 Suppose that (8.6.14) holds. What can be said about dependency between observations?

8.6.15 Consider a continuous random variable Z with density f^Z. What is the density of $Y := aZ + b$? Note that we are dealing with a scale-location transformation of Z.

8.6.16 Explain the motivation behind estimator (8.6.15). Why do we use such a complicated density estimator?

8.6.17* The density estimator (8.6.15) tells us that the support of X is $[X_{(1)}, X_{(n)}]$. Is this also the case for the empirical cumulative distribution function?

8.6.18* Consider a sample of size n from X. Find the probability $\mathbb{P}(X \geq X_{(n)})$. Use your result to improve the approach (8.6.15).

8.6.19 Explain formulas (8.6.16) and (8.6.17).

8.6.20 Explain the simulation that creates top diagrams in Figure 8.7.

8.6.21 Explain how estimates, shown in Figure 8.7, are calculated.

8.6.22 What is the difference, if any, between left and right columns of diagrams in Figure 8.7?

8.6.23 Repeat Figure 8.7 a number of times, and make a conclusion about which type of ARMA processes benefits estimation of the density.

8.6.24 Repeat Figure 8.7 with different parameters σ and σ_1. Report on how they affect estimation of the density.

8.6.25 Find better parameters of the E-estimator used in Figure 8.7.

8.6.26 Explain the simulation that creates the top diagram in Figure 8.8.

8.6.27 Explain the estimate shown in the second from the top diagram in Figure 8.8.

8.6.28 Explain the simulation that creates the third from the top diagram in Figure 8.8.

8.6.29 Explain the estimate shown in the bottom diagram in Figure 8.8.

8.6.30 Evaluate the mean and variance of the available number N of observations in simulation in Figure 8.8.

8.6.31 Suggest better parameters of the E-estimator used in Figure 8.8.

8.6.32* Using Figure 8.8, as well as your understanding of the theory, comment on the effect of missing data on density estimation for processes with long memory. Hint: Think about the case when the sample size of a hidden sample and the number of complete cases in a larger sample with missing observations are the same.

8.6.33 Explain the underlying idea of the estimator (8.6.18).

8.6.34* Find the mean and variance of the estimator (8.6.18).

8.6.35* Suppose that you have a sample of observations. What test would you suggest for independence of observations?

8.6.36* Propose a generator of a time series with long memory.

8.7.1 Explain the model of nonparametric autoregression.

8.7.2 Why can the model (8.7.1) be referred to as the prediction (forecasiting) model?

8.7.3* Consider a model $X_t = m_1(X_{t-1}) + m_2(X_{t-2}) + \sigma(X_{t-1}, X_{t-2})W_t$. Propose E-estimators for functions $m_1(x)$, $m_2(x)$ and $\sigma(x_1, x_2)$.

8.7.4 Is the process, defined by (8.7.1), second-order stationary? Hint: Think about assumptions.

8.7.5 Explain (8.7.2).

8.7.6 Verify (8.7.3).

8.7.7 Explain how the nonparametric autoregression model is converted into a nonparametric regression model.

8.7.8 Describe the simulation used in Figure 8.9.

8.7.9 How does a Markov-Bernoulli missing mechanism perform?

8.7.10 Is there any useful information that may be gained from analysis of Diagram 1 in Figure 8.9?

8.7.11 How was Diagram 2 in Figure 8.9 created?

8.7.12* Explain all steps in construction of E-estimator of $m(x)$. Then conduct a number of simulations, using Figure 8.9, and comment on performance of the estimator.

8.7.13* Use the theory and Figure 8.9 to explain how parameters of the ARMA process affect estimation of $m(x)$. Hint: Check the assumption $\mathbb{E}\{U_t|X_{t-1}\} = 0$.

8.7.14 Scale function in model (8.7.1) is an important function on its own, and it is often referred to as the volatility. Explain how it may be estimated using E-estimator. Then check its performance using Figure 8.9.

8.7.15 Repeat Figure 8.9 several times, make hard copies of figures, and then write down a report that explains performance of the E-estimators.

8.7.16 Find better parameters of the E-estimator used in Figure 8.9.

8.7.17 Explain how parameters of the E-estimator affect estimation of $m(x)$. Then test your conclusion using Figure 8.9.

8.7.18 Explain how parameters of the E-estimator affect estimation of $\sigma(x)$. Then test your conclusion using Figure 8.9.

8.7.19* Using Figure 8.9, find how parameters of the underlying model for $m(x)$ affect estimation of $m(x)$. Then explain your conclusion theoretically.

8.7.20* Using Figure 8.9, find how parameters of the underlying model for $m(x)$ affect estimation of $\sigma(x)$. Then explain your conclusion theoretically.

8.7.21 Using your understanding of the theory and simulations conducted by Figure 8.10, explain how parameters of the Markov-Bernoulli missing affect estimation of $m(x)$.

8.7.22 Scale function in model (8.7.1) is an important function on its own, and it is often referred to as the volatility. Explain how it may be estimated using E-estimator.

8.7.23 Using your understanding of the theory and simulations conducted by Figure 8.10, explain how parameters of the Markov-Bernoulli missing affect estimation of $\sigma(x)$.

8.7.24 Repeat Figure 8.10 several times, make hard copies of figures, and then write a report which explains shapes of the recorded E-estimates.

8.7.25* Model (8.7.1) is often referred to as a nonlinear one-step prediction. What will be the definition of a nonlinear two-step prediction model? Suggest an E-estimator.

8.7.26* Consider a functional-coefficient autoregression model

$$X_t = a_1(X_{t-d})X_{t-1} + \ldots + a_p(X_{t-d})X_{t-p} + \sigma(X_{t-d})W_t. \qquad (8.8.1)$$

Propose an E-estimator for the functions $a_1(x), \ldots, a_p(x)$.

8.9 Notes

There is a number of excellent books devoted to time series analysis, for example Anderson
(1971), Dzhaparidze (1985), Diggle (1990), Brockwell and Davis (1991), Fan and Yao (2003,
2015), Bloomfield (2004), Györfi et al. (2013), Box et al. (2016), Montgomery, Jennings and
Kulahci (2016), Shumway and Stoffer (2017). The last three books contain the introduction
to R and many R-examples. A large number of examples in applied economics can be found
in the book Greiner, Semmler and Gong (2005). A book-length treatment of hidden Markov
models for time series using R can be found in Zucchini, MacDonald and Langrock (2016).
Chapter 5 in Efromovich (1999a) is devoted to a number of topics in nonparametric time
series analysis, and the review of classical time series results is based on that chapter.

8.1 Discussion of ARMA processes, function class (8.1.8) and a wide spectrum of related
results can be found in Chapter 3 of Brockwell and Davis (1991), Efromovich (1998b, 1999a),
Montgomery, Jennings and Kulahci (2016). Mixing processes and weak dependence are
discussed in Doukhan (1994) and Dedecker et al. (2007).

8.2 The spectral density is defined via a cosine series approximation (8.2.2), and hence
the use of E-estimator becomes natural and attractive. For all other applications, like estimation of regression, hazard rate or probability density, one may argue for and against
a cosine-series approach in particular and an orthogonal series approach in general (versus
other approaches like kernel or spline). And nonetheless, kernel smoothing of the periodogram was the first proposed and so far the most popular method of estimation of the
spectral density. Lag-window spectral density estimates are discussed in Wu and Zaffaroni
(2017) where further references may be found. Parts of Sections 8.1 and 8.2, devoted to a
review of classical results of the time series theory, are based on Chapter 5 in Efromovich
(1999a), Efromovich and Pinsker (1981) and Efromovich (1984), where more discussion
and examples may be found. Asymptotic justification of the series estimation methodology can be found in Levit and Samarov (1978), Efromovich and Pinsker (1981, 1986), and
Efromovich (1984, 1998b, 1999a).

8.3 The problem of nonparametric spectral density estimation for discrete time series
in the presence of missing observations has a long history. Parzen (1963) has coined the
term amplitude-modulated for the time series $\{A_t X_t\}$ and the term amplitude-modulating
for the time-series $\{A_t\}$. Among first and most prominent theoretical results devoted to
the spectral density estimation with stochastically missing observations, let us mention
Scheinok (1965) who studied the case of A_t, $t = 1, 2, \ldots$ being independent and identically
distributed Bernoulli random variables, and Bloomfield (1970) who considered the case
of a stationary $\{A_t\}$. For these settings, consistent spectral density estimators have been
proposed. Further references and developments in time-series analysis with missing data
can be found in Dunsmuir and Robinson (1981a,b), Baisch and Bokelmann (1999), Jiang
and Hui (2004), Lee (2004), Tarczynski and Allay (2004), Robinson (2008), Vorotniskaya
(2008), Baby and Stoica (2010), Butcher and Gillard (2016), and Jiang et al. (2016). The
book by Ross (2014) gives a good introduction to Markov chains.

First consistent estimators of the spectral density have been developed at about the
same time as consistent estimators for nonparametric regression. At the same time, while the
theory of efficient and adaptive nonparametric regression and probability density estimation
is known for more than four decades, the theory of efficient estimation of the spectral density
with missing data was developed only recently, see Efromovich (2014b, 2017).

8.4 Examples of amplitude-modulated times series may be found in Parzen (1963),
Dunsmuir and Robinson (1981a,b), Efromovich (1999a), Bloomfield (2004), Jiang and Hui
(2004), and Vorotniskaya (2008). The asymptotic analysis of the proposed estimation procedure, which proves efficiency of the E-estimation approach, may be found in Efromovich
(2014d). An interesting and practically important topic is the sequential estimation with
assigned risk, see a discussion and corresponding results in Efromovich (2016b).

8.5 An interesting analysis of econometric methods and examples can be found in Li and Racine (2007) and Robinson (2008). Helsel (2011) provides a number of environmental examples and a discussion of Minitab and R softwares. Another good time series book to read is Box et al. (2016). Sun and Zhao (2013) provides a book-length treatment of survival topics including analysis of interval-censored failure time data. Among recent papers devoted to the analysis of censored time series, a special case of censored autoregressive model is considered in Wang and Chan (2017) where an elaborate system of unbiased estimating equations is proposed, see also Choi and Portnoy (2016) where quantile autoregression is explored for censored time series data.

8.6 The book by Beran (1994) is devoted to long-memory processes. Probability density estimation and conditional density estimation for dependent observations is considered in the book by Fan and Yao (2003) where a local polynomial technique is used.

8.7 Nonparametric autoregression is a classical topic in nonlinear time series. Books by Fan and Gijbels (1996), Efromovich (1999a), Fan and Yao (2003, 2015) and Tsay (2005) discuss different aspects of this topic and present interesting examples and extensions. The discussion of nonparametric functional autoregression may be found in Zhu and Politis (2017).

Chapter 9

Dependent Observations

This chapter is a continuation of Chapter 8 and it uses its notions and notations. The primary interest is in the study of nonstationary processes where the joint distribution of studied time series and/or the modification mechanism may change in time. The nonstationarity should test the limits of the E-estimation methodology.

The chapter begins with the analysis of a nonparametric regression with dependent regression errors, in particular long-memory ones. We will learn in Section 9.1 that the design of predictors, which may be either fixed or random, makes a dramatic effect on quality of estimation. Recall that this phenomenon does not exist for independent regression errors. Section 9.2 discusses classical continuous time processes, including Brownian motion and white noise, and we are learning how to filter a continuous signal from a white noise. In Section 9.3 we consider a nonstationary discrete-time series and learn how to detrend, deseasonalize and descale it so we may estimate the spectral density of an underlying stationary time series. The case of missing observations is considered as well. Section 9.4 considers a classical decomposition of amplitude-modulated time series. Section 9.5 explains how to deal with a missing mechanism that changes in time. Section 9.6 considers the case of a nonstationary time series with changing in time spectral density. Section 9.7 introduces us to a Simpson's paradox which explains the importance of paying attention to lurking variables that may dramatically change our opinion about data. Finally, Section 9.8 introduces us to a sequential estimation. Here we are exploring the potential of a controlled design of predictors in a regression.

9.1 Nonparametric Regression

In this section our aim is to understand how dependency affects estimation of a regression function. We begin with recalling some basic results for the case of independent observations. Then the case of dependent regression errors is considered. The main learning objective is to understand that regressions with fixed-design and random-design of predictors are affected differently by dependency between regression errors.

First, let us recall a classical regression model discussed in Section 2.3,

$$Y_l = m(X_l) + \sigma\varepsilon_l, \quad l = 1, 2, \ldots, n. \tag{9.1.1}$$

We observe n pairs of observations $(X_1, Y_1), \ldots, (X_n, Y_n)$ where X_l is called the predictor and Y_l is called the response. A problem, considered in Section 2.3, was to estimate a smooth (say differentiable with bounded derivative) regression function $m(x)$ on the unit interval $[0, 1]$ given that random errors $\varepsilon_1, \ldots, \varepsilon_n$ are mutually independent, zero mean, unit variance and independent of predictors.

The underlying idea of the proposed E-estimation is to write the regression function as a series expansion using the cosine bases on $[0, 1]$ with elements $\varphi_0(x) := 1$, $\varphi_j(x) :=$

$2^{1/2}\cos(\pi jx)$, $j = 1, 2, \ldots$,

$$m(x) = \sum_{j=0}^{n} \theta_j \varphi_j(x), \qquad (9.1.2)$$

where Fourier coefficients θ_j are

$$\theta_j := \int_0^1 m(x)\varphi_j(x)dx. \qquad (9.1.3)$$

Assume that predictors have either the uniform fixed-design $X_l = l/n$, $l = 1, 2, \ldots, n$ or the Uniform on $[0,1]$ random design with the density $f^X(x) = I(x \in [0,1])$. Then for both designs we can estimate Fourier coefficients θ_j of the regression function by a sample mean estimator

$$\hat{\theta}_j := n^{-1}\sum_{l=1}^{n} Y_l\varphi_j(X_l) = n^{-1}\sum_{l=1}^{n} m(X_l)\varphi_j(X_l) + \sigma n^{-1}\sum_{l=1}^{n}\varepsilon_l\varphi_j(X_l). \qquad (9.1.4)$$

Let us also assume that the regression function is differentiable with bounded derivative on $[0,1]$. Then the variance $\mathbb{V}(\hat{\theta}_j)$ and the mean squared error of the Fourier estimator decrease with the classical parametric rate n^{-1}.

The Fourier estimator (9.1.4) yields the regression E-estimator $\hat{m}(x)$ introduced in Section 2.3.

Now let us explore the same characteristics of Fourier estimator (9.1.4) for the new case when regression errors are a realization of a zero-mean and second-order stationary time series $\{\varepsilon_t\}$. The dependence does not change the mean of the Fourier estimator (9.1.4) due to the zero-mean property of regression errors. Let us evaluate the variance of $\hat{\theta}_j$.

We begin with analysis of the variance of $\hat{\theta}_0$. Using (8.2.14) and the zero-mean property of the errors we may write,

$$\mathbb{V}(\hat{\theta}_0) = \mathbb{E}\{[n^{-1}\sum_{l=1}^{n}(m(X_l) - \mathbb{E}\{m(X_l)\})]^2\} + \sigma^2\mathbb{E}\{(n^{-1}\sum_{l=1}^{n}\varepsilon_l)^2\}$$

$$= \left[\mathbb{E}\{[n^{-1}\sum_{l=1}^{n}(m(X_l) - \mathbb{E}\{m(X_l)\})]^2\}\right] + \left[\sigma^2 n^{-2}\sum_{l=-n}^{n}(n - |l|)\gamma^\varepsilon(l)\right], \qquad (9.1.5)$$

where $\gamma^\varepsilon(l) := \mathbb{E}\{\varepsilon_0\varepsilon_l\}$ is the autocovariance function of the regression errors. There are two terms on the right side of (9.1.5) highlighted by square brackets. The first one is either zero for the fixed design or it decreases with the rate not slower than n^{-1} for the random design. Further, the regression errors have no effect on the first term.

Regression errors affect only the second term in (9.1.5), and it is important to stress that the design of predictors has no effect on the term. On the other hand, the dependence in regression errors may dramatically change the rate of convergence to zero of the second term. Indeed, if the dependence has short memory (say the errors are generated by an ARMA process or m-dependent) then $\sum_{l=0}^{\infty}|\gamma^\varepsilon(l)| < \infty$ and the second term, as well as the variance of $\hat{\theta}_0$, are proportional to n^{-1}. For a long-memory dependence of order β with $\beta \in (0,1)$, when the autocovariance function $\gamma^\varepsilon(j)$ is proportional to $j^{-\beta}$, the rate of convergence to zero of the second term slows down to $n^{-\beta}$. The latter yields the same slowing down for the rate of convergence to zero of the variance of $\hat{\theta}_0$. Further, according to the Parseval identity, we may expect lower rates for the MISE convergence.

The conclusion is that while the design of predictors does not affect the rate of convergence of the variance of estimator $\hat{\theta}_0$, the dependence in regression errors may have a pronounced effect on the variance. Furthermore, as we will see shortly, a long-memory dependence in regression errors may make a reliable estimation of θ_0 impossible for small samples.

Now let us consider the effect of dependent regression errors on the variance of Fourier estimator $\hat{\theta}_j$ for $j \geq 1$. Write,

$$\mathbb{V}(\hat{\theta}_j) = \mathbb{E}\{(\hat{\theta}_j - \mathbb{E}\{\hat{\theta}_j\})^2\} = n^{-2}\mathbb{E}\Big\{\Big[\sum_{l=1}^{n}(Y_l\varphi_j(X_l) - \mathbb{E}\{\hat{\theta}_j\})\Big]^2\Big\}$$

$$= n^{-2}\mathbb{E}\Big\{\Big[\sum_{l=1}^{n}\Big((m(X_l)\varphi_j(X_l) - \mathbb{E}\{\hat{\theta}_j\}) + \sigma\varepsilon_l\varphi_j(X_l)\Big)\Big]^2\Big\}$$

$$= n^{-2}\sum_{l,t=1}^{n}\mathbb{E}\{(m(X_l)\varphi_j(X_l) - \mathbb{E}\{\hat{\theta}_j\})(m(X_t)\varphi_j(X_t) - \mathbb{E}\{\hat{\theta}_j\})\} \qquad (9.1.6)$$

$$+2n^{-2}\sum_{l,t=1}^{n}\mathbb{E}\{(m(X_l)\varphi_j(X_l) - \theta_j)\sigma\varepsilon_t\varphi_j(X_t)\} + n^{-2}\sigma^2\sum_{l,t=1}^{n}\mathbb{E}\{\varepsilon_l\varepsilon_t\varphi_j(X_l)\varphi_j(X_t)\}. \quad (9.1.7)$$

Because the time series $\{\varepsilon_t\}$ is zero-mean, the term (9.1.6) does not depend on the distribution of regression errors. Further, for the fixed design this term is zero while for the random design it decreases with the rate not slower than n^{-1}. To verify the last assertion for the random design we use $\mathbb{E}\{m(X)\varphi_j(X)\} = \mathbb{E}\{\hat{\theta}_j\} = \theta_j$, and for the fixed design we use $X_l = l/n$ and $n^{-1}\sum_{l=1}^{n}m(l/n)\varphi_j(l/n) = \mathbb{E}\{\hat{\theta}_j\}$.

The first sum in (9.1.7) is zero because $\{\varepsilon_t\}$ is zero-mean and independent of predictors. The main term to explore is the second sum in (9.1.7), and this is where the design of predictors becomes critical. Note that a particular expectation in that sum can be written as

$$\mathbb{E}\{\varepsilon_l\varepsilon_t\varphi_j(X_l)\varphi_j(X_t)\} = \mathbb{E}\{\varepsilon_l\varepsilon_t\}\mathbb{E}\{\varphi_j(X_l)\varphi_j(X_t)\} = \gamma^{\varepsilon}(t-l)\mathbb{E}\{\varphi_j(X_l)\varphi_j(X_t)\}. \quad (9.1.8)$$

Further, we may write for the both designs of predictors that

$$\mathbb{E}\{\varphi_j(X_l)\varphi_j(X_t)\} = \mathbb{E}\{\varphi_j(X_l)\}\mathbb{E}\{\varphi_j(X_t)\}I(l \neq t) + \mathbb{E}\{[\varphi_j(X_l)]^2\}I(l = t). \qquad (9.1.9)$$

For the random design we have

$$\mathbb{E}\{\varphi_j(X_l)\} = \int_0^1 \varphi_j(x)dx = 0, \quad j \geq 1, \qquad (9.1.10)$$

while for the fixed design we have

$$\mathbb{E}\{\varphi_j(l/n)\} = \varphi_j(l/n). \qquad (9.1.11)$$

Relations (9.1.10) and (9.1.11) explain the pronounced difference between the effect of dependence in the time series $\{\varepsilon_t\}$ on regressions with random and fixed designs. For the random design we use (9.1.10) to evaluate the second term in (9.1.7) and get

$$n^{-2}\sigma^2\sum_{l,t=1}^{n}\mathbb{E}\{\varepsilon_l\varepsilon_t\varphi_j(X_l)\varphi_j(X_t)\} = n^{-2}\sigma^2\sum_{l=1}^{n}\mathbb{E}\{\varepsilon_l^2\}\mathbb{E}\{\varphi_j^2(X_l)\}$$

$$= n^{-1}\sigma^2\mathbb{E}\{[\varphi_j(X)]^2\}. \qquad (9.1.12)$$

As a result, regardless of dependence between regression errors, for the random-design predictors the variance of the sample mean estimator $\hat{\theta}_j$ converges with the parametric rate

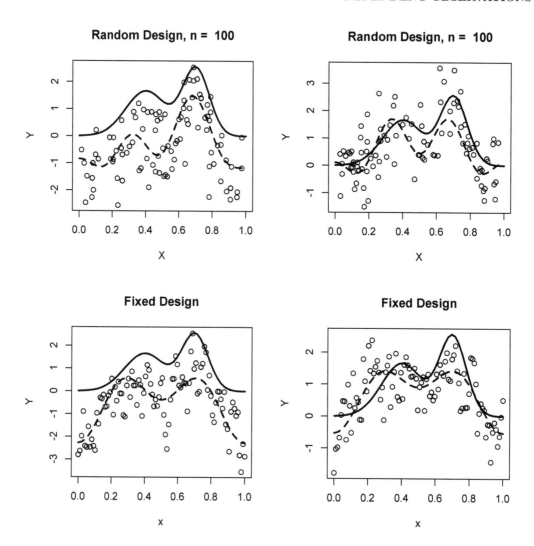

Figure 9.1 *Regression with long-memory errors of order $\beta \in (0,1)$. Two columns correspond to different simulations of the same experiment. The same regression errors are used in generating random and fixed design regressions. The predictors are uniform in both designs. Circles show a scattergram of observations overlaid by the underlying regression function (the solid line) and its E-estimate (the dashed line). {Parameter β is controlled by the argument beta.} [corn = 3, n = 100, beta = 0.2, sigma = 1, cJ0 = 4, cJ1 = 0.5, cTH = 4]*

n^{-1} for all $j \geq 1$. Of course, this is not the case for the $\hat{\theta}_0$, but if we are interested solely in estimation of the shape $\sum_{j=1}^{\infty} \theta_j \varphi_j(x)$ of the regression function $m(x)$, then this is a remarkable statistical outcome.

Unfortunately, there is no similarly nice conclusion for the fixed-design case because (9.1.11) does not allow us to eliminate the effect of dependency.

The teachable moment from our theoretical analysis is that a random design may attenuate the effect of dependency if we are interested in estimating the shape of a regression function.

Figure 9.1 allows us to understand the setting and appreciate performance of the E-estimator for the case of long-memory regression errors and the uniform random and fixed designs of predictors. Let us look at the left column of diagrams. Here we have the same regression function, the same time series of long-memory regression errors, and the only difference is in the design of predictors. What we see here is a typical long-memory series that begins with negative values and, even after 100 realizations, it is still negative. Note that the errors are zero mean, that is, eventually they will be positive and then they will stay positive over a long period of time. Nonetheless, we do see the two modes, and even the relation between the modes is shown reasonably well. In other words, while the regression is shifted down (because we cannot reliably estimate θ_0 for this sample size), the shape of the regression is clearly visualized. The outcome is worse for the case of the fixed design, and we know why. The right column of diagrams presents another realization of the same underlying experiment. Here again we can see how the design, together with dependent regression errors, affects both the scattergram and the E-estimator.

Let us make two more remarks. The first one is that there is no way for us to reliably estimate parameter $\theta_0 = \int_0^1 m(x)dx$ for the considered sample size $n = 100$. It will be a teachable moment to repeat Figure 9.1 with different sample sizes and parameter β and analyze chances of feasible estimation of θ_0. The second remark is about the observed time series $\{Y_t\}$ of responses. Consider the case of a stationary time series $\{\varepsilon_t\}$ of regression errors. Then, under the random-design of predictors the time series $\{Y_t\}$ of responses is also stationary, and it is *nonstationary* for the fixed-design predictors whenever the regression function $m(x)$ is not constant. To see the latter, note that the mean of the time series $\{Y_t\}$ changes in time.

We may conclude that for a regression problem the dependence between responses may dramatically affect quality of estimation. Further, there is a striking difference between random- and fixed-design regressions, and knowing this fact may help in designing an experiment and choosing an appropriate methodology of estimation. Finally, the E-estimator still may be used and the asymptotic theory asserts optimality of the series methodology of estimation.

9.2 Stochastic Process

So far we have discussed discrete-time series (processes), with the main examples being ARMA processes. Recall that ARMA processes are generated by a special modification of a series of independent and identically distributed variables called a discrete-time white noise. In many applications an observation is continuous in time, and it may be referred to as a continuous-time series, or a continuous-time stochastic process, or a continuous signal, and in the terminology continuos-time or continuous may be skipped whenever no confusion occurs.

Information theory, theory of stochastic processes, and signal processing theory are examples of sciences with emphasis on exploring continuous signals. In this section we will explore a classical Brownian motion, a white noise and the statistical problem of filtering a signal from white noise. Typically these notions and methods of filtering are studied in advance statistical classes, but for us they will not be too complicated because their discussion is based on a series approach and using the E-estimation methodology.

We begin with introducing a continuous in time $t \in [0, 1]$ stochastic process

$$W(t,k) := \sum_{j=0}^{k-1} W_j \varphi_j(t), \quad 0 \leq t \leq 1. \tag{9.2.1}$$

Here $W_0, W_1, \ldots$ are independent standard Gaussian random variables, k is a positive integer, and $\varphi_0(t) = 1$, $\varphi_j(t) = 2^{1/2} \cos(\pi j t)$ are elements of the cosine basis on $[0, 1]$. Recall that in (8.1.5) a discrete-time series $\{W_t\}$, called a discrete-time Gaussian white noise, was used to define a stationary discrete-time ARMA process, and in (9.2.1) $\{W_t\}$ is used to create a continuous-time stochastic process. The name "white" reflects the fact that the spectral density of $\{W_t\}$ is constant. Further, we know from the Fourier theory of Section 2.1 that in the sum (9.2.1) W_j is the jth Fourier coefficient of a square-integrable on $[0,1]$ processes $W(t,k)$, and $k-1$ is the cutoff in the frequency domain. This is why the process (9.2.1) is called a continuous-time *frequency-limited white noise*.

There is one specific issue with the process $W(t,k)$ that should be explained and then addressed. A frequency-limited white noise (9.2.1) is zero-mean, and this is a very attractive property. Further, using the Parseval identity (recall Section 1.3) we can also write,

$$\mathbb{E}\{\int_0^1 [W(t,k)]^2 dt\} = \sum_{j=0}^{k-1} \mathbb{E}\{W_j^2\} = k. \tag{9.2.2}$$

As we see, the power of a frequency-limited white noise increases to infinity as its frequency domain increases. There is nothing wrong with dealing with such a process, at least mathematically, but this fact is necessary to know.

There is a simple way to overcome the last complication via taking an integral of the frequency-limited white noise. Indeed, introduce its integral

$$B(t,k) := \int_0^t W(u,k)du = \sum_{j=0}^{k-1} W_j \int_0^t \varphi_j(u)du$$

$$= W_0 t + \sum_{j=1}^{k-1} W_j[(\pi j)^{-1} 2^{1/2} \sin(\pi jt)], \ 0 \le t \le 1. \tag{9.2.3}$$

The process $B(t,k)$ is called the frequency-limited standard Brownian motion. Let us look at its distribution. For a fixed time t, the frequency-limited standard Brownian motion $B(t,k)$ is a Gaussian variable with zero mean and variance

$$\mathbb{V}(B(t,k)) = \sum_{j=0}^{k-1}[\int_0^t \varphi_j(u)du]^2 = t^2 + 2\sum_{k=1}^{k-1}(\pi j)^{-2}\sin^2(\pi jt), \ 0 \le t \le 1. \tag{9.2.4}$$

We can immediately say that the variance is bounded by a constant for all k, and we also will establish shortly that $\mathbb{V}(B(t,k)) \le t$. Hence, at least formally, we can introduce a *standard Brownian motion* (also known as a *Wiener process*)

$$B(t) := \sum_{j=0}^{\infty} W_j \int_0^t \varphi_j(u)du. \tag{9.2.5}$$

Brownian motion plays a central role in the theory of stochastic processes, similarly to the role of a Gaussian variable in the classical theory of probability. Let us look at some of basic properties of the Brownian motion. First, $B(0) = 0$. Second, at any moment t a Brownian motion $B(t)$ is a Gaussian variable with zero mean. Let us calculate its variance, and this is a very nice exercise on using the Parseval identity. Note that $\int_0^t \varphi_j(u)du$ is the jth Fourier coefficient of function $I(0 \le u \le t)$, and then the Parseval identity implies

$$t = \int_0^1 [I(0 \le u \le t)]^2 du = \sum_{j=0}^{\infty}[\int_0^t \varphi_j(u)du]^2. \tag{9.2.6}$$

The conclusion is that $B(t)$ is a Gaussian variable with zero mean and variance t. Third, the random process $B(t_0 + t) - B(t_0)$ is again a standard Brownian motion that starts at time

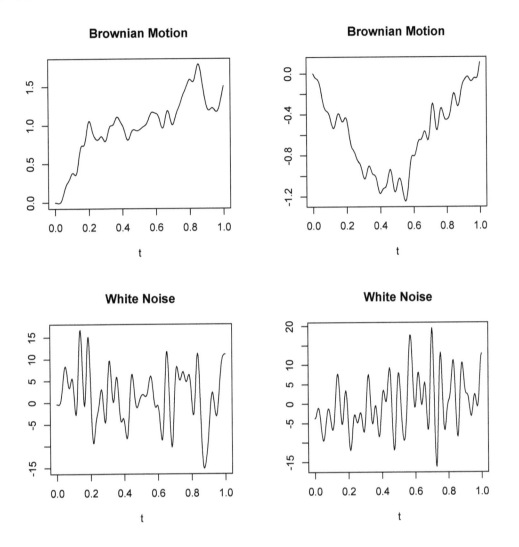

Figure 9.2 *Examples of a Brownian motion $B(t, k)$ and a corresponding white noise $W(t, k)$. {The signals are frequency-limited by a cutoff k.} [k = 50]*

t_0. Finally, we have an interesting property that for any $0 \leq t_1 < t_2 \leq t_3 < t_4 \leq 1$ Gaussian variables $B(t_2) - B(t_1)$ and $B(t_4) - B(t_3)$ are independent. To prove this assertion, we note that these two variables have a bivariate Gaussian distribution, and then we may use the Parseval identity (1.3.39) and get,

$$0 = \int_0^1 I(t_1 \leq u \leq t_2) I(t_3 \leq u \leq t_4) du = \sum_{k=0}^{\infty} \int_{t_1}^{t_2} \varphi_j(u) du \int_{t_3}^{t_4} \varphi_j(u) du$$

$$= \mathbb{E}\{[B(t_2) - B(t_1)][B(t_4) - B(t_3)]\}. \tag{9.2.7}$$

This implies that the two Gaussian variables are uncorrelated and hence independent. We now know all major properties of the Brownian motion.

It is of a special interest to look at several realizations of a standard Brownian motion and the corresponding Gaussian white noise. The top row of diagrams in Figure 9.2 shows two particular realizations of a frequency-limited ($k = 50$) Brownian motion. The realizations are completely stochastic and nonetheless it looks like they contain some important information about the trend in data. The reader familiar with curves of stock prices may easily recall similar curves and plenty of speculations of why there is this or that particular shape of a stock price. This is why many Wall Street pundits believe that Brownian motion is an excellent (and the only realistic) model for stock prices. Indeed, we see a bull market in the left diagram with some recent short-lived correction, and in the right diagram we see a bear market that reached the bottom and then roared back with vengeance. But again, recall that these realizations are purely stochastic, and here we know this for sure. It is also possible to say that a Brownian motion resembles behavior of a long-memory time series. The bottom row of diagrams shows us the corresponding frequency-limited white noise, and here it is simply the derivative of the Brownian motion. The latter is correct only for the considered frequency-limited processes.

Now we are ready to formulate a classical statistical problem of filtering a signal from noise. Let $m(t)$, $0 \le t \le 1$ be a signal of interest (estimand), and we observe a continuous in time signal $Y(t)$ such that

$$Y(t) = \int_0^t m(u)du + \sigma B(t), \quad 0 \le t \le 1. \tag{9.2.8}$$

Here σ is a positive constant. Another often used equation for $Y(t)$ is

$$dY(t) = m(t)dt + \sigma dB(t), \quad 0 \le t \le 1. \tag{9.2.9}$$

Equation (9.2.9) explains why the problem of estimating $m(t)$ is called filtering a signal from a white Gaussian noise.

As we already know, a white noise is a pure mathematical notion. Indeed, a white noise $W(t) := \sum_{j=0}^{\infty} W_j \varphi_j(t)$ has the same power at all frequencies (this explains the name "white"). Thus its total power is infinity, and no physical system can generate a white noise. On the other hand, its frequency-limited version $W(t, k) = \sum_{j=0}^{k-1} W_j \varphi_j(t)$ has a perfect physical sense, and at least theoretically, $W(t, k)$ may be treated as $W(t)$ passed through an ideal low-pass rectangular filter. This explains why a white noise is widely used in communication theory.

Our E-estimation methodology perfectly fits the problem of filtering a signal from white noise. Indeed, as usual we write the signal $m(t)$ as a Fourier series,

$$m(t) = \sum_{j=0}^{\infty} \theta_j \varphi_j(t), \ 0 \le t \le 1, \tag{9.2.10}$$

where

$$\theta_j := \int_0^1 m(t)\varphi_j(t)dt \tag{9.2.11}$$

are Fourier coefficients of the signal. Then using (9.2.8) or (9.2.9) we may introduce the statistic

$$\hat{\theta}_j := \int_0^1 \varphi_j(t)dY(t) = \int_0^1 \varphi_j(t)m(t)dt + \sigma \int_0^1 \varphi_j(t)dB(t) = \theta_j + \sigma W_j. \tag{9.2.12}$$

Recall that $W_0, W_1, \ldots$ are independent standard Gaussian variables. As a result, the statistic $\hat{\theta}_j$ is unbiased estimate of parameter θ_j. Hence, we need only to understand how to estimate the parameter σ which is the standard deviation of the statistic $\hat{\theta}_j$. Up until now, in all our problems the used estimator was a sample mean estimator, and then its standard deviation was naturally estimated via classical square root of the sample variance. Here we cannot use this approach because $\hat{\theta}_j$ is like a sample mean estimate, only here we do not

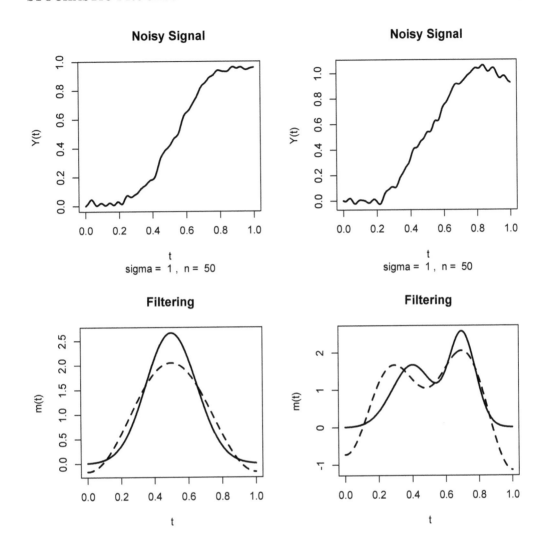

Figure 9.3 *Filtering a signal from a white Gaussian noise by the E-estimator. Two columns of diagrams correspond to the Normal and the Bimodal underlying signals, respectively. Top diagrams show processes $Y(t)$ simulated according to (9.2.8) with $B(t)$ replaced by $B(t,k)$. A bottom diagram shows an underlying signal $m(t)$ and its E-estimate by the solid and dashed lines, respectively. {Parameter $\sigma := \sigma^*/n^{1/2}$, and σ^* is controlled by the argument sigma. The argument J controls the parameter J in (9.2.13). Choosing of two underlying signals is controlled by the argument set.c.} [set.c = c(2,3), sigma = 1, n = 50, k = 50, cJ0 = 4, cJ1 = .5, cTH = 4, J = 20]*

know variables that were used to calculate it. What can be done in this case? A possible solution is based on the result of Section 2.1 that Fourier coefficients of a smooth signal decrease fast. Recall that in Section 2.2 we also used this fact in restricting our attention to estimating only first $c_{J0} + c_{J1}\ln(n)$ Fourier coefficients. As a result, we may introduce the following estimator of the unknown parameter σ,

$$\hat{\sigma} := \left[J^{-1} \sum_{j=J+1}^{2J} \hat{\theta}_j^2 \right]^{1/2} \tag{9.2.13}$$

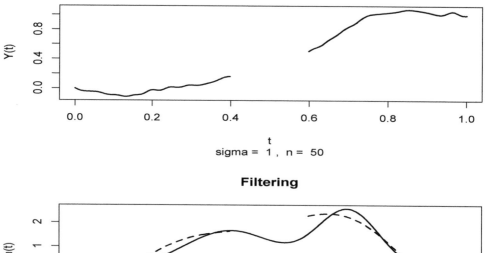

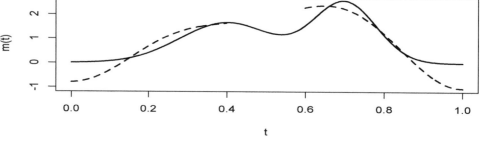

Figure 9.4 *Filtering a signal from a white Gaussian noise where a part of the noisy signal is missed. The simulation is identical to the one in the right column of Figure 9.3. In the bottom diagram the solid line shows the underlying signal and the dashed line shows the filtered signal over the two periods when the noisy signal is observed. [corn = 3, sigma = 1, n = 50, k = 50, cJ0 = 4, cJ1 = 0.5, cTH = 4, J = 20]*

with some reasonably large integer J. Statistics (9.2.12) and (9.2.13) allow us to use the E-estimator of Section 2.2 with an artificial parameter n defined below.

Let us shed light on the parameter σ and how it is related to the sample size n in the density estimation problem discussed in Section 2.2. Using notation of that section, $\theta_j := \int_0^1 \varphi_j(x) f^X(x) dx$ is the jth Fourier coefficient of the density $f^X(x)$ supported on $[0, 1]$. The Fourier coefficients are estimated by the sample mean estimator

$$\hat{\theta}_j = n^{-1} \sum_{l=1}^{n} \varphi_j(X_l) = \theta_j + n^{-1} \sum_{j=1}^{n} [\varphi_j(X_l) - \theta_j]. \qquad (9.2.14)$$

Then, if n is relatively large and in our case $n \geq 30$ is sufficient, the Central Limit Theorem asserts that $n^{-1} \sum_{j=1}^{n} [\varphi_j(X_l) - \theta_j]$ is a zero-mean Gaussian variable with the standard deviation $[\mathbb{V}(\varphi_j(X))/n]^{1/2}$. Further, according to (2.2.19), $\mathbb{V}(\varphi_j(X)) \to 1$ as j increases.

This is a pivotal result for understanding the meaning of parameter σ. Parameter σ in models (9.2.8) and (9.2.9) is equivalent to parameter $n^{-1/2}$ in the density estimation problem. This allows us to introduce an artificial $n := \sigma^{-2}$, and then use its estimator $\hat{n} := [\hat{\sigma}]^{-2}$, together with (9.2.12) and (9.2.13) in the E-estimator of Section 2.2. The asymptotic theory supports this approach, and the so-called *principle of equivalence* says that if a risk with a bounded loss function is considered, then, under mild assumptions on the smoothness of $m(x)$, results are valid for the filtering model if and only if they are valid for other statistical models including probability density estimation, regression, and spectral density estimation. In particular, for a regression model $Y_l = m(X_l) + \sigma(X_l)\varepsilon_l$, $l = 1, \ldots, n$ discussed in Section 2.3, parameter σ is equivalent to $(\int_0^1 [\sigma^2(x)/f^X(x)]dx)^{1/2}$.

Figure 9.3 illustrates the problem of filtering a signal $m(t)$ from a Gaussian white noise. First, let us look at the top diagrams. Here we see two realizations of the signal (9.2.8) corresponding to different underlying signals $m(t)$. Can you recognize the underlying signals? The hint is that they are two of our corner functions. Does the hint help? The answer is probably "no," and indeed it is a very complicated problem to manually filter the signal from $Y(t)$. The bottom diagrams show the underlying signals and their E-estimates. Overall, they do a reasonable job under the circumstances, and the principle of equivalence helps us to appreciate the complexity. In the simulation $\sigma = 1/n^{1/2}$ with $n = 50$. Hence, the filtering problem is equivalent to the problem of density estimation based on a sample of size $n = 50$. We know from Section 2.2 that this is a challenging sample size for estimation of our corner densities, and specifically the Bimodal. This understanding allows us to conclude that the particular estimates are reasonable. The reader is advised to repeat Figure 9.3 with different parameters and get a feeling of the filtering problem.

In the filtering (communication) theory literature one may find a literature devoted to the case of a signal missed over some periods of time. Figure 9.4 illustrates such a situation where no observation is available for the period $0.4 < t < 0.6$. In this case we can use our E-estimator for denoising an underlying signal over the periods of time when we have observations. Note that for each of these periods we are dealing with the same filtering problem according to the above-presented properties of a Brownian motion. The critical issue here is that without any additional information we cannot restore an underlying signal $m(t)$ over the missed period of time $(0.4, 0.6)$. This should be clear from the bottom diagram. As a result, this missing is destructive. The conclusion may be confusing a bit because no such destruction occurs in a regression problem with missing responses. The explanation is that in the regression we have considered the case of random predictors, while here the time is deterministic and we miss a large interval in time. This is a nice learning moment to realize that the nature of predictor (and here it is the time) plays a critical role in missing data. We will return to this issue in the next section.

9.3 Nonstationary Time Series With Missing Data

An observed discrete-time series $\{Y_t\}$ is rarely stationary and zero-mean. At the same time, the time series analysis is traditionally dealing with stationary and zero-mean series of observations. As a result, the first step in the statistical analysis of a time series is to extract from $\{Y_t\}$ an underlying stationary time series $\{X_t\}$. To do this, the following classical *decomposition model* of a nonstationary time series $\{Y_t\}$ is assumed,

$$Y_t := m(t) + S(t) + \sigma(t)X_t. \tag{9.3.1}$$

This decomposition resembles a fixed-design regression model, only now the aim is to evaluate statistical properties of the time series $\{X_t\}$ while all other functions are considered as a nuisance. The following terminology is used. In (9.3.1) $m(t)$ is a slowly changing function known as a *trend* component, $S(t)$ is a periodic function with period T (that is, $S(t + T) = S(t)$ for all t), known as a *seasonal* (cyclical) component (it is also customarily assumed that the integral or sum of the values of the seasonal component over the period is zero), $\sigma(t)$ is called a *scale* function (it is also often referred to, especially in finance and econometrics literature, as a volatility), and $\{X_t\}$ is a zero-mean and unit-variance stationary time series. Of course, the nuisance components may be of a practical interest on their own, and we will discuss their estimation.

Another typical complication in analysis of time series is that some observations may be missed, and instead of (9.3.1) we observe a time series

$$A_t Y_t = A_t[m(t) + S(t) + \sigma(t)X_t]. \tag{9.3.2}$$

Here $\{A_t\}$ is a Bernoulli time series discussed in Section 8.3. In particular, we will consider examples of Markov-Bernoulli and batch-Bernoulli time series. It is assumed that time series $\{A_t\}$ and $\{X_t\}$ are independent. The main aim is again to restore the modified time series $\{X_t\}$ and evaluate its statistical characteristics. In particular, we are interested in the spectral density of $\{X_t\}$.

The underlying idea of solving the problem is to estimate the three nuisance components by statistics $\tilde{m}(t)$, $\tilde{S}(t)$ and $\tilde{\sigma}(t)$, then use the detrended, deseasoned and rescaled statistics $A_t[Y_t - \tilde{m}(t) - \tilde{S}(t)][\tilde{\sigma}(t)]^{-1}$ as a proxy for $A_t X_t$, and finally invoke the technique of Section 8.3 of the analysis of stationary time series with missing observations. This is the plan that we will follow in this section. In the meantime, it will be also explained how to estimate a trend, a seasonal component and a scale function. This is the reason why it is convenient to divide the rest of this section into subsections devoted to a particular statistical component of the proposed solution with the last subsection being a detailed example explaining all steps of the analysis. The reader may also first look at the example in Subsection 9.3.6 and then return to subsections of interest.

In what follows, it is always assumed that the (hidden) time series of interest $\{X_t\}$ is zero-mean, unit-variance and second-order stationary.

9.3.1 Estimation of a Trend. Estimation of a trend is a regression problem, and as usual it is convenient to rescale times of observations onto the unit interval $[0, 1]$. As a result, in this subsection it is convenient to consider times $t = 1/n, 2/n, \ldots, 1$ where n is the number of available observations.

We begin with the case of no missing data, that is, the model (9.3.1). In that model the sum $m(t) + S(t)$ is deterministic, and we distinguish between the trend $m(t)$ and the seasonal component $S(t)$ in frequency domain. Namely, it is assumed that the trend is a slowly changing component. As a result, it is natural to use an orthogonal series approach to find Fourier coefficients of $m(t) + S(t)$ and then discuss how to separate the trend from the seasonal component. (In some cases this separation may be a tricky issue and we postpone this discussion until particular examples). In what follows $c_1, c_2, \ldots$ denote finite positive constants whose specific values are not of interest. Set $q(t) := m(t) + S(t)$, $t \in [0, 1]$ and, with some abuse of the notation, rewrite our observations of the time series (9.3.1) as

$$Y_l = q(l/n) + \sigma(l/n)X_l, \quad l = 1, 2, \ldots, n. \tag{9.3.3}$$

Our first step is to estimate Fourier coefficient $\theta_j := \int_0^1 q(t)\varphi_j(t)dt$, $j = 0, 1, \ldots$ where $\varphi_0(t) = 1$ and $\varphi_j(t) = 2^{1/2}\cos(\pi j t)$, $j > 0$ are elements of the cosine basis on $[0, 1]$. So far we have used the sample mean methodology to motivate an estimation procedure, and we still may do this here. At the same time, it will be a nice learning experience to look at another approach. The idea is to write down an integral as a corresponding Riemann sum,

$$\theta_j := \int_0^1 q(t)\varphi_j(t)dt = n^{-1}\sum_{l=1}^{n}\int_{(l-1)/n}^{l/n} q(t)\varphi_j(t)dt = \sum_{l=1}^{n} Y_l \int_{(l-1)/n}^{l/n} \varphi_j(t)dt$$

$$+\Big[\sum_{l=1}^{n}\int_{(l-1)/n}^{l/n} [q(t) - q(l/n)]\varphi_j(t)dt - \sum_{l=1}^{n} X_l \sigma(l/n)\int_{(l-1)/n}^{l/n} \varphi_j(t)dt\Big]$$

$$=: \tilde{\theta}_j + [\nu_j - \eta_j]. \tag{9.3.4}$$

On the right side of (9.3.4) the term

$$\tilde{\theta}_j := n^{-1} \sum_{l=1}^{n} Y_l \int_{(l-1)/n}^{l/n} \varphi_j(t)dt \qquad (9.3.5)$$

is a Fourier estimator of θ_j. Let us explore its properties via the analysis of ν_j and η_j.

In (9.3.4) the term ν_j is deterministic and defines the bias of estimator $\tilde{\theta}_j$, while the term η_j is random, has zero mean and defines the variance of the Fourier estimator. For ν_j we need to evaluate its absolute value. Let us assume that $\max_{t \in [0,1]} |dq(t)/dt| \le c_1$, that is the trend and the seasonal components are differentiable and their derivatives are bounded on $[0, 1]$. Then the mean value theorem allows us to write

$$|\nu_j| \le \sum_{l=1}^{n} \int_{(l-1)/n}^{l/n} c_1 n^{-1} |\varphi_j(t)| dt = n^{-1}[c_1 \int_0^1 |\varphi_j(t)| dt]. \qquad (9.3.6)$$

We conclude that the deterministic term ν_j is of order n^{-1}. Recall that a sample mean estimate, based on a sample of independent and identically distributed variables, is unbiased and its variance is of order n^{-1}. Estimator $\tilde{\theta}_j$ is biased but the squared bias decreases in order faster than n^{-1}. As a result, we may conclude that the biased nature of $\tilde{\theta}_j$ has no effect on its statistical properties as long as we are dealing with reasonably large samples. Further, (9.3.6) explains why in (9.3.5) it is better to use $\int_{(l-1)/n}^{l/n} \varphi_j(t)dt$ in place of $n^{-1}\varphi_j(l/n)$. Indeed, in the latter case we would have an extra factor j in (9.3.6) because the derivative of $\varphi_j(t)$ is proportional to j.

Now let us consider the stochastic component η_j. Write,

$$\mathbb{E}\{\eta_j\} = \mathbb{E}\left\{ \sum_{l=1}^{n} X_l \sigma(l/n) \int_{(l-1)/n}^{l/n} \varphi_j(t)dt \right\} = 0. \qquad (9.3.7)$$

Here we used the assumed zero-mean property of $\{X_t\}$. For the variance we get,

$$\mathbb{V}(\eta_j) = \sum_{l,s=1}^{n} \mathbb{E}\{X_l X_s\} \sigma(l/n)\sigma(s/n) [\int_{(l-1)/n}^{l/n} \varphi_j(t)dt][\int_{(s-1)/n}^{s/n} \varphi_j(t)dt]$$

$$= \sum_{l,s=1}^{n} \gamma^X(l-s) \sigma(l/n)\sigma(s/n) [\int_{(l-1)/n}^{l/n} \varphi_j(t)dt][\int_{(s-1)/n}^{s/n} \varphi_j(t)dt]. \qquad (9.3.8)$$

Assume that $\{X_t\}$ is a short-memory time series, with the main example of interest being an ARMA process. Then we have

$$\sum_{l=-\infty}^{\infty} |\gamma^X(l)| \le c_2 < \infty. \qquad (9.3.9)$$

Let us also assume that $\max_{t \in [0,1]} |\sigma(t)| \le c_3 < \infty$. Then we can continue evaluation of the right side of (9.3.8),

$$\mathbb{V}(\eta_j) \le \sum_{l,s=1}^{n} |\gamma^X(l-s)| n^{-2} 2 c_3^2 \le n^{-1}[2c_2 c_3^2]. \qquad (9.3.10)$$

Combining the obtained results in (9.3.4) we conclude that

$$|\mathbb{E}\{\tilde{\theta}_j\} - \theta_j| \le n^{-1}[2^{1/2} c_1], \quad \mathbb{V}(\tilde{\theta}_j) \le n^{-1}[2c_2 c_3^2]. \qquad (9.3.11)$$

These results justify the use of Fourier estimator $\tilde{\theta}_j$ for the regression E-estimator $\hat{q}(t)$.

Now let us propose an estimator of θ_j for the case of n observations of a time series (9.3.2) with missing observations, that is when we observe $A_l Y_l$, $l = 1, \ldots, n$ and Y_l is defined in (9.3.3). In what follows it is assumed that $\mathbb{P}(Y_l = 0) = 0$, and hence we have $\mathbb{P}(A_l = I(A_l Y_l \neq 0)) = 1$ (otherwise recall our discussion in Chapter 4 that we need to keep track of $\{A_t\}$). Let us check that again, similarly to (9.3.5), we can use a Riemann sum for construction of a Fourier estimator. Only to simplify the following formulae, we introduce several new notations. Denote by $Z_s, s = 1, \ldots, N$ moments when the underlying series is observed, in other words integer-valued random variables Z_s are defined as the solution of equation $A_{Z_s} = 1$, $1 \leq Z_s \leq n$. We also set $Z_0 := 0$ and $Z_{N+k} := n$ for $k \geq 1$. Here $N := \sum_{l=1}^n A_l = \sum_{l=1}^n I(A_l Y_l \neq 0)$ is the number of available observations of the underlying time series, it is random, and $N \leq n$. In what follows we consider only cases when $N > 1$ to avoid some trivial complications with notation, and recall our discussion in Section 4.1 that in missing data N plays the role of n for the case of direct data, and hence nonparametric estimation should be used only for samples with relatively large N.

Then we set $A_0 Y_0 := A_{Z_1} Y_{Z_1}$, $X_0 := X_{Z_1}$, $q_n(l/n) := I(l = 0)q(1/n) + I(l > 0)q(l/n)$, and $\sigma_n(l/n) := I(l = 0)\sigma(1/n) + I(l > 0)\sigma(l/n)$.

A natural generalization of (9.3.5) for the case of missing observations is the estimator

$$\hat{\theta}_j := \sum_{s=0}^N A_{Z_s} Y_{Z_s} \int_{Z_s/n}^{Z_{s+1}/n} \varphi_j(t)dt. \qquad (9.3.12)$$

Note that the only reason why this estimator may perform poorly is if the intervals of integration $(Z_{s+1} - Z_s)/n$ do not vanish. This is where our discussion of properties of Markov-Bernoulli and batch-Bernoulli time series $\{A_t\}$ becomes handy because we know how to evaluate the probability of large gaps in observation of the underlying time series.

Let us explore the proposed Fourier estimator (9.3.12). Similarly to (9.3.4) and using the above-introduced notation we can write,

$$\theta_j - \hat{\theta}_j = \sum_{s=0}^N \int_{Z_s/n}^{Z_{s+1}/n} [q(t) - q_n(Z_s/n)]\varphi_j(t)dt$$

$$- \sum_{s=0}^N X_{Z_s}\sigma_n(Z_s/n) \int_{Z_s/n}^{Z_{s+1}/n} \varphi_j(t)dt. \qquad (9.3.13)$$

Consider the first sum in (9.3.13). Using the assumption that the absolute value of the derivative of $q(t)$ is bounded by a constant c_1, we write,

$$\Big| \sum_{s=0}^N \int_{Z_s/n}^{Z_{s+1}/n} [q(t) - q_n(Z_s/n)]\varphi_j(t)dt \Big|$$

$$\leq \sum_{s=0}^n I(s \leq N) \int_{Z_s/n}^{Z_{s+1}/n} |q(t) - q_n(Z_s/n)||\varphi_j(t)|dt$$

$$\leq 2^{1/2}c_1 n^{-2} \sum_{s=0}^n I(s \leq N)[Z_{s+1} - Z_s]^2. \qquad (9.3.14)$$

Note how we were able to replace the random number $N + 1$ of terms in the sum by the fixed $n + 1$, and this will allow us to use our traditional methods for calculation of the expectation and the variance. As we already know from Section 8.3, for the considered time series $\{A_t\}$ we have $\mathbb{E}\{[Z_{s+1} - Z_s]^4\} \le c_4 < \infty$, and this yields (compare with (9.3.11))

$$\mathbb{E}\{|\sum_{s=0}^{N} \int_{Z_s/n}^{Z_{s+1}/n} [q(t) - q_n(Z_s/n)]\varphi_j(t)dt|\} \le c_5 n^{-1}. \tag{9.3.15}$$

Using the same technique we can evaluate the second moment of the sum,

$$\mathbb{E}\{[\sum_{s=0}^{N} \int_{Z_s/n}^{Z_{s+1}/n} [q(t) - q_n(Z_s/n)]\varphi_j(t)dt]^2\}$$

$$\le 2c_1^2 n^{-4} \sum_{s,r=0}^{n} \mathbb{E}\{(Z_{s+1} - Z_s)^2 (Z_{r+1} - Z_r)^2\} \le c_6 n^{-2}. \tag{9.3.16}$$

Again, it is of interest to compare (9.3.16) with (9.3.11). We conclude that the effect of the first sum in (9.3.13) on the Fourier estimator is negligible.

Now we are considering the second sum in (9.3.13) which is the main term. Because $\{X_t\}$ is zero-mean and independent of $\{A_t\}$ (and hence of $\{Z_s\}$), we get

$$\mathbb{E}\{\sum_{s=0}^{N} X_{Z_s} \sigma_n(Z_s/n) \int_{Z_s/n}^{Z_{s+1}/n} \varphi_j(t)dt\} = 0. \tag{9.3.17}$$

Now let us consider the variance of the sum. Write,

$$\mathbb{V}\left(\sum_{s=0}^{N} X_{Z_s} \sigma_n(Z_s/n) \int_{Z_s/n}^{Z_{s+1}/n} \varphi_j(t)dt\right)$$

$$= \mathbb{V}\left(\sum_{s=0}^{n} I(s \le N) X_{Z_s} \sigma_n(Z_s/n) \int_{Z_s/n}^{Z_{s+1}/n} \varphi_j(t)dt\right)$$

$$= E\left\{\sum_{s,r=0}^{n} I(s \le N) I(r \le N) X_{Z_s} X_{Z_r}\right.$$

$$\left. \times \sigma_n(Z_s/n) \sigma_n(Z_r/n) \int_{Z_s/n}^{Z_{s+1}/n} \varphi_j(t)dt \int_{Z_r/n}^{Z_{r+1}/n} \varphi_j(t)dt\right\}$$

$$\le 2c_3^2 n^{-2} \sum_{s,r=0}^{n} \mathbb{E}\{|\gamma^X(Z_s - Z_r)|(Z_{s+1} - Z_s)(Z_{r+1} - Z_r)\} \le c_7 n^{-1}. \tag{9.3.18}$$

In the last inequality we used the assumed (9.3.9), and recall that c_i are positive constants whose specific values are not of interest.

Combining the results we get that $\hat{\theta}_j$, suggested for time series $\{A_t Y_t\}$ with missing observations, has the same property as $\tilde{\theta}_j$ for $\{Y_t\}$, namely

$$|\mathbb{E}\{\hat{\theta}_j\} - \theta_j| \le c_5 n^{-1}, \quad \mathbb{V}(\hat{\theta}_j) \le c_8 n^{-1}. \tag{9.3.19}$$

We conclude that, depending on a given time series, we may use either Fourier estimator $\tilde{\theta}_j$ or Fourier estimator $\hat{\theta}_j$ to construct a corresponding regression E-estimator of function $q(t) := m(t) + S(t)$.

Recall that in this subsection our aim is to estimate the trend $m(t)$, and to do this we simply bound the largest frequency J in the E-estimator. How to choose a feasible frequency bound is explained in the next subsection.

9.3.2 Separation of Trend from Seasonal Component. These two components may be separated using either frequency or time domains. By the latter it is understood that a deterministic function with the period less than $T_{\max}$ is referred to as a seasonal component, and as a trend component otherwise. In some applications it may be easier to think about a seasonal component in terms of periods and the choice of $T_{\max}$ comes naturally. For instance, for a long-term money investor $T_{\max}$ is about several years, while for an active stock trader it may be just several hours or even minutes.

If $T_{\max}$ is specified, then in the above-proposed regression E-estimator, the largest frequency used by the estimator should not exceed $J_{\max}$ which is defined as the minimal integer such that $\varphi_{J_{\max}}(x + T_{\max}) \approx \varphi_{J_{\max}}(x)$ for all x. For instance, for the cosine basis on $[0, n]$ with the elements $\varphi_0(t) := n^{-1/2}$, $\varphi_j(t) := (n/2)^{-1/2} \cos(\pi jt/n)$, $j = 1, 2, \ldots, 0 \le t \le n$, we get

$$J_{\max} = \lfloor 2n/T_{\max} \rfloor. \tag{9.3.20}$$

Recall that $\lfloor x \rfloor$ denotes the rounded down x.

Using (9.3.20) in the E-estimator, proposed for $q(t)$ in subsection 9.3.1, yields the E-estimator $\hat{m}(t)$ of the trend.

9.3.3 Estimation of a Seasonal Component. If T is the period of a seasonal component $S(t)$, then by definition of a seasonal component, we have $S(t+T) = S(t)$ for any t, and if a time series is defined at integer points (and in this subsection this convention is convenient), then $\sum_{l=1}^{T} S(l) = 0$ (a seasonal component should be zero-mean because the mean of a time series is a part of the trend).

A classical time series theory assumes that the period T of an underlying seasonal component $S(x)$ is given. And indeed, in many practical examples, such as monthly housing starts, hourly electricity demands, migration of birds, or monthly average temperatures, periods of possible cyclical components are apparent. If the period is unknown, we will use spectral density to estimate it.

For now, let us assume that the period T is known and $\tilde{m}(t)$ is an estimate of the trend. We begin with the case when observations of $\{Y_t\}$ are not missed. Set $\tilde{Y}_t := Y_t - \tilde{m}(t)$ and introduce the estimator

$$\tilde{S}(t) := (\lfloor (n-t)/T \rfloor + 1)^{-1} \sum_{r=0}^{\lfloor (n-t)/T \rfloor} \tilde{Y}_{t+rT}, \quad t = 1, 2, \ldots, T. \tag{9.3.21}$$

The estimator is nonparametric because no underlying model of $S(t)$ is assumed.

To shed light on performance of $\tilde{S}(t)$, consider a simple example when $\tilde{Y}_l = S(l) + \sigma W_l$, $l = 1, 2, \ldots, n$, where $n = kT$, k is integer, and $W_1, W_2, \ldots$ are independent standard normal variables. Then

$$\tilde{S}(t) = S(t) + \sigma k^{-1} \sum_{r=0}^{k-1} W_{t+rT} =: S(t) + \sigma k^{-1/2} W'_t, \quad t = 1, 2, \ldots, T, \tag{9.3.22}$$

where $W'_t := k^{-1/2} \sum_{r=0}^{k-1} W_{t+rT}$ are again independent standard normal variables. Thus, if k is large enough (that is, if n is large and T is relatively small), then the estimator should perform well.

Estimation of the seasonal component for the case of missed observations (9.3.2) is similar,

$$\hat{S}(t) := \frac{\sum_{r=0}^{\lfloor (n-t)/T \rfloor} I(A_{t+rT} Y_{t+rT} \ne 0)[A_{t+rT} Y_{t+rT} - \tilde{m}(t+rT)]}{\sum_{r=0}^{\lfloor (n-t)/T \rfloor} I(A_{t+rT} Y_{t+rT} \ne 0)}, t = 1, 2, \ldots, T. \tag{9.3.23}$$

9.3.4 Estimation of Scale Function. Estimation of the scale $\sigma(x)$ in models (9.3.1) and (9.3.2) is not a part of a classical time series analysis. The primary concern of the classical time series theory is that the stochastic term $\{X_t\}$ should be second-order stationary, that is, the scale function $\sigma(x)$ should be constant. Since this is typically not the case, the usually recommended approach is to transform a dataset at hand in order to produce a new data set that can be successfully modeled as a stationary time series. In particular, to reduce the variability (volatility) of data, Box–Cox transformations are recommended when the original positive observations $Y_1, \ldots, Y_n$ are converted to $\psi_\lambda(Y_1), \ldots, \psi_\lambda(Y_n)$, where two popular choices are $\psi_\lambda(y) := (y^\lambda - 1)/\lambda$, $\lambda \neq 0$ and $\psi_\lambda(y) := \log(y)$, $\lambda = 0$. By a suitable choice of λ, the variability may be significantly reduced.

Our aim is twofold. First, we would like to estimate the scale function because in a number of applications, specifically in finance where the scale is called the volatility, it is important to know this function. Second, when the scale is estimated, this gives us an access to the hidden $\{X_t\}$.

Section 3.6 explains, for the case of complete observations, how to convert the problem of estimation of the scale function into a classical regression one. Furthermore, here again an E-estimator $\hat{\sigma}(t)$ may be used. The same conclusion holds for the case of missed data. We will see shortly how the scale E-estimator performs.

9.3.5 Estimation of Spectral Density Function. This is our final step. As soon as we have estimated trend, seasonal component and scale, we can use a plug-in estimator for values of $X_1, \ldots, X_n$ or $A_1 X_1, \ldots, A_n X_n$ depending on the data, and then utilize the corresponding spectral density E-estimator of Chapter 8.

9.3.6 Example of the Nonparametric Analysis of a Time Series. This subsection presents a simulated example of a nonstationary time series (9.3.2) and an explanation of how the spectral density of an underlying zero-mean and second-order stationary time series $\{X_t\}$ can be estimated. The example is simulated by Figure 9.6, it contains 10 diagrams and to improve their visualization the first four diagrams are shown in Figure 9.5 and the rest is in Figure 9.6. Captions to these figures present explanation of corresponding diagrams.

The used model of nonstationary time series is (9.3.2). Namely, we observe the product $\{A_t Y_t\}$ of two time series at times $t = 1, 2, \ldots, n$. The stationary time series $\{A_t\}$ is Markov-Bernoulli, its definition may be found in Section 8.3, and it creates missing observations whenever $A_t = 0$. The nonstationary time series $\{Y_t\}$ is defined as

$$Y_t := m(t) + S(t) + \sigma(t)X_t. \tag{9.3.24}$$

Here $\{X_t\}$ is a zero-mean second-order stationary Gaussian time series of interest, in the simulation it is an ARMA(1,1) process similar to those in Figure 8.1. The main aim is to estimate its spectral density. Deterministic function $m(t)$ is a trend, in the simulation it is one of our corner functions with domain $[1, n]$. Deterministic function $S(t)$ is a periodic seasonal component. In the simulation it is a trigonometric function $S(t) := s_s \sin(2\pi t/T) + s_c \cos(2\pi t/T)$ with the period T. Deterministic function $\sigma(t)$ is the scale, and it is defined as $\sigma_{sc}(1 + f(l/n))$, where $f(x)$ is one of our corner functions. How to choose specific underlying parameters and functions is explained in the caption.

Now let us look at a realization of the underlying time series $\{X_t\}$ shown in Diagram 1 of Figure 9.5. Here, crosses show the time series. The diagram is congested due to the sample size $n = 200$, but overall the time series looks like a reasonable stationary realization. Diagram 2 shows us the available time series with missing data. The horizontal line $y = 0$ helps us to see times when missing occurs. Note that the diagram is dramatically less congested because only $N = 144$ observations are available and more than a quarter of observations are missed. Note that the time series in Diagram 2 is all that we have for the statistical analysis. Can you visualize the underlying trend, seasonal component, scale

340

DEPENDENT OBSERVATIONS

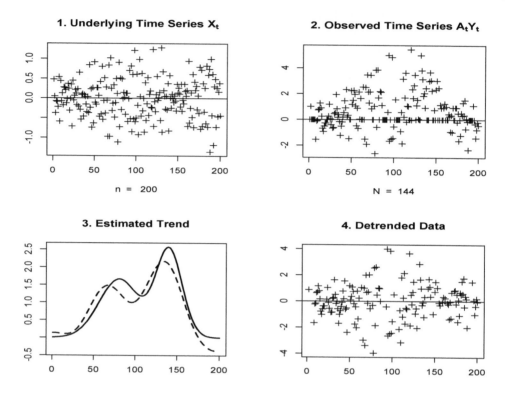

Figure 9.5 *Analysis of a nonstationary time series. Figure 9.6 creates ten diagrams and here the first four are shown. Diagram 1 shows the underlying (hidden) time series $\{X_t\}$ which is an ARMA(1,1) similar to those in Figure 8.1. The sample size $n = 200$ is shown in the subtitle. Diagram 2 shows the observed time series $\{A_tY_t\}$ defined in (9.3.2). The time series $\{A_t\}$ is generated as in Figure 8.2. The horizontal line helps to recognize missing observations, and the number of available observations $N := \sum_{l=1}^n A_l = 144$ is shown in the subtitle. The underlying trend and its E-estimate are shown in Diagram 3 by the solid and dashed lines, respectively. Diagram 4 shows the detrended data. Note that detrended data may be calculated only when $A_t = 1$, and hence only $N = 144$ observations are shown in this diagram.*

function, and the spectral structure of the stochastic noise from the data? The answer is probably "no," so let us see how the nonparametric data-driven procedure, discussed earlier, handles the data.

The first step is the nonparametric estimation of the trend. This is done by the regression E-estimator whose largest frequency is bounded by $2n/T_{\max}$ where the possible largest period of seasonal component $T_{\max}$ must be chosen manually. For the considered data we choose the default $T_{\max} = 35$, which implies $J_{\max} = 7$. The E-estimate of the trend (the dashed line) is shown in the third diagram. Note that it is based on the time series with missed data. The estimate is relatively good, it undervalues the modes but otherwise nicely shows the overall shape of the Bimodal corner function. The right tail goes down too much, but this is what the data indicate (look at the right tail of the observed time series). The reader is advised to repeat Figure 9.6 and get used to possible regression estimates because here we are dealing with a very complicated setting and a rather delicate procedure of estimation of the trend.

As soon as the trend is estimated, we can detrend the data (subtract the estimated trend from observations shown in Diagram 2), and the result is exhibited in Diagram 4.

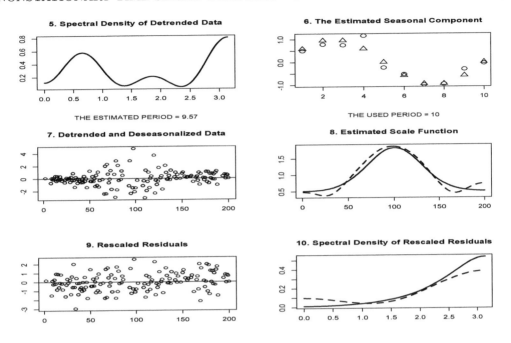

Figure 9.6 *Analysis of a nonstationary time series. Here the last 6 diagrams of Figure 9.6 are presented while the first four are shown in Figure 9.5. Time series in Diagram 4 (see Figure 9.5) is used to estimate its spectral density. Diagram 5 shows the estimated spectral density whose mode is used to estimate a possible period of a seasonal component shown in the subtitle. Then the rounded period, shown in the subtitle of Diagram 6, is used to estimate an underlying seasonal component. The seasonal component (the triangles) and its estimate (the circles) are shown in Diagram 6. The estimated seasonal component is subtracted from the detrended data of Diagram 4, and the result is shown by circles in Diagram 7. Again, only available observations (when $A_t = 1$) are exhibited. These statistics are used to estimate the scale function, and Diagram 8 shows the underlying scale function (the solid line) and its E-estimate (the dashed line). Data, shown in Diagram 9, are the observations in Diagram 7 divided by the scale E-estimate. Finally, the time series of Diagram 9 is used by the spectral density E-estimate. Diagram 10 shows the underlying spectral density (the solid line) of the underlying time series $\{X_t\}$ and its E-estimate (the dashed line). {Choosing the trend and scale functions is controlled by arguments trendf and scalef. Parameter σ is controlled by sigmasc. The estimate of the scale is bounded below by lbscale. The seasonal component is $S(t) = s_s \sin(2\pi t/T) + s_c \sin(2\pi t/T)$ whose parameters are controlled by arguments ss, sc and Tseas. ARMA(1,1) time series $\{X_t\}$ is generated as in Figure 8.1 and it is controlled by arguments a and b. Time series $\{A_t\}$ is generated as in Figure 8.2 and it is controlled by arguments alpha and beta. Intervals for the search of an underlying period are controlled by arguments set.period (in the time domain) and set.lambda (in the spectral domain). Argument TMAX separates trend from seasonal component. Setting ManualPer=T allows the user to manually choose a period, and this option is illustrated in Figure 9.8. A warning is issued if the estimated period is beyond a wished range.} [n = 200, trendf = 3, scalef = 2, sigmasc = 0.5, ss = 1, sc = 1, a = -0.4, b = -0.5, alpha = 0.4, beta = 0.8, TMAX = 35, Tseas = 10, ManualPer = F, set.period = c(8,12), set.lambda = c(0,2), lbscale = 0.1, cJ0 = 4, cJ1 = 0.5, cTH = 4, cJ0sp = 2, cJ1sp = 0.5, cTHsp = 4]*

Let us look at it. First, if a majority of observations in the second diagram are positive, here it is fair to say that the sample mean is likely zero. Second, note that only available observations are shown, and their number is $N = 144$. Third, based on this time series with missed data, we need to recover an underlying seasonal component if one exists. Can you recognize a seasonal component in the detrended data? Even if you know that this is a smooth function with period 10, it is difficult to see it in the data. Furthermore, the pronounced scale function complicates visualization of a seasonal component.

Estimation of the seasonal component is explained in diagrams of Figure 9.6 (recall that it presents continuation of the analysis of data shown in Figure 9.5). Diagram 5 shows us the E-estimate of the spectral density of the detrended time series with missing data. (Recall that as in the previous sections, arguments of the spectral density E-estimator have the attached string *sp*, for instance, *cJ0sp* is the argument that controls the coefficient c_{J0} of the spectral density E-estimator. This allows us to use separate arguments for the regression estimator, which recovers the trend and scale functions, and the spectral density estimator.) Diagram 5 indicates that the detrended data have a spectral density with a pronounced mode at the frequency about 0.6. The period 9.62 (the estimated period), calculated according to the formula $T = 2\pi/\lambda$, is given in the subtitle. The corresponding rounded (to the nearest integer) period is 10, and this is exactly the underlying period. What we see is one of the important practical applications of the spectral density that allows us to find periods of seasonal components.

While for this particular simulated time series the rounded estimated period has been determined correctly, this is not always the case. The small sample sizes and large errors may take their toll and lead to an incorrect estimate of the period. We will return to this issue shortly.

The rounded estimated period is used to estimate the underlying seasonal component. Here the estimator (9.3.23) is used and recall that it uses the fact that $S(t + T) = S(t)$. Circles in the sixth diagram show the estimate while triangles show the underlying seasonal component. The estimate is not perfect, but it is not chaotic or unreasonable. Note that its magnitude is fair, and the phase is shown absolutely correctly. Keep in mind that each point is the average of about 14 observations, so even for a parametric setting this would be considered a small sample size.

It is fair to say that for this particular data the nonparametric technique produced a remarkable outcome keeping in mind complexity of data in Diagram 2.

As soon as the seasonal component is estimated, we can subtract it from the detrended time series, and the resulting time series with missing observations is shown in Diagram 7. Now we may almost feel the shape of the scale functions which still makes the series nonstationary. The E-estimate of the scale function (the dashed line) is shown in Diagram 8, and we may compare it with the underlying scale (the solid line). Overall the E-estimate is good, and note that it is based on a time series where more than a quarter of observations are missed.

The data shown in Diagram 7, using the regression terminology, may be referred to as the time series of residuals. As soon as the scale function is estimated, we divide the residuals by the scale. To avoid a zero divisor, the estimate is truncated from below by the argument *lbscale*; the default value is 0.1. The resulting time series, called rescaled residuals, is shown in Diagram 9 for times t when $A_t = 1$. Note that the rescaled residuals are our plug-in E-estimates $\hat{X}_t$ of the underlying stationary time series X_t. Visual analysis shows that there is no apparent trend, or a seasonal component, or a scale function.

Finally, we arrive at the last step of estimation of the spectral density of the rescaled residuals. The E-estimate (the dashed line) and the underlying spectral density (the solid line) are shown in Diagram 10. Let us explain what to look upon here. The spectral density estimate is clearly different from the one in Diagram 5. The main issue is that we do not have a pronounced mode near frequency 0.6. This tells us that a seasonal component was successfully removed from the data. Otherwise we would again see a pronounced mode in the spectral density. The spectral density E-estimate is relatively good keeping in mind the complexity of the problem.

This finishes our analysis of this particular time series. Repeated simulations may show different outcomes, and Figure 9.6 allows us to address some issues. In Diagram 5 the mode correctly indicates the period of seasonal component, but in general this may not be the case. First, there may be several local modes created by both a seasonal component and a stochastic component, and large errors also may produce a wrong global mode. As a result, the period may be estimated incorrectly. One of the possibilities to avoid such a complication is to use prior information about the domain of possible periods. To play around with this possibility, two arguments are added to Figure 9.6, namely, *set.period* and *set.lambda*. The first one, *set.period = c(T1,T2)*, allows one to skip estimation of a seasonal component whenever an estimated period is beyond the interval $[T1, T2]$. The second argument, *set.lambda = c(\lambda_1, \lambda_2)$*, allows one to restrict the search for the mode to this particular frequency interval. While these two arguments do a similar job, they are good tools for gaining the necessary experience in dealing with the time and frequency domains. Note that Diagrams 6 and 7 are skipped if the estimated period is beyond the interval [T1,T2] or the frequency is beyond its interval, and then a warning statement is issued.

The second reason for the failure of the estimation of the period is that due to large noise and small sample size, the mode of an estimated spectral density may be relatively flat. To understand why, consider, as an example, frequencies $\lambda_1^* = 0.6$, $\lambda_2^* = 0.59$, and $\lambda_3^* = 0.54$. Then, the corresponding periods (recall the formula $T = 2\pi/\lambda$) are $T_1^* = 2\pi/0.6 = 10.47$, $T_2^* = 2\pi/0.59 = 10.64$, and $T_3^* = 2\pi/0.54 = 11.63$, which imply the rounded periods 10, 11, and 12, respectively. We conclude that a relatively small error in the location of a mode may imply a significant error in the estimated period.

Two questions immediately arise: how to detect such a case and how to correct the mistake. An incorrect period will not remove the mode in the spectral estimate shown in Diagram 10. Further, an incorrect period will not show a reasonable seasonal component in Diagram 6. These are the key points to check. The obvious method to deal with a wrong period is to use a manual period, and Figure 9.6 allows us to do this. Set the argument *ManualPer = T* (in R-language "T" stands for "True" and "F" for "False"). This stops the calculations at Diagram 5. Then the program prompts for entering a wished period from the keyboard. At the prompt 1: enter a period (here it should be 10, but any integer period may be tried) from the keyboard and then press Return; then at the prompt 2: just press Return. This completes the procedure, and the seasonal component will be calculated with the period entered. The period will be shown in the subtitle of Diagram 6. This option will be illustrated in Figure 9.8.

One more comment is due. In some cases a spectral density estimate of rescaled residuals has a relatively large left tail, as in Diagram 10, while an underlying theoretical spectral density does not. One of the typical reasons for such a mistake is a poorly estimated trend. Unfortunately, for the cases of small sample sizes and relatively large errors there is no cure for this "disease," but knowledge of this phenomenon may shed light on a particular outcome.

9.4   Decomposition of Amplitude-Modulated Time Series

The main aim of this section is to estimate the spectral density of an underlying zero-mean and unit-variance stationary time series $\{X_t\}$ which is not observed directly. Instead, we observe the result of a two-stage modification. First, the stationary time series is modified by a location-scale transformation,

$$Y_t := m(t) + S(t) + \sigma(t)X_t. \tag{9.4.1}$$

Here, as in Section 9.3, $m(t)$ is the trend, $S(t)$ is the seasonal component, and $\sigma(t)$ is the scale. As a result, the time series $\{Y_t\}$ is neither zero-mean, nor unit-variance, nor stationary. Second, $\{Y_t\}$ is not observed directly, and instead we observe its modification by an amplitude-modulating process $\{U_t\}$ (recall Section 8.4). Namely, the available observation is a time series with elements

$$V_t := U_t Y_t = U_t[m(t) + S(t) + \sigma(t)X_t]. \qquad (9.4.2)$$

Here $\{U_t\}$ is a time series of independent and identically distributed Poisson random variables with an unknown mean λ. Note that $\{U_t\}$ creates both the missing and the amplitude modulation of the time series $\{Y_t\}$. Let us additionally assume that $\mathbb{P}(Y_t = 0) = 0$, then it is sufficient to observe only V_t to recognize if the underlying Y_t is missed ($U_t = 0$) or scaled ($U_t > 0$). Indeed, using the assumption we get $\mathbb{P}(I(U_t = 0) = I(V_t = 0)) = 1$, and hence $I(V_t \neq 0)$ may be used as the availability.

The main aim is to estimate the spectral density of $\{X_t\}$ based on observations $V_1, \ldots, V_n$ of the time series $\{V_t\}$, and we also would like to estimate the trend, the seasonal component, the scale and the parameter λ.

The problem looks similar to the one considered in Section 9.3, but here we have an additional complication that not only some observations of time series $\{Y_t\}$ are missed but they are also amplitude-modulated. In other words, we are dealing with a sophisticated modification of the hidden time series of interest $\{X_t\}$. Nonetheless, the only way to solve the problem is to estimate all nuisance functions and get an access to the underlying time series of interest.

To shed light on a possible solution, similarly to Section 9.3 we begin with estimation of $q(t) := m(t) + S(t)$ and only for estimation of this function assume that the times of observations are $t_l := l/n$, $l = 1, 2, \ldots, n$. In a general case we may rescale all observations on the unit interval, and we do this to use our traditional regression approach of estimation over the unit interval.

To employ a regression E-estimator for estimation of $q(t)$, we need to estimate a Fourier coefficient

$$\theta_j := \int_0^1 q(t)\varphi_j(x)dx. \qquad (9.4.3)$$

To do this, set $N := \sum_{l=1}^n I(V_l \neq 0)$ for the number of available observations of the underlying time series, assume that $N > 1$, and then similarly to Section 9.3 we denote by Z_s, $s \in \{1, 2, \ldots, N\}$ random variables such that $V_{Z_s} \neq 0$. Recall our discussion in Sections 4.1 and 9.3 about cases with small N and that for a feasible estimation we need to have N comparable with sizes n used for directly observed time series.

Set $Z_0 := 0$, $Z_{N+k} := n$ for $k \geq 1$, with some plain abuse of notation set $V_0 := V_1$, and introduce a statistic

$$\check{\theta}_j := \sum_{s=0}^N V_{Z_s} \int_{Z_s/n}^{Z_{s+1}/n} \varphi_j(t)dt. \qquad (9.4.4)$$

We can continue (9.4.4) and replace the random number of terms in the sum by deterministic one,

$$\check{\theta}_j := \sum_{s=0}^n I(s \leq N)V_{Z_s} \int_{Z_s/n}^{Z_{s+1}/n} \varphi_j(t)dt. \qquad (9.4.5)$$

Consider the expectation of (9.4.5). Write,

$$\mathbb{E}\{\check{\theta}_j\} := \sum_{s=0}^n \mathbb{E}\{I(s \leq N)U_{Z_s}q(Z_s/n) \int_{Z_s/n}^{Z_{s+1}/n} \varphi_j(t)dt\}$$

$$= \sum_{s=0}^n \mathbb{E}\{\mathbb{E}\{U_{Z_s}|Z_s\}I(s \leq N)q(Z_s/n) \int_{Z_s/n}^{Z_{s+1}/n} \varphi_j(t)dt\}. \qquad (9.4.6)$$

Note that for a given Z_s the Poisson variable U_{Z_s} is positive, and hence

$$\mathbb{E}\{U_{Z_s}|Z_s\} = \frac{\mathbb{E}\{U_t I(U_t > 0)\}}{\mathbb{P}(U_t > 0)} = \frac{\lambda}{1 - e^{-\lambda}}. \qquad (9.4.7)$$

We conclude that the statistic $\check{\theta}_j$ cannot be used for estimation of θ_j, but there is a simple remedy. Assume that λ is known and $q(t)$ is differentiable and the absolute value of the derivative is bounded. Introduce a Fourier estimator

$$\tilde{\theta}_j := \frac{1 - e^{-\lambda}}{\lambda} \sum_{s=0}^{N} V_{Z_s} \int_{Z_s/n}^{Z_{s+1}/n} \varphi_j(t)dt. \qquad (9.4.8)$$

Then similarly to (9.4.6)–(9.4.7) and (9.3.13)–(9.3.19) we conclude that

$$\mathbb{E}\{|\tilde{\theta}_j - \theta_j|\} \le c_1 n^{-1}, \quad \mathbb{V}(\tilde{\theta}_j) \le c_2 n^{-1}, \qquad (9.4.9)$$

where c_1 and c_2 are finite constants.

These results justify using $\tilde{\theta}_j$ in the regression E-estimator. If parameter λ is unknown, then using the assumed $\mathbb{P}(Y_t = 0) = 0$, we may replace λ by its estimate

$$\hat{\lambda} := -\ln(n^{-1} \sum_{l=1}^{n} I(V_l = 0)). \qquad (9.4.10)$$

Here we used $\mathbb{P}(U_t = 0) = e^{-\lambda}$ and note that $n^{-1} \sum_{l=1}^{n} I(V_l = 0)$ is the sample mean estimate of $\mathbb{P}(U_t = 0)$.

Then the plug-in Fourier estimator yields the regression E-estimator $\hat{q}(t)$. All other steps in separating $m(t)$ from $S(t)$, estimation of the seasonal component, the scale function and the spectral density $g^X(\lambda)$ follow similarly along lines of the previous section.

Figure 9.8 illustrates the problem and its solution via ten diagrams, and its first four diagrams are shown in Figure 9.7 Diagram 1 in Figure 9.7 shows us an underlying time series $\{X_t\}$ which is simulated by the same ARMA process as in Figure 9.5. The observed amplitude-modulated time series $\{U_t Y_t\} =: \{V_t\}$ is shown in the second diagram. The underlying nonstationary process Y_t has the same trend, seasonal component and scale as in Figure 9.5. The only difference in the simulations conducted by Figure 9.6 and Figure 9.8 is that here $\{Y_t\}$ is multiplied by a white Poisson process with mean $\lambda = 1.5$. The amplitude modulation by the Poisson process creates a complicated time-series $\{V_t\}$ which hides the trend, seasonal component and the scale. Further, pay attention to the relatively large, with respect to Diagram 1, amplitude of the observed series, and that only $N = 231$ observations from hidden $n = 300$ are available while 69 are missed. As a result, the nonparametric problem at hand is extremely complicated and a possible poor estimation should not be a surprise.

The estimated trend $\hat{m}(t)$ is shown in Diagram 3. This estimate is good, and look at the correctly shown magnitude and location of the main mode and a good exhibition of the second mode. Compare magnitudes of the modes with what may be seen in Diagram 2, and then conclude that our formulas (9.4.7), (9.4.8) and (9.4.10) performed well in recovery of the hidden process $\{Y_t\}$ modified by the Poisson amplitude-modulation.

Our next step is to look at detrended data that then will be used for estimation of a seasonal component. In the previous Section 9.3, where the case of observed time series $\{A_t Y_t\}$ with missing observations was considered, this was a trivial step. Indeed, detrended data are the differences $A_t Y_t - \hat{m}(t)$ for times t when $A_t = 1$. Here, because of the amplitude-modulation by Poisson processes, we cannot simply subtract the trend. Instead, in Diagram 4 we show $V_t/[\hat{\lambda}/(1 - \exp(-\hat{\lambda}))] - \hat{m}(t)$ for times t when $V_t \ne 0$. It is of interest to compare Diagrams 2 and 4 and notice that the detrending did not remove largest observations.

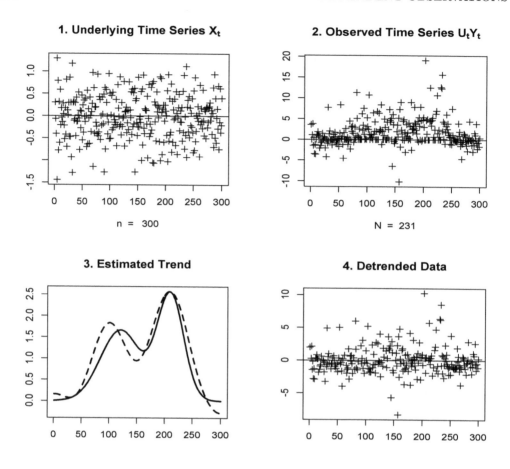

Figure 9.7 *Analysis of amplitude-modulated nonstationary time series. Here the first four diagrams, created by Figure 9.8, are shown. The underlying simulation of $Y_t := m(t) + S(t) + \sigma(t)X_t$ is identical to Figure 9.6. The only difference is that here the observed time series is $\{U_tY_t\}$ where U_t are independent and identically distributed Poisson variables with mean $\lambda = 1.5$. The structure of diagrams is similar to Figure 9.5.*

Further, the magnitude of detrained observations is still relatively large with respect to the hidden data in Diagram 1.

We continue our analysis of the data in Figure 9.8. Diagram 5 shows us E-estimate of the spectral density of the detrended time series. Look how flat is the left mode which is in the range of frequencies of interest. We know from our discussion in Section 9.3 that this may lead to inconsistency in estimation of the period of a possible seasonal component. Because Figure 9.8 uses the default argument ManualPer=T, the program stops after Diagram 5 and allows us to enter a wished period (the caption explains how to do this). Here the period 10 was entered, and Diagram 6 shows the estimated seasonal component by circles and the underlying seasonal component by triangles. The estimate is reasonable given complexity of the problem. A wrongly chosen period would not produce a reasonable shape of seasonal component, and also the left mode of the spectral density, observed in Diagram 5, will be again seen in Diagram 10. These two facts may help to choose a correct period for Diagram 6.

Detrended and deseasonalized time series is shown in Diagram 7. It is clear that the

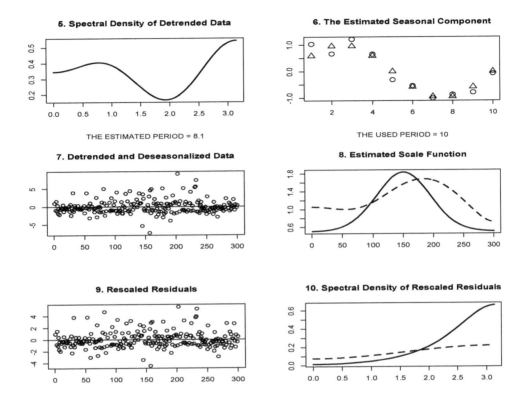

Figure 9.8 *Analysis of amplitude-modulated nonstationary time series* $\{V_t\} := \{U_tY_t\}$. *Here* $\{U_t\}$ *is the Poisson time series of independent variables with the mean* λ, *time series* Y_t *and* $\{U_t\}$ *are independent, and* $\{Y_t\}$ *is generated as in Figure 9.6. The case of a manually chosen period* $T = 10$ *of seasonal component is presented. This option is set by the argument* $ManualPer = T$ *which stops the program after exhibiting Diagram 5. Then the program prompts for entering a wished period. At prompt 1: enter a period (here it is 10, but any integer period may be tried) from the keyboard and press Return, then at the prompt 2: press Return. This completes the procedure, the seasonal component will be calculated with the manual period which will be also shown in the subtitle of Diagram 6. Apart of this, the structure of diagrams is identical to those in Figure 9.6.* [n = 200, lambda = 1.5, trendf = 3, scalef = 2, sigmasc = 0.5, ss = 1, sc = 1, a = -0.3,b = -0.5, TMAX = 35, Tseas = 10, ManualPer = T, set.period = c(8,12), set.lambda = c(0,2), lbscale = 0.1, cJ0 = 4, cJ1 = 0.5, cTH = 4, cJ0sp = 2, cJ1sp = 0.5, cTHsp = 4]

scale is not constant and it is larger for t around 200. This is what we see in the scale estimate (the dashed line) shown in Diagram 8. The estimate is skewed to the right with respect to the underlying scale (the solid line), and its tails are also wrong. At the same time, let us note that the E-estimator knows only the time series shown in Diagram 7, and the data support conclusion of the E-estimator. The main issue here is the large volatility created by Poisson amplitude-modulation, recall the discussion in Section 8.4.

We are almost done with the decomposition. Diagram 9 shows us the rescaled residuals, and Diagram 10 shows the spectral density E-estimate (the dashed line) and the true spectral density (the solid line) of time series $\{X_t\}$ shown in Diagram 1. First of all, note that the left mode, observed in the spectral density of detrended data in Diagram 5, is completely gone. This tells us that the underlying time series $\{Y_t\}$ has a seasonal component with period 10, the estimated seasonal component (see Diagram 6) is very reasonable, and that the procedure of removing the seasonal component was successful. The spectral density estimate by itself is far from being perfect, but at least it correctly shows that $\{X_t\}$ has a larger power at high frequencies.

The explored model of amplitude-modulated nonstationary time series is complicated, and the reader is encouraged to repeat Figure 9.8 with different parameters, appreciate its complexity, and get a good training experience. Let us stress one more time that here we are dealing with compounded effects of the scale-location modification $\{Y_t\} = \{m(t) + S(t) + \sigma(t)X_t\}$ of the process of interest $\{X_t\}$, missing observations of $\{Y_t\}$ created by zero realizations of a Poisson process, and amplitude-modulation of nonstationary $\{Y_t\}$ by the Poisson process. Figure 9.8 helps to shed light on all these modifications.

9.5   Nonstationary Amplitude-Modulation

So far we have considered stationary missing mechanisms. What will be if the missing mechanism is not stationary? Can we still use E-estimation?

To answer these questions, consider the case of an amplitude-modulated time series $\{Y_t\} := \{U_t X_t\}$. Here $\{X_t\}$ is a zero-mean and second-order stationary time series of interest for which we would like to estimate its spectral density, and it is additionally assumed that $\mathbb{P}(X_t = 0) = 0$. $\{U_t\}$ is a time series of independent Poisson variables with the mean that depends on t. This yields a nonstationary missing mechanism as well as a nonstationary amplitude-modulation of $\{X_t\}$.

In what follows it is assumed that time series $\{X_t\}$ and $\{U_t\}$ are independent. It is also convenient to assume that $\{Y_t\}$ is observed on the unit interval of time $[0,1]$ at moments $1/n, 2/n, \ldots, n/n$, that is we observe $Y_{1/n}, \ldots, Y_{n/n}$, otherwise we simply rescale times of n observations onto the unit interval.

To describe the proposed solution, we begin with the autocovariance function of the observed time series $\{Y_t\}$,

$$\gamma^Y(j) := \mathbb{E}\{Y_0 Y_{j/n}\}, \tag{9.5.1}$$

which may be estimated by a sample autocovariance

$$\hat{\gamma}^Y(j) := (n-j)^{-1} \sum_{l=1}^{n} Y_{l/n} Y_{(l+j)/n}. \tag{9.5.2}$$

Let us evaluate the expectation of the sample autocovariance. Write,

$$\mathbb{E}\{\hat{\gamma}^Y(j)\} = \gamma^X(j)\left[(n-j)^{-1} \sum_{l=1}^{n-j} \mathbb{E}\{U_{l/n}\}\mathbb{E}\{U_{(l+j)/n}\}\right]. \tag{9.5.3}$$

Set $\lambda(t) := \mathbb{E}\{U_t\}$, $t \in [0,1]$ and assume that the function $\lambda(t)$ is differentiable and the absolute value of the derivative is bounded on $[0,1]$. Then we can continue (9.5.3) and, to simplify formulae, we are considering separately cases $j = 0$ and $j > 0$. Using relation $\mathbb{E}\{U^2(l/n)\} = \lambda(l/n) + [\lambda(l/n)]^2$, we can continue (9.5.3) for the case $j = 0$,

$$\mathbb{E}\{\hat{\gamma}^Y(0)\} = \gamma^X(0)[n^{-1} \sum_{l=1}^{n} \lambda(l/n)(1 + \lambda(l/n))]$$

$$= \gamma^X(0) \int_0^1 \lambda(t)(1 + \lambda(t))dt$$

$$+ \left\{\gamma^X(0)[n^{-1} \sum_{l=1}^{n} \lambda(l/n)(1 + \lambda(l/n)) - \int_0^1 \lambda(t)(1 + \lambda(t))dt]\right\}. \tag{9.5.4}$$

The absolute value of the term in the square brackets, due to the assumed smoothness of function $\lambda(t)$, is not larger than $c_1 n^{-1}$ (note that we are evaluating the remainder for a Riemann sum and we have done a similar calculation in Section 9.3). Here and in what follows c_i are some positive finite constants whose specific values are not of interest to us. We conclude that

$$\left| \gamma^X(0) - \frac{\mathbb{E}\{\hat{\gamma}^Y(0)\}}{\int_0^1 \lambda(t)(1+\lambda(t))dt} \right| \leq c_2 \gamma^X(0) n^{-1}. \tag{9.5.5}$$

For $j > 0$ we can continue (9.5.3) and write,

$$\mathbb{E}\{\hat{\gamma}^Y(j)\} = \gamma^X(j)[(n-j)^{-1} \sum_{l=1}^{n-j} \lambda(l/n)\lambda((l+j)/n)]$$

$$= \gamma^X(j) \int_0^1 (\lambda(t))^2 dt + \left\{ \gamma^X(j)[(n-j)^{-1} \sum_{l=1}^{n-j} \lambda(l/n)\lambda((l+j)/n) - \int_0^1 (\lambda(t))^2 dt] \right\}. \tag{9.5.6}$$

The absolute value of the term in the square brackets, due to the assumed smoothness of function $\lambda(t)$, is not larger than $c_3 j n^{-1}$, and we get

$$\left| \gamma^X(j) - \frac{\mathbb{E}\{\hat{\gamma}^Y(j)\}}{\int_0^1 (\lambda(t))^2 dt} \right| \leq c_4 |\gamma^X(j)| j n^{-1}. \tag{9.5.7}$$

We conclude that if function $\lambda(t)$, defining the missing mechanism, is known then $\hat{\gamma}^Y(j) / \int_0^1 (\lambda(t))^2 dt$ may be used as a Fourier estimator of the autocovariance function of interest $\gamma^X(j)$.

In general function $\lambda(t)$ is unknown. Recall the assumption $\mathbb{P}(X_t = 0) = 0$, and using it we can write,

$$\mathbb{E}\{I(Y_l = 0)\} = \mathbb{E}\{I(U_l X_l = 0)\} = \mathbb{P}(U_l = 0) = e^{-\lambda(l/n)}. \tag{9.5.8}$$

Set

$$m(t) := e^{-\lambda(t)}. \tag{9.5.9}$$

Then (9.5.8) implies that $m(t)$ is the regression function in the fixed design Bernoulli regression with the response $I(Y_l = 0)$ and the predictor l/n, $l = 1, 2, \ldots, n$. Hence we can use Bernoulli regression E-estimator $\hat{m}(t)$ and propose the plug-in estimator of $\lambda(t)$,

$$\hat{\lambda}(t) := -\ln(\hat{m}(t)). \tag{9.5.10}$$

Obtained relations allow us to propose the following estimator of the autocovariance function $\gamma^X(j)$,

$$\hat{\gamma}^X(0) := \frac{n^{-1} \sum_{l=1}^n Y_{l/n}^2}{\int_0^1 [\hat{\lambda}(t)]^2 dt}, \tag{9.5.11}$$

and

$$\hat{\gamma}^X(j) := \frac{(n-j)^{-1} \sum_{l=1}^{n-j} Y_{l/n} Y_{(l+j)/n}}{\int_0^1 \hat{\lambda}(t)(1+\hat{\lambda}(t))dt}, \quad j \geq 1. \tag{9.5.12}$$

Further, note how simple the two integrals in (9.5.11) and (9.5.12) may be expressed via Fourier coefficients of $\hat{\lambda}(t)$ (to realize that recall the Parseval identity).

This autocovariance estimator allows us to construct the spectral density E-estimator introduced in Section 8.2.

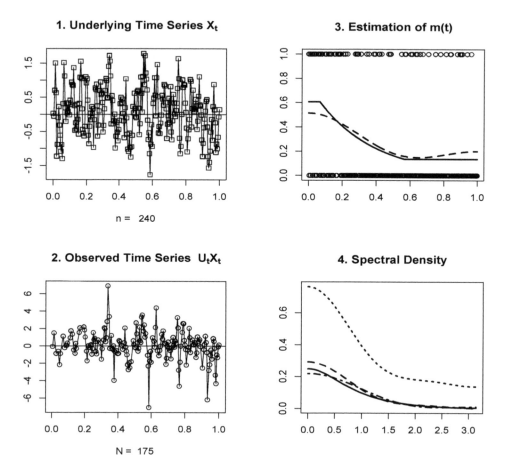

Figure 9.9 *Nonstationary amplitude modulation. Observations are* $Y_t := U_t X_t$, $t = 1/n, 2/n \ldots, n/n$, *where* $\{X_t\}$ *is the Gaussian ARMA(1,1) process defined in the caption of Figure 8.2, and* $U_{1/n}, U_{2/n}, \ldots, U_{n/n}$ *are independent Poisson random variables with the mean* $\mathbb{E}\{U_t\} = \lambda(t) = \max(L_{min}, \min(L_{max}, A + Bt))$. *Diagram 3 shows estimation of the regression function* $m(t) := \mathbb{E}\{I(U_t X_t = 0)\}$ *based on n pairs of observations* $(1/n, I(U_{1/n} X_{1/n} = 0)), \ldots, (n/n, I(U_{n/n} X_{n/n} = 0))$. *The underlying regression and its E-estimate are shown by the solid and dashed lines, respectively. In Diagram 4 the curves are: the underlying spectral density (the sold line), the proposed E-estimate (the long-dashed line), the oracle E-estimate based on the hidden* $\{X_t\}$ *(the dot-dashed line), and the naïve E-estimate based on* $\{U_t X_t\}$ *(the dotted line). [n = 240, sigma = 0.5, a = 0.4, b = 0.5, LMIN = 0.5, LMAX = 2, A = 0.3, B = 3, cJ0 = 4, cJ1 = 0.5, cTH = 4, cJ0sp = 2, cJ1sp = 0.5, cTHsp = 4]*

Figure 9.9 illustrates the problem and the proposed solution. Diagram 1 shows realization of an underlying stationary ARMA(1,1) time series, the sample size $n = 240$ is shown in the subtitle. Diagram 2 shows us observed realizations of the nonstationary amplitude-modulated time series $\{U_t X_t\}$. Here missed realizations, when $U_t = 0$, are skipped. This presentation allows us to conclude that the availability likelihood increases as t increases because as t increases the number of missing realizations decreases. We also clearly observe the amplitude-modulated structure of the time series, which is in strike contrast to the time series in Diagram 1. Diagram 3 shows us the scattergram used to estimate the regression function $m(t)$ defined in (9.5.9). The solid line shows the underlying regression and the dashed line shows the regression E-estimate. Note that the function $\lambda(t)$ is not differentiable but still it is possible to show that all our conclusions hold for this function because it is piecewise differentiable. As we see from Diagram 3, the E-estimate is not perfect but it does follow the data. Further, Diagram 3 is a nice place to realize that the amplitude-modulating time series is not stationary.

The final result of the estimation procedure is exhibited in Diagram 4. Here the solid line shows the spectral density of the underlying (hidden) process $\{X_t\}$. We see that it has a larger power on lower frequency, and returning to Diagram 1 we realize why. The dotted line shows us the estimated spectral density of $\{U_t X_t\}$. Note that it has much larger power, thanks to the Poisson amplitude-modulation. Clearly using the naïve estimate would be misleading. Nonetheless, please note that its overall decreasing shape is correct. The interested reader may return to the above-presented formulae to test this conclusion theoretically. The dot-dashed line is the oracle E-estimate based on the hidden (but known to us from the simulation) time series $\{X_t\}$ shown in Diagram 1. This estimate is good and it indicates that the underlying time series is reasonable to begin with. Finally, the proposed E-estimate is shown by the dashed line. This is a fair estimate keeping in mind complexity of the nonstationary modification and that almost 27% of initial observations are missed, and note that the E-estimate is dramatically better than the naïve estimate.

9.6 Nonstationary Autocovariance and Spectral Density

Consider a zero-mean time series $\{X_t\}$ and suppose that we observe it on the unit interval of time $[0,1]$ at moments $1/n, 2/n, \ldots, n/n$, that is we observe $X_{1/n}, X_{2/n}, \ldots, X_{n/n}$. So far we have considered only cases of stationary or at least second-order stationary time series for which the spectral density is well defined. What will be if $\{X_t\}$ is not second-order stationary? Can we introduce the notion of a spectral density that changes over the time? How can this problem be formulated and then solved? This section explores possible answers to these questions.

First of all, let us recall that for a zero-mean and second-order stationary time series $\{X_t\}$ the spectral density is defined as

$$g^X(\lambda) := (2\pi)^{-1}\gamma^X(0) + \pi^{-1}\sum_{j=1}^{\infty}\gamma^X(j)\cos(j\lambda), \quad -\pi < \lambda \leq \pi. \tag{9.6.1}$$

Here $\gamma^X(j)$ is the autocovariance function which for any $l = 1, 2, \ldots, n-j$ can be defined as

$$\gamma^X(j) := \gamma_l^X(j) := \mathbb{E}\{X_{l/n}X_{(l+j)/n}\}. \tag{9.6.2}$$

Because the time series is second-order stationary, in (9.6.2) the autocovariance does not depend on a particular l. In general this may not be the case.

Let us present an example where (9.6.2) does not hold. Consider a causal ARMA(1,1) process $X_t - aX_{t-1/n} = \sigma(W_t + bW_{t-1/n})$, $|a| < 1$. If a and b are constants, we know from Section 8.2 that

$$\gamma^X(0) = \frac{\sigma^2[(a+b)^2 + 1 - a^2]}{(1-a^2)}, \quad \gamma^X(1) = \frac{\sigma^2(a+b)(1+ab)}{(1-a^2)},$$

$$\gamma^X(j) = a^{j-1}\gamma(1), \quad j \geq 2, \tag{9.6.3}$$

and the corresponding spectral density is

$$g^X(\lambda) = \frac{\sigma^2|1 + be^{i\lambda}|^2}{2\pi|1 - ae^{i\lambda}|^2}. \tag{9.6.4}$$

Note that if $a > 0$ and $b > 0$, then $\gamma(1) > 0$, and a realization of the time series will "slowly" change over time. On the other hand, if $a+b < 0$ and $1+ab > 0$, then a realization of the time series may change its sign almost every time. Thus, depending on parameters a and b, we may see either slow or fast oscillations in a realization of an ARMA(1,1) process. Figures 8.1 and 8.2 allow us to visualize ARMA(1,1) processes with $(a = -0.3, b = -0.6)$ and $(a = 0.4, b = 0.5)$, respectively. What will be if the parameters of ARMA(1,1) process change from those in Figure 8.1 to those in Figure 8.2 when time t increases from 0 to 1? In other words, suppose that

$$a := a(t) := a_0 + a_1 t, \quad b := b(t) := b_0 + b_1 t, \ t \in [0,1]. \tag{9.6.5}$$

Then, according to (9.6.3) and (9.6.4), we have changing in time (dynamic, nonstationary) autocovariance function and spectral density.

This example motivates us to introduce the following characteristics of a zero-mean time series $\{X_t, \ t = 1/n, 2/n, \ldots\}$ with bounded second moments $\mathbb{E}\{X_t^2\} \le c^* < \infty$. We introduce a changing in time (dynamic, nonstationary) autocovariance function

$$\gamma_t^X(j) := \mathbb{E}\{X_t X_{t+j/n}\}, \quad 0 \le t \le 1. \tag{9.6.6}$$

In its turn, the changing in time autocovariance yields the dynamic (nonstationary) spectral density

$$g_t^X(\lambda) := (2\pi)^{-1}\gamma_t^X(0) + \pi^{-1}\sum_{j=1}^{\infty}\gamma_t^X(j)\cos(j\lambda), \quad -\pi < \lambda \le \pi, \ 0 \le t \le 1. \tag{9.6.7}$$

These characteristics explain a possible approach for the spectral analysis of nonstationary time series. Suppose that we are dealing with a realization $X_{1/n}, X_{2/n}, \ldots, X_{n/n}$ of a zero-mean time series $\{X_t\}$ with a bounded second moment. Then the problem is to estimate the dynamic autocovariance $\gamma_t^X(j)$ and the dynamic spectral density $g_t^X(\lambda)$.

Note that a dynamic autocovariance is a bivariate function in t and j. As a result, the proposed solution is to fix $j < n$ and consider (9.6.6) as a nonparametric regression function with t being the predictor and $X_t X_{t+j/n}$ being the response. Indeed, we may formally write

$$X_{l/n} X_{(l+j)/n} = \gamma_{l/n}^X(j) + \varepsilon_{l,j,n}, \quad l = 1, 2, \ldots, n-j, \tag{9.6.8}$$

and use the regression E-estimator of Section 2.3 to estimate $\gamma_t^X(j)$ as a function in t. Then we can use these estimators to construct E-estimator of the dynamic spectral density. Because the dynamic spectral density is a bivariate function, it is better to visualize the corresponding dynamic autocovariance functions, and this is the recommended approach.

Now, when we know the model and the estimation procedure, let us analyze several realizations of a nonstationary time series.

Figure 9.10 allows us to look at a particular realization of the above-presented example of ARMA(1,1) process with changing in time coefficients. Its caption explains the simulation and the diagrams. The top diagram shows us a particular realization of the nonstationary Gaussian ARMA process. Note how, over the time of observation of the process, initial high-frequency oscillations transfer into low-frequency oscillations. It may be instructive to look one more time at the corresponding stationary oscillations shown in Figures 8.1 and 8.2 and then compare them with the nonstationary one.

Now let us see what the discussed in Section 8.2 traditional methods of spectral density estimation, developed for stationary processes, produce. The periodogram is clearly confused and exhibits several possible periodic components (look at the modes and try to evaluate possible periods using the rule $T = \lambda/(2\pi)$ discussed in Section 8.2). The bottom diagram shows the initial underlying spectral density (the solid line), the final underlying spectral density (the dotted line), and the E-estimate of Section 8.2 (the dashed line). Note that

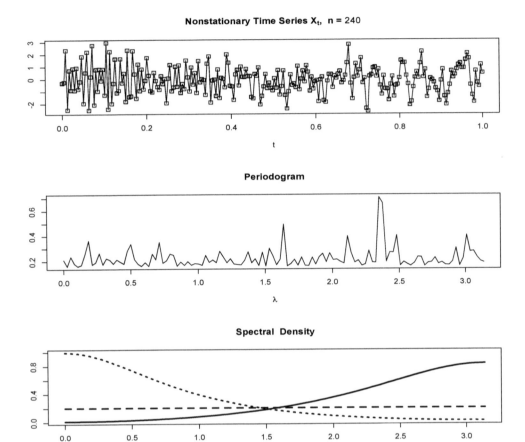

Figure 9.10 *Nonstationary ARMA(1,1) time series and its naïve spectral analysis. The top diagram shows a particular realization of a nonstationary Gaussian ARMA(1,1) time series $X_{l/n} - a(l/n)X_{(l-1)/n} = \sigma(W_{l/n} + b(l/n)W_{(l-1)/n})$, $l = 1, 2, \ldots, n$, where $a(l/n) = a_0 + (l/n)a_1$, $b(l/n) = b_0 + (l/n)b_1$, and $\{W_t\}$ is a standard Gaussian white noise. The middle diagram shows the periodogram. The bottom diagram shows by the solid and dotted lines spectral densities of stationary Gaussian ARMA(1,1) processes with parameters (a_0, b_0, σ) and (a_1, b_1, σ), respectively. In other words, these are initial $(t = 0)$ and final $(t = 1)$ spectral densities of the underlying nonstationary ARMA time series. The dashed line shows the spectral density E-estimate of Section 8.2 which is developed for a second-order stationary time series. {Parameters a_0, b_0, a_1, b_1 are controlled by arguments a0, b0, a1, and b1, respectively.} [n = 240, sigma = 1, a0 = -0.3, b0 = -0.6, a1 = 0.4, b1 = 0.5, cJ0sp = 2, cJ1sp = 0.5, cTHsp = 4]*

the E-estimate, similarly to the periodogram, is confused and indicates that the observed process is a white noise.

What we see in Figure 9.10 is in no way the fault of the estimators because they are developed for stationary processes and cannot be used for nonstationary ones.

Now let us check how the proposed methodology of dynamic autocovariances estimated via the regression E-estimator works out. Figure 9.11 illustrates the proposed approach. Here the top diagram shows us another simulation of the same Gaussian nonstationary process used in Figure 9.10. We again see a familiar pattern of changing the dynamic of the process over time from high to low frequency oscillations. The three other diagrams show underlying autocovariances (the solid line) and the E-estimates (the dashed line). As

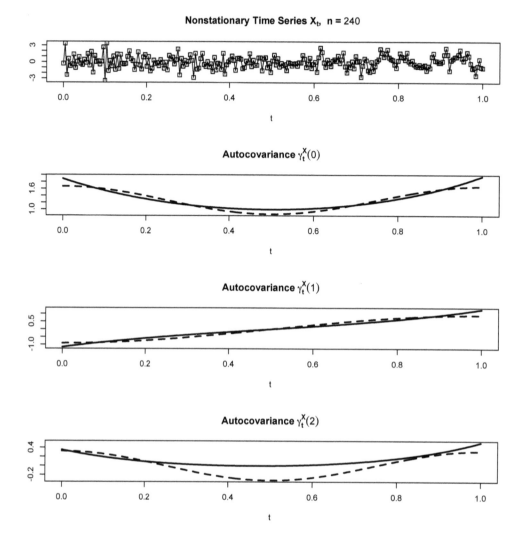

Figure 9.11 *Estimation of the first three autocovariance functions for a nonstationary time series. The time series is generated as in Figure 9.10. The solid and dashed lines show an underlying autocovariance $\gamma_t^X(j)$ and its regression E-estimate, respectively. [n = 240, sigma = 1, a0 = -0.3, b0 = -0.6, a1 = 0.4, b1 = 0.5, cJ0 = 4, cJ1 = 0.5, cTH = 4]*

we see, the E-estimates are not perfect but they correctly show the dynamic of underlying autocovariances. Note that for a stationary process these three curves should be constant in time. Hence we can use the developed methodology for testing second-order stationarity.

We do not show here the corresponding dynamic spectral density $g_t^X(\lambda)$ because it is a bivariate function in t and λ, and we know that it is not easy to visualize a bivariate function. On the other hand, recall that at each moment t the dynamic autocovariances are coefficients in Fourier expansion (9.6.7). Hence, at each moment in time we can reconstruct the spectral density E-estimate using E-estimates of dynamic autocovariances.

Let us finish this section by a remark about more general formulation of the problem. Consider a set of time series $\{X_{t,\tau}\}$ where t is a discrete time of moments when we can observe these time series and τ is a continuous parameter that controls second-order characteristics like the autocovariance function and the spectral density. Note that we are dealing with infinitely many underlying time series of interest that are synchronized in time. We do not observe all these time series simultaneously. Instead, we may observe the trajectory (realization) of a process $X_{t,t}$, $t = 1, 2, \ldots, n$. Note that every time we observe realization of a different underlying time series. The latter is a very special modification of underlying time series because each time we "jump" from observing one time series to observing another time series. Similarly to the previous setting with a dynamic ARMA process, we would like to infer about second-order characteristics of the underlying set of time series. The main assumption is that if $\tau = t$, then the second-order characteristics change slowly in time t. Under this assumption, the above-presented solution is applicable.

9.7   The Simpson Paradox

The Simpson paradox is an eye-opening statistical phenomenon which is traditionally used to introduce multi-way tables and then explain why it is important to keep track of lurking variables. The paradox shows that aggregation of data and ignoring lurking variables can reverse our conclusion/opinion about data and make it debatable. This section shows how the paradox can be used to enrich our understanding of the linear regression and the usefulness of conditional density in searching after possible lurking variables.

We begin with a simple, and at the same time, classical example of Simpson's paradox. Suppose that we want to make an informative decision about the salary of college A and college B graduates with a master's degree who graduated 10 years ago (let us agree to refer to those graduates as graduates A and B). We may randomly survey 100 A and 100 B former students who graduated 10 years ago and get data which includes their current salary and college GPA (grade point average). For sensitivity reasons, we agree to deal only with rescaled onto $[0, 1]^2$ data. Table 9.1 presents sample means (of rescaled to $[0, 1]$) salaries for A and B graduates.

Table 9.1 *Sample mean salaries*

A Graduates	B Graduates
0.47	0.31

Our conclusion from the survey is clear: the mean salary of B graduates is only 66% of the mean salary of the A graduates.

Now let us look more closely at the available data and take into account the lurking variable "field of concentration" which can be either science or engineering. Table 9.2 presents the corresponding data.

Table 9.2 *Sample mean salaries*

Graduates	Engineers	Scientists
A	0.56	0.18
B	0.61	0.21

Table 9.2 sheds an absolutely different light on the same data and dramatically changes our opinion about salaries of A and B graduates. The data indicates that B graduates have larger salaries in both fields. Note that Table 9.2 is a classical two-way table which takes into account the lurking variable "field of concentration."

How can B graduates do better than A graduates in every field according to Table 9.2 yet still fall far behind A graduates according to Table 9.1? The explanation is in the larger number of B graduates concentrating in science where salaries are lower than in engineering. When salaries from both concentrations are lumped together, the B graduates place lower because the fields they favor pay less.

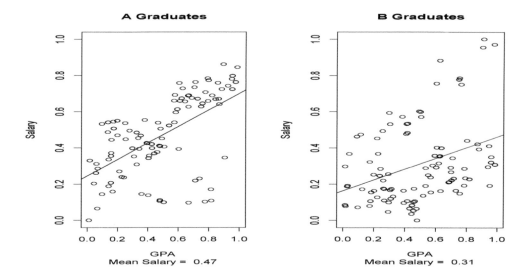

Figure 9.12 *Two scattergrams overlaid by linear regressions. The left diagram presents data for graduates with a master's degree from college A, and the right diagram for graduates with a master's degree from college B. The simulation and its parameters are explained at the end of Section 9.7. {This figure is the first part of the triplet generated by Figure 9.12. To proceed to Figure 9.13, at the prompt* **Browser[1]** *> enter letter c and then press Return. The same procedure is used to proceed to Figure 9.14.}* [n = 100, k = 10, sigma = 0.1, a = 0.8, b = 0.2, eta = 0.4]

The original one-way Table 9.1 is misleading because it does not take into account the lurking categorical variable "field." This misleading constitutes Simpson's paradox when an association or comparison that holds for all of several groups can reverse direction when the data are combined to form a single group.

Of course, there are many other examples that are similar to the above-presented case of two colleges. For instance, you may want to compare two hospitals with the lurking variable being the proportion of elderly patients, or you may want to compare returns of two mutual funds with different required allocations in bonds and stocks, etc.

The main teaching moment is that Simpson's paradox helps us to recognize the importance of paying attention to lurking variables. It truly teaches us to look "inside" the data and challenge "obvious" conclusions.

Can the paradox be useful in understanding other familiar statistical topics? Let us continue the discussion of the salary example, and we will be able to recognize Simpson's paradox in a linear regression and learn how to investigate it using the conditional density. Another interesting feature of the example is that we will be using a continuous stochastic process in an underlying model.

Figure 9.12 presents data and linear regressions with the salary, 10 years after graduation, being the response and GPA being the predictor. The two diagrams correspond to graduates from schools A and B, and there are $n = 100$ graduates from each school. The solid lines, overlaying the scattergrams, exhibit the least squares linear regressions for A and B graduates. The mean salaries are shown in the subtitles.

What we see in Figure 9.12 is a definite testament to the superiority of School A. Not only the mean salary of A graduates is significantly larger, the better learning at school A, reflected by the GPA, is dramatically more rewarding in terms of the salary as we can see from the larger slope in the linear regression. Interestingly, even A graduates with the worst grades do better than their peers from school B (compare the y-intercepts). Note that Figure 9.12 is the extended regression analog of Table 9.1, and it allows us to make much

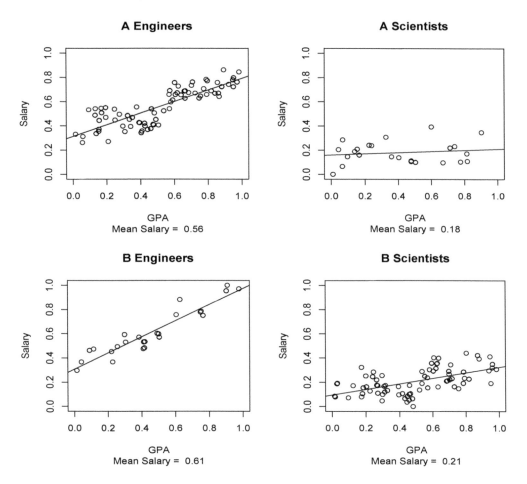

Figure 9.13 *Scattergrams taking into account the field of specialization. Solid lines are least-squares regressions. This is a continuation of Figure 9.12.*

stronger conclusion about rather dramatic differences in the salary patterns of A and B graduates. After analysis of the data, it is absolutely clear that school A does a superb job in educating its students and preparing them for the future career.

Now, similar to Table 9.2, let us take into account the lurking variable field (science or engineering). Figure 9.13 presents corresponding scatterplots overlaid by linear regressions. Note that the four diagrams give us much better visualization and understanding of data than a two-way table like Table 9.2. These diagrams also completely change our opinion about the two schools. What we see is that in each field B graduates do better and the success in learning, measured by the GPA, is more rewarding at school B for both fields. This is a complete reversal of the previous conclusion based on the analysis of Figure 9.12. Just think about a similar example when performances of two mutual funds are compared and how wrong conclusions that do not take into account all pivotal information may be.

How can B graduates do better in all aspects of the salary in every field, yet fall far behind when we look at all engineers and scientists? The answer is the same as in the previous discussion of the multi-way tables. When salaries are lumped together, the B graduates are placed lower because the fields they favor pay less. What makes the presented example so attractive is that the lurking variable "field" dramatically changes our opinion not only

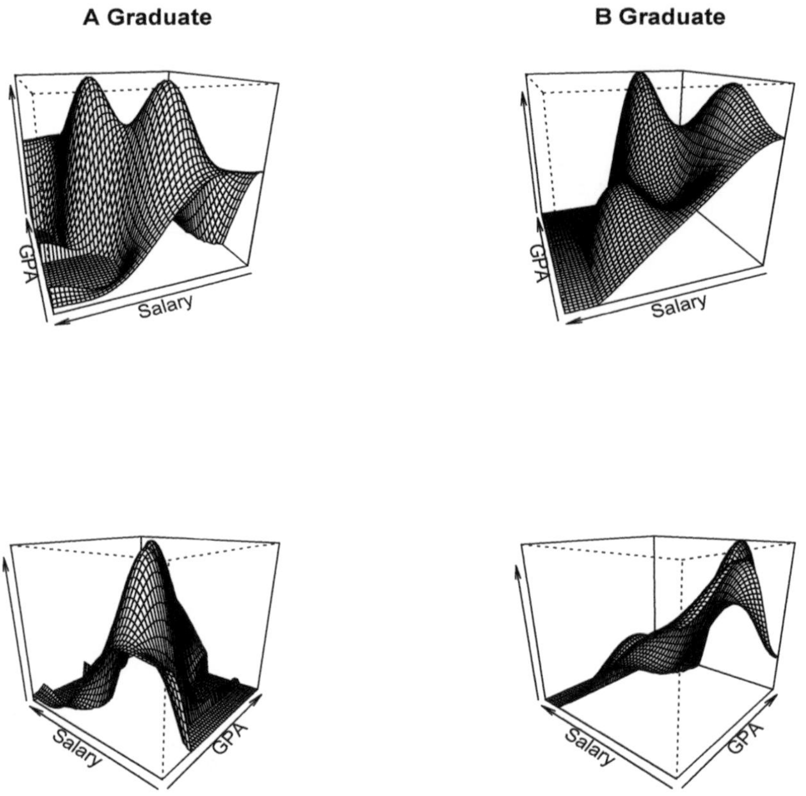

A Graduate   **B Graduate**

Figure 9.14 *Estimated conditional densities of salary given GPA. Each column of diagrams shows the same E-estimate using two different "eye" locations. This is a final figure created by Figure 9.12.*

about average salaries, but also about the effect of being a good-standing (higher GPA) student.

As we see, Simpson's paradox for a regression may be even more confusing and challenging than its classical multi-way-table counterpart where only mean values are compared. As a result, the paradox motivates us to think about data and to not rush with conclusions based on employing standard statistical tools.

Can a statistical science suggest a tool for recognizing a possible Simpson's paradox and/or the presence of an important lurking variable? The answer is "yes," and the tool is the conditional density. Remember that if $f^{XY}(x,y)$ is the joint density of the pair (X,Y) of random variables, then the conditional density of Y given X is defined as $f^{Y|X}(y|x) := f^{XY}(x,y)/f^{X}(x)$ assuming that the marginal density $f^{X}(x)$ of X is positive. Of course, the conditional density is a bivariate function and, as we know from Section 4.4, this presents its own complications caused by the curse of dimensionality. Nonetheless, it may help us in visualization of a potential issue with a lurking variable.

Figure 9.14 allows us to look at nonparametric E-estimates of the conditional density of salary given GPA for the two schools based on data exhibited in Figure 9.12. The two "eye" locations help us to visualize the surfaces. Note that intersection of a vertical slice (with constant GPA) with the surface shows us the estimated conditional density of the salary given a specific GPA. The two pronounced ridges in conditional densities for both schools clearly indicate the dependence of the salary on the GPA, and they also indicate two possible strata in the data that may be explained by a lurking variable. After making this observation, it may be a good learning experience to return to the available data shown in Figure 9.12 and realize that the indicated by the conditional density strata may be also seen in the scattergrams.

Now let us explain the underlying simulation used in Figure 9.12. For both schools the rescaled GPA is a uniform random variable on $[0,1]$. For an lth graduate from a school A with a field of concentration being F_l, which is either engineering when $F_l = 1$ or science if $F_l = 0$, the model for the salary $S_l(A)$ as a function of the GPA G is defined as a rescaled onto $[0,1]$ random variable $S_l'(A)$ which is defined as

$$S_l'(A) = 2 + G_l + F_l(A)(1 + 2G_l) + \sigma W(A, G_l, k) + \eta Z_l(A), \ l = 1, \dots, n. \quad (9.7.1)$$

Here $F_l(A)$ is a Bernoulli random variable with $\mathbb{P}(F_l(A) = 1) = a$, $W(A, t, k) = \sum_{j=0}^{k-1} W_j(A)\varphi_j(t)$ is a standard frequency-limited white Gaussian noise with $W_j(A)$ being independent standard Gaussian variables, $Z_l(A)$ are also independent standard Gaussian variables, and a, k, σ, η are parameters that may be changed while using Figure 9.12. Similarly for the school B,

$$S_l'(B) = 2 + 1.2G_l + F_l(B)(1 + 3G_l) + \sigma W(B, G_l, k) + \eta Z_l(B), \ l = 1, \dots, n. \quad (9.7.2)$$

Here $F_l(B)$ is a Bernoulli random variable with $\mathbb{P}(F_l(B) = 1) = b$, it is the indicator that the lth graduate from school B is engineer, and b is the parameter that may be changed while using Figure 9.12. The process $W(B, t, k)$ is defined as explained above for school A.

The reader is advised to repeat Figure 9.12 with different parameters and get a better understanding of the underlying idea of Simpson's paradox for regression and how the conditional density may shed light on the paradox. Another teachable moment is to visualize data in Figure 9.12 and learn how to search after possible strata directly from scattergrams.

9.8 Sequential Design

So far we have considered problems where a sample is given and then the problem is to estimate a nonparametric function of interest. In some situations it may be up to a statistician to design an experiment. For instance, consider a familiar nonparametric regression,

$$Y_l = m(X_l) + \sigma(X_l)\varepsilon_l, \ l = 1, 2, \dots, n. \quad (9.8.1)$$

In a standard setting it is assumed that $X_1, \dots, X_n$ are independent and identically distributed random variables with a design density $f^X(x)$ supported on $[0,1]$, $\sigma(x)$ is a scale function (by definition it is nonnegative) and $\varepsilon_1, \dots, \varepsilon_l$ are independent zero-mean and unit-variance variables. Then the asymptotic theory asserts that, under a mild assumption, the MISE (mean integrated squared error) of an optimal regression estimator is proportional to the following functional,

$$d(f^X, \sigma) := \int_0^1 \frac{\sigma^2(x)}{f^X(x)} dx. \quad (9.8.2)$$

This is an interesting theoretical result because it shows how the design and the scale affect the quality of regression estimation. Further, it immediately raises the following question. What is the optimal design of predictors that minimizes the MISE? To answer the question, we need to find a density which minimizes (9.8.2) for a given scale function $\sigma(x)$. It is a straightforward minimization problem that implies optimality of the design density

$$f_*^X(x) := \frac{\sigma(x)}{\int_0^1 \sigma(x)dx}. \quad (9.8.3)$$

In its turn, this design implies that the minimal functional $d(f^X, \sigma)$ is

$$d_*(\sigma) := d(f^X_*, \sigma) = \int_0^1 \sigma^2(x)[\sigma(x)/\int_0^1 \sigma(u)du]^{-1}dx = [\int_0^1 \sigma(x)dx]^2. \qquad (9.8.4)$$

It is a nice exercise in probability to check, using Cauchy-Schwarz inequality (1.3.32), that $f_*(x)$ indeed minimizes (9.8.2). Write,

$$d(f^X_*, \sigma) = [\int_0^1 \sigma(x)dx]^2 = [\mathbb{E}\{\sigma(X)/f^X(X)\}]^2$$

$$\leq \mathbb{E}\{[\sigma(X)/f^X(X)]^2\}\mathbb{E}\{1\} = d(f^X, \sigma). \qquad (9.8.5)$$

It is also possible to establish (9.8.5) using a functional form (1.3.33) of Cauchy-Schwarz inequality or the inequality $[\mathbb{E}\{Z\}]^2 \leq \mathbb{E}\{Z^2\}$.

In some applications it is possible to collect data sequentially, that is first get a pair (X_1, Y_1), then (X_2, Y_2), etc. Further, in a controlled design it is possible to employ different design densities for each predictor. As a result, in a sequential controlled design it is possible after each observation to choose a design density for the next predictor. For instance, after obtaining the first $k < n$ observations $(X_1, Y_1), \ldots, (X_k, Y_k)$, it is possible to decide what distribution to choose for generating the next predictor X_{k+1}. Then we observe a new pair (X_{k+1}, Y_{k+1}), and based on the available $k+1$ pairs decide what distribution to use for the next predictor, etc.

It follows from (9.8.3) that the optimal design density is proportion to the scale function. On one hand, this makes sense because the larger the scale, the more difficult to estimate an underlying regression function. On the other hand, the result is pure theoretical. As a result, the following question is worthy of exploration. Can the asymptotic result (9.8.3) guide us in designing a controlled regression for relatively small sample sizes? This is not a simple question. First of all, the approach is based on the asymptotic theory. This means that in a series expansion a large number of Fourier coefficients is estimated and used, and this is clearly not the case for small samples. Further, to use the underlying idea of implementing (9.8.3) we need to estimate the scale function and then plug it in (9.8.3). We know from Section 3.6 that estimation of the scale is a challenging problem on its own because it involves two steps, with the first step being estimation of the regression function and with the second step being estimation of the scale based on squared residuals. Further, we cannot use all n observations for estimation of the scale and then use the optimal design for n new observations, we have a total of n observations to deal with. The asymptotic theory asserts that a two-stage sequential procedure is asymptotically optimal when the first $k := \lfloor b_n n \rfloor$, $0 < b_n < 1$, $b_n = o_n(1)$ observations are used to estimate the scale and the corresponding design density (9.8.3), and then the estimated design density is used for the remaining $n - k$ observations. This result sheds a new light on the problem because it may look more attractive to constantly adapt, after each observation, the design to an underlying scale function. The two-stage approach is a significant simplification, but it also stresses the challenges. Indeed, only $k < n$ observations are used to estimate the scale, and then only $n - k < n$ observations have an optimal design, and it is clear that asymptotically k should be in order smaller than n to attain the optimal outcome. Finally, the ratio between (9.8.2) and (9.8.4) may be close to 1, and we cannot evaluate it prior to an experiment.

Despite all these potential complications for small samples, it is of interest to test this appealing asymptotic idea and to learn about a sequential controlled design. One of the typical approaches in testing ideas, based on asymptotic results, is via simulations, and this approach is used in this section. We divide the test into two parts. First of all, we check that using the design density $f^X_*(x)$, defined in (9.8.3), improves quality of estimation measured by the MISE. Then we test a sequential controlled design.

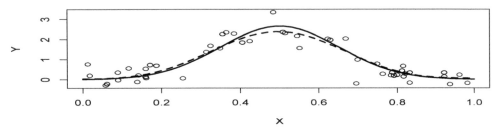

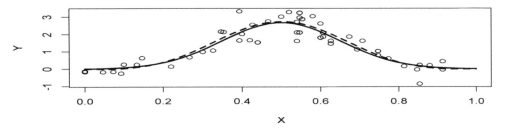

Figure 9.15 *Testing the idea of the optimal design. Regression scattergrams, simulated according to (9.8.1), are shown by circles overlaid by the underlying regression function (the solid line) and its E-estimate (the dashed line). The Uniform design density is used in the top diagram, the optimal design density $f_*^X(x)$, proportional to the scale, is used in the bottom diagram. The same errors ε_l, $l = 1, \ldots, n$ are used in the two regressions. The scale function is defined as $\sigma(a + f_2(x))$ where $f_2(x)$ is the Normal density. In a title, ISE is the empirical integrated squared error, AISE is the sample mean of ISEs obtained for nsim simulations. $R := \int_0^1 [\sigma(x)]^2 dx / [\int_0^1 \sigma(x) dx]^2$ is the ratio between the functionals (9.8.2) for the Uniform and optimal designs, it is shown in the title of the bottom diagram. {The argument corn controls an underlying regression function, n controls the sample size, sigma controls σ, nsim controls the number of repeated simulations used to calculate AISE.} [n = 50, corn = 2, sigma = 0.1, a = 3, nsim = 100, cJ0 = 4, cJ1 = 0.5, cTH = 4]*

Figure 9.15 allows us to conduct the first test, and its caption explains the diagrams. Here the scale function is $\sigma(x) = \sigma(a + f_2(x))$ where $f_2(x)$ is the Normal density, $\sigma = 0.1$ and $a = 3$. The top diagram uses the Uniform density as the design density, and the bottom diagram uses the design density $f_*^X(x)$ proportional to the underlying scale function. The same regression errors ε_l are used in the both simulations, so the only difference is in the design of predictors. What we see here is that indeed the optimal design placed more observations in the middle of the unit interval. On the other hand, note that in the bottom diagram we have less observations to estimate tails of the regression function. In particular, under the optimal design there are no observations near the end of the right tail.

Let us stress that we are considering random designs, so each simulation may have its own specific issues. Further, because the scale is larger near the center of the interval of estimation, in the bottom diagram we observe more cases with larger regression errors. Nonetheless, for this particular simulation the optimal design implied a smaller integrated squared error (ISE) shown in the title.

Of course, we have analyzed a single simulation, and it is of interest to repeat the simulation a number of times and then compare the sample means of corresponding ISEs. Figure 9.15 allows us to do this. Namely, it repeats the outlined simulation $nsim = 100$ times, for each simulation calculates ISEs for the Uniform and optimal designs, then averages

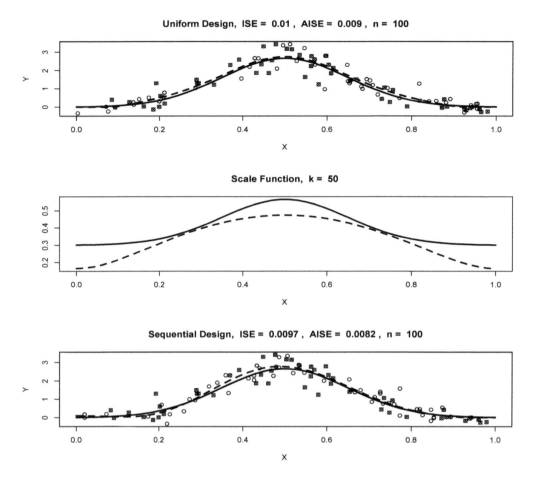

Figure 9.16 *Testing a two-stage sequential design. The underlying regression is the same as in Figure 9.15. Regression scattergrams are shown by circles overlaid by the underlying regression function (the solid line) and its estimate (the dashed line). The top diagram shows a simulation (9.8.1) with the Uniform design. The middle diagram shows estimation of the scale function based on first k observations shown by crossed circles in the top diagram. Here $k := \lfloor bn \rfloor$ where b is the parameter. The bottom diagram shows results for a particular sequential simulation. Here the first k pairs of observations are the same as in the top diagram (generated according to the Uniform design density) and indicated by crossed circles, and the next $n - k$ are generated according to the density proportional to the estimated scale function shown in the middle diagram. Indicated ISE and AISE are the integrated squared errors for the shown regression estimates and averaged ISEs over nsim simulations. [n = 100, corn = 2, sigma = 0.1, a = 3, b = 0.5, nsim = 100, cJ0 = 4, cJ1 = 0.5, cTH = 4]*

them and shows AISEs in the titles. The above-described numerical study is a traditional statistical tool to compare different statistical procedures.

The AISEs, shown in the titles of the corresponding diagrams, are relatively close to each other but do benefit the optimal design. It is of interest to compare them with the theoretical ratio $R := d(1, \sigma)/d_*(\sigma) = 1.06$ shown in the title of the bottom diagram. As we see, the potential for improvement is rather modest, but nonetheless it does exist. Of course, the small ratio stresses complexity of the problem of a sequential controlled design for small samples because this design involves more estimation procedures that may go wrong. Nonetheless, the outcome of Figure 9.15 is encouraging, and we may proceed to exploring a sequential design.

A sequential design, explored in Figure 9.16, is a two-stage design. First, we conduct $k = \lfloor bn \rfloor$ simulations according to the Uniform design. These observations are used to estimate the scale function by the regression E-estimator of Section 3.6. This procedure constitutes the first stage of the design. The second stage is to generate the remaining $n - k$ pairs of observations using the design density calculated according to (9.8.3) with the plug-in scale E-estimate. Note that in the available sample the last $n - k$ pairs of observations are mutually independent, but observations collected during the first and second stages are dependent.

Figure 9.16 shows us a particular two-stage sequential simulation. Its top diagram is simulated identically to the top diagram in Figure 9.15. The only difference here is that the first k pairs are highlighted by crossed circles. These are the pairs used to estimate the underlying scale function, and the result is shown in the middle diagram. The E-estimate of the scale (the dashed line) is far from being perfect, but note that only its shape is of interest (recall formula (9.8.3)), and the E-estimate does indicate that the scale function is symmetric about 0.5 and has a pronounced mode. Sequential design is shown in the bottom diagram. Here the first k pairs are the same as in the top diagram, and they are highlighted by crossed circles. The remaining $n - k$ pairs are generated using the estimated optimal design. The particular outcomes and the AISEs favor the sequential design but they are too close to each other. Because of this, another simulation may reverse the outcome. The issue here is that the size n is too small for the considered challenging regression problem. Nonetheless, this outcome may be considered as a success. Indeed, with the ratio $R = 1.06$ (recall Figure 9.15), the sequential controlled design has a tiny margin for error. Further, the studied sequential regression estimation is a dramatically more complicated procedure than the regression E-estimator for the Uniform design. And the fact that the AISEs are close to each other for the relatively small sample size $n = 100$ is encouraging.

We may conclude that a sequential controlled design is an interesting and promising idea, but in general it is a challenging one for small samples when initial observations are used for estimation of nuisance functions and then obtained estimates define an optimal design for next observations. This is when a numerical study becomes a pivotal tool. For the considered regression problem, Figures 9.15 and 9.16, repeated with different parameters, may help to learn more about this interesting problem.

9.9   Exercises

9.1.1 Explain the problem of regression estimation with dependent responses.

9.1.2 Suppose that in the regression model (9.1.1) the predictor X_l is uniformly distributed and ε_l are realizations of a stationary time series. Under this assumption, are observations (X_l, Y_l) stationary or nonstationary?

9.1.3 Consider the previous exercise only now $X_l = l/n$. Are observations (X_l, Y_l) stationary or nonstationary?

9.1.4 Explain how regression E-estimator performs for the case of a model (9.1.1) with independent errors.

9.1.5* Suppose that regression errors are a zero-mean and second-order stationary time series. What is the mean and the variance of the sample mean estimator (9.1.4)?

9.1.6 Verify every step in establishing (9.1.6). Comment about used assumptions.

9.1.7* Explain why the second sum in (9.1.7) is sensitive to the design of predictors.

9.1.8 Verify (9.1.8).

9.1.9 Explain why (9.1.9) holds or does not hold for both designs of predictors.

9.1.10 Verify (9.1.12) for a random design and then explain why this result is a good news for a random design regression.

9.1.11 Is (9.1.12) valid for a fixed design regression?

9.1.12* What is the MISE of the regression E-estimator for the case of Gaussian ARMA regression errors and a fixed design of predictors? Make any additional assumptions that may be helpful.

9.1.13 Explain the underlying simulation used in Figure 9.1.

9.1.14 What is the definition of long-memory errors?

9.1.15 Do long-memory errors affect regression estimation for the case of a random design? Explain your answer theoretically and then support by simulations using Figure 9.1.

9.1.16* Consider the case of regression with fixed design of predictors and long-memory errors. Calculate the rate of the MISE convergence for the E-estimator, and then complement your answer by simulations.

9.1.17* Explore the effect of σ on regression estimation with dependent errors. Use both theoretical and empirical approaches.

9.1.18 How well may the shape of a regression function be estimated?

9.1.19* Figure 9.1 indicates that the E-estimator has difficulties with estimation of the mean of regression function. Explain why, and then answer the following question. Is this problem specific for the E-estimator?

9.1.20 Using Figure 9.1 for different sample sizes, write a report on how the sample size affects estimation of regression function in the presence of long-memory errors.

9.1.21 Using Figure 9.1, explore the effect of parameter β on quality of estimation.

9.1.22 Propose better parameters for the E-estimator used in Figure 9.1.

9.2.1 Give several examples of continuous stochastic processes.

9.2.2 Present several examples of stationary and nonstationary stochastic processes.

9.2.3 Find the mean and the variance of the continuous stochastic process (9.2.1). Hint: Calculate these characteristics for a particular time t.

9.2.4 Is the process (9.2.1) stationary or nonstationary?

9.2.5 What is the distribution of process (9.2.1) at time t?

9.2.6 Why is the limit, as $k \to \infty$, of the process $W(t, k)$ called a white process?

9.2.7 Is it possible to simulate a process $W(t, \infty)$?

9.2.8 Give the definition of a Brownian (Wiener) process. What is the underlying idea behind its definition?

9.2.9 Verify (9.2.4).

9.2.10* Consider two independent Brownian motions. What can be said about their difference and sum? What will change if the Brownian motions are dependent?

9.2.11 Consider times $t_1 < t_2$. What is the distribution of $B(t_2) - B(t_1)$?

9.2.12* Consider times $t_1 < t_2$. What is the distribution of $B(t_1) + B(t_2)$?

9.2.13* Explain the meaning of infinite sum in (9.2.5).

9.2.14 Verify (9.2.6).

9.2.15 Verify each equality in (9.2.7).

9.2.16 Explain the simulation used by Figure 9.2.

9.2.17* Explain theoretically, and then test empirically using Figure 9.2, how parameter k affects the shape of curves.

9.2.18 Explain the problem of filtering a continuous signal from white noise.

9.2.19 Are the models (9.2.8) and (9.2.9) equivalent?

9.2.20* Explain how the E-estimator performs filtering a signal from white noise.

9.2.21* Consider the case of a signal from a Sobolev class $\mathcal{S}_{\alpha,Q}$ defined in (2.1.11). First, find the MISE of the E-estimator for a given cutoff J. Second, find a cutoff that minimizes the MISE. Finally, calculate the corresponding minimal MISE. Note that the obtained rate of the MISE convergence is the fastest and no other estimator can produce a better rate.

9.2.22* Consider the case of a signal from an analytic class $\mathcal{A}_{r,Q}$ defined in (2.1.12). First, find the MISE of the E-estimator for a given cutoff J. Second, find a cutoff that minimizes the MISE. Finally, calculate the corresponding minimal MISE. Note that the obtained rate of the MISE convergence is the fastest and no other estimator can produce a better rate as $n \to \infty$. Further, even the constant of the MISE is minimal among all possible estimators.

9.2.23 Explain the relations (9.2.12).

9.2.24 Consider (9.2.12). Why are W_j independent and have a standard Gaussian distribution?

9.2.25 Explain the underlying idea of estimator (9.2.13).

9.2.26* Find the mean and variance of estimator (9.2.13).

9.2.27 Explain the simulation used in Figure 9.3.

9.2.28 How is parameter σ chosen in Figure 9.3?

9.2.29 Repeat Figure 9.3 with different n and explain how this parameter affects performance of the E-estimator.

9.2.30 Using Figure 9.3, find better parameters of the E-estimator.

9.2.31 Explain relation (9.2.14) and its implications.

9.2.32 Explain the principle of equivalence.

9.2.33 Explain the underlying experiment of Figure 9.4.

9.2.34 Consider the available realization of the process shown in Figure 9.4. Can you propose a method for recovery the missed portion of the process?

9.2.35* Consider a filtering problem. Assume that, similarly to Figure 9.4, a portion of the noisy signal is missed. Can you suggest a situation when an underlying signal may be consistently estimated? Hint: Think about a parametric signal (like a linear regression) or a seasonal component.

9.3.1 What is the definition of a stationary and second-order stationary time series?

9.3.2 Explain a classical decomposition model of a nonstationary time series. Present an example.

9.3.3 What is the definition of a seasonal component?

9.3.4 What is the scale function?

9.3.5 Present and discuss several examples of time series with missed data.

9.3.6 Explain the idea of estimation of the trend.

9.3.7 Verify (9.3.4). Describe terms in the right side of (9.3.4).

9.3.8 What is the underlying idea of estimator (9.3.5)?

9.3.9 Explain (9.3.6). What are the assumptions needed for the validity of this inequality?

9.3.10 Verify (9.3.7).

9.3.11 Prove all equalities in (9.3.8).

9.3.12 Why do we need assumption (9.3.9) for evaluation of the variance of η_j? What will be if it does not hold?

9.3.13* What do relations (9.3.11) tell us about the estimator $\tilde{\theta}_j$? Note that the estimator is biased. Can an unbiased estimator decrease the rate of the mean squared error (MSE) convergence?

9.3.14 Explain the idea of estimation of Fourier coefficients for the case of model (9.3.2) with missing data.

9.3.15* Evaluate the mean and the variance of the estimator (9.3.12).

9.3.16* Explain how the sum, with the random number of addends on the left side of (9.3.14), is replaced by the sum with fixed number of addends. Then prove (9.3.14).

9.3.17* Verify inequality (9.3.15). What is the assumption sufficient for its validity?

9.3.18* Prove (9.3.16).

9.3.19* Estimator $\hat{\theta}_j$, defined in (9.3.12), is studied under the assumption that the function $q(t)$ has a bounded derivative. Is it possible to relax this assumption and still prove that $\mathbb{E}\{(\hat{\theta}_j - \theta_j)^2\} \le cn^{-1}$?

9.3.20* Verify (9.3.17). Note that here N is a random variable.

9.3.21* Verify every relation in (9.3.18). Pay attention to the fact that N is a random variable.

9.3.22 Based on (9.3.19), what can be said about the mean squared error of the estimator $\hat{\theta}_j$?

9.3.23 Explain the underlying idea of estimation of a seasonal component.

9.3.24 Explain formula (9.3.22).

9.3.25* How can a trend be separated from a seasonal component?

9.3.26 Explain formula (9.3.23).

9.3.27 Explain the simulation used in Figure 9.6.

9.3.28* Explain all steps in the analysis of nonstationary time series. Then repeat Figure 9.6 and present analysis of all ten diagrams.

9.4.1 Is the time series $\{Y_t\}$, defined in (9.4.1), stationary or nonstationary?

9.4.2 Define and explain all components of the classical decomposition model (9.4.1).

9.4.3 Is there a difference, if any, between the trend and the seasonal component?

9.4.4 Present several examples of a time series (9.4.1) with a pronounced trend, seasonal component and the scale.

9.4.5 Explain the model (9.4.2) of a nonstationary amplitude-modulated time series.

9.4.6 What is the underlying idea of estimation of the trend?

9.4.7 Explain the motivation behind estimator (9.4.4).

9.4.8* Evaluate the mean and the variance of estimator (9.4.4).

9.4.9 Does the statistic (9.4.5) use unavailable V_{Z_s} for $s > N$?

9.4.10 Suppose that $\mathbb{P}(Y_t = 0) = 0$. Show that in this case from an observation of V_t we can conclude if Y_t is missed, that is, if $U_t = 0$.

9.4.11* Formula (9.4.4) is a naïve numerical integration. Propose a more accurate formula and then evaluate its mean and variance.

9.4.12 Verify (9.4.6).

9.4.13 Prove (9.4.7).

9.4.14 Explain the underlying idea of estimator (9.4.8).

9.4.15* Calculate the mean and variance of estimator (9.4.8).

9.4.16* Prove (9.4.9).

9.4.17* Evaluate the mean and variance of estimator (9.4.10).

9.4.18 Explain the underlying idea of estimator (9.4.10).

9.4.19* Propose another feasible estimator of λ and compare it with estimator (9.4.10).

9.4.20 Explain the simulation used in Figure 9.8.

9.4.21 Explain the difference between Diagrams 1 and 2 in Figure 9.7.

9.4.22* Explain how the E-estimator calculates the E-estimate of the trend in Diagram 3.

9.4.23 Explain how the detrended data in Diagram 4 are obtained. Hint: Pay attention to missing data in Diagram 2.

9.4.24 How was the estimated period in Diagram 5 calculated?

9.4.25 Explain how the estimate of seasonal component is constructed. Repeat Figure 9.8 several times, get a wrong estimate of the seasonal component, and then explain why this happened.

9.4.26 The E-estimate in Diagram 8 is poor, it does not resemble the underlying scale function. Why did this happen?

9.4.27 Does the time series in Diagram 9 look stationary? Do you see any seasonal component or trend?

9.4.28 Diagram 10 shows the estimated spectral density of the rescaled residuals. Does it indicate a seasonal component?

9.4.29 Repeat Figure 9.8 several times using different periods for the seasonal component. Then write a report on how the period affects estimation of the spectral density.

9.4.30 Propose better parameters for the E-estimator of the trend and scale function. Use Figure 9.8 to check your suggestions.

9.4.31 Propose better parameters for the E-estimator of the spectral density. Use Figure 9.8 to check your suggestions.

9.4.32* Explain, both theoretically and using simulations, how parameter λ affects estimation of the trend, scale and spectral density.

9.4.33* Consider a Poisson variable U. Calculate $\mathbb{E}\{U^2|(U > 0)\}$. Then explain how this result can be used in estimation of the scale function in the decomposition of $\{U_t Y_t\}$.

9.5.1 Explain a model of time series with nonstationary missing. Present several examples.

9.5.2 Why do we consider a model of time series on the unit time interval?

9.5.3 Explain formula (9.5.1) for the autocovariance function.

9.5.4* Evaluate the mean and variance of the sample autocovariance function (9.5.2).

9.5.5* Prove every equality in (9.5.3) and (9.5.4).

9.5.6* Is it possible to relax the assumption about bounded derivative of $\lambda(t)$ and still have (9.5.5)?

9.5.7 Verify (9.5.6).

9.5.8 Prove (9.5.7)

9.5.9* Is it possible to relax the assumption of bounded derivative of $\lambda(t)$ for validity of (9.5.7)?

9.5.10 Explain how function $\lambda(t)$ may be estimated based on observations of the time series $\{U_t X_t\}$.

9.5.11 Explain the estimator (9.5.10).

9.5.12* Suppose that the MISE of the regression E-estimator $\hat{m}(t)$ is known. Evaluate the MISE of estimator $\hat{\lambda}(t)$ defined in (9.5.10).

9.5.13* Evaluate the mean and variance of estimator (9.5.11).

9.5.14* Evaluate the mean and variance of estimator (9.5.12).

9.5.15 Repeat Figure 9.9 with different sample sizes n. Based on the experiment, what is the minimal sample size that may be recommended for a reliable estimation?

9.5.16 Using different parameters of an underlying ARMA(1,1) process in Figure 9.9, explore the problem of how these parameters affect estimation of the spectral density.

9.5.17* In Section 8.3 a stationary batch-Bernoulli missing mechanism was studied. Consider a nonstationary batch-Bernoulli missing mechanism and propose a spectral density estimator for an underlying time series $\{X_t\}$.

9.5.18* Explore the case of nonstationary Markov–Bernoulli missing mechanism. Hint: Recall Section 8.3. Think about how many parameters are needed to define a stationary Markov chain, then make them changing in time.

9.5.19* For the setting of the previous exercise, suggest E-estimator for the spectral density.

9.6.1 Give definition of the spectral density of a stationary time series.

9.6.2 Give definition of the autocovariance function of a stationary time series.

9.6.3* Prove (9.6.3) for a causal ARMA(1,1) process.

9.6.4* Verify (9.6.4).

9.6.5 Suppose that (9.6.5) holds. Explain how a typical realization of the corresponding nonstationary ARMA(1,1) process will look like.

9.6.6 Present several practical situations when (9.6.5) occurs.

9.6.7 Why is (9.6.6) called a dynamic autocovariance?

9.6.8 Explain the underlying idea of the dynamic spectral density.

9.6.9 Is the dynamic spectral density a univariate or bivariate function?

9.6.10 What is the simulation used in Figure 9.10?

9.6.11 Explain the time series of observations in the top diagram in Figure 9.10. Does it look like a stationary time series? Explain.

9.6.12 Explain how the periodogram in Figure 9.10 is calculated. What do its modes tell us?

9.6.13 Explain the three curves in the bottom diagram in Figure 9.10.

9.6.14 Can changing parameters of the E-estimator, used in Figure 9.10, help us to realize that the underlying time series is nonstationary?

9.6.15 Explain the diagrams in Figure 9.11.

9.6.16 What are $\gamma_t^X(j)$ shown in Figure 9.11?

9.6.17 Does Figure 9.11 alert us about nonstationarity of the time series?

9.6.18 Suppose that you would like to simulate a stationary time series using Figure 9.11. How can this be done? Then what type of curves could be expected for $\gamma_t^X(j)$?

9.6.19 Explain why (9.6.6) can be considered as a nonparametric regression. Hint: Use (9.6.8).

9.6.20* Explain how E-estimator $\hat{\gamma}_t^X(j)$ may be constructed.

9.6.21* Propose an estimator of $\gamma_t^X(j)$, and then evaluate its mean and variance.

9.6.22* What may define the quality of estimation of a dynamic autocovariance function?

9.6.23* Propose a spectral density E-estimator for a nonstationary time series. Hint: Use E-estimators of dynamic autocovariances.

9.6.24* At the end of Section 9.6, a general setting of a set of time series $\{X_{t,\tau}\}$ is presented. Propose a feasible approach for a corresponding dynamic spectral density and how it may be estimated.

9.7.1 What is the definition of a one-way table? Give several examples.

9.7.2 What is the definition of a two-way table? Give several examples.

9.7.3 What is a plausible definition of a three-way table? Give several examples.

9.7.4 It looks like Tables 9.1 and 9.2 contradict each other. Explain why it is possible that they are based on the same data.

9.7.5 Suggest an example of Simpson's paradox for performance of two mutual funds with lurking variable being the allocation between bonds and stocks.

9.7.6 Explain the underlying simulation used in Figure 9.12.

9.7.7 Figures 9.12 and 9.13 are based on the same data and nonetheless they imply different conclusions about the two schools. How is this possible?

9.7.8 Based on the simulation, what is the mean number of graduates with a major in science?

9.7.9 Based on the simulation, what is the variance of the number of graduates with a major in science?

9.7.10* In Figure 9.12, parameters a and b control the probability of graduates from schools A and B being an engineer. Using both the theory and simulations, for what values of these parameters may the Simpson's paradox no longer be observed?

9.7.11 Explain how E-estimator of the conditional density is constructed.

9.7.12 In model (9.7.1) the GPA is considered as a continuous "time" variable in a stochastic process. Does this make sense for a salary model?

9.7.13 Suggest a stochastic model for weakly returns of a mutual fund for the last ten years with lurking variable being the asset allocation.

9.7.14* Explain how dependence between observations in model (9.7.1) affect estimation of the linear regression and the conditional density.

9.8.1 Explain the idea of a sequential design in a controlled regression experiment.

9.8.2 Suppose that in a controlled regression the next predictor is generated according to a density whose choice is based on previous observations. In this case, are available observations dependent or independent?

9.8.3* Propose a Fourier estimator of $\theta_j := \int_0^1 m(x)\varphi_j(x)dx$ whose mean squared error is $n^{-1}d(f^X, \sigma)[1 + o_j(1) + o_n(1)]$. Hint: Recall our discussion in Section 2.3.

9.8.4 Show that the design density that minimizes $d(f^X, \sigma)$ is proportional to the scale function $\sigma(x)$.

9.8.5 Verify (9.8.4).

9.8.6 Proof inequality (9.8.5) using the functional form (1.3.33) of the Cauchy-Schwarz inequality.

9.8.7* Prove the Cauchy-Schwarz inequality used in (9.8.5). Hint: Use the Cauchy inequality $2|ab| \leq a^2 + b^2$ and then think about choosing appropriate a and b.

9.8.8 Explain the underlying simulation used in the top diagram of Figure 9.15.

9.8.9 Explain the underlying simulation used in the bottom diagram of Figure 9.15.

9.8.10* Repeat Figure 9.15 a number of times and make your own conclusion about the opportunity of using an optimal design. Then explain all possible complications in using the idea of optimal design by a sequential estimator.

9.8.11 Repeat Figure 9.15 and notice that AISEs vary rather significantly from one experiment to another. Explain the variation.

9.8.12 Find better parameters of the used E-estimator.

9.8.13 Using Figure 9.15 for other regression functions and sample sizes, make your own conclusion about feasibility of using an optimal design.

9.8.14 Repeat Figure 9.16 a number of times and write a report about sensitivity of the sequential design to the scale's estimate.

9.8.15* Consider the problem of choosing the size k of a sample on the first stage. Asymptotically, should k be of the same order as n or it may be smaller in order than n?

9.8.16 Use Figure 9.16 and explore the effect of k on estimation. Hint: The size k of the first stage is controlled by argument b.

9.10   Notes

There are a number of excellent books devoted to time series analysis like Anderson (1971), Diggle (1990), Brockwell and Davis (1991), Fan and Yao (2003), Bloomfield (2004). Among more recent ones that discuss nonstationary time series, let us mention Box et al. (2016), De Gooijer (2017), and Tanaka (2017).

   9.1 The book by Beran (1994) covers a number of topics devoted to dependent variables including long-memory processes, see also Ibragimov and Linnik (1971) and Samorodnitsky (2016). The books Dedecker et al. (2007) and Rio (2017) cover a wide spectrum of topics on weak dependence. Hall and Hart (1990) established that dependent regression errors may significantly slow down the MISE convergence for a fixed-design regression. The theory of regression with dependent errors is discussed in Efromovich (1997c, 1999a, 1999c) and Yang (2001). The books by Dryden and Mardia (1998) and Efromovich (1999a) discuss estimation of shapes.

   9.2 White noise, Brownian motion and filtering a signal are classical statistical topics with a rich literature. See, for instance, books by Ibragimov and Khasminskii (1981), Mallat (1998), Efromovich (2009a), Tsay (2005), Tsybakov (2009), Fan and Yao (2003, 2015), Del Moral and Penev (2014), Pavliotis (2014), Durrett (2016) and Samorodnitsky (2016). The book Dobrow (2016) uses R to present introduction to stochastic processes.

   Asymptotic theory of efficient filtering a signal from white Gaussian noise was pioneered by Pinsker (1980) for the case of a known class of signals, efficient adaptation was proposed in Efromovich and Pinsker (1984, 1989), and multidimensional settings were considered in Efromovich (1994b, 2000b). It is worthwhile to note that typically a sine-cosine basis is used in a series expansion.

   The principle of equivalence between filtering model and other classical statistical models has been initiated by Brown and Low (1996) for regression and Nussbaum (1996) for the probability density. It is important to stress that the equivalence is based on some specific assumptions, and this implies limits on its applications; see more in Efromovich and Samarov (1996) and Efromovich (1999a, 2003a).

Missing data is discussed in Tsay (2005) and Box et al. (2016).

9.3 The decomposition model is discussed in a number of books, see Brockwell and Davis (1991), Fan and Gijbels (1996), Efromovich (1999a), Fan and Yao (2003, 2015), Tsay (2005) and Box et al. (2016). The typically recommended method of estimation of the trend is the liner regression. A thorough discussion of Fourier estimators using the idea of approximation of an integral by a Riemann sum can be found in Efromovich (1999a) and Efromovich and Samarov (2000).

9.4 Amplitude-modulated time series are discussed in Efromovich (2014d) where further references may be found. Wavelet and multiwavelet analysis may be of a special interest for the considered problem. See some relevant results in Efromovich (1999a, 2001c, 2004e, 2007b, 2009b, 2017), Efromovich et al. (2004), Efromovich et al. (2008), Efromovich and Valdez-Jasso (2010), Efromovich and Smirnova (2014a,b), as well as the asymptotic theory in the monograph Johnstone (2017).

9.5 Matsuda and Yajima (2009) consider a time series $\{A_t X_t\}$ with missing observations where the availability variables A_t are independent Bernoulli variables and they are also independent from the time series of interest $\{X_t\}$, but A_t are not identically distributed Bernoulli variables with $\mathbb{P}(A_t = 1) = w(t)$ being unknown. This model was suggested for time series with irregularly spaced data. The case of a stationary Poisson amplitude-modulation was studied in Vorotniskaya (2008), and Efromovich (2014d) proved efficiency of the E-estimation approach.

9.6 Time-varying nonstationary processes is a popular topic due to numerous practical applications, see a discussion in books by Tsay (2005), Stoica and Moses (2005), Sandsten (2016) and Tanaka (2017). One of the main ideas to deal with nonstationarity, complementary to the presented one, is the segmentation of the period of observation, assuming that the process is stationary on each segment, and then using a smoothing procedure. This, and other interesting approaches may be found in Priestly (1965), Kitagawa and Akaike (1978), Zurbenko (1991), Dahlhaus (1997), Adak (1998), and Rosen, Wood and Stoffer (2012). As an example of application, see Chen et al. (2016) and Efromovich and Wu (2017).

9.7 Simpson's paradox is an extreme example showing that observed relationships and associations can be misleading when there are lurking variables, see a discussion in Moore, McGabe and Craig (2009). Due to its confusing nature, it is an excellent pedagogical tool to attract our attention to statistical data analysis, see Gou and Zhang (2017). It is a common practice to use the paradox in conjunction with explanation of multi-way tables. At the same time, as we have seen from Figures 9.12–9.14, it is also useful in understanding of regressions, continuous processes and conditional densities. It is also important to stress that as these figures indicate, the conditional density, and not a regression, is the ultimate description of the relationship between the predictor and the response. As such, the conditional density may uncover the "mystery" of Simpson's paradox and may teach us a valuable lesson about the practical value of the conditional density. A nice introduction to multi-way tables can be found in Moore, McGabe and Craig (2009), to linear models in Kutner et al. (2005), and the theory of estimation of the conditional density in Efromovich (2007g).

9.8 Stein (1945) and Wald (1947,1950) pioneered principles of sequential estimation, and more good reading can be found in Prakasa Rao (1983), Wilks (1962), Mukhopadhyay and Solanky (1994). There are a number of good books that discuss controlled sampling procedures, see Pukelsheim (1993), Thompson and Seber (1996), Arnab (2017), and Dean, Voss and Draguljic (2017).

High-dimensional two-stage procedures are considered in Aoshima and Yata (2011). Asymptotic issues of nonparametric sequential estimation are considered in Efromovich (1989; 2004b,d; 2007d; 2008c; 2012b; 2015; 2017). Quickest detection problem is another interesting topic, see Efromovich and Baron (2010).

Chapter 10

Ill-Posed Modifications

A modification of underlying data is called ill-posed if a large change in an estimand (something that we are interested in estimation of) yields a relatively small change in the modified data which, in its turn, slows down the rate of a risk (for instance the MISE) convergence with respect to estimation based on the underlying data. Let us stress that so far studied modifications of data, including missing, truncation and censoring, affected only a constant of the MISE convergence, while ill-posed problems are characterized by slower rates of the MISE convergence and hence require dramatically larger sample sizes to attain the same quality of estimation. A classical example of ill-posed modification is contaminating data by additive errors, traditionally referred to as measurement errors. In particular, normal measurement errors may decrease the MISE convergence to logarithmic rates. Further, in some cases measurement errors imply a destructive modification which makes consistent estimation impossible. As it will be explained shortly, the problem of density estimation for data contaminated by measurement errors may be also referred to as a deconvolution problem, and there is a rich mathematical and statistical literature exploring deconvolution problems. Regression with predictors contaminated by additive errors is another classical example of ill-posed modification.

Ill-posedness may also occur if a directly observed sample does not correspond to the problem at hand. For instance, suppose that we have a sample from a variable X or a sample from a pair (X, Y). If we are interested in estimation of the density of X or the regression of Y on X, then we get traditional rates of the MISE convergence discussed in Chapter 2. At the same time, if we are interested in the derivative of density or regression function, then the problem becomes ill-posed. In other words, directly observed data correspond to underlying density or regression function but not to the derivatives.

Further, it is important to know that it is accurate to call a well-known problem ill-posed if there exists a related problem where a better quality of estimation is available. In other words, everything depends on a benchmark. Let us illustrate the above-made remark by the following example. Based on a sample of size n from X, the cumulative distribution function (CDF) $F^X(x)$ can be estimated with the classical parametric rate n^{-1} regardless of the CDF smoothness, while we know from Chapter 2 that the probability density $f^X(x)$ can be estimated with a slower rate which depends on smoothness of the density. As a result, with respect to the CDF estimation it is possible to say that density estimation is an ill-posed problem. And indeed, because first theoretical results on density estimation were obtained at the time when the theory of estimation of the CDF was well developed, nonparametric density estimation was initially considered as an ill-posed problem. Only with the progress of the theory of nonparametric estimation and increased interest in its practical applications, the density estimation is no longer considered as an ill-posed (while there is nothing wrong in doing so with respect to the CDF estimation).

Another feature to learn about is that some of our previous conclusions about optimal procedures for missing/survival data may no longer hold. For instance, this is the case for regression with missing responses and measurement errors in predictors when a complete case approach may no longer be consistent.

While measurement errors, deconvolution and estimation of derivatives are probably the better known example of ill-posed settings, survival analysis has its own number of interesting examples. In particular, we will consider a practically important current status censoring (CSC) where the censoring affects the rate of the MISE convergence.

The context of the chapter is as follows. Density estimation with measurement errors, which is a particular example of a deconvolution problem, is studied in the first three sections. Section 10.1 considers the case of a random variable whose observations are contaminated by measurement errors. Section 10.2 makes the setting more complicated by considering missing data with measurement errors. Section 10.3 adds another layer of modification via censoring. Let us note that the proposed deconvolution is based on estimation of the characteristic function which is an important statistical problem on its own, and this is the first time when this function is discussed. The characteristic function, similar to the cumulative distribution function, completely defines a random variable, and for a random variable X it is defined as $\phi^X(t) := \mathbb{E}\{e^{itX}\} = \mathbb{E}\{\cos(tX)\} + i\mathbb{E}\{\sin(tX)\}$, where $t \in (-\infty, \infty)$ and i is the imaginary unit, that is $i^2 = -1$. Current status censoring, also referred to as case 1 interval censoring, is discussed in Section 10.4. Regression with measurement errors in predictors is discussed in Sections 10.5 and 10.6. Estimation of derivatives is discussed in Section 10.7.

10.1   Measurement Errors in Density Estimation

Measurement errors occur in many practical problems. They may be due to a defective measurement tool, or negligence, or may be the intrinsic part of a problem at hand. For instance, measurement errors often present in studies of the behavior of insects or animals when only indirect measurements are possible. Another classical example is a score on an IQ test that should measure the IQ of a person. Let us also note that quantities that cannot be directly measured are sometimes called latent.

The aim is to estimate the probability density $f^X(x)$ of a random variable X. An underlying sample $X_1, \ldots, X_n$ from X is not available, and instead a sample $Y_1 := X_1 + \varepsilon_1, \ldots, Y_n := X_n + \varepsilon_n$ from $Y := X + \varepsilon$ is available where ε is a measurement error. It is assumed that X and ε are independent continuous variables and X is supported on $[0, 1]$.

If $f^\varepsilon(u)$ is the density of the measurement error, then the density of the sum $Y = X + \varepsilon$ is defined by the *convolution* formula

$$f^Y(y) = \int_{-\infty}^{\infty} f^\varepsilon(u) f^X(y - u) du. \tag{10.1.1}$$

To verify (10.1.1), we begin with the corresponding CDF and write,

$$F^Y(y) := \mathbb{P}(Y \le y) = \mathbb{E}\{I(Y \le y)\} = \mathbb{E}\{\mathbb{E}\{I(X + \varepsilon \le y)|\varepsilon\}\}$$

$$= \mathbb{E}\{\mathbb{E}\{I(X \le y - \varepsilon)|\varepsilon\}\} = \int_{-\infty}^{\infty} f^\varepsilon(u) F^X(y - u) du. \tag{10.1.2}$$

Taking derivative with respect to y verifies (10.1.1).

Formula (10.1.1) explains why the problem of estimation of the density f^X is often referred to as a *deconvolution* problem.

Convolution, as a data modification, occurs in a number of practical settings, for instance in signal processing, pattern recognition, etc. This explains the importance of the deconvolution problem and its special place in the statistics. Further, the problem may be also referred to as an *inverse* problem because the density of interest should be reconstructed from the results of its convolution with another density.

A proposed solution of the deconvolution problem (estimation of the density f^X) is based on utilizing a characteristic function which completely describes distribution of any random variable and hence of interest on its own. For a random variable Z its *characteristic function* is defined as

$$\phi^Z(t) := \mathbb{E}\{e^{itZ}\} := \mathbb{E}\{\cos(tZ)\} + i\mathbb{E}\{\sin(tZ)\}, \qquad (10.1.3)$$

where i is the imaginary unit, that is $i^2 = -1$. The characteristic function resembles the moment generating function; to see this, just remove the imaginary unit from the exponential function in (10.1.3) and this yields the moment generating function. While the characteristic function exists for any random variable and its absolute value does not exceed 1 (it is a bounded function), the moment generating function exists only for a number of variables. If the distribution of Z is symmetric about zero, then its characteristic function becomes real (because $\mathbb{E}\{\sin(tZ)\} = 0$), that is we have

$$\phi^Z(t) = \mathbb{E}\{\cos(tZ)\} \quad \text{if} \quad \mathbb{P}(Z < -z) = \mathbb{P}(Z > z), \, z \in [0, \infty). \qquad (10.1.4)$$

Another important property of a characteristic function $\phi^Z(t)$ is that if it is integrable, then Z is a continuous variable with the probability density

$$f^Z(z) = \frac{1}{2\pi} \int_{-\infty}^{\infty} e^{-itx} \phi^Z(t) dt. \qquad (10.1.5)$$

Let us stress that the characteristic function of a random variable always exists while the density may not exist (say for a discrete random variable). Overall, estimation of the characteristic function, considered shortly, is of interest on its own and it is an important statistical problem.

For the deconvolution problem, the pivotal property of the characteristic function is that

$$\phi^{X+\varepsilon}(t) = \phi^X(t)\phi^\varepsilon(t) \text{ whenever } X \text{ is independent of } \varepsilon. \qquad (10.1.6)$$

As a result, if a sample $Y_1, \dots, Y_n$ from $Y := X + \varepsilon$ is available and the distribution of the measurement error ε is known and the absolute value of its characteristic function is positive, we can estimate the characteristic function of an underlying random variable of interest X. Indeed, we may estimate the characteristic function of Y by the *empirical characteristic function* (which is another wording for the sample mean estimator)

$$\hat{\phi}^Y(t) := n^{-1} \sum_{l=1}^{n} e^{itY_l}, \qquad (10.1.7)$$

and then, using (10.1.6), estimate the characteristic function of an underlying X by the sample mean estimator

$$\tilde{\phi}^X(t) := \frac{\hat{\phi}^Y(t)}{\phi^\varepsilon(t)} = \frac{n^{-1} \sum_{l=1}^{n} e^{itY_l}}{\phi^\varepsilon(t)}. \qquad (10.1.8)$$

The only (and extremely series) complication here is that in (10.1.8) $h^\varepsilon(t)$ is used in the denominator. The following examples shed light on the complexity. The most common measurement error is normal. If ε has a normal distribution with zero mean and variance σ^2, then its characteristic function is $\phi^\varepsilon(t) = e^{-t^2\sigma^2/2}$. Another often used distribution for ε is Laplace (double exponential) with zero mean and variance $2b^2$, and its characteristic function is $\phi^\varepsilon(t) = [1 + b^2t^2]^{-1}$. The fast decrease in the characteristic function of the measurement error makes estimation of $\phi^X(t)$ for large t problematic and causes an acute ill-posedness. Indeed, the good news is that the estimator (10.1.8) is unbiased, the bad news is that its variance is proportional to $|\phi^\varepsilon(t)|^{-2}$, namely

$$\mathbb{V}(\tilde{\phi}^X(t)) = \frac{\mathbb{E}\{|\hat{\phi}^Y(t) - \phi^Y(t)|^2\}}{|\phi^\varepsilon(t)|^2} = n^{-1}\frac{1 - |\phi^Y(t)|^2}{|\phi^\varepsilon(t)|^2}. \qquad (10.1.9)$$

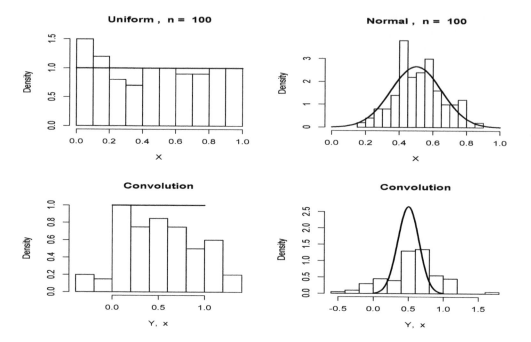

Figure 10.1 *Simulated hidden data and the same data contaminated by normal measurement errors (the convolution). Hidden samples of size $n = 100$ are simulated according to the Uniform and the Normal densities and are shown in the two columns. The same measurement errors are used for both samples and the errors are generated according to a normal distribution with zero mean and standard deviation σ. Samples are shown by histograms, underlying densities f^X are shown by solid lines. Note that each of the bottom diagrams shows the histogram of a sample from Y and the underlying density $f^X(x)$, $x \in [0,1]$. The latter is stressed by the horizontal axis label "Y, x". {The sample size is controlled by the argument n, the set of underlying densities by set.c, and σ by the argument sigma.} [n = 100, set.c = c(1,2), sigma = 0.3]*

We conclude that the variance may dramatically (exponentially for the case of a normally distributed measurement error) increase in t.

These results shed light on the ill-posedness caused by measurement errors. According to (10.1.6), for a large t a change in an underlying $\phi^X(t)$ may cause a relatively small change in the observed $\phi^Y(t)$, and this is what causes ill-posedness. Further, (10.1.9) shows that the variance of $\tilde{\phi}^X(t)$ may be dramatically larger than the variance of $\hat{\phi}^Y(t)$.

Figure 10.1 allows us to visualize the effect of measurement errors. The left-top diagram shows us a simulated sample from the Uniform density (the caption explains the simulation). This is a direct sample, and an appropriate smoothing of the histogram should produce a good estimate. The diagram below shows the same sample contaminated by independent normal measurement errors with zero mean and standard deviation $\sigma = 0.3$. No longer the histogram even remotely resembles the Uniform density shown by the solid line. Further, over the interval of interest $[0,1]$, the histogram clearly indicates a decreasing density. Furthermore, please look at the larger support of Y and the asymmetric about 0.5 data, and note that the distribution of Y is symmetric about 0.5. Can you visualize the underlying Uniform density in the histogram? The answer is "no." Here only a rigorous statistical approach may help. The right column of diagrams shows a similar experiment only with the underlying Normal density and the same measurement errors. Here the direct histogram (in the right-top diagram) is good and we may visualize the underlying Normal density shown by the solid line. On the other hand, the histogram of the same sample with measurement errors does not resemble the Normal and it is clearly skewed and asymmetric about 0.5. We may conclude that no longer visualization is a helpful step in the analysis of data modified by measurement errors. The reader is advised to repeat this figure, use different arguments, and get a feeling of the problem and its complexity.

Now we are in a position to explain how our E-estimator, described in Section 2.2 for the case of direct observations of X, can be used for estimating an underlying density $f^X(x)$ when data is modified by measurement errors. Recall that X is supported on $[0,1]$, and then the density of interest can be written as

$$f^X(x) = 1 + 2^{1/2} \sum_{j=1}^{\infty} \text{Re}\{\phi^X(\pi j)\}\varphi_j(x), \quad x \in [0,1], \qquad (10.1.10)$$

where $\varphi_j(x) := 2^{1/2}\cos(\pi j x)$ are elements of the cosine basis on $[0,1]$ and $\text{Re}\{z\}$ is the real part of a complex number z. Then, assuming that the distribution of the measurement error is known, the characteristic function may be estimated by the sample mean estimator (10.1.8). This characteristic function estimator yields the deconvolution E-estimator.

Furthermore, according to (10.1.4), the E-estimator simplifies and looks more familiar if the measurement error is symmetric about zero. Indeed, in this case it is convenient to rewrite (10.1.9) as

$$f^X(x) = 1 + \sum_{j=1}^{\infty} \theta_j \varphi_j(x), \qquad (10.1.11)$$

where $\theta_j := \int_0^1 f^X(x)\varphi_j(x)dx$, and then use the sample mean Fourier estimator

$$\hat{\theta}_j := n^{-1} \sum_{l=1}^{n} \varphi_j(Y_l)/\phi^\varepsilon(\pi j). \qquad (10.1.12)$$

Estimator (10.1.12) resembles the traditional Fourier estimator of Section 2.2, with one pronounced difference. Here we divide by the quantity $\phi^\varepsilon(\pi j)$ which may be small and close to zero. The latter also complicates plugging in an estimate of the characteristic function. The asymptotic theory shows that estimation on frequencies where the characteristic function is too small must be skipped. In particular, we may restrict our attention to frequencies where $|\phi^\varepsilon(\pi j)|^2 > C_H n^{-1} \log(n)$, and if m is the sample size of an extra sample of measurement errors used to estimate ϕ^ε, then n is replaced by m. Here C_H is a positive constant which is used by the E-estimator.

Figure 10.2 illustrates performance of the E-estimator for the case of a known distribution of measurement errors. Here the distribution of measurement errors is Laplace with parameter $b = 0.2$. The distribution is symmetric about zero, the density is $f^\varepsilon(u) = (2b)^{-1}\exp(-|u|/b)$ and the characteristic function is $\phi^\varepsilon(t) = (1 + b^2 t^2)^{-1}$. Note that Laplace distribution may be also referred to as double exponential or Gumble. Laplace distribution has heavier tails than a Normal distribution and a slower decreasing characteristic function. Now let us look at particular simulations. The E-estimate for the case of direct observation and the Normal density is almost perfect despite a rough histogram reflecting the underlying sample from the Normal. The corresponding histogram for the modified by measurement errors data (the left-bottom diagram) indicates a skewed and asymmetric sample. This yields a far from perfect E-estimate. On the other hand, keeping in mind the underlying data, the E-estimate correctly indicates the symmetric about 0.5 and bell-shaped density as well as the almost perfect support. For the case of the Bimodal underlying distribution (see the right column of diagrams) the E-estimates are worse, and this reflects the more challenging shape of the density. We see in the right-top diagram that the histogram (and hence the sample) does not indicate the shape of the Bimodal, and this is also reflected by

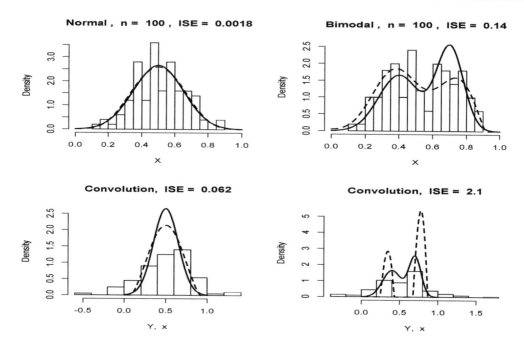

Figure 10.2 *Performance of the density E-estimator for direct data and the deconvolution E-estimator for data modified by Laplace measurement errors. Parameter of the Laplace distribution is b = 0.2. Data are shown by histograms, underlying densities by solid lines and E-estimates by dashed lines. {Argument CH controls parameter C_H.} [n = 100, b = 0.2, set.c = c(2,3), cJ0 = 3, cJ1 = 0.8, cTH = 4, CH = 0.1]*

the E-estimate. At the same time, the E-estimate does indicate two modes with the only caveat that the magnitude of the right one is too small. Measurement errors caused an interesting modification of the sample from X that may be observed in the right-bottom diagram. Surprisingly, measurement errors "corrected" the underlying sample in terms of the relative magnitudes of the modes. On the other hand, the E-estimate shows sharper modes and incorrectly indicates two strata. This is a teachable outcome because it shows how the deconvolution performs. Finally, note how heavier (with respect to normal) tails of Laplace distribution affect the modified data. The reader is advised to repeat Figures 10.1 and 10.2 and compare data modified by these two typical measurement errors.

If the distribution of measurement errors is unknown, then in general the deconvolution is impossible. In what follows we are considering two possible scenarios that allow us to overcome this complication. The first one is when an extra sample of measurement errors of size m is available. Then the characteristic function ϕ^ε may be estimated and used in place of an unknown ϕ^ε.

Figure 10.3 illustrates this situation. Simulations are similar to those in Figure 10.2. We begin analysis of estimates with the left column corresponding to the Normal density $f^X(x)$. The particular sample of direct observations, shown by the histogram in the left-top diagram, is clearly skewed and asymmetric. The E-estimate improves this drawback of the histogram but still it is far from being good. The latter is also reflected by the large ISE (compare with Figure 10.2). The extra sample of measurement errors, shown by the histogram in the left-middle diagram, is also far from being perfect. It is used to estimate the characteristic function $\phi^\varepsilon(\pi j)$. The left-bottom diagram shows the modified data and the plug-in density E-estimate which is relatively good. We observe a rare outcome when

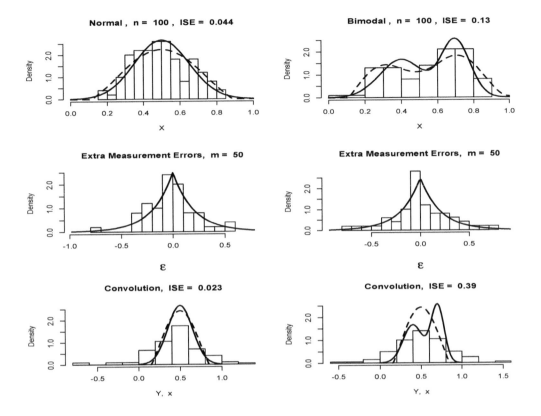

Figure 10.3 *Performance of the deconvolution E-estimator for the case of an unknown distribution of measurement errors when an extra sample of size m of the errors is available. The underlying distribution of errors is Laplace with parameter b = 0.2. Solid and dashed lines show underlying densities and E-estimates, respectively. [n = 100, m = 50, b = 0.2, set.c = c(2,3), cJ0 = 3, cJ1 = 0.8, cTH = 4, CH = 0.1]*

measurement errors improved the E-estimate. No such pleasant surprise for the Bimodal density shown in the right column of diagrams. The unimodal shape of the plug-in E-estimate for the modified data is typical because the Bimodal shape is too complicated for the deconvolution and small sample sizes. At the same time, we may conclude that a relatively small size of extra samples of measurement errors is feasible for our purposes. It is highly advisable to repeat Figures 10.3 with different parameters and learn more about this interesting problem and the case of small samples.

Another possibility to solve the deconvolution problem without knowing the distribution of measurement errors is as follows. Consider the case of *repeated observations* when available observations are

$$Y_{lk} = X_l + \varepsilon_{lk}, \quad l = 1, 2, \ldots, n, \quad k = 1, 2. \tag{10.1.13}$$

Here ε_{lk} are iid observations (the sample of size $2n$) from ε, and it is known that the distribution of ε is symmetric about zero and its characteristic function is positive (note that this is the case for the Normal, Laplace, Cauchy, their mixtures and a number of other popular distributions).

To understand how the characteristic function $\phi^\varepsilon(t)$ of the measurement error may be estimated, let us note that for any independent and identically distributed ε_{l1} and ε_{l2} we can write,

$$\mathbb{E}\{e^{it(Y_{l1}-Y_{l2})}\} = \mathbb{E}\{e^{it(\varepsilon_{l1}-\varepsilon_{l2})}\} = \phi^\varepsilon(t)\phi^\varepsilon(-t) = |\phi^\varepsilon(t)|^2. \tag{10.1.14}$$

As a result, $n^{-1} \sum_{l=1}^{n} e^{it(Y_{l1}-Y_{l2})}$ is the unbiased estimator of $|\phi^{\varepsilon}(t)|^2$. Furthermore, if the distribution of ε is symmetric about zero and its characteristic function is positive, then

$$\hat{\phi}^{\varepsilon}(t) := [n^{-1} \sum_{l=1}^{n} e^{it(Y_{l1}-Y_{l2})}]^{1/2} \qquad (10.1.15)$$

is a consistent (as well as asymptotically rate optimal) estimator of $\phi^{\varepsilon}(t)$. This characteristic function estimator yields the plug-in deconvolution E-estimator.

So far we have considered the case of a known finite support of X. In a general case of an unknown support, similarly to the case of directly observed X, the density is estimated over the range of observed sample from $Y = X + \varepsilon$.

We are finishing this section by presenting an interesting and practically important case of *directional (angular, circular)* data where data are measured in the form of angles. Such data may be found almost everywhere throughout science. Typical examples include wind and ocean current directions, times of accident occurrence, and energy demand over a period of 24 hours. It is customary to measure directions in radians with the range $[0, 2\pi)$ radians. In this case the mathematical procedure of translation of any value onto this interval by *modulo 2π* (the shorthand notation is $[\mod 2\pi]$) is useful. As an example, $5\pi[\mod 2\pi] = 5\pi - 2(2\pi) = \pi$, and $-3.1\pi[\mod 2\pi] = -3.1\pi + 4\pi = 0.9\pi$. In words, you add or subtract $j2\pi$ (where j is an integer) to get a result in the range $[0, 2\pi)$.

The corresponding statistical setting is as follows. The data are n independent and identically distributed realizations (so-called directions) Y_l, $l = 1, 2, \ldots, n$, of a circular random variable Y that is defined by $Y := (X + \varepsilon)[\mod 2\pi]$ (or $Y := (X[\mod 2\pi] + \varepsilon[\mod 2\pi])[\mod 2\pi]$), where the random variable ε is independent of X. The variable ε is referred to as the measurement error. The problem is to estimate the probability density $f^X(x)$, $0 \le x < 2\pi$, of the random variable $X[\mod 2\pi]$.

Before explaining the solution, several comments about circular random variables should be made. Many examples of circular probability densities are obtained by *wrapping* a probability density defined on the line around the circumference of a circle of unit radius (or similarly one may say that a continuous random variable on the line is wrapped around the circumference). In this case, if Z is a continuous random variable on the line and X is the corresponding wrapped random variable, then

$$X = Z[\mod 2\pi], \quad f^X(x) = \sum_{k=-\infty}^{\infty} f^Z(x + 2\pi k). \qquad (10.1.16)$$

While the notion of a wrapped density is intuitively clear, the formulae are not simple. For instance, a wrapped normal $N(\mu, \sigma^2)$ random variable has the circular density (obtained after some nontrivial simplifications)

$$f^X(x) = (2\pi)^{-1} \Big(1 + 2 \sum_{k=1}^{\infty} e^{-k^2 \sigma^2/2} \cos(k(x - \mu)) \Big). \qquad (10.1.17)$$

Fortunately, for the problem at hand these complications with wrapped densities are not crucial because for the case of a wrapped distribution we get the following simple formulae for the characteristic function

$$\phi^X(j) = \int_0^{2\pi} \sum_{k=-\infty}^{\infty} f^Z(x + 2\pi k) e^{ijx} dx = \phi^{Z[\mod 2\pi]}(j) = \phi^Z(j). \qquad (10.1.18)$$

Hence for independent X and ε formula (10.1.6) holds and, in its turn, this implies that the proposed deconvolution E-estimator may be used for directional data modified by measurement errors.

10.2   Density Deconvolution with Missing Data

Here we are considering the same model as in the previous section only when some observations may be missed. In a sense, this setting is a combination of models considered in Sections 10.1, 4.1 and 5.1 only now a missing mechanism may be more complicated due to a larger number of underlying random variables and the absence of direct observations of a random variable of interest.

Let us formally describe the setting. There are two underlying (and hidden) samples of size n from the continuous random variable of interest X and the continuous measurement error ε. It is assumed that X is supported on $[0,1]$ and independent of ε. The element-wise sum of these two samples $Y_1 := X_1 + \varepsilon_1, \ldots, Y_n := X_n + \varepsilon_n$, which may be considered as a sample from $Y := X + \varepsilon$, is also hidden due to a missing. The missing mechanism is described by a Bernoulli random variable A, called the availability, which also generates a sample $A_1, \ldots, A_n$, and the availability likelihood is

$$\mathbb{P}(A = 1 | X = x, \varepsilon = u, Y = y) =: w(x, u, y) \geq c_0 > 0. \tag{10.2.1}$$

It is understood that in (10.2.1) we consider only $y = x + u$.

The observed (available) sample is $A_1 Y_1, A_2 Y_2, \ldots, A_n Y_n$. Note that $Y_l = X_l + \varepsilon_l$ is observed only if the availability $A_l = 1$ and otherwise it is missed (not available). The problem is to estimate the density of interest $f^X(x)$, that is, to solve the deconvolution problem for the case of an observed sample from $A(X + \varepsilon)$.

Because $\mathbb{P}(Y = 0) = 0$, we get $\mathbb{P}(A = I(AY \neq 0)) = 1$. Hence, even if we do not observe A directly, we know it.

In what follows, following Section 10.1, we are assuming that the distribution of a continuous ε is symmetric about zero and its characteristic function $\phi^\varepsilon(\pi j)$ is known and $|\phi^\varepsilon(\pi j)| > 0$, $j = 1, 2, \ldots$ It is also assumed that the availability likelihood $w(x, u, y)$ is known (it is explained in Chapter 5 when and how it is possible to relax this assumption).

We are considering in turn several possible availability likelihood functions.

Case 1. Availability likelihood is a positive constant. This is the simplest case of the MCAR (missing completely at random) when missing occurs purely at random and independently of the underlying random variables. As we already know from Section 4.1, in this case using complete cases (available observations of $Y = X + \varepsilon$) implies an optimal solution. Section 10.1 explains how to construct in this case the E-estimator. A serious complication is that the size $N := \sum_{l=1}^n A_l$ of available observations of Y may be small with respect to n. Apart of this complication, there are no new issues to comment on.

Case 2. Availability likelihood depends on Y. Here the missing mechanism is defined by the value of $Y := X + \varepsilon$, that is,

$$\mathbb{P}(A = 1 | X = x, \varepsilon = u, Y = y) =: w(y) \geq c_0 > 0. \tag{10.2.2}$$

The missing resembles the classical MNAR (missing not at random) of Section 5.1 where Y is the variable of interest. For instance we may think that the missing occurs after an observations of Y is generated, and then given Y the missing does not depend on hidden variables X and ε. The obvious complication, with respect to the classical MNAR of Section 5.1, is that the variable of interest X is modified by the measurement error and hence a special deconvolution is needed to recover the density f^X.

If the availability likelihood $w(y)$ is known, then for any t, such that $|\phi^\varepsilon(t)| > 0$, the sample mean estimator of the characteristic function $\phi^X(t)$ of the random variable of interest X is

$$\hat{\phi}^X(t) := n^{-1} \sum_{l=1}^n \frac{I(A_l Y_l \neq 0) e^{it A_l Y_l}}{\phi^\varepsilon(t) w(A_l Y_l)}. \tag{10.2.3}$$

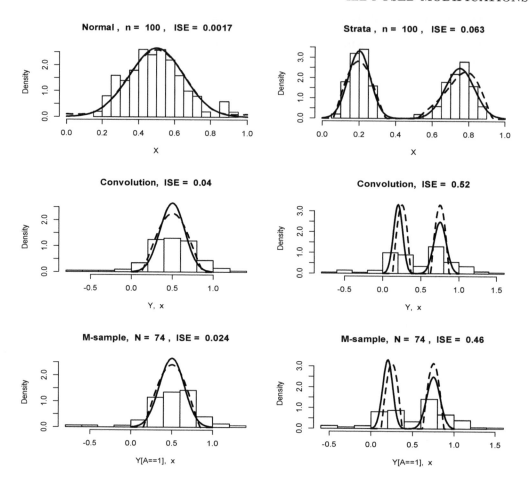

Figure 10.4 *Performance of the E-estimator for direct data, the data contaminated by Laplace measurement errors, and M-sample where some observations with measurement errors are missed according to the availability likelihood* $w(y)$. *Samples are shown by histograms, underlying densities and E-estimates are shown by solid and dashed lines, respectively.* {*The used availability likelihood is* $w^*(y) := \max(d_{wL}, \min(d_{wU}, w(y)))$ *where* $w(y) = 0.3 + 0.5exp\{(1+4y)/(1+exp(1+4y))\}$, *measurement errors are Laplace with parameter b. Parameters* d_{wL} *and* d_{wU} *are controlled by arguments dwL and dwU, function* $w(y)$ *is defined by the string w.*} *[n = 100, set.c = c(2,4), b = 0.2, w = "0.3+0.5*exp((1+4*y)/(1+exp(1+4*y)))", dwL = 0.3, dwU = 0.9, cJ0 = 3, cJ1 = 0.8, cTH = 4, cH = 0.1]*

Let us check that the estimator is unbiased. Write,

$$\mathbb{E}\{\hat{\phi}^X(t)\} = \mathbb{E}\Big\{\frac{Ae^{itAY}}{\phi^\varepsilon(t)w(AY)}\Big\}$$

$$= \mathbb{E}\Big\{\frac{e^{itY}}{\phi^\varepsilon(t)w(Y)}\mathbb{E}\{A|Y\}\Big\} = \mathbb{E}\Big\{\frac{e^{itY}}{\phi^\varepsilon(t)w(Y)}w(Y)\Big\}. \qquad (10.2.4)$$

In the last equality we used relation $\mathbb{E}\{A|Y\} = \mathbb{P}(A = 1|Y) = w(Y)$ which holds for the Bernoulli variable A due to the assumption (10.2.2). Using independence of X and ε we continue (10.2.4),

$$\mathbb{E}\{\hat{\phi}^X(t)\} = \mathbb{E}\Big\{\frac{e^{it(X+\varepsilon)}}{\phi^\varepsilon(t)}\Big\} = \frac{\phi^X(t)\phi^\varepsilon(t)}{\phi^\varepsilon(t)} = \phi^X(t). \qquad (10.2.5)$$

This proves that the estimator (10.2.3) is unbiased. The variance of this sample mean estimator decreases with the classical rate n^{-1}. These results allow us to use the characteristic function estimator (10.2.3) for construction of the deconvolution density E-estimator of Section 10.1.

Figure 10.4 illustrates the setting and performance of the deconvolution E-estimator for the case of missing data. We begin our visualization of Figure 10.4 with the left column of diagrams where the underlying density is the Normal. The left-top diagram shows us the hidden sample from X. The sample is good, and this is reflected by the histogram, the E-estimate and the small ISE. The middle diagram shows us the same observations of X but contaminated by measurement errors. Here we can see the deconvolution estimator of Section 10.1. Note that measurement errors make the histogram practically flat around 0.5. Nonetheless, the deconvolution E-estimate preserves the pronounced unimodal shape of the Normal density because it takes into account the known characteristic function of measurement errors. The bottom diagram shows us the histogram of M-sample (sample with missing data), which is a subsample from $Y := X + \varepsilon$ corresponding to $A = 1$ (in R language this subsample would be written as $Y[A == 1]$ where Y and A denote vectors of observations). The availability likelihood skews available data to the right, and we observe this in the histogram of the M-sample. The E-estimate does take the missing into account and correctly shows the symmetric about 0.5 bell-shaped density. Another effect of the missing is the decreased sample size when 26% of observations are lost. Nonetheless, here we have an interesting (and rare) case when the E-estimate based on missing data is better (both visually and in terms of the ISE) than the E-estimate for the data with measurement errors. The latter is explained by the small size of samples and particular structure of the hidden samples from X and ε.

The right column of diagrams in Figure 10.4 presents a similar simulation for the Strata density. As we see in the top diagram, the histogram and the E-estimate, based on the underlying sample from X, are fair. The middle digram shows that Laplace measurement errors made the right stratum higher, also look at the larger support of $Y = X + \varepsilon$. In the bottom diagram, as it could be expected, the missing further skewed data to the right, and this is reflected by the histogram. The E-estimate takes the missing mechanism into account and corrects the right-skewness. Further, note the decreased ISE with respect to the deconvolution estimate in the middle diagram. This improvement is atypical, and the particular simulations are chosen to stress the possible phenomena of a small sample. It is advisable to repeat Figure 10.4 and get used to possible outcomes for different sample sizes, measurement errors and availability likelihood functions.

So far the availability likelihood $w(y)$ was supposed to be known. If it is unknown then, without additional information about an underlying missing mechanism, no consistent estimation of the density f^X is possible. Different scenarios for obtaining a sufficient information for consistent estimation were discussed in Chapter 5. Further, if the characteristic function ϕ^ε of the measurement error is unknown, then again no consistent estimation is possible without an extra information about the measurement error, and the possibilities to get such information were discussed in Section 10.1.

Case 3. Availability likelihood depends on X. In some settings missing may occur before contamination of X by a measurement error ε. This is one of the examples when the availability likelihood depends on the variable of interest X, and

$$\mathbb{P}(A = 1 | X = x, \varepsilon = u, Y = y) = \mathbb{P}(A = 1 | X = x) =: w(x) \geq c_0 > 0. \qquad (10.2.6)$$

This is a new case for us because the missing is MNAR and it is defined by a hidden variable whose observations are contaminated by measurement errors. This makes the missing mechanism more complicated.

Our aim is to show that if the nuisance functions $\phi^\varepsilon(\pi j)$ and $w(x)$ are known, then consistent estimation of the density of interest $f^X(x)$ is possible.

Let us explain the proposed solution. We begin with a preliminary calculation of the expectation of the product Ae^{itAY}. Using the technique of conditional expectation we can write,

$$\mathbb{E}\{Ae^{itAY}\} = \mathbb{E}\{e^{itX}e^{it\varepsilon}\mathbb{E}\{A|X,\varepsilon\}\} = \mathbb{E}\{w(X)e^{itX}\}\phi^\varepsilon(t). \qquad (10.2.7)$$

Here we used (10.2.6) and the independence of X and ε. In its turn (10.2.7) implies that

$$\mathbb{E}\{Ae^{itAY}\} = \phi^\varepsilon(t)\int_0^1 [f^X(x)w(x)]e^{itx}dx. \qquad (10.2.8)$$

Equation (10.2.8) is the pivot that explains how the density of interest $f^X(x)$ may be estimated. Recall the cosine basis $\varphi_0(x) := 1$, $\varphi_j(x) = 2^{1/2}\cos(\pi jx)$, $j = 1, 2, \ldots$ on $[0, 1]$, set $g(x) := f^X(x)w(x)$ for the product of the density of interest and the availability likelihood, and introduce Fourier coefficients of the function $g(x)$,

$$\kappa_j := \int_0^1 g(x)\varphi_j(x)dx. \qquad (10.2.9)$$

Then, according to (10.2.8) and the assumed symmetry about zero of the distribution of ε, the sample mean Fourier estimator of κ_j is

$$\hat{\kappa}_j := n^{-1}\sum_{l=1}^n I(A_lY_l \neq 0)\varphi_j(AY_l)/\phi^\varepsilon(\pi j). \qquad (10.2.10)$$

The Fourier estimator yields the E-estimator $\hat{g}(x)$, $x \in [0, 1]$.

Using (10.2.11), the proposed plug-in deconvolution E-estimator of the density of interest is

$$\hat{f}^X(x) := \frac{\hat{g}(x)}{w(x)}. \qquad (10.2.11)$$

Figure 10.5 allows us to understand the setting and check performance of the deconvolution E-estimator. We begin with the left column corresponding to the Normal density. The top diagram shows us a simulation from the Normal density which is clearly skewed to the left, and this is reflected by the histogram, the E-estimate and the relatively large ISE. The middle diagram indicates that measurement errors "helped" to restore the symmetry, and here the E-estimate does a good deconvolution of the underlying Normal density from the Laplace measurement errors. The bottom diagram exhibits the same data with measurement errors (shown in the middle diagram) where some observations are missed according to the availability likelihood $w(x) = 0.3 + 0.5x$. This skews data to the right and we clearly see this in the histogram. The E-estimator, by taking into account the convolution and the MNAR, does a good job in restoring the Normal. Its performance is especially impressive keeping in mind the sample size $N = 51$ of available observations of Y. The right column of diagrams corresponds to the Strata density of X. The top diagram indicates that the simulated sample from X is skewed to the right, and this is reflected in the E-estimate which exhibits identical magnitudes of the modes. Measurement errors, as it could be expected, make estimation of the density of X even more difficult, see the middle diagram. Nonetheless, the deconvolution E-estimate clearly indicates two strata. Very interesting and atypical outcome for the missing data is exhibited in the bottom diagram. As we can observe from the E-estimate, the missing helped to restore the shape of the Strata.

Let us make a final remark about sample sizes in the missing data. In both cases they are the same ($N = 51$), and we observe a similar outcome in Figure 10.4. This happened

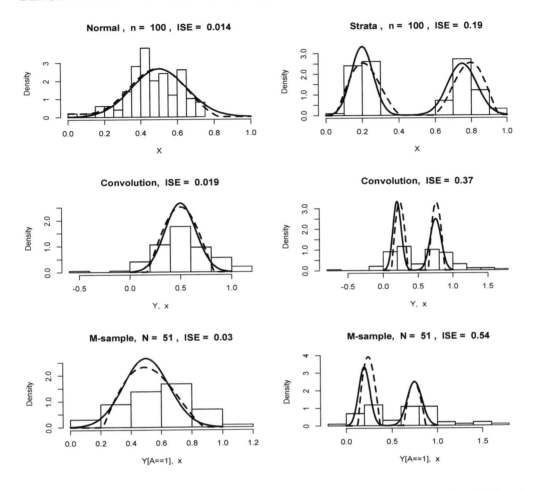

Figure 10.5 *Deconvolution with missing data when the availability likelihood depends on X. Samples are shown by histograms, underlying densities and E-estimates are shown by solid and dashed lines, respectively. {The availability likelihood is $w^*(x) := \max(d_{wL}, \min(d_{wU}, w(x)))$ where $w(x) = 0.3 + 0.5x$, measurements errors are Laplace with parameter b. Parameters d_{wL} and d_{wU} are controlled by arguments dwL and dwU, function $w(x)$ is defined by the string w.} [n = 100, set.c = c(2,4), b = 0.2, w = "0.3+0.5*x", dwL = 0.3, dwU = 0.9, cJ0 = 3, cJ1 = 0.8, cTH = 4, cH = 0.1]*

just by chance, the interested reader may repeat these figures and check this. At the same time, it is a nice exercise in probability to calculate the likelihood of this event.

Case 4. Availability likelihood depends on ε. This is the case when missing is defined by value of the measurement error, and the availability likelihood is

$$\mathbb{P}(A = 1 | X = x, \varepsilon = u, Y = y) = \mathbb{P}(A = 1 | \varepsilon = u) = w(u) \geq c_0 > 0. \qquad (10.2.12)$$

As a result, we again consider a deconvolution problem with MNAR observations. Furthermore, the missing depends on a hidden random variable which is never observed.

Let us see if we can find a way to estimate the underlying density of interest $f^X(x)$. We begin with a preliminary result. Write,

$$\mathbb{E}\{Ae^{itAY}\} = \mathbb{E}\{e^{itX}e^{it\varepsilon}\mathbb{E}\{A | X, \varepsilon\}\}$$

$$= \mathbb{E}\{e^{itX}\}\mathbb{E}\{w(\varepsilon)e^{it\varepsilon}\} = \phi^X(t)\mathbb{E}\{w(\varepsilon)e^{it\varepsilon}\}. \qquad (10.2.13)$$

Introduce a new function

$$g(t) := \mathbb{E}\{w(\varepsilon)e^{it\varepsilon}\}. \tag{10.2.14}$$

The distribution of ε and the availability likelihood $w(u)$ are assumed to be known, and hence the function $g(t)$ is also known. Let us additionally assume that $|g(\pi j)| > 0$, $j = 1, 2, \ldots$ Then (10.2.13) implies the following sample mean estimator of the characteristic function of X,

$$\hat{\phi}^X(\pi j) := \frac{n^{-1}\sum_{l=1}^n I(A_lY_l \neq 0)e^{i\pi j A_l Y_l}}{g(\pi j)}. \tag{10.2.15}$$

In its turn, $\hat{\phi}^X(\pi j)$ yields the deconvolution E-estimator $\hat{f}^X(x)$, $x \in [0,1]$ of Section 10.1.

Note that the estimator (10.2.15) of the characteristic function of X is of interest on its own, and if $|g(t)| > 0$ then we can calculate $\hat{\phi}^X(t)$.

Case 5. Availability likelihood depends on an auxiliary variable Z. This is a relatively simple case when we may observe a sample from a pair (AY, Z) and

$$\mathbb{P}(A = 1|X = x, \varepsilon = u, Y = y, Z = z) = w(z) \geq c_0 > 0. \tag{10.2.16}$$

This setting is similar to the one considered in Section 5.3, and given $|\phi^\varepsilon(t)| > 0$ we may introduce the following sample mean estimator of the characteristic function of X,

$$\hat{\phi}^X(t) = \frac{n^{-1}\sum_{l=1}^n [I(A_lY_l \neq 0)e^{itA_lY_l}/w(Z_l)]}{\phi^\varepsilon(t)}. \tag{10.2.17}$$

Let us check that this estimator is unbiased. Using independence of X and ε, together with (10.2.16) and the rule of calculation of the expectation via conditional expectation, we can write,

$$\mathbb{E}\{\hat{\phi}^X(t)\} = \mathbb{E}\{\mathbb{E}\{Ae^{itAY}[w(Z)\phi^\varepsilon(t)]^{-1}|X, \varepsilon, Z\}\}$$
$$= \mathbb{E}\{e^{itX}e^{it\varepsilon}[w(Z)\phi^\varepsilon(t)]^{-1}\mathbb{E}\{A|X, \varepsilon, Z\}\}$$
$$= \mathbb{E}\{e^{itX}e^{it\varepsilon}[w(Z)\phi^\varepsilon(t)]^{-1}w(Z)\}$$
$$= \mathbb{E}\{e^{itX}\}\mathbb{E}\{e^{it\varepsilon}\}/\phi^\varepsilon(t) = \phi^X(t). \tag{10.2.18}$$

We conclude that the proposed estimator of the characteristic function is unbiased.

In its turn, the characteristic function estimator (10.2.17) yields the deconvolution estimator $\hat{f}^X(x)$, $x \in [0,1]$ of Section 10.1.

If the function $w(z)$ is unknown, then we can estimate it from the sample $(Z_1, I(A_1Y_1 \neq 0)), \ldots, (Z_n, I(A_nY_n \neq 0))$. Indeed,

$$w(z) := \mathbb{P}(A = 1|Z = z) = \mathbb{E}\{I(AY \neq 0)|Z = z\}, \tag{10.2.19}$$

and hence we can use the Bernoulli regression estimator $\hat{w}(z)$ of Section 2.4.

Figure 10.6 allows us to understand the setting and test performance of the proposed deconvolution E-estimator $\hat{f}^X(x)$. The experiment is explained in the caption. We begin with the left column of diagrams where the case of the Normal density f^X is considered. For the hidden sample from X, shown in the top diagram, the E-estimator does a very good job. The middle diagram shows by circles the scattergram for the Bernoulli regression of A on Z, triangles show us values of $w(Z_l)$ for Z_l corresponding to $A_l = 1$. Note that according to (10.2.17) only for these values we need to know the availability likelihood. The E-estimate is not perfect, and we know from Section 2.4 that Bernoulli regression is a complicated problem for small sample sizes. Indeed, the right tail of the estimate is wrong, but note that it does correspond to the scattergram. The bottom diagram shows us an E-estimate

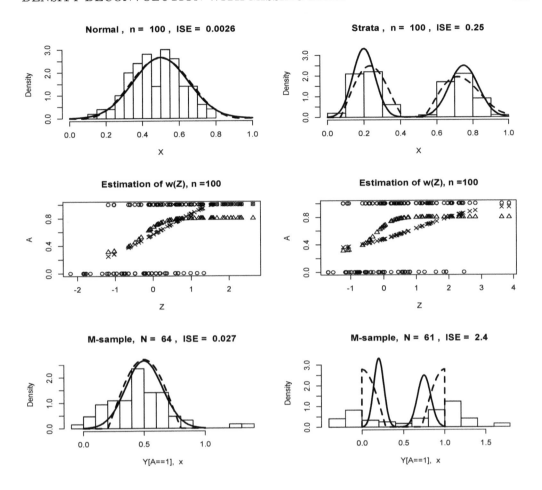

Figure 10.6 *Deconvolution with missing data when the availability likelihood depends on auxiliary variable Z. In the experiment $Z = \beta_0 + \beta_1 X + \sigma\eta$ with $\beta_0 = 0.2$, $\beta_1 = 0.4$, $\sigma = 1$, and η being standard normal random variable and independent of X and ε. Measurement errors are Laplace with parameter $b = 0.2$. Solid and dashed lines show underlying densities of interest and their E-estimates, respectively. In the middle diagrams observations of the pair (Z, A) are shown by circles, and for Z corresponding to $A = 1$ the likelihood function and its E-estimate are shown by triangles and crosses, respectively. {The availability likelihood is $w^*(z) := \max(d_{wL}, \min(d_{wU}, w(z)))$ where $w(z) = 0.3 + 0.5exp((1+4z)/(1+exp(1+4z)))$. Parameters d_{wL} and d_{wU} are controlled by arguments dwL and dwU, function $w(z)$ is controlled by the string w, set.beta = (β_0, β_1).} [n = 100, set.c = c(2,4), b = 0.2, set.beta = c(0.2,0.4), sigma = 1, w = "0.3+0.5*exp((1+4*z)/(1+exp(1+4*z)))", dwL = 0.3, dwU = 0.3, cJ0 = 3, cJ1 = 0.8, cTH = 4, cH=0.1]*

of $f^X(x)$, $x \in [0, 1]$. The estimate is good keeping in mind complexity of the deconvolution problem and when the estimate of the availability likelihood is far from perfect. As before, the E-estimate shows a smaller support but otherwise its shape is very good. Note how the underlying data, shown in the top diagram, is skewed to the right by the missing, and nonetheless the E-estimate correctly indicates a symmetric about 0.5 density.

The case of the underlying density Strata is considered in the right column of diagrams. The E-estimate for the hidden sample from X, shown in the top diagram, is reasonable for this sample size. The E-estimate of the availability likelihood, shown in the middle diagram by crosses, is far from being perfect but it does reflect the data at hand. The deconvolution E-estimate of the density of interest, shown in the bottom diagram, correctly shows two strata but shifts the modes. The estimate does not look attractive, but if we carefully look at the histogram, which presents available (not missed) observations of Y, and note its size $N = 61$, it becomes clear that the deconvolution E-estimator does a good job in recovering the Strata. After all, we are dealing with a complicated ill-posed deconvolution problem with missing data, and a relatively poor estimation must be expected.

It is a prudent advice to repeat Figures 10.4-10.6 with different parameters and get used to the deconvolution problem with missing data.

10.3   Density Deconvolution for Censored Data

We are considering a practically important problem of density estimation based on a sample that is first contaminated by additive errors and then the contaminated observations are right censored (RC). This is an interesting combination of the deconvolution problem discussed in Section 10.1 and the RC problem discussed in Section 6.5.

Let us formally define the setting. There is a hidden sample $(X_1, \varepsilon_1, C_1), \ldots, (X_n, \varepsilon_n, C_n)$ from the triplet (X, ε, C) of continuous and mutually independent variables. Here X is the variable of interest supported on $[0, 1]$, ε is the measurement error which is symmetric about zero and whose characteristic function $\phi^\varepsilon(t)$ is known and $|\phi^\varepsilon(\pi j)| > 0$, $j = 1, 2, \ldots$, and C is a nonnegative censoring variable.

There are two modifications of the hidden sample of interest from X. First, it is modified by measurement errors and this creates a new sample $Y_1 := X_1 + \varepsilon_1, \ldots, Y_n := X_n + \varepsilon_n$ from $Y := X + \varepsilon$. Second, the contaminated Y is modified by a censoring variable C which creates a pair (V, Δ) of random variables where $V := \min(Y, C)$ and $\Delta := I(Y \leq C)$. As a result of the second modification, the observed sample is $(V_1, \Delta_1), \ldots, (V_n, \Delta_n)$ where $V_l := \min(Y_l, C_l)$ and $\Delta_l := I(Y_l \leq C_l)$. The aim is to estimate the density $f^X(x)$ over its support $[0, 1]$.

Following the E-estimation methodology, we need to propose a sample mean estimator of Fourier coefficients

$$\theta_j := \int_0^1 \varphi_j(x) f^X(x) dx, \ \ j = 1, 2, \ldots \tag{10.3.1}$$

where $\varphi_j(x) = 2^{1/2} \cos(\pi j x)$ are elements of the cosine basis on $[0, 1]$. Because the characteristic function $\phi^\varepsilon(t)$ of the measurement error is a known real function and $\phi^\varepsilon(\pi j) \neq 0$, using (10.1.6) we can continue (10.3.1) and get,

$$\theta_j = \frac{\mathbb{E}\{\varphi_j(Y)\}}{\phi^\varepsilon(\pi j)}. \tag{10.3.2}$$

Now we need to understand how the expectation in (10.3.2) can be estimated based on a sample from (V, Δ). We note that

$$f^{V,\Delta}(v, 1) = f^Y(v) G^C(v), \tag{10.3.3}$$

where $G^C(x)$ is the survival function of the censoring variable C. Suppose that $\beta_C \geq \beta_Y$ (recall that β_Z denotes the upper bound of the support of Z). This allows us to continue (10.3.2) and write

$$\theta_j = \frac{\mathbb{E}\{\Delta \varphi_j(V)/G^C(V)\}}{\phi^\varepsilon(\pi j)}. \tag{10.3.4}$$

It was explained in Section 6.5 that $G^C(v)$ may be estimated by

$$\hat{G}^C(v) := \exp\left\{-\frac{\sum_{l=1}^n (1 - \Delta_l) I(V_l \leq v)}{\sum_{l=1}^n I(V_l \geq v)}\right\}. \tag{10.3.5}$$

1. Hidden Sample from Normal , n = 100 , ISE = 0.022

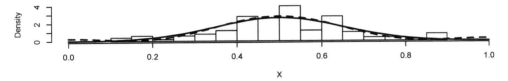

2. Hidden Sample Contaminated by Measurement Errors, ISE = 0.036

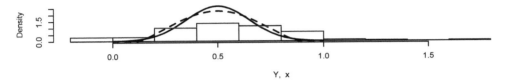

3. Censored Data, cens = Unif , n = 100 , N = 66 , uC = 1.5

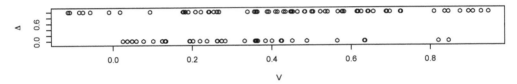

4. E-estimate and Confidence Bands, ISE = 0.052

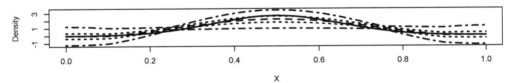

Figure 10.7 *Density estimation for the case when observations are first contaminated by measurement errors and then right censored. In the simulation the variable of interest X is the Normal, measurement error ε is Laplace with parameter $b = 0.2$, and C is Uniform$(0, U_C)$ with $U_C = 1.5$. Diagram 1 shows the histogram of a simulated hidden sample from the random variable of interest X, the solid and dashed lines exhibit the underlying Normal density and its E-estimate. Diagram 2 shows the histogram of the hidden sample contaminated by Laplace measurement errors. The solid and dashed lines exhibit the underlying Normal density and the deconvolution E-estimate. Diagram 3 shows the observed censored data which is a sample from (V, Δ). Here $N = \sum_{l=1}^{n} \Delta_l$ is the number of uncensored observations of $Y := X + \varepsilon$. Diagram 4 exhibits the underlying density (the solid line), the E-estimate (the dashed line), $(1 - \alpha)$ pointwise (the dotted lines) and simultaneous (the dot-dashed lines) confidence bands. {Use cens = "Expon" for exponential censoring variable whose mean is controlled by argument lambdaC. Argument alpha controls α.} [n = 100, corn = 2, b = 0.2, cens = "Unif", uC = 1.5, lambdaC = 1.5, alpha = 0.05, c = 1, cJ0 = 3, cJ1 = 0.8, cTH = 4]*

Combining the results, we may propose the following plug-in sample mean estimator of θ_j,

$$\hat{\theta}_j = \frac{n^{-1} \sum_{l=1}^{n} \Delta_l \varphi_j(V_l) / \max(\hat{G}^C(V_l), c/\ln(n))}{\phi^\varepsilon(\pi j)}. \tag{10.3.6}$$

The Fourier estimator (10.3.6) yields the density E-estimator $\hat{f}^X(x)$. Further, let us recall that, according to Section 2.6, E-estimators allow us to visualize their pointwise and simultaneous confidence bands.

Figure 10.7 illustrates the setting and the estimation procedure (see comments in the caption). Diagram 1 shows the underlying sample from X. It is atypical for the Normal, and this is reflected in the E-estimate and relatively large ISE. Then this sample is modified by Laplace measurement errors. Diagram 2 shows the modified sample $Y_1 = X_1 + \varepsilon_1, \ldots, Y_n = X_n + \varepsilon_n$. Note that the sample is heavily skewed and also check the larger support. The E-estimate does a good job in deconvolution of f^X. We see the symmetric about 0.5 E-estimate (the dashed line) and there is only a relatively small increase, for the ill-posed problem, in the ISE with respect to the case of direct observations. Then the sample, contaminated by measurement errors, is right censored by a random variable C with uniform on $[0, 1.5]$ distribution. The available observations of the pair (V, Δ) are shown in Diagram 3. As we see, only $N = 66$ observations of Y are uncensored.

Because the support $[0, 1.5]$ of the censoring variable C is smaller than the support $(-\infty, \infty)$ of Y, we known from Section 6.5 that consistent estimation of the survival function G^C is possible. On the other hand, consistent estimation of the density f^Y is impossible because there are no observations of Y larger than 1.5. Nonetheless, let us continue our analysis of the censored data shown in Diagram 3. As we see, only $N = 66$ from $n = 100$ of the underlying observations of Y are available in the censored sample. This is the bad news. The good news is that, as we know from Diagram 2, a majority of observations of Y are smaller than $\beta_C = 1.5$, and this tells us that the E-estimate of f^X may be relatively good (under the circumstances). And indeed, Diagram 4 exhibits symmetric about 0.5 and bell-shaped E-estimate whose ISE is comparable in order with the ones for the case of the convoluted and directly observed data. Furthermore, the confidence bands perform well.

Figure 10.7 also allows us to simulate experiments with exponential censoring variable C, and it is highly advisable to analyze this case, as well as to repeat Figure 10.7 with different parameters. The simulations and their analysis will allow the reader to gain first-hand experience in dealing with this complicated statistical problem when underlying data are subject to the two-fold modification by measurement errors and censoring.

10.4   Current Status Censoring

A theoretically interesting and practically important current status censoring (CSC), also known as "case I" interval censoring, is new to us. CSC yields an ill-posed data modification, and this is why it was not introduced in previous chapters.

The CSC is less known than the right or left censoring, and it is worthwhile to begin with a number of examples.

A general example, which explains the CSC, is as follows. Suppose that we would like to know the distribution of a time X when an event of interest occurs, but we cannot constantly monitor the time. Instead, there is a possibility to check the status of the event at some random moment of time Z, called the *monitoring* time. Then the available observation is a pair of random variables (Z, Δ) where Z is the monitoring time and $\Delta := I(X \leq Z)$ is the status of the event of interest, namely the status (indicator) is equal to 1 if the event of interest already occurred and 0 otherwise. Note that we never observe the time when the event occurred, and only know that the event occurred before or after the monitoring time Z. A sample from (Z, Δ) is called current status censored.

It is also possible that some of CSC observations are missed. For instance, we may observe a realization of (Z, Δ) only when $\Delta = 1$ and otherwise Z is missed. This means that in addition to the current status censoring, we observe the monitoring time Z only if the event of interest occurs before the monitoring time. Using the traditional missing data notation, in this case we observe a sample from (AZ, Δ, A), and $A := \Delta$ is the availability.

Of course, a missing mechanism with $A := 1 - \Delta$ is another possibility. We will refer to this modification as missing CSC.

Further, similarly to a truncation, another possible additional modification of CSC data is that a realization of (Z, Δ) is observed given $\Delta = 1$ (or $\Delta = 0$) and otherwise the realization is skipped and we even do not know that it occurred. This modification of CSC data is a truncation, and we may refer to it as truncated CSC.

Now let us consider several specific examples.

Our first example is actuarial. Suppose that we may get information from an insurance company about the deductible Z in each insurance policy and the indicator of payment to the policyholder for an insurable loss. The aim is to estimate the distribution of the loss X by a policyholder. In this case the available data is a sample from (Z, Δ) where $\Delta := I(X \leq Z)$ is the status of the policy with $\Delta = 0$ implying a payment on the insurance policy. This is a classical example of CSC data. Further, let us consider the case when we may get information only about deductibles for insurance policies with reported payments to policyholders (and no information is provided about policies with no payments and we even do not know the total number of policies). In this case we observe a sample from (Z, Δ) given that $\Delta = 0$, and this is the truncated CSC. Indeed, we observe the deductible Z only if the loss X exceeds the deductible Z, and vice versa if Z is observed then we know that the loss exceeded the deductible. Further, if we know the total number of insurance policies, then we have the missing CSC. Note that under all these possible scenarios we never observe the underlying insurable loss. Let us shed additional light on this CSC example with missing data. As we know from Chapter 6, deductible is a classical truncation variable, and if we additionally could get information about X for policies with reported payments, then we would deal with a left truncated sample from (Z, X) given that $X \geq Z$. This would be a dramatically more informative sample than the above-presented CSC with missing data, because in the truncated sample we would observe losses. Of course, the observed truncated losses are biased, but this is a relatively mild modification with respect to the CSC when we never observe losses.

The second example is epidemiological. During an outbreak of an infectious disease, a particular random variable of interest is the time X it takes for a subject to be in a contact with the source of infection until becoming infected. A doctor can examine the subject at a time Z, counted from the time of contact, and conclude that the subject is either infected ($\Delta := I(X \leq Z) = 1$) or not ($\Delta := I(X \leq Z) = 0$). As a result, the available information about the variable of interest X is a sample from (Z, Δ) and not X. Further, if a doctor is required to submit information only about infected patients, we get a sample from (Z, Δ) given $\Delta = 1$, and this is a truncated CSC.

The third example is about monitoring the time X of a migration of a subject when the only available information is the last time Z when there is a record on the subject. Historically this example was one of the first that raised the question about the CSC.

Finally, let us finish our set of examples with one that clarifies the difference between the CSC and censoring. Suppose that we have only a period of time T to watch until a just hatched baby eagle makes a first flight. If it is possible to constantly monitor the baby eagle, then the time X from hatching until making first flight is known exactly unless this time is larger than the censoring time T, and in the latter case we only know that the time of interest is larger than T. This is a classical right censoring model. On the other hand, if the baby eagle is observed only at end of the available period, then T is the monitoring time and we know only that either the baby eagle already made the first flight or the flight will occur later, and this is a classical CSC model.

After all these examples, let us stress that in many controlled studies CSC sampling is simpler and less expensive to arrange than a direct sampling of the variable of interest, but unfortunately, as we will see shortly, a CSC modification is ill-posed and dramatically affects quality of estimation. As a result, the choice should be accurately thought through, and this section will help to shed light on the CSC.

Now let us formulate the considered CSC problem. There is an underlying (hidden) sample $X_1, \ldots, X_n$ from a random variable of interest X. There is also a sample $Z_1, \ldots, Z_n$ from an independent monitoring variable Z. Available observations are $(Z_1, \Delta_1), \ldots, (Z_n, \Delta_n)$ where $\Delta_l := I(X_l \leq Z_l)$. We may say that (Z, Δ), where $\Delta := I(X \leq Z)$, is the CSC modification of (X, Z). Note that the variable of interest X is hidden and *never* observed, and the only information about a hidden realization X_l is that it is either not larger or larger than an independently generated monitoring Z_l. Further, in some cases even the CSC observations may be missed and we may observe either cases corresponding to $\Delta = 0$ or $\Delta = 1$.

On first glance, it is even difficult to believe that estimation of the distribution of X based on CSC modified data is possible, but as we will see shortly, while the problem is indeed complicated and ill-posed, a consistent estimation is possible under an appropriate assumption. Let us explain how the density $f^X(x)$ for a continuous X may be estimated via the E-estimation methodology.

We begin with a pivotal probability formula. Suppose that X and Z are continuous and independent random variables with smooth densities. Then the joint (mixed) density of the observed pair (Z, Δ) is

$$f^{Z,\Delta}(z, \delta) = f^Z(z)[\mathbb{P}(X \leq z)]^\delta [\mathbb{P}(X > z)]^{1-\delta}$$

$$= f^Z(z)[F^X(z)]^\delta [1 - F^X(z))]^{1-\delta}, \quad \delta \in \{0, 1\}. \qquad (10.4.1)$$

Formula (10.4.1) immediately implies that consistent estimation of the distribution of X is possible if and only if the support S^X of X is the subset of the support S^Z of Z, that is

$$S^X \subset S^Z. \qquad (10.4.2)$$

This assertion is based on the fact that both $f^{Z,\Delta}(z, \delta)$ and $f^Z(z)$, $z \in S^Z$ can be consistently estimated by the density E-estimator of Section 2.2. At the same time, $F^X(z)$ cannot be restored for $z \notin S^Z$. The conclusion should be absolutely clear from the actuarial example. Indeed, if deductible Z cannot be larger than a constant c_0 and the loss X may be larger than c_0, namely $\mathbb{P}(X > c_0) > 0$, then the CSC modification is destructive and we cannot restore the right tail of the distribution of the loss. The latter is a series complication of the CSC that we should be aware of.

Now let us return to the problem of estimation of the density of interest $f^X(x)$, and we are considering a number of possible CSC models in turn.

We begin with a more complicated case of missing CSC when the available sample is from (AZ, Δ, A) and the availability $A := \Delta$. The latter is equivalent to observing a sample from $(\Delta Z, \Delta)$ or observing Z only if $X \leq Z$ and knowing the sample size n of a hidden sample from X. (Recall that in some examples $A := 1 - \Delta$ and this setting will be explored shortly.) In what follows, to be specific about the above-discussed effect of supports on the possibility of consistent estimation, we suppose that X is supported on $[0, 1]$, Z is supported on $[a_1, a_2]$, and

$$a_1 \leq 0 \quad \text{and} \quad 1 \leq a_2. \qquad (10.4.3)$$

Note that (10.4.3) is the special case of (10.4.2).

Now several remarks are due. First of all, let us consider the following question. Is this missing mechanism MAR or MNAR? While it is tempting to say that it is MNAR (because the missing is defined by the value of Z), it is actually MAR. Indeed, we have

$$\mathbb{P}(A = 1 | X = x, Z = z, \Delta = \delta) = \mathbb{P}(A = 1 | \Delta = \delta) = I(\delta = 1). \qquad (10.4.4)$$

The value of Δ is always observed and this yields the MAR.

To be specific, assume that the observed sample is $(\Delta_1 Z_1, \Delta_1), \ldots, (\Delta_n Z_n, \Delta_n)$. The (mixed) joint density of $(\Delta Z, \Delta)$ is (compare with (10.4.1))

$$f^{\Delta Z, \Delta}(z, 1) = f^Z(z) F^X(z), \qquad (10.4.5)$$

and

$$f^{\Delta Z, \Delta}(0, 0) = \mathbb{P}(\Delta = 0) = \mathbb{P}(X > Z) = 1 - \int_{-\infty}^{\infty} f^Z(z) F^X(z) dz. \qquad (10.4.6)$$

It immediately follows from (10.4.5) and (10.4.6) that no consistent estimation of the distribution of X is possible unless $f^Z(z)$ is known. Indeed, note that the right sides of (10.4.5) and (10.4.6) depend solely on the product $f^Z(z) F^X(z)$. As a result, based on missed CSC data we may estimate only the product $f^Z(z) F^X(z)$. This implies that to estimate F^X, one needs to know f^Z. Of course, if no missing occurs then f^Z may be estimated based on n direct observations of Z. On the learning side of the story, we now have an example when MAR implies destructive missing.

Now, after all these preliminary remarks, we are ready to propose an E-estimator for the density $f^X(x)$, $x \in [0, 1]$ based on the above-described sample and known density $f^Z(z)$ of the monitoring variable. Further, let us assume that the monitoring variable is also supported on $[0, 1]$ and $f^Z(z) \geq c_* > 0$, $z \in [0, 1]$ and leave the case of a larger support as an exercise. Following the E-estimation methodology, we are using the cosine basis $\{1, \varphi_j(x) := 2^{1/2} \cos(\pi j x), j = 0, 1, \ldots\}$ on $[0, 1]$ and write,

$$f^X(x) = 1 + \sum_{j=1}^{\infty} \theta_j \varphi_j(x), \quad x \in [0, 1]. \qquad (10.4.7)$$

Let us write down a Fourier coefficient θ_j as the expectation of a function of observed random variables. Using integration by parts, we may write for $j \geq 1$,

$$\theta_j := \int_0^1 f^X(x) \varphi_j(x) dx = 2^{1/2} \int_0^1 \cos(\pi j x) f^X(x) dx$$

$$= 2^{1/2} [\cos(\pi j) F^X(1) - \cos(0) F^X(0)] + (2^{1/2} \pi j) \int_0^1 \sin(\pi j x) F^X(x) dx$$

$$= 2^{1/2} \cos(\pi j) + (2^{1/2} \pi j) \mathbb{E} \Big\{ \Delta \frac{\sin(\pi j \Delta Z)}{f^Z(\Delta Z)} \Big\}. \qquad (10.4.8)$$

In the last equality we used $F^X(0) = 0$, $F^X(1) = 1$ and (10.4.5). The expression (10.4.8) allows us to propose the sample mean estimator of θ_j,

$$\tilde{\theta}_j := 2^{1/2} \cos(\pi j) + n^{-1} (2^{1/2} \pi j) \sum_{l=1}^{n} \Delta_l \frac{\sin(\pi j \Delta_l Z_l)}{f^Z(\Delta_l Z_l)}. \qquad (10.4.9)$$

The Fourier estimator is unbiased, and it yields the density E-estimator $\tilde{f}^X(x)$, $x \in [0, 1]$. The asymptotic theory asserts that no other estimator can improve the rate of the MISE convergence achieved by the E-estimator. But the asymptotic theory also tells us that the problem is ill-posed and the MISE convergence slows down with respect to the case of direct observations of X.

Let us present two explanations of why the CSC modification is ill-posed. The first one is to directly evaluate the variance of the sample mean estimator (10.4.9) and get the following asymptotic expression,

$$\mathbb{V}(\tilde{\theta}_j) = n^{-1} (\pi j)^2 \int_0^1 \frac{F^X(z)}{f^Z(z)} dz [1 + o_j(1)]. \qquad (10.4.10)$$

Recall that $o_s(1)$ denotes a generic sequence in s that tends to zero as s increases. Formula (10.4.10) shows that the variance increases as j increases, and this is what slows down the MISE convergence (compare with the discussion in Section 10.1). At the same time, for small sample sizes, when the E-estimator uses only a few first Fourier coefficients, that increase in the variances may not be too large and we still may get relatively fair E-estimates (and recall that this is always the hope with ill-posed problems).

Another way to realize the ill-posedness is via direct analysis of our probability formulas. First of all, according to (10.4.5), the density of available observations is proportional to the CDF $F^X(x)$. In other words, the likelihood of a particular observation is defined by the CDF. On the other hand, the estimand of interest is not the CDF but the density $f^X(x)$. Let us check how a change in the density of interest affects the CFD and correspondingly the density of available observations. Set $\kappa_j := \int_0^1 [2^{1/2} \sin(\pi j x)] F^X(x) dx$ for the jth Fourier coefficient of the CDF $F^X(x)$ for the sine basis on $[0, 1]$. Then, following (10.4.7), introduce a new probability density $f'(x) := 1 + \sum_{s=1}^{\infty} \theta'_s \varphi_s(x)$ with $\theta'_s := \theta_s$, $s \neq j$ and $\theta'_j := \theta_j + \gamma$. Note that the difference between densities $f^X(x)$, defined in (10.4.7), and $f'(x)$ is only in their jth Fourier coefficients. Then the CDF $F'(x) := \int_0^x f'(u) du$ has Fourier coefficients $\kappa'_s = \kappa_s$, $s \neq j$ and $\kappa'_j = \kappa_j + \gamma/(\pi j)$. As a result, the effect of changing the estimand (here $f^X(x)$) on the density (10.4.5) of observations vanishes as j increases. The latter is exactly what ill-posedness is about when observations (data) become less sensitive to changes in the estimand. Note that no such phenomenon occurs for the case of direct observations of X when the density of observations $f^X(x)$ and the estimand $f^X(x)$ coincide.

Now let us consider another model of missing CSC data when Z is observed only when $\Delta = 0$, that is when $X > Z$. In this case we observe a sample $((1 - \Delta_1)Z_1, \Delta_1), \ldots, ((1 - \Delta_n)Z_n, \Delta_n)$ from pair $((1 - \Delta)Z, \Delta)$, and

$$f^{(1-\Delta)Z, \Delta}(z, 0) = f^Z(z)(1 - F^X(z)). \tag{10.4.11}$$

Let us explain how to construct a sample mean estimator of Fourier coefficients θ_j of $f^X(x)$ in the expansion (10.4.7). Note that the two top lines in (10.4.8) are still valid, and we can write using (10.4.11) for $j \geq 1$,

$$\theta_j = 2^{1/2} \cos(\pi j) + (2^{1/2} \pi j) \int_0^1 \sin(\pi j x)[1 - \frac{f^{(1-\Delta)Z, \Delta}(x, 0)}{f^Z(x)}] dx$$

$$= 2^{1/2} \cos(\pi j) + (2^{1/2} \pi j) \int_0^1 \sin(\pi j x) dx$$

$$- (2^{1/2} \pi j) \mathbb{E}\left\{ (1 - \Delta) \frac{\sin(\pi j (1 - \Delta) Z)}{f^Z((1 - \Delta)Z)} \right\}$$

$$= 2^{1/2} - (2^{1/2} \pi j) \mathbb{E}\left\{ (1 - \Delta) \frac{\sin(\pi j (1 - \Delta) Z)}{f^Z((1 - \Delta)Z)} \right\}. \tag{10.4.12}$$

In the last line we used $\int_0^1 \sin(\pi j x) dx = [-\cos(\pi j) + 1]/(\pi j)$.

Relation (10.4.12) implies the following sample mean estimator of Fourier coefficients,

$$\check{\theta}_j := 2^{1/2} - n^{-1} (2^{1/2} \pi j) \sum_{l=1}^{n} (1 - \Delta_l) \frac{\sin(\pi j (1 - \Delta_l) Z_l)}{f^Z((1 - \Delta_l) Z_l)}. \tag{10.4.13}$$

This Fourier estimator is unbiased and it yields the corresponding density E-estimator $\check{f}^X(x)$, $x \in [0, 1]$.

The third considered model is the classical CSC. We observe a sample $(Z_1, \Delta_1), \ldots, (Z_n, \Delta_n)$ from (Z, Δ) (recall that $\Delta := I(X \leq Z)$) with the joint density (10.4.1). Here we no longer need to know density $f^Z(z)$ of the monitoring variable Z because we can use the E-estimator $\hat{f}^Z(z)$ of Section 2.2 based on the direct observations $Z_1,\ldots,Z_n$. Then we may plug this estimator in (10.4.9) and (10.4.13) in place of $f^Z(z)$ and get two estimators of $f^X(x)$. And this leads us to a new statistical problem of *aggregation* of the two estimators.

Aggregation of several estimators into a better one is a classical topic in statistics, and for us it is a right time and place to explore it. We begin with a parametric estimation problem. Suppose that there are two unbiased and independent estimators $\tilde{\gamma}_1$ and $\tilde{\gamma}_2$ of parameter γ whose variances are σ_1^2 and σ_2^2, respectively. We are interested in finding an aggregation

$$\tilde{\gamma}(\lambda) := \lambda\tilde{\gamma}_1 + (1-\lambda)\tilde{\gamma}_2, \quad \lambda \in [0,1], \tag{10.4.14}$$

of the two estimators that minimizes the variance $\mathbb{V}(\tilde{\gamma}(\lambda))$. To solve the problem, we first find an explicit expression for the variance,

$$\mathbb{V}(\tilde{\gamma}(\lambda)) = \mathbb{E}\{[\lambda(\tilde{\gamma}_1 - \gamma) + (1-\lambda)(\tilde{\gamma}_2 - \gamma)]^2\}$$

$$= \lambda^2\mathbb{E}\{(\tilde{\gamma}_1 - \gamma)^2\} + (1-\lambda)^2\mathbb{E}\{(\tilde{\gamma}_2 - \gamma)^2\} = \lambda^2\sigma_1^2 + (1-\lambda)^2\sigma_2^2. \tag{10.4.15}$$

The right side of (10.4.15) is minimal when λ is equal to

$$\lambda^* := \frac{\sigma_2^2}{\sigma_1^2 + \sigma_2^2}. \tag{10.4.16}$$

Using this λ^* in (10.4.14) and (10.4.15) we find that the optimal aggregated estimator is

$$\hat{\gamma} := \frac{\sigma_2^2}{\sigma_1^2 + \sigma_2^2}\tilde{\gamma}_1 + \frac{\sigma_1^2}{\sigma_1^2 + \sigma_2^2}\tilde{\gamma}_2, \tag{10.4.17}$$

and

$$\mathbb{V}(\hat{\gamma}) = \frac{\sigma_1^2\sigma_2^2}{\sigma_1^2 + \sigma_2^2} < \min(\sigma_1^2, \sigma_2^2). \tag{10.4.18}$$

The inequality in (10.4.18) sheds light on performance of the aggregation.

Importance of this unbiased aggregation procedure for parametric statistics is explained by the fact that it is optimal (no other unbiased estimator can have a smaller variance) whenever γ is the mean of a normal random variable and we need to combine sample mean estimators based on two independent samples.

Formula (10.4.17) gives us a tool for aggregation of Fourier estimators $\tilde{\theta}_j$ and $\check{\theta}_j$ for the considered CSC model. The aggregated Fourier estimator $\hat{\theta}_j$ yields the density E-estimator $\hat{f}^X(x)$, $x \in [0,1]$.

Now let us look at simulations and test performance of the proposed E-estimators. We begin with the case of CSC data illustrated in Figure 10.8. Its top diagram shows us a sample of size $n = 100$ from the Normal variable X. This is the classical case of direct observations and it will serve us as a reference. The lines are explained in the caption. As we see, the E-estimate nicely smooths the histogram. At the same time, note that the relatively small sample size $n = 100$ precludes us from having observations near the boundary points. The empirical integrated squared error (ISE) of the E-estimate is equal to 0.0017, and it will be our benchmark. Also, the diagram shows the underlying cumulative distribution function $F^X(x)$ and the empirical cumulative distribution function (ECDF)

$$\hat{F}^X(x) := n^{-1}\sum_{l=1}^{n} I(X_l \le x). \tag{10.4.19}$$

Note that ECDF is a classical sample mean estimator because $F^X(x) := \mathbb{P}(X \le x) =$

Direct Sample from X, n = 100 , ISE = 0.0017 , ISEF = 0.00036

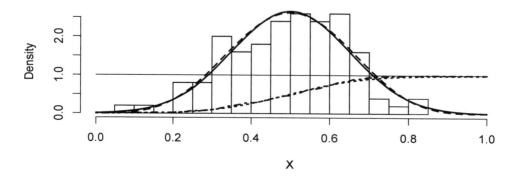

CSC Data, n = 100 , ISE = 0.023

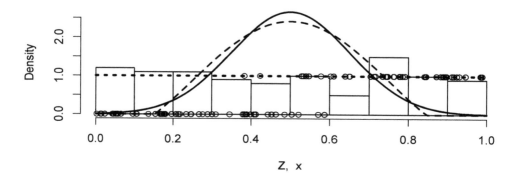

Figure 10.8 *Density estimation based on a direct sample from X and its current status censored (CSC) modification. Distributions of independent X and Z are the Normal and the Uniform. The top diagram presents direct observations via the histogram, the solid and dashed lines show the underlying density and its E-estimate, and the dotted and dot-dashed lines show the underlying cumulative distribution function and the empirical cumulative distribution function (ECDF). ISE and ISEF are the integrated squared errors for the E-estimate and the ECDF. The bottom diagram shows CSC data. Here the histogram shows the sample from Z, and the circles show pairs (Z_l, Δ_l), $l = 1, 2, \ldots, n$. The solid and dashed lines are the underlying density of X and its E-estimate, the ISE is shown in the title. The dotted line shows E-estimate of the density $f^Z(z)$. [n = 100, corn = 2, cJ0 = 3, cJ1 = 0.8, cTH = 4]*

$\mathbb{E}\{I(X \leq x)\}$. This immediately implies that the estimator is unbiased and $\mathbb{V}(\hat{F}^X(x)) = n^{-1}F^X(x)(1 - F^X(x))$. For the particular simulation the integrated squared error of the ECDF is ISEF = 0.00036. Note that the ISEF is in order smaller than the ISE, and this reflects the fact that estimation of the cumulative distribution function is possible with the parametric rate n^{-1} while estimation of the density is a nonparametric problem with a slower rate. The latter also explains why historically density estimation was considered as an ill-posed problem.

The bottom diagram in Figure 10.8 is devoted to the current status censoring (CSC). The data are created by the above-presented sample from X and an independent sample from the Uniform Z shown by the histogram. The circles show the observed sample

$(Z_1, \Delta_1), \ldots, (Z_n, \Delta_n)$ where $\Delta_l = I(X_l \leq Z_l)$. The E-estimate of the density f^Z (the dotted line) is perfect, and this may lead to a good E-estimate of the density of interest f^X. And indeed, its E-estimate (the dashed line) is very respectful, it correctly indicates the symmetric and unimodal shape of the Normal. Of course, the tails are shown incorrectly and indicate a smaller support, but this is also what we have seen in the histogram for direct observations. Note that the E-estimator knows only the scattergram of circles, and from this data it reconstructs the density. Also, look at the ISE which is in order larger than the benchmark. This is what defines the ill-posedness of the current status censoring. At the same time, the interested reader can repeat Figure 10.8 and find simulations where the E-estimate based on CSC data is better than the one based on direct observations. These are rare but realistic outcomes for small samples, and the likelihood of such an outcome diminishes for larger sample sizes.

Figure 10.9 is devoted to the case of missing CSC data. Here, to have as a reference the hidden underlying data, we use the sample from (X, Z) shown in Figure 10.8. The top diagram is devoted to the case when only observations Z_l satisfying $X_l \leq Z_l$ are available, that is when $\Delta_l = 1$. If we return for a moment to the bottom diagram in Figure 10.8, then the circles corresponding to $\Delta_l = 1$ are observed and others are missed. In Figure 10.9 the histogram of available observations of Z clearly indicates that only larger observations of Z are present. Nonetheless, the E-estimate of $f^X(x)$ is perfectly symmetric, unimodal and its ISE is good. It is truly impressive how, based on the right-skewed observations of the monitoring variable Z, the E-estimator has recovered the correct shape of the underlying Normal. The diagram also shows us the underlying CDF $F^X(x)$ of the Normal and its estimate by the dotted and dot-dashed lines, respectively.

Let us explain how the CDF estimator is constructed because it is of interest on its own. According to (10.4.5),

$$F^X(z) = \frac{f^{\Delta Z, \Delta}(z, 1)}{f^Z(z)} \quad \text{whenever } f^Z(z) > 0. \tag{10.4.20}$$

Note that the mixed density $f^{\Delta Z, \Delta}(z, 1) = f^{\Delta Z | \Delta}(z|1)\mathbb{P}(\Delta = 1)$ can be estimated by the E-estimator and $f^Z(z)$ is assumed to be known for the case of missing data. This yields a naïve ratio CDF E-estimator. Note that accuracy of estimation of the CDF is defined by the nonparametric accuracy of estimation of the density $f^{\Delta Z | \Delta}(z|1)$, and hence the MSE converges with a slower rate than the classical n^{-1}. The asymptotic theory asserts that no other estimator can improve the MSE convergence and that the proposed estimator is rate-optimal. The same conclusion holds for the case of CSC data. This yields that estimation of the CDF, based on CSC data, is ill-posed. This conclusion is both important and teachable because now we have an example of the modification by censoring which makes estimation of the CDF ill-posed. Further, comparing Figures 10.8 and 10.9, it is possible to explore the effect of missing CSC data on estimation of the CDF.

The bottom diagram in Figure 10.9 is constructed identically to the top one, only here we observe Z_l corresponding to $\Delta_l = 0$; recall that this is another missing scenario when Fourier estimator (10.4.13) is used. Note that the number $N = 53$ of available observations complements the number $N = 47$ in the top diagram, and the small sample sizes stress how difficult the estimation problems are. The reader is advised to repeat Figure 10.9 with different parameters and realize that the problem of missing CSC data is indeed extremely complicated due to the combined effects of missing and ill-posedness.

Now let us finish our discussion of the CSC by considering an even more complicated case when (10.4.3), and respectively (10.4.2), does not hold, meaning that the support of X is no longer a subset of the support of Z. In this case no consistent estimation of the density of interest f^X is possible because no inference about tail of f^X can be made. At the same time, is it possible to gain any insight into the distribution of X?

Missing CSC, n = 100 , N = 47 , ISE = 0.026 , ISEF = 0.0036

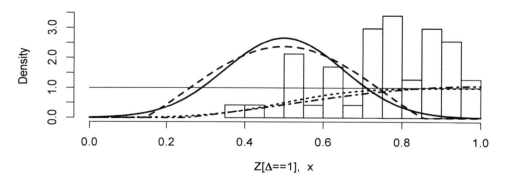

Z[Δ==1], x

Missing CSC, n = 100 , N = 53 , ISE = 0.021 , ISEF = 0.003

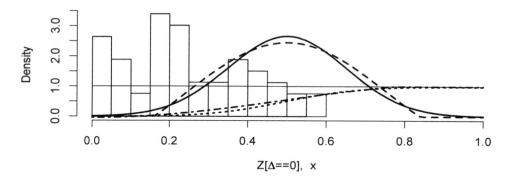

Z[Δ==0], x

Figure 10.9 *Density estimation for missing current status censored (CSC) data. The density of the monitoring time Z is assumed to be known. The underlying hidden sample from (X, Z) is the same as in Figure 10.8 (any other simulation by this figure will produce its own independent sample). In both diagrams the observed sample is from (AZ, Δ), only in the top diagram $A := \Delta$ and in the bottom $A := 1 - \Delta$. As a result, in the top diagram only observations of Z_l satisfying $Z_l \geq X_l$ are available and their number N is shown in the title. The available observations of the monitoring time Z are shown via the histogram, the solid and dashed lines show the underlying density $f^X(x)$ and its E-estimate, respectively. The dotted and dot-dashed lines show the underlying $F^X(x)$ and the naïve estimate $\tilde{F}^X(z) := \hat{f}^{Z,\Delta}(z,1)/f^Z(z)$ where $\hat{f}^{Z,\Delta}(z,1)$ is the E-estimate based on N available observations of Z. ISE and ISEF are the (empirical) integrated squared errors for the density and the cumulative distribution function estimates, respectively. The bottom diagram is similar to the top one, only here observations Z_l satisfying $Z_l < X_l$ are available. [n = 100, corn = 2, cJ0 = 3, cJ1 = 0.8, cTH = 4]*

We are exploring this issue for the case of missing CSC when only observations Z_l corresponding to $\Delta_l = 1$ are available. We are also relaxing our assumption that the support of $f^X(x)$ is known. Denote by $\hat{a}$ and $\hat{b}$ the smallest and largest available monitoring time, by $\hat{c} := \hat{b} - \hat{a}$ the range, and by $\psi_0(x) := [1/\hat{c}]^{1/2}$, $\psi_j(x) := [2/\hat{c}]^{1/2} \cos(\pi j(x - \hat{a})/\hat{c})$ the cosine basis on $[\hat{a}, \hat{b}]$.

Let us check the possibility to estimate $f^X(x)$ over the interval $[\hat{a}, \hat{b}]$, which is the empirical support for the monitoring time Z given $X \leq Z$. Set

$$\kappa_j := \int_{\hat{a}}^{\hat{b}} \psi(x) f^X(x) dx, \quad j = 0, 1, \ldots \tag{10.4.21}$$

for Fourier coefficients of $f^X(x)$ on $[\hat{a}, \hat{b}]$. Because the density of observations is directly related to F^X and not to the density f^X (recall (10.4.1)), it is prudent to rewrite (10.4.21) via F^X. We do this separately for $j = 0$ and $j \geq 1$. For $j = 0$ we can write that

$$\kappa_0 = [F^X(\hat{b}) - F^X(\hat{a})]/\hat{c}^{1/2}. \tag{10.4.22}$$

For $j \geq 1$, using integration by parts, we get

$$\kappa_j = (2/\hat{c})^{1/2} \int_{\hat{a}}^{\hat{b}} \cos(\pi j(x - \hat{a})/\hat{c}) f^X(x) dx$$

$$= (2/\hat{c})^{1/2} [\cos(\pi j) F^X(\hat{b}) - F^X(\hat{a})]$$

$$+ [2^{1/2} \pi j/\hat{c}^{3/2}] \int_{\hat{a}}^{\hat{b}} \sin(\pi j(x - \hat{a})/\hat{c}) F^X(x) dx. \tag{10.4.23}$$

It is a teachable moment to compare (10.4.23) with the second line in (10.4.8) that holds for the simpler setting (10.4.3) and known support $[0, 1]$ of X. We can clearly see the difference and the challenges of the considered setting. First of all, and this is very important, we no longer know values of $F^X(\hat{b})$ and $F^X(\hat{a})$, and they affect every Fourier coefficient. Recall that estimation of F^X is an ill-posed problem, and this is a very serious complication. A possible approach of estimating F^X is based on using (10.4.5) which still holds in our case. It follows from (10.4.5) that

$$F^X(x) = \frac{f^{\Delta Z, \Delta}(x, 1)}{f^Z(x)}, \quad x \in [a, b], \tag{10.4.24}$$

where $[a, b]$ is the support of Z given $\Delta = 1$, and let us assume that $f^{Z|\Delta}(z|1) \geq c_0 > 0$ on $[a, b]$. Recall that $f^Z(z)$ is supposed to be known for the case of missing CSC, and we already have discussed how to construct the mixed density E-estimator $\hat{f}^{\Delta Z, \Delta}(x, 1)$. This immediately yields a plug-in estimator

$$\tilde{F}^X(x) := \frac{\hat{f}^{\Delta Z, \Delta}(x, 1)}{f^Z(x)}, \quad x \in [\hat{a}, \hat{b}]. \tag{10.4.25}$$

We can also use the bona fide properties of the CDF to improve the estimator by considering its projection $\hat{F}^X(z)$ on a class of monotone and not exceeding 1 functions, see the Notes. Further, similarly to the motivation of the sample mean estimator (10.4.9) by (10.4.8), we may use (10.4.23), as well as (10.4.22), and propose the following sample mean estimator of Fourier coefficients κ_j, $j = 0, 1, \ldots$

$$\hat{\kappa}_j := \hat{c}^{-1/2} [I(j = 0) + 2^{1/2} I(j \geq 1)][\cos(\pi j) \hat{F}^X(\hat{b}) - \hat{F}^X(\hat{a})]$$

$$+ n^{-1} [2^{1/2} \pi j/\hat{c}^{3/2}] \sum_{l=1}^{n} \Delta_l \frac{\sin(\pi j(\Delta_l Z_l - \hat{a})/\hat{c})}{f^Z(\Delta_l Z_l)}. \tag{10.4.26}$$

This Fourier estimator implies the density E-estimator $\hat{f}^X(x)$, $x \in [\hat{a}, \hat{b}]$.

Now let us look at how the proposed E-estimator performs. Figure 10.10 sheds light on the setting and the density estimator. Here the random variable of interest X has the Strata

Hidden CSC, a1 = 0.1 , a2 = 0.6 , n = 200

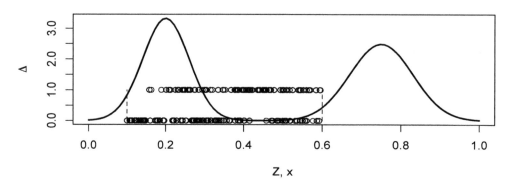

Missing CSC, N = 90

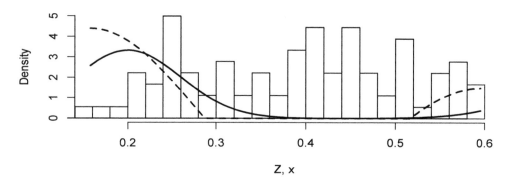

Figure 10.10 *Density estimation for missing CSC data when the support of monitoring time Z is the subset of an unknown support of X. The case of missing with the availability $A := \Delta := I(X \leq Z)$ is considered. The distributions of X and Z are the Strata and Uniform(a1, a2), respectively. The top diagram shows by circles a hidden CSC sample from (Z, Δ) which is overlaid by the density of interest $f^X(x)$ on its support $[0, 1]$. The support of the monitoring variable Z is $[a1, a2]$, it is shown in the title and highlighted by the two vertical dashed lines. The bottom diagram shows the available observations of Z by the histogram, as well as the underlying density $f^X(x)$ (the solid line) and its E-estimate (the dashed line) over the range of available monitoring times. Note that the histogram corresponds to z-coordinates of the top circles in the upper diagram. The number of available observations is $N := \sum_{l=1}^{n} \Delta_l$. [n = 200, corn = 4, a1 = 0.1, a2 = 0.6, cJ0 = 3, cJ1 = 0, cTH = 4]*

distribution while the monitoring variable (time) Z has uniform distribution on $[0.1, 0.6]$. The top diagram allows us to understand the setting. Density $f^X(x)$ is shown by the solid line. The monitoring variable Z is uniform on the interval indicated by two vertical dashed lines. Hidden CSC observations are shown by the scatterplot of pairs (Z, Δ). Note that in the missing CSC we are observing only values of Z corresponding to $\Delta = 1$, that is we know only the top circles, and their total number is $N = 90$. In addition to the small number of available monitoring times, note that the empirical support is also significantly smaller than the support $[0, 1]$ of the variable of interest X. The bottom diagram shows the histogram of available observations of Z, and this is another visualization of the data at hand. The histogram is overlaid by the solid line which indicates the underlying density $f^X(x)$ over the empirical support $[\hat{a}, \hat{b}]$. Note that in no way the histogram resembles or hints upon the underlying density $f^X(x)$. The density E-estimate $\hat{f}^X(x)$, $x \in [\hat{a}, \hat{b}]$ is shown by the dashed line. The density estimate is relatively good, but it is highly advisable to repeat Figure 10.10 a number of times with different parameters and to realize that here we are dealing with an extremely complicated ill-posed problem where a chance to get a fair estimate is slim.

Our final remark is about the possibility to estimate $f^X(x)$ by taking the derivative of an estimate of $F^X(x)$. This approach is natural keeping in mind that the density of observations is expressed via the cumulative distribution function $F^X(x)$ and not via the density. Assuming that $f^Z(z)$ is known, this approach, as it follows from (10.4.5) or (10.4.24), is equivalent to estimating the derivative of $f^{\Delta Z, \Delta}(z, 1)$. Estimation of derivatives is another classical ill-posed problem and it will be considered in Section 10.7.

10.5   Regression with Measurement Errors in Predictors

We begin with discussion of related mathematical and statistical problems that shed a new light on the phenomenon of ill-posedness in nonparametric regression.

Suppose that we know some values of a continuous function $m(x)$, $x \in [0, 1]$, namely we know values

$$m_l := m(x_l), \quad l = 1, 2, \ldots, n \tag{10.5.1}$$

where $x_l = (l-1)/(n-1)$ are equidistant points on $[0, 1]$. This is equivalent to say, using the regression terminology, that we observe n pairs $(m_1, x_1), \ldots, (m_n, x_n)$. The aim is to evaluate the function $m(x)$ for $x \in [0, 1]$. This is a fundamental and classical mathematical problem called the *interpolation*. Note that in (10.5.1) the function of interest $m(x)$ is observed without errors and at deterministic points x_l. As a result, under a mild assumption on smoothness of $m(x)$, we can find an interpolating function $\breve{m}(x)$ such that $|\breve{m}(x) - m(x)| < c_0 n^{-1}$. Further, if the function has a bounded second derivative, then we may even get the rate n^{-2} in the last inequality.

Now let us add some stochasticity to the interpolation model (10.5.1). Suppose that function $m(x)$ can be observed only with errors. Then (10.5.1) transforms into a classical regression model

$$Y_l = m(x_l) + \varepsilon_l, \quad l = 1, 2, \ldots, n. \tag{10.5.2}$$

Here $\varepsilon_1, \ldots, \varepsilon_n$ are called regression errors, and they are realizations of a zero mean and finite variance random variable (regression error) ε. The problem is again to estimate function $m(x)$, $x \in [0, 1]$. For us this is a familiar nonparametric regression problem where $m(x)$ is called the regression function, x_l is called the predictor and Y_l is called the response. While a regression problem may look similar to an interpolation problem, and they indeed share a lot in common, the consequences of measuring a function $m(x)$ with errors are dramatic because the best that we can get for α-fold differentiable functions is $\mathbb{E}\{|\breve{m}(x) - m(x)|\} \leq c_1 n^{-\alpha/(2\alpha+1)}$, $c_1 < \infty$. This tells us that the nonparametric regression is ill-posed with respect to the interpolation, or equivalently we may say that the modification of interpolation data (10.5.1) by regression errors is ill-posed.

Further, we may consider even a more general model. First, we replace deterministic predictors by random X_l that are realizations of a random variable X supported on $[0, 1]$. Second, we assume that not only an underlying function of interest may be observed with errors, but also predictors may be observed with errors. Under this scenario we observe a sample $(U_1, Y_1), \ldots, (U_n, Y_n)$ from (U, Y) where

$$Y_l = m(X_l) + \varepsilon_l, \quad U_l = X_l + \xi_l, \quad l = 1, 2, \ldots, n. \tag{10.5.3}$$

The problem is again to estimate the regression function $m(x)$, $x \in [0,1]$, and now this problem is referred to as a regression problem with *measurement errors in predictors* or simply a MEP regression. Note that a MEP regression is "symmetric" in terms of affecting both the predictor and the response by additive errors. As we shall see shortly, errors in predictors slow down the MISE convergence with respect to a standard regression and make estimation of the regression function extremely complicated. For instance, if ξ in (10.5.3) is normal, then the MISE decreases in n only logarithmically, and if the distribution of the measurement error is unknown, then consistent regression estimation is impossible (note that there is no such outcome for the classical regression model). This is an important information to keep in mind because errors in the response and the predictor yield dramatically different effects on regression estimation. We conclude that a MEP regression is ill-posed with respect to a standard regression, or we may say that the modification of predictors by measurement errors is ill-posed.

After this bad news, let us mention several good news. First, E-estimation is still optimal and dominates any other methodology. Second, for small samples and simple regression functions, we may get reasonable results.

Let us explain when and how an E-estimator can be constructed for the MEP model (10.5.3). The best way to explain the proposed solution is first to look at statistic

$$\check{\kappa}_j := n^{-1} \sum_{l=1}^n Y_l \varphi_j(U_l), \quad j = 0, 1, \ldots \tag{10.5.4}$$

This statistic is a naïve mimicking of the estimator $n^{-1} \sum_{l=1}^n Y_l \varphi_j(x_l)$ of Fourier coefficients $\theta_j := \int_0^1 \varphi_j(x) m(x) dx$ for the regression model (10.5.2), and recall our notation $\varphi_0(1) := 1$, $\varphi_j(x) := 2^{1/2} \cos(\pi j x)$, $j = 1, 2, \ldots$ for the cosine basis on $[0,1]$.

Now we step-by-step calculate the expectation of statistic (10.5.4). For $j = 0$ we have

$$\mathbb{E}\{\check{\kappa}_0\} = \mathbb{E}\{Y\} = \int_0^1 [f^X(x) m(x)] dx. \tag{10.5.5}$$

We continue for $j \geq 1$,

$$\mathbb{E}\{\check{\kappa}_j\} = \mathbb{E}\Big\{n^{-1} \sum_{l=1}^n Y_l \varphi_j(U_l)\Big\} = \mathbb{E}\{Y \varphi_j(U)\}$$

$$= \mathbb{E}\{m(X)\varphi_j(U)\} + \mathbb{E}\{\varepsilon \varphi_j(U)\}. \tag{10.5.6}$$

Assuming that $\mathbb{E}\{\varepsilon\} = 0$ and ε is independent of U, we can continue,

$$\mathbb{E}\{\check{\kappa}_j\} = \sqrt{2} E\{m(X) \cos(\pi j(X + \xi))\}. \tag{10.5.7}$$

There is a helpful trigonometric equality to use in (10.5.7),

$$\cos(\alpha + \beta) = \cos(\alpha)\cos(\beta) - \sin(\alpha)\sin(\beta). \tag{10.5.8}$$

Using it and additionally assuming that the predictor X and the measurement error ξ are independent, we continue (10.5.7),

$$\mathbb{E}\{\check{\kappa}_j\} = \mathbb{E}\{m(X)\varphi_j(X)\}\mathbb{E}\{\cos(\pi j \xi)\} - 2^{1/2} \mathbb{E}\{m(X)\sin(\pi j X)\}\mathbb{E}\{\sin(\pi j \xi)\}. \tag{10.5.9}$$

If we additionally assume that the distribution of measurement error ξ is symmetric about zero (Normal, Cauchy, Laplace are particular examples), then we get the identity $\mathbb{E}\{\sin(\pi j \xi)\} \equiv 0$, and hence can continue (10.5.9),

$$\mathbb{E}\{\check{\kappa}_j\} = \mathbb{E}\{m(X)\varphi_j(X)\}\mathbb{E}\{\cos(\pi j \xi)\}$$

$$= \left[\int_0^1 \varphi_j(x)[f^X(x)m(x)]dx \right] E\{\cos(\pi j\xi)\}. \tag{10.5.10}$$

Relation (10.5.10) is the pivot for our understanding of how to construct a sample mean estimator of Fourier coefficients $\theta_j := \int_0^1 \varphi_j(x)m(x)dx$ and hence the corresponding regression E-estimator for the MEP regression. But first let us combine together the above-made assumptions. It is assumed that X, ε and ξ are mutually independent, ε has zero mean and finite variance, and the distribution of ξ is symmetric about zero. Under these assumptions (10.5.10) holds, and the characteristic function $\phi^\xi(t)$ of the measurement error ξ can be written as

$$\phi^\xi(t) := \mathbb{E}\{e^{it\xi}\} = \mathbb{E}\{\cos(t\xi)\} + i\mathbb{E}\{\sin(t\xi)\} = \mathbb{E}\{\cos(t\xi)\}. \tag{10.5.11}$$

This allows us to continue (10.5.10),

$$\mathbb{E}\{\check{\kappa}_j\} = \phi^\xi(\pi j) \int_0^1 \varphi_j(x)[f^X(x)m(x)]dx. \tag{10.5.12}$$

A conclusion from (10.5.5) and (10.5.12) is as follows. If values $\phi^\xi(\pi j)$, $j = 1, 2, \ldots$ of the characteristic function of measurement error ξ are known and not equal to zero, then we can estimate the product

$$g(x) := f^X(x)m(x) \tag{10.5.13}$$

by the E-estimator $\check{g}(x)$ based on Fourier estimator $\check{\kappa}_j/\phi^\xi(\pi j)$. Further, as we know from Section 10.1, we can also estimate the design density $f^X(x)$ by the deconvolution E-estimator $\hat{f}^X(x)$. Then the regression function of interest $m(x)$ can be estimated by the corresponding ratio $\check{g}(x)/\hat{f}^X(x)$. Two important remarks are due. First, the estimation of $f^X(x)$ is an ill-posed deconvolution problem. Second, if the characteristic function of ξ is unknown, then consistent regression estimation is impossible and the MEP modification becomes destructive. In the latter case an additional information, like an extra sample from ξ, should be searched after.

After this discussion and heuristic explanation of the approach, let us formally describe three steps in construction of a MEP regression estimator. In what follows it is assumed that the function $\phi^\xi(\pi j)$ is known and not equal to zero, and the above-formulated assumptions about the MEP model hold. First, we need to estimate Fourier coefficients of function $g(x)$ defined in (10.5.13),

$$g_j := \int_0^1 g(x)\varphi_j(x)dx. \tag{10.5.14}$$

According to (10.5.5) and (10.5.12), the corresponding sample mean estimator is

$$\hat{g}_j := [n\phi^\xi(\pi j)]^{-1} \sum_{l=1}^n Y_l \varphi_j(U_l). \tag{10.5.15}$$

This Fourier estimator yields the E-estimator $\hat{g}(x)$, $x \in [0, 1]$.

The second step is to estimate the design density $f^X(x)$. Here we use the deconvolution density E-estimator $\hat{f}^X(x)$, $x \in [0, 1]$ of Section 10.1.

The final step is to use (10.5.13) and construct the plug-in regression E-estimator. Let us, as usual, assume that the design density $f^X(x)$ is bounded below from zero on its support. Then the proposed MEP regression estimator is

$$\hat{m}(x) := \frac{\hat{g}(x)}{\max(\hat{f}^X(x), c/\ln(n))}, \quad x \in [0, 1], \tag{10.5.16}$$

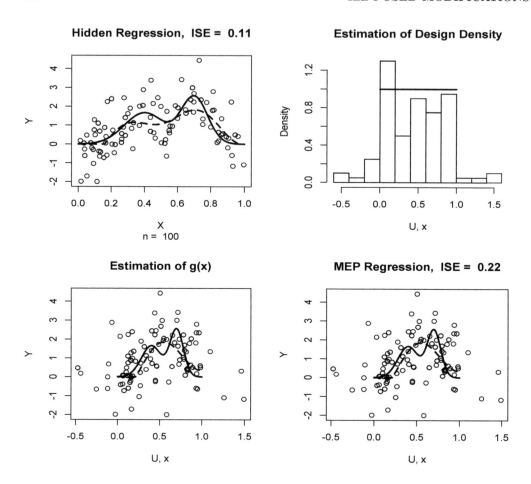

Figure 10.11 *Regression with measurement errors in predictors (MEP regression). The underlying model is (10.5.3) with the regression function being the Bimodal, regression error ε being normal with zero mean and variance σ^2, measurement error ξ being Laplace with parameter b (the same measurement error is used in Figure 10.2), and the predictor X being the Uniform. These variables are mutually independent. The left-top diagram shows a hidden sample of size n from (X, Y) via the scattergram. The underlying regression (the solid line) and the regression E-estimate of Section 2.3 (the dashed line) are also shown. The left-bottom diagram illustrates estimation of function $g(x)$ defined in (10.5.13). The circles show the available sample from (U, Y) where $U := X + \xi$, and this scattergram is overlaid by the underlying function $g(x)$ (the solid line) and its E-estimate (the dashed line) over the interval of interest $[0, 1]$. The right-top diagram shows the histogram of the sample from U, and by solid and dashed lines the underlying density $f^X(x)$, $x \in [0, 1]$ and its deconvolution E-estimate, respectively. (In this simulation the deconvolution E-estimate is perfect and is hidden by the solid line.) Finally, the right-bottom diagram again shows us the scattergram of available observations overlaid by the underlying regression (the solid line) and the MEP regression estimate (the dashed line). {Argument c controls the parameter c in (10.5.16).} [n = 100, sigma = 1, b = 0.2, corn = 3, c = 1, cJ0 = 3, cJ1 = 0, cTH = 4, c = 1]*

where $c > 0$ is the parameter of the estimator.

Of course, instead of (10.5.16) we may directly estimate Fourier coefficients of $m(x)$ and then use the regression E-estimator of Section 2.3. This approach is left as an exercise.

Figure 10.11 illustrates the studied MEP regression and the proposed regression estimator. The caption of Figure 10.11 explains the simulation and the diagrams. The left-top diagram shows us the hidden scattergram of the response Y with respect to the predictor X, and it corresponds to model (10.5.3). Note that we can observe the data only due to the simulation. The underlying regression is the Bimodal, and we know how difficult it is to estimate this function even for the case of a regular regression. The underlying regression function, shown by the solid line, helps us to see the pattern in the cloud of $n = 100$ observations, and overall we may agree that it is possible to visualize a bimodal-type regression. The E-estimate (the dashed line) is far from being perfect and it barely indicates the left mode. The not perfect estimation is also stressed by the empirical ISE = 0.11. But overall, for the case of the Bimodal regression, the outcome is satisfactory. Finally, it is reasonable to conclude from the scattergram that the underlying distribution of the predictor X is the Uniform.

Now let us look at the left-bottom diagram. It shows the scattergram of the MEP regression (10.5.3) with Laplace measurement errors in predictors. Here we use the same distribution for these errors as in Figure 10.2 so we have some experience with this modification of predictors. As we know, the distribution of the predictor is the Uniform, and hence according to (10.5.13) the function $g(x) = f^X(x)m(x) = m(x)$ is equal to the regression function. This is handy here because the function $g(x)$, which should be estimated, is the Bimodal. Can you see it in the cloud of circles? The solid line is the underlying function $g(x)$, does it help you to visualize the Bimodal in the scattergram? The likely answer is "no," and the reason for this is that measurement errors in predictors blur the Bimodal pattern by a stochastic movement of each response along the horizontal line. The E-estimate $\hat{g}(x)$ is shown by the dashed line, it indicates only one mode but at least it correctly indicates a smoothed shape of the Bimodal.

Recall that the E-estimation of $g(x)$ is the first part of the proposed estimation procedure. The next one is to estimate the design density f^X. The right-top diagram illustrates estimation of the design density $f^X(x)$. This is a familiar deconvolution procedure discussed in Section 10.1, the only new moment here is that in that section we did not consider an example of deconvolution of the Uniform density. The histogram is based on the sample from $U := X + \xi$. The histogram does not help us in visualization of the underlying density f^X, but the E-estimate performs a perfect deconvolution. For the Uniform this outcome is not surprising because this function is the first element of the cosine basis, but this is still an ill-posed problem and another simulation may imply a worse outcome.

The right-bottom diagram shows us the final step when the plug-in regression estimate (10.5.16) is calculated. Note that the estimate is the ratio between the estimates shown in the left-bottom and right-top diagrams. If we compare it with the regression estimate for the hidden regression in the left-top diagram, the quality of estimation is dramatically worse. Nonetheless, it is fair to say that the outcome is very reasonable for the MEP setting, the relatively small sample size, and large errors in the response and the predictor. It is advisable to repeat Figure 10.11 with different parameters and learn more about this complicated regression problem.

Let us finish our discussion of the ill-posed MEP regression with the following remark. In the beginning of this section, three settings (10.5.1)-(10.5.3) were formulated and possible outcomes for evaluation of an underlying function $m(x)$ were presented. In terms of accuracy, we may say that the classical regression (10.5.2) is ill-posed with respect to the interpolation (10.5.1), and the MEP regression (10.5.3) is ill-posed with respect to the classical regression (10.5.2). In other words, the notion of ill-posedness is defined by its benchmark. And indeed, before becoming one of the "mainstream" problems in statistics, nonparametric regression was considered as an ill-posed modification of the interpolation.

10.6 MEP Regression with Missing Responses

An underlying MEP (measurement errors in predictors) regression model is the same as in the previous section,

$$Y = m(X) + \sigma\varepsilon, \ \ U = X + \xi. \tag{10.6.1}$$

Only now a sample $(U_1, Y_1), \dots, (U_n, Y_n)$ from (U, Y) is hidden. Instead we observe a sample $(U_1, A_1Y_1, A_1), \dots, (U_n, A_nY_n, A_n)$ from the triplet (U, AY, A) where the availability A has a Bernoulli distribution with the availability likelihood

$$\mathbb{P}(A = 1|Y = y, U = u, X = x) = \mathbb{P}(A = 1|U = u) =: w(u) \geq c_0 > 0. \tag{10.6.2}$$

This is the MAR case because the missing is defined by the always observed variable U. According to Chapter 4, for a standard regression (10.5.2) (regression with $\mathbb{P}(\xi = 0) = 1$) the MAR is not destructive and further a complete-case approach is optimal.

In what follows all assumptions of the previous section hold, namely X is supported on $[0, 1]$, the design density $f^X(x)$ is positive on the support, variables X, ε and ξ are mutually independent, ε is zero mean and has a finite variance, the distribution of ξ is symmetric about zero and values $\phi^\xi(j\pi)$, $j = 1, 2, \dots$ of its characteristic function are known and not equal to zero. The problem is to estimate the regression function $m(x)$ based on MAR data.

We begin with the case of a known availability likelihood $w(u)$. Introduce a statistic (compare with (10.5.4))

$$\tilde{\kappa}_j := n^{-1} \sum_{l=1}^{n} A_l Y_l \varphi_j(U_l)/w(U_l), \tag{10.6.3}$$

and calculate its expectation. Write,

$$\mathbb{E}\{\tilde{\kappa}_j\} = \mathbb{E}\{AY\varphi_j(U)/w(U)\}$$

$$= \mathbb{E}\{\mathbb{E}\{[AY\varphi_j(U)/w(U)]|Y, U\}\}$$

$$= E\{[Y\varphi_j(U)/w(U)]\mathbb{E}\{A|Y, U\}\}$$

$$= \mathbb{E}\{[Y\varphi_j(U)/w(U)]w(U)\} = \mathbb{E}\{Y\varphi_j(U)\}. \tag{10.6.4}$$

In the last line we used relation $\mathbb{E}\{A|Y, U\} = \mathbb{P}(A = 1|Y, U) = w(U)$ which holds because A is Bernoulli and due to (10.6.2). The right side of (10.6.4) has been already evaluated in (10.5.6)-(10.5.12) and we conclude that

$$\mathbb{E}\{\tilde{\kappa}_j\} = \phi^\xi(\pi j) \int_0^1 \varphi_j(x)[f^X(x)m(x)]dx. \tag{10.6.5}$$

This is the same expression as in (10.5.12) and hence we can use the proposed in Section 10.5 plug-in regression E-estimator which performs as follows. First, the E-estimator $\tilde{g}(x)$ of $g(x) := m(x)f^X(x)$ is constructed using the Fourier estimator

$$\tilde{g}_j := (n\phi^\xi(\pi j))^{-1} \sum_{l=1}^{n} A_l Y_l \varphi_j(U_l)/w(U_l) \tag{10.6.6}$$

of Fourier coefficients

$$g_j := \int_0^1 \varphi_j(x)[f^X(x)m(x)]dx. \tag{10.6.7}$$

Second, the deconvolution density E-estimator $\hat{f}^X(x)$ of Section 10.1 is calculated, and it is based on the sample $U_1, \ldots, U_n$ and the characteristic function $\phi^\xi(\pi j)$. Finally, the plug-in regression estimator is defined as (compare with (10.5.16))

$$\tilde{m}(x) := \frac{\tilde{g}(x)}{\max(\hat{f}^X(x), c/\ln(n))} \ , \quad x \in [0,1]. \tag{10.6.8}$$

Let us stress that if the characteristic function $\phi^\xi(\pi j)$ is unknown then, similarly to the previous section, consistent estimation of regression function $m(x)$ is impossible. If the latter is the case, then an extra sample from ξ may be used to estimate the characteristic function.

So far it was assumed that the availability likelihood $w(u)$ is known and it was used in (10.6.3). If this function is unknown, and this is a typical case, then it can be estimated from the sample $(U_1, A_1), \ldots, (U_n, A_n)$. Indeed, we may write that $w(u) = \mathbb{E}\{A|U = u\}$, and hence $w(u)$ is a Bernoulli regression function and the regression E-estimator $\hat{w}(u)$ of Section 3.7 may be used. The only issue here is that no longer the design density $f^U(u)$ is necessarily separated from zero, and this may affect the estimation.

Figure 10.12 illustrates the model and the proposed estimator. The underlying regression model is identical to the one in Figure 10.11, only here the Normal regression function is used. The left-top diagram shows us the underlying regression of Y on X, and we can visualize it thanks to the simulation (in a practical example it would not be available). The underlying regression function is the Normal, and due to relatively large noise and small sample size the estimate is not perfect but correctly indicates the unimodal shape of the regression. Further, note that the estimate does reflect the scattergram. It may be instructive to compare this diagram with the left-top diagram in Figure 10.11. The new diagram in Figure 10.12 is the left-bottom one. It shows us a classical Bernoulli regression of A on U. According to (10.6.3), we need to know values of the availability likelihood function $w(u)$ only at points U_l. This is why the triangles show values of $w(U_l)$ and the crosses show values of its estimate $\hat{w}(U_l)$. The estimate is clearly not good, but it does follow the scattergram and at least correctly shows that the availability likelihood increases in u. The right-top diagram shows that the design density estimate is perfect. Finally, the right-bottom diagram shows the underlying regression and its estimate. For this particular simulation only $N = 75$ responses from $n = 100$ are available. The outcome is curious because the stochasticity, together with the small sample size, created a simulation where the estimate based on the missing MEP regression is more accurate than the one based on the hidden regression. It is advisable to repeat Figure 10.12 and realize that, despite this particular optimistic outcome, the problem is ill-posed, it is complicated by missing responses, and hence another simulation may create a worse outcome. Further, repeated simulations indicate that the effect of inaccurate estimation of $w(u)$ is less severe than of the design density $f^X(x)$.

Now let us consider a more complicated, and at the same time very interesting, case when the missing is defined by the hidden predictor X, and the availability likelihood is

$$\mathbb{P}(A = 1|Y = y, U = u, X = x) = \mathbb{P}(A = 1|X = x) =: w(x) \geq c_0 > 0. \tag{10.6.9}$$

In particular, this is the situation when missing of the response occurs before the predictor X is contaminated by the measurement error ξ. In other words, we have a standard MAR missing in a classical regression, and only then the predictor is contaminated by the measurement error.

Let us see what can be done in this case. Introduce a statistic

$$\hat{\kappa}_j := n^{-1} \sum_{l=1}^{n} A_l Y_l \varphi_j(U_l). \tag{10.6.10}$$

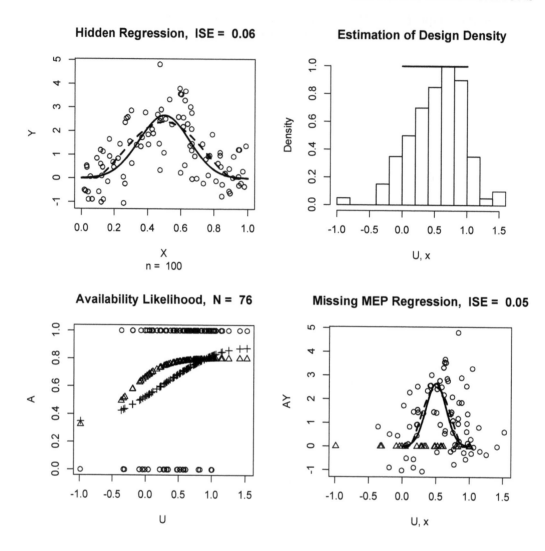

Figure 10.12 *MEP regression (10.6.1) with missing responses, the case when the missing is defined by always observable U. The underlying model is the same as in Figure 10.11, only here the particular regression function is the Normal. The missing mechanism is (10.6.2) with the availability likelihood $w^*(u) := \max(d_{wL}, \min(d_{wU}, w(u)))$, and for this particular simulation $w(u) = 0.3 + 0.5e^{1+4u}/(1 + e^{1+4u})$, $d_{wL} = 0.3$ and $d_{wU} = 0.9$. With respect to Figure 10.11, the new diagram is the left-bottom one that shows estimation of the availability likelihood. Here circles show the Bernoulli scattergram, the triangles and crosses show values $w(U_l)$ and $\hat{w}(U_l)$, respectively. In the right-bottom diagram the circles show complete pairs and triangles show incomplete ones with missed responses. {Function $w(u)$ is defined by a string w.} [n = 100, sigma = 1, b = 0.2, corn = 2, w = "0.3 + 0.5 *exp(1+4*u)/(1+exp(1+4*u))", dwL = 0.3, dwU = 0.9, cJ0 = 3, cJ1 = 0.8, cTH = 4, c = 1]*

Its expectation can be evaluated as follows,

$$\mathbb{E}\{\hat{\kappa}_j\} = \mathbb{E}\{AY\varphi_j(U)\} = \mathbb{E}\{\mathbb{E}\{AY\varphi_j(U)|Y, U, X\}\}$$

$$= E\{Y\varphi_j(U)\mathbb{E}\{A|Y, U, X\}\} = \mathbb{E}\{w(X)Y\varphi_j(U)\}. \tag{10.6.11}$$

In the last equality we used $\mathbb{E}\{A|Y, U, X\} = \mathbb{P}(A = 1|Y, U, X) = w(X)$. Then following

(10.5.6)-(10.5.12) we continue (10.6.11),

$$\mathbb{E}\{\hat{\kappa}_j\} = \mathbb{E}\{w(X)m(X)\varphi_j(X)\cos(\pi j\xi)\}$$

$$= \mathbb{E}\{w(X)m(X)\varphi_j(X)\}\phi^\xi(\pi j)$$

$$= \phi^\xi(\pi j)\int_0^1 [f^X(x)w(x)m(x)]\varphi_j(x)dx. \tag{10.6.12}$$

Set

$$q(x) := [f^X(x)w(x)]m(x), \ \ x \in [0,1] \tag{10.6.13}$$

and note that this is the same function that we see in the square brackets in (10.6.12). Further, the integral in (10.6.12) is the jth Fourier coefficients of this function,

$$q_j := \int_0^1 [f^X(x)w(x)m(x)]\varphi_j(x)dx \tag{10.6.14}$$

We can estimate this Fourier coefficient using the sample mean estimator

$$\hat{q}_j = (n\phi^\xi(\pi j))^{-1}\sum_{l=1}^n A_l Y_l \varphi_j(U_l). \tag{10.6.15}$$

In its turn, this Fourier estimator yields the E-estimator $\hat{q}(x)$ of $q(x)$.

Returning to (10.6.13), we now have a clear path for estimating the regression function $m(x)$ if we are able to estimate the product $f^X(x)w(x)$. Let us explain a possible approach. There exists a useful formula,

$$f^{X|A}(x|1) = \frac{f^{X,A}(x,1)}{\mathbb{P}(A=1)} = \frac{f^X(x)w(x)}{\mathbb{P}(A=1)}. \tag{10.6.16}$$

The numerator on the right side of (10.6.16) is the product $f^X(x)w(x)$, which we would like to know, and the probability $\mathbb{P}(A=1)$ is straightforwardly estimated by the sample mean estimator $n^{-1}\sum_{l=1}^n A_l$.

As a result, we need to understand how to estimate the conditional density $f^{X|A}(x|1)$ based on a sample $(U_1, A_1Y_1, A_1), \ldots, (U_n, A_nY_n, A_n)$. Is it possible to use here the deconvolution density estimator of Section 10.1? To answer this question, consider a subsample of (U_l, A_l) with $A_l = 1$, that is a subsample of MEP predictors in complete cases. Because X and ξ are independent, this, together with (10.6.9), implies the equality $f^{\xi|A}(z|1) = f^\xi(z)$. In its turn, this equality allows us to deduce the following convolution formula,

$$f^{U|A}(u|1) = \int_0^1 f^{X|A}(x|1)f^\xi(u-x)dx. \tag{10.6.17}$$

As a result, we indeed can use the subsample of MEP predictors in complete cases and construct the deconvolution density E-estimator $\hat{f}^{X|A}(x|1)$ of Section 10.1.

Now returning to (10.6.16), we can estimate the product $f^X(x)w(x)$ by the estimator $\hat{f}^{X|A}(x|1)n^{-1}\sum_{l=1}^n A_l$. This, together with (10.6.13) and the above-constructed estimator $\hat{q}(x)$, yields the following regression estimator for the case when the availability likelihood is defined by the value of X,

$$\hat{m}(x) := \frac{\hat{q}(x)}{[n^{-1}\sum_{l=1}^n A_l]\max(\hat{f}^{X|A}(x|1), c/\ln(n))}, \ \ x \in [0,1]. \tag{10.6.18}$$

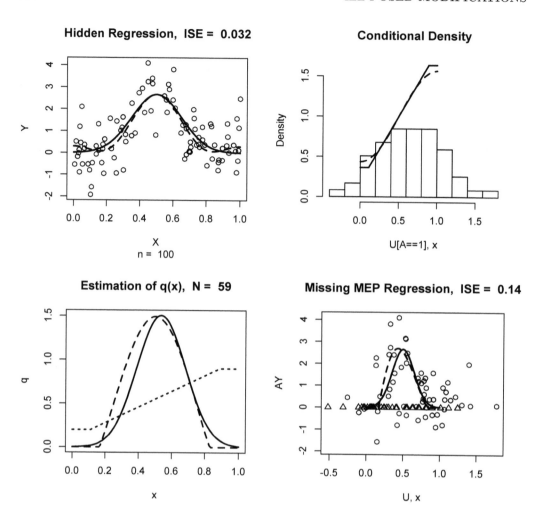

Figure 10.13 *MEP regression with missing responses when the missing is defined by the hidden predictor X. The underlying model is the same as in Figure 10.12 only here the missing mechanism is (10.6.9). With respect to Figure 10.12, the new diagrams are the left-bottom and right-top ones. The left-bottom diagram shows estimation of the function q(x) defined in (10.6.13). Here the solid and dashed lines are q(x) and its E-estimate, respectively. Also, the dotted line shows the underlying availability likelihood function w(x). The right-top diagram shows the histogram of U_l corresponding to $A_l = 1$, as well as the conditional density $f^{X|A}(x|1)$ (the solid line) and its deconvolution E-estimate $\hat{f}^{X|A}(x|1)$ (the dashed line) for $x \in [0,1]$. {In the default simulation the availability likelihood is $w^*(x) := \max(d_{wL}, \min(d_{wU}, w(x)))$, $w(x) = 0.1+0.9x$, $d_{wL} = 0.2$ and $d_{wU} = 0.9$. The function w(x) is defined by the string w, and parameters d_{wL} and d_{wU} are controlled by arguments dwL and dwU.} [n = 100, sigma = 1, b = 0.2, corn = 2, w = "0.1 + 0.9 *x", dwL = 0.2, dwU = 0.9, cJ0 = 3, cJ1 = 0.8, cTH = 4, c = 1]*

Of course, instead of developing the ratio estimator (10.6.18) we could use our traditional approach of estimating Fourier coefficients of the regression function; developing of this approach is left as an exercise.

Figure 10.13 illustrates the setting and performance of the regression estimator. The underlying MEP regression model is the same as in Figure 10.12, and the left-top diagram shows us the scattergram of the hidden regression of Y on X. The solid and dashed lines are the underlying Normal regression function and its E-estimate. It is a nice exercise to use your imagination and try to propose a better fit for this particular scattergram (and, of course, repeat the figure and visualize more simulations). The missing MEP data are shown in the right-bottom diagram. Here the circles show complete pairs $(U_l, A_l Y_l)$ corresponding to $A_l = 1$ and the triangles show incomplete pairs $(U_l, 0)$ corresponding to $A_l = 0$. You may see how the MEP spreads observations along corresponding horizontal lines and how the missing, defined by an underlying value of X, causes incomplete cases. The underlying availability likelihood function $w(x)$ is shown by the dotted line in the left-bottom diagram.

Note that only $N := \sum_{l=1}^{n} A_l = 59$ pairs are complete, and hence we are dealing with a simulation which is difficult even for the case of a regular regression. There are three steps in construction of the regression estimator for the MEP regression with missing responses. The first one is to estimate the function $q(x) := f^X(x) w(x) m(x)$. We do this by the E-estimator based on Fourier estimator (10.6.15). The function and the E-estimate are shown in the left-bottom diagram. The outcome is good. At the same time, let us stress that estimation of the function $q(x)$ is an ill-posed problem with missing data, this is a difficult problem, and it is advisable to repeat Figure 10.13 to realize the complexity.

The second step is to estimate the conditional density $f^{X|A}(x|1)$. Recall that the estimation is based on observations of U_l from complete pairs. According to (10.6.16), this density is proportional to $f^X(x) w(x)$, and note that in the simulation the design density $f^X(x)$ is the Uniform and the availability likelihood $w(x)$ is shown by the dotted line in the left-bottom diagram. Function $w(x)$ is only piecewise differentiable, and this creates a complication in estimation of the conditional density $f^{X|A}(x|1)$. The right-top diagram illustrates the estimation. The histogram shows us $N = 59$ observations of U_l in complete cases. Note that the histogram resembles neither $f^X(x)$, nor $w(x)$, nor their product. The underlying function $f^{X|A}(x|1)$ is shown by the solid line. Note that it is about twice in magnitude as the function $w(x)$ shown in the left-bottom diagram, and this is due to the denominator $\mathbb{P}(A = 1)$ in (10.6.16). The deconvolution E-estimate $\hat{f}^{X|A}(x|1)$ is shown by the dashed line, and it is very good keeping in mind the small sample size and ill-posed nature of the deconvolution problem.

The final step is (10.6.18) where the estimate of $q(x)$ (the dashed line in the left-bottom diagram) is divided by the estimate of the conditional density (the dashed line in the right-top diagram) and then divided by the sample mean estimate of $\mathbb{P}(A = 1)$. Note that (10.6.18) involves the ratio of two nonparametric estimates, and each of them is for an ill-posed problem. This is what makes the problem of MEP regression with missing data so complicated. The particular estimate, shown in the right-bottom diagram, is good, but keeping in mind complexity of the problem, another simulation may imply a worse outcome.

It is highly recommended to repeat Figures 10.12 and 10.13 with different parameters and compare simulations and performance of the estimators. Also, pay attention to the nuisance functions that may be of interest in practical applications.

10.7 Estimation of Derivative

So far we have considered ill-posed problems that are based on relatively complicated models like convolution or current status censoring. This may create an illusion that an ill-posed problem is simply a complicated problem. The conclusion is correct if a "complication" means slower rates of a risk convergence, and it is wrong if it is understood in how difficult a problem looks like. For instance, truncated data is a complicated notion but it is not ill-posed.

In this section we are considering a ladder of simply formulated and well understood ill-posed problems of estimation of the derivatives. We begin with estimation of the derivative of the cumulative distribution function of a continuous random variable, and recall that this derivative is called the probability density, and then explore the problem of estimation of the derivative of the density, or equivalently the problem of estimation of the second derivative of the cumulative distribution function.

Suppose that we have a sample $X_1, \ldots, X_n$ from a continuous random variable X supported on $[0, 1]$.

As we know, the cumulative distribution function $F^X(x) := \mathbb{P}(X \le x)$ completely describes the random variable X as well as the sample. This is why estimation of the cumulative distribution function is a classical statistical problem. Note that in general, when it is not assumed that $F^X(x)$ is from a particular family, say Gaussian, Cauchy or Laplace, estimation of $F^X(x)$ is a nonparametric problem. The most popular nonparametric estimator is an *empirical cumulative distribution function*

$$\hat{F}^X(x) := n^{-1} \sum_{l=1}^{n} I(X_l \le x), \qquad (10.7.1)$$

where $I(\cdot)$ is the indicator. Note that (10.7.1) is the sample mean estimator because

$$F^X(x) := \mathbb{P}(X \le x) = \mathbb{E}\{I(X \le x)\}. \qquad (10.7.2)$$

This immediately yields that the estimator (10.7.1) is unbiased and its variance decreases with the parametric rate n^{-1}, namely

$$\mathbb{V}(\hat{F}^X(x)) = n^{-1} F(x)(1 - F(x)), \quad x \in [0, 1]. \qquad (10.7.3)$$

Further, using the Hoeffding inequality (1.3.30) we conclude that

$$\mathbb{P}(|\hat{F}(x) - F(x)| > tn^{-1/2}) \le 2e^{-2t^2}, \quad t > 0. \qquad (10.7.4)$$

We may conclude that, despite being a nonparametric estimation problem, the cumulative distribution function can be estimated with the parametric rate n^{-1}. This is an important conclusion on its own because now we have an example of nonparametric estimation with the parametric rate. Further, as we will see shortly, it allows us to understand that the notion of ill-posedness depends on a chosen benchmark.

Now let us consider the problem of estimation of the derivative of $F^X(x)$ which is called the probability density of X,

$$f^X(x) := dF^X(x)/dx. \qquad (10.7.5)$$

As we know from Section 2.2, a probability density cannot be estimated with the parametric rate n^{-1} even if it is a smooth function. If the density is α-fold differentiable and belongs to a Sobolev class $S_{\alpha,Q}$ defined in (2.1.11) (the latter is assumed in this section), then the best result that can be guaranteed is

$$\mathbb{E}\{(\tilde{f}^X(x) - f^X(x))^2\} \le c_1 n^{-2\alpha/(2\alpha+1)}, \qquad (10.7.6)$$

and the same rate of convergence holds for the MISE. Here and in what follows $c_1, c_2, \ldots$ denote generic positive constants whose exact values are not of interest.

We may conclude that density estimation, with respect to estimation of the cumulative distribution function, is an ill-posed problem. And indeed, this is how this problem was treated when the research on nonparametric probability density estimation was initiated. With time passing by, more statisticians have been involved in the nonparametric research, the literature has grown rapidly, and then slowly but surely the nonparametric density estimation, characterized by slower rates, has become a classical statistical problem on its own. As a result, for now, papers and books that consider nonparametric density estimation as ill-posed problem are next to none. At the same time, of course the problem of nonparametric density estimation is ill-posed with respect to estimation of the cumulative distribution function.

Let us comment on why we have slower rates of the MISE convergence for the density estimation. Consider a cosine series estimator

$$\check{f}^X(x) := 1 + \sum_{j=1}^{J} \hat{\theta}_j \varphi_j(x), \ x \in [0,1]. \tag{10.7.7}$$

Here

$$\hat{\theta}_j := n^{-1} \sum_{l=1}^{n} \varphi_j(X_l) \tag{10.7.8}$$

is the sample mean estimator of Fourier coefficients

$$\theta := \int_0^1 \varphi_j(x) f^X(x) dx = \mathbb{E}\{\varphi_j(X)\}, \tag{10.7.9}$$

and parameter J in (10.7.7) is called the cutoff. Assuming that the density is square integrable, we can write

$$f^X(x) = 1 + \sum_{j=1}^{\infty} \theta_j \varphi_j(x), \ x \in [0,1]. \tag{10.7.10}$$

Using the Parseval identity (1.3.40) we get the following expression for the mean integrated squared error (MISE) of the estimator $\check{f}^X(x)$ (compare with (2.2.10))

$$\mathrm{MISE}(\check{f}^X, f^X) := \mathbb{E}\{\int_0^1 (\check{f}^X(x) - f^X(x))^2 dx\}$$

$$= \sum_{j=1}^{J} \mathbb{E}\{(\hat{\theta}_j - \theta_j)^2\} + \sum_{j>J} \theta_j^2. \tag{10.7.11}$$

Estimator $\hat{\theta}_j$ is the sample mean estimator, and hence it is unbiased, its variance is equal to the mean squared error, and the variance is easily evaluated (the calculation is left as an exercise and recall notation $o_s(1)$ for vanishing in $s \to \infty$ sequences),

$$\mathbb{E}\{(\hat{\theta}_j - \theta_j)^2\} = n^{-1}(1 + o_j(1)). \tag{10.7.12}$$

Further, according to (2.1.11) we have the following inequality

$$\sum_{j>J} \theta_j^2 \leq c_2 J^{-2\alpha}. \tag{10.7.13}$$

Using (10.7.12) and (10.7.13) in (10.7.11) we conclude that the MISE of the estimator (10.7.7) can be bounded from above,

$$\mathrm{MISE}(\check{f}^X, f^X) \leq c_3[Jn^{-1} + J^{-2\alpha}]. \tag{10.7.14}$$

A cutoff J_n^*, which minimizes the right side of (10.7.14), is

$$J_n^* := (2\alpha)^{2\alpha+1} n^{1/(2\alpha+1)}(1 + o_n(1)), \tag{10.7.15}$$

and it implies the above-mentioned rate $n^{-2\alpha/(2\alpha+1)}$ of the MISE convergence. Further, the asymptotic theory asserts that this rate of the MISE convergence is optimal in the sense that no other estimator can achieve a faster rate uniformly over the Sobolev class (2.1.11) of α-fold differentiable densities.

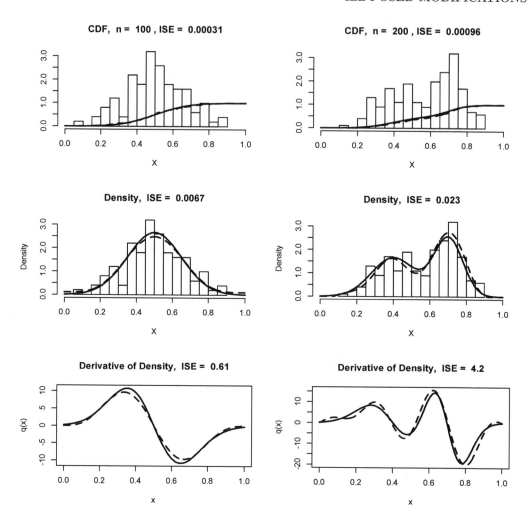

Figure 10.14 *Estimation of the cumulative distribution function (CDF), its first derivative (the probability density), and its second derivative (derivative of the probability density). Two simulations with different underlying distributions and different sample sizes are shown in the two columns. The solid and dashed lines show underlying functions and their estimates, respectively. Samples are shown by histograms. {Argument set.n allows to choose sample sizes for the two experiments.} [set.n = c(100,200), set.corn = c(2,3), cJ0 = 3, cJ1 = 0.8, cTH = 4]*

We may conclude that nonparametric estimation of the derivative of the cumulative distribution function, or equivalently nonparametric estimation of the density, is an ill-posed problem which implies slower rates of a risk convergence with respect to the benchmark problem of estimation of the cumulative distribution function. There is also another interesting conclusion to point upon. The smoothness of $F^X(x)$, say it is piece-wise continuous or twice differentiable, has no effect on the rate n^{-1} of the variance or the MISE convergence, see (10.7.3). At the same time, (10.7.6) indicates that smoothness of the cumulative distribution function dramatically affects accuracy of estimation of the derivative of the cumulative distribution function.

Now we are in a position to continue our investigation of estimation of derivatives. Consider the problem of estimation of the second derivative of the cumulative distribution function, or equivalently the problem of estimation of the derivative of the density. In what follows, we are using the latter formulation to simplify notation.

Several questions immediately arise. Is the problem of estimation of the density's derivative ill-posed with respect to the density estimation? Will the smoothness of density affect estimation of its derivative?

To answer these questions, set

$$q^X(x) := df^X(x)/dx \qquad (10.7.16)$$

for the density's derivative. Note that because we are dealing with only one random variable X, we may skip the superscript X in the density's derivative $q^X(x)$ and write $q(x)$ instead.

We are going to consider a very simple estimator of $q^X(x)$, which is the derivative of the series estimator (10.7.7),

$$\check{q}^X(x) = d\check{f}^X(x)/dx = -\sum_{j=1}^{J} \hat{\theta}_j(\pi j)\psi_j(x), \quad \psi_j(x) := 2^{1/2}\sin(\pi jx). \qquad (10.7.17)$$

Note that $\{\psi_j(x), j = 1, 2, \ldots\}$ is the sine orthonormal basis on $[0, 1]$.

Let us calculate the MISE of estimator (10.7.17). Using the Parseval identity we may write,

$$\text{MISE}(\check{q}^X, q^X) := \mathbb{E}\{\int_0^1 (\check{q}^X(x) - q^X(x))^2 dx\}$$

$$= \sum_{j=1}^{J}(\pi j)^2 \mathbb{E}\{(\hat{\theta}_j - \theta_j)^2\} + \sum_{j>J}(\pi j)^2\theta_j^2. \qquad (10.7.18)$$

Recall that we are considering $f^X(x)$ from the Sobolev class (2.1.11) of α-fold differentiable densities. Assume that $\alpha > 1$ and evaluate the second sum on the right side of (10.7.18),

$$\sum_{j>J}(\pi j)^2\theta_j^2 < c_4 J^{-2\alpha+2}. \qquad (10.7.19)$$

Using this and (10.7.12) in (10.7.18), we get the following upper bound for the MISE of the density's derivative estimator,

$$\text{MISE}(\check{q}^X, q^X) < c_5[J^3 n^{-1} + J^{-2\alpha+2}]. \qquad (10.7.20)$$

A cutoff J_n', which minimizes the right side of (10.7.20), is

$$J_n' := [(2\alpha - 2)/3]^{2\alpha+1} n^{1/(2\alpha+1)}(1 + o_n(1)). \qquad (10.7.21)$$

It is of interest to compare this cutoff for the density's derivative estimator with the optimal cutoff J_n^* for the density estimator defined in (10.7.15). As we see, they are both proportional to $n^{1/(2\alpha+1)}$, that is estimation of the derivative does not affect how the optimal cutoff increases in n.

Using (10.7.21) in (10.7.20) we conclude that

$$\text{MISE}(\check{q}^X, q^X) < c_6 n^{-2(\alpha-1)/(2\alpha+1)}. \qquad (10.7.22)$$

The asymptotic theory asserts that no other estimator can improve the above-presented rate of the MISE convergence uniformly over densities from the Sobolev class of α-fold differentiable densities. The latter means that our estimator of the derivative is rate optimal. Further, using the cutoff J_n^* in place of J_n' in the estimator of derivative also yields the optimal rate of the MISE convergence. We conclude that the derivative of the density E-estimator of Section 2.2 may be used as the estimator of the density's derivative.

The optimal rate $n^{-2(\alpha-1)/(2\alpha+1)}$ for estimation of the derivative is slower than the optimal rate $n^{-2\alpha/(2\alpha+1)}$ for estimation of the density, and this implies that the problem of estimation of the derivative is ill-posed with respect to the problem of density estimation. In short, estimation of derivative is ill-posed.

The same conclusion holds for other classical nonparametric problems like regression, filtering a signal or spectral analysis. This does not come as a surprise to us because we know that all these nonparametric problems have a lot in common, and moreover there exists a theory of equivalence between these problems.

Figure 10.14 illustrates the problem of estimation of the density and its derivative. The top diagrams show us two simulations via histograms and estimation of the cumulative distribution function. Note that the sample sizes of simulations are different to take into account the more complicated shape of the Bimodal. Pay attention to the small integrated squared errors of the empirical cumulative distribution function.

The middle diagrams illustrate the familiar density estimation problem. Of course, we may equivalently say that this is the problem of estimation of the derivative of the cumulative distribution function. In both cases E-estimates are good and this is reflected by the corresponding empirical ISEs. At the same time, compare these ISEs with those in the corresponding top diagrams and get the feeling of ill-posedness.

The bottom diagrams show derivatives of the underlying densities and their E-estimates. Overall, the estimated derivatives are good but look at the large empirical ISEs. The latter, at least in part, is explained by the fact that the derivatives are larger functions (look at the scales), but it also reflects the ill-posed nature of the problem. Further, repeated simulations indicate a relatively large number of poor estimates, and it is highly recommended to repeat Figure 10.14 with different parameters and get used to the problem of estimation of derivatives.

Let us also note that, using (10.7.17), it is possible to develop a sine series E-estimator of estimation of the derivative. Further, it is possible to consider more complicated settings when data are missed and/or modified by truncation, censoring, measurement errors, etc. These extensions are done straightforwardly using the E-methodology and they are left as exercises.

10.8   Exercises

10.1.1 Give several examples of a problem with measurement errors. In those examples, what can be said about the distribution of errors?

10.1.2* Consider a pair of random variables (X, ε). If the joint cumulative distribution function is given, what is the cumulative distribution function of their sum? Hint: Note that no information about the independence is provided.

10.1.3 Consider a pair of continuous random variables (X, ε). What is the density of their sum? Hint: Begin with the case of independent variables and then look at a general case.

10.1.4 Give definition of a characteristic function. Does it always exist?

10.1.5 Calculate characteristic functions of uniform, normal and Laplace random variables.

10.1.6 What is the value of a characteristic function of X at point zero, that is $\phi^X(0)$? Hint: Note that the distribution of X is not specified.

10.1.7 Prove (10.1.4).

10.1.8 Verify assertion (10.1.5).

10.1.9 Find the mean and the variance of an empirical characteristic function. Do the mean and the variance always exist?

10.1.10 Explain how the deconvolution problem can be solved.

10.1.11 Verify formula (10.1.9) for the variance of estimator (10.1.8).

10.1.12 Using formula (10.1.9), present examples of measurement errors when the variance increases exponentially and as a power in t.

10.1.13 Explain the simulation used by Figure 10.1.

10.1.14 Repeat Figure 10.1 using different arguments and comment on obtained result.

10.1.15 For which density, among our four corner ones, the effect of additive noise is more visually devastating in terms of "hiding" an underlying distribution? Hint: Begin with the theory and then confirm your conclusion using Figure 10.1.

10.1.16 Explain the estimate (10.1.12).

10.1.17 Find the mean and variance of the estimate (10.1.12).

10.1.18 Explain the simulation used in Figure 10.2.

10.1.19 Using Figure 10.2, find better values for parameters of the E-estimator. Hint: Use empirical ISEs.

10.1.20 In Figure 10.2, what is the role of parameter CH? Repeat Figure 10.2 for different sample sizes and suggest a better value for the parameter.

10.1.21 Explain the purpose of using an extra sample in Figure 10.3.

10.1.22 Explain the E-estimator used in Figure 10.3.

10.1.23 Suggest better parameters for E-estimator used in Figure 10.3. Hint: Use empirical ISEs.

10.1.24 Consider sample sizes $n = 100$ and $n = 200$. Then, using Figure 10.3, explore minimal sample sizes m of extra samples that imply a reliable estimation.

10.1.25 Explain how the case of repeated observations can be used for solving the deconvolution problem with unknown distribution of the measurement error.

10.1.26* Find the mean and variance of the estimator (10.1.15).

10.1.27 What is the definition of a circular random variable? Give examples.

10.1.28 Explain formula (10.1.16)

10.1.29* Verify (10.1.17).

10.1.30* Explain why the proposed deconvolution E-estimator may be also used for circular data.

10.1.31* Evaluate the MISE of the proposed deconvolution E-estimator. Hint: Begin with the case of a known nuisance function.

10.1.32* Consider the case of unknown distribution of ε. Suppose that an extra sample from ε may be obtained. Suggest an E-estimator for this case and analyze its MISE.

10.1.33* Suppose that the measurement error ε and the variable of interest X are dependent. Suggest a deconvolution estimator.

10.2.1 Explain the problem of density deconvolution with missing data.

10.2.2 What type of observations is available for density deconvolution with missing data?

10.2.3 Explain a possible solution for the case of MCAR.

10.2.4 Explain a setting when the availability likelihood depends on Y. Present several examples.

10.2.5 What is an underlying idea of estimator (10.2.3)?

10.2.6* Find the mean and variance of estimator (10.2.3). Is it unbiased?

10.2.7 Explain the underlying simulation used by Figure 10.4.

10.2.8 Repeat Figure 10.4 with different sample sizes, and comment on outcomes.

10.2.9 Present several examples when the availability likelihood depends on the underlying variable of interest X.

10.2.10 Explain how E-estimator, proposed for the case when the availability likelihood depends on X, is constructed.

10.2.11* Find the mean and variance of estimator (10.2.10). Then evaluate the MISE of E-estimator (10.2.11).

10.2.12* Use Figure 10.5 and analyze the effect of the availability likelihood $w(x)$ on estimation. Which functions are less and more favorable for the E-estimator? Hint: Theoretically analyze the MISE and then test your conclusion using empirical ISEs. Pay attention to the number of available observations.

10.2.13* Explain the case when the availability likelihood depends on the measurement error ε. Present several examples. Then propose a deconvolution E-estimator and evaluate its MISE.

10.2.14 What is the motivation of the estimator (10.2.17)?

10.2.15 Find the mean and variance of estimator (10.2.17).

10.2.16 How can the availability likelihood $w(z)$ be estimated?

10.2.17 Explain the underlying simulation in Figure 10.6.

10.2.18* Using Figure 10.6, analyze the effect of the estimator of the availability likelihood on performance of the E-estimator. Hint: Develop a numerical study and use empirical ISEs.

10.2.19* Does the shape of the availability likelihood function affect the deconvolution? If the answer is "yes," then explain how. Hint: Use both the theory and Figure 10.6.

10.2.20 Consider sample sizes $n = 100$ and $n = 400$ for Figure 10.6. Do you see a significant difference in quality of estimation? Hint: Support your conclusion by analysis of empirical ISEs.

10.2.21 Suggest better values for parameters of the E-estimator used in Figure 10.6.

10.2.22 Verify each equality in (10.2.18).

10.2.23* Explain the idea of estimation of the availability function $w(z)$. Propose an E-estimator. Evaluate the MISE of that E-estimator.

10.2.24* Repeat Figures 10.4-10.6, compare quality of estimation, analyze the results, and write a report.

10.2.25* Consider the case of unknown distribution of ε. Suppose that an extra sample from ε may be obtained. Suggest a deconvolution E-estimator for this case and analyze its MISE.

10.3.1 Explain the model of density deconvolution for censored data, and also present several examples. Further, is it necessary to assume that the variable Y is nonnegative (lifetime)?

10.3.2 Verify (10.3.2).

10.3.3 Explain why (10.3.3) is valid. Hint: Think about a condition needed for the validity of (10.3.3).

10.3.4 Why can (10.3.1) be written as (10.3.4)?

10.3.5 Explain the underlying idea of estimator (10.3.5).

10.3.6* Find the mean and variance of estimator (10.3.5) of the survival function.

10.3.7 Explain the estimator (10.3.6).

10.3.8 Find the mean of statistic (10.3.6). Hint: Begin with the case of known G^C.

10.3.9* Find the mean squared error of estimator (10.3.6) when the estimand is θ_j.

10.3.10* Describe the proposed deconvolution E-estimator of density f^X. Then calculate its MISE.

10.3.11 Explain the underlying simulation of Figure 10.7.

10.3.12 Using Figure 10.7, explain data and how the E-estimator performs.

10.3.13* Using Figure 10.7, explain how measurement errors and censoring affect the estimation. Hint: Propose a numerical study and then make a statistical analysis of empirical ISEs.

10.3.14 Using Figure 10.7, explain how parameters of this figure affect the quality of estimation.

10.3.15 Find better arguments of the E-estimator used in Figure 10.7. Do your recommendations depend on the underlying censoring distribution and sample size?

10.3.16* Explain, using both the theory and simulations of Figure 10.7, how the support of censoring variable C affects the studied estimation.

10.3.17* Consider the case of an unknown distribution of ε. Suppose that an extra sample from ε may be obtained. Suggest an E-estimator for this case and analyze its MISE.

10.3.18* Explain how the confidence bands are constructed.

10.3.19* Evaluate the MISE of proposed E-estimator. Hint: Begin with the case of known nuisance functions.

10.3.20* Consider the case of truncated data and propose a consistent estimator.

10.4.1 Describe several examples of the current status censoring (CSC).

10.4.2 What is the statistical model of CSC?

10.4.3 Suppose that you would like to simulate CSC data. Explain how this may be done.

10.4.4 Explain formula (10.4.1) for the joint (mixed) density of (Z, Δ). What are the assumptions?

10.4.5* Consider the case when support of the monitoring variable Z is the subset of the support of the random variable of interest X. In other words, suppose that (10.4.2) does not hold. Is it possible in this case to consistently estimate the density of X? Explain your answer. Then explore an estimand that may be consistently estimated.

10.4.6 The studied CSC modification is ill-posed. What does this mean?

10.4.7 There are two considered models of missing CSC data. Explain them. Further, what is the difference between truncated CSC and missing CSC?

10.4.8 Explain and prove expression (10.4.5) for the joint density of $(\Delta Z, \Delta)$.

10.4.9 Verify (10.4.6).

10.4.10 Explain and prove each equality in (10.4.8).

10.4.11 Explain the motivation behind the estimator (10.4.9).

10.4.12* Find the mean and variance of the estimator (10.4.9). Is it unbiased? What is the effect of the nuisance function?

10.4.13 Use formula (10.4.10) to explain ill-posedness.

10.4.14* Consider a series density estimator $\tilde{f}^X(x, J) := 1 + \sum_{j=0}^{J} \tilde{\theta}_j \varphi_j(x)$. Find its MISE. Then propose an estimator of the cutoff J and evaluate the MISE of the plug-in density estimator.

10.4.15 Explain a missing mechanism which implies (10.4.11).

10.4.16 Verify each equality in (10.4.12).

10.4.17* Explain the underlying idea of estimator (10.4.13). Is it unbiased? Then calculate its mean squared error.

10.4.18 Explain the idea of an aggregated estimator (10.4.14). Is it unbiased? Why may we refer to it as linear? Hint: Recall why a regression is called linear.

10.4.19* Verify (10.4.15). What is the used assumption? Then consider the case of dependent estimators $\tilde{\gamma}_1$ and $\tilde{\gamma}_2$, and repeat calculation of the variance.

10.4.20 What is the value of parameter λ which minimizes the variance of estimator (10.4.14)? Prove your assertion.

10.4.21 Explain why the weights in (10.4.17) for the two aggregated estimators have sense. Hint: Think about σ_1^2 significantly smaller than σ_2^2 and vice versa.

10.4.22* Prove (10.4.18). Does the inequality justify the aggregation? Then consider a sample of size $3n$ from X. Suppose that the estimand is the expectation of X. Propose an estimator based on first n observations and another estimator based on remaining $2n$ observations. Aggregate the two estimators and analyze the outcome.

10.4.23 Suppose that you have a sample of size n. What is the empirical cumulative distribution function?

10.4.24 Describe statistical properties of the empirical cumulative distribution function.

10.4.25 Explain the simulation used by Figure 10.8.

10.4.26 Explain all curves in Figure 10.8.

10.4.27 How are ISE and ISEF calculated in Figure 10.8?

10.4.28 Using Figure 10.8, consider different sample sizes and comment about performance of the E-estimator.

10.4.29 Using Figure 10.8, consider different corner functions. Write down a report about your observations.

10.4.30 Explain the simulation used by Figure 10.9.

10.4.31 What is the difference between top and bottom diagrams in Figure 10.9?

10.4.32 How are ISE and ISEF calculated in Figure 10.9?

10.4.33 Consider Figure 10.9. Explain why the histogram in the top diagram is skewed to the right and in the bottom diagram to the left.

10.4.34 Repeat Figure 10.9 with different sample sizes. Present a report about your conclusions.

10.4.35 Repeat Figure 10.9 with different corner function. Present a report about your conclusions.

10.4.36 Using Figures 10.8 and 10.9, propose optimal parameters for the E-estimator. Explain your choice.

10.4.37* Evaluate the MISE of proposed estimator.

10.4.38* Consider the case of missing CSC and propose a consistent estimator.

10.4.39 Explain the approach used for the case when (10.4.3) does not hold.

10.4.40 Are Fourier coefficients (10.4.21) deterministic or stochastic?

10.4.41 Verify each equality in (10.4.23).

10.4.42 Explain formula (10.4.25).

10.4.43* Find the mean and variance of the estimator (10.4.26).

10.4.44 Explain the simulation used in Figure 10.10.

10.4.45 Repeat Figure 10.10 with different parameters and write a report about your findings.

10.4.46* Consider the case when the support of Z is larger than the support of X. Suggest an E-estimator.

10.5.1 Explain the model of regression with measurement errors in predictors.

10.5.2 Present several examples of a regression with measurement errors in predictors. In those examples, is it possible to avoid the errors?

10.5.3 What is the interpolation problem? Give an example and explain how accurately a differentiable function may be interpolated.

10.5.4 What is the accuracy of estimation of a differentiable regression function? Why is it worse than for the interpolation?

10.5.5* Modify statistic (10.5.4) by using $n^{-1} \sum_{l=1}^{n} Y_l \varphi_j(U_l)/f^X(U_l)$. The underlying idea of this modification is to mimic a known estimator for the case of a standard regression. Calculate its expectation and make a conclusion about usefulness of this modification.

10.5.6 Modify statistic (10.5.4) by using $n^{-1} \sum_{l=1}^{n} Y_l \varphi_j(U_l)/f^U(U_l)$. Calculate its expectation and make a conclusion about usefulness of this modification.

10.5.7 Verify (10.5.6).

10.5.8* To get (10.5.7) it is assumed that ε is independent of U. First, is this a reasonable assumption? Second, what will be if it does not hold?

10.5.9* To get (10.5.9), it is assumed that X and ξ are independent. Consider the case when this assumption does not hold and propose a solution.

10.5.10 What is the assumption needed for validity of (10.5.10)?

10.5.11 Does the characteristic function completely define a random variable? If the answer is "yes," show how it can be used to find the second moment.

10.5.12 Verify (10.5.12) and formulate all used assumptions.

10.5.13 Explain how E-estimator can be used to estimate function $g(x) := f^X(x)m(x)$. Is this a regular or ill-posed problem?

10.5.14* Explain the estimator (10.5.15). Find its mean and variance.

10.5.15 Describe the underlying simulation of Figure 10.11.

10.5.16 Explain the procedure of regression E-estimation using diagrams in Figure 10.11.

10.5.17 Suggest better parameters of the E-estimator used in Figure 10.11.

10.5.18* Figure 10.11 uses the same parameters of the E-estimator for estimation of all involved functions. Is this a good idea? Is it better to use different parameters for each function? Support your conclusion via a numerical study based on repeated simulations.

10.5.19 To understand the effect of measurement errors, it is possible to compare ISEs for the case of regular and MEP regressions. Use Figure 10.11 to conduct such a numerical study.

10.5.20 Repeat Figure 10.11 a number of times and answer the following question. Based on the simulations, is estimation of $g(x)$ or $f^X(x)$ more critical for E-estimation?

10.5.21 Consider the same zero-mean normal distribution for ε and ξ. Assume that the standard deviation is σ. Explain how σ affects the regression estimation.

10.5.22* Evaluate the MISE of proposed estimator.

10.5.23* Consider the case of a linear regression with errors in predictors. Propose a consistent estimator.

10.5.24* Consider the case of unknown distribution of measurement errors ξ. Suggest a possible solution of the regression problem.

10.5.25* Propose a regression E-estimator following the methodology of Section 2.3.

10.6.1 Explain the MEP regression with responses missing according to (10.6.2).

10.6.2 Present several examples of the MEP regression given (10.6.2).

10.6.3 Explain motivation of the statistic (10.6.3).

10.6.4 Verify (10.6.4). What are the used assumptions?

10.6.5 What are the conditions used to get (10.6.5)?

10.6.6 How can formula (10.6.5) be used to suggest a regression E-estimator?

10.6.7 How can the function $g(x) := m(x)f^X(x)$ be estimated? What are required assumptions?

10.6.8* Explain the E-estimator (10.6.8). Evaluate the MISE.

10.6.9 How is the estimator $\hat{f}^X(x)$, used in (10.6.8), constructed? What are the assumptions?

10.6.10* Estimator (10.6.8) is based on the assumption that the characteristic function $\phi^\xi(\pi j)$ is known. What can be done if it is unknown?

10.6.11 How can the availability likelihood function, used in (10.6.8), be estimated? Is this an ill-posed problem?

10.6.12 Explain the simulation used in Figure 10.12.

10.6.13 Explain diagrams in Figure 10.12.

10.6.14 Using Figure 10.12, make a suggestion about better parameters for the E-estimator.

10.6.15 Conduct a numerical study with the purpose of understanding the effect of missing (10.6.2) in the MEP regression on quality of estimation. Hint: Use Figure 10.12.

10.6.16 Explain missing mechanism (10.6.9) and present several examples.

10.6.17 What is the underlying idea of the statistic (10.6.10)?

10.6.18 Verify every equality in (10.6.11).

10.6.19 Explain each step in establishing (10.6.12). What are the used assumptions?

10.6.20 Why do we introduce a new function $q(x)$ in (10.6.13)?

10.6.21 Explain why (10.6.15) is a sample mean estimator.

10.6.22 What is the variance of estimator (10.6.15)?

10.6.23 Verify (10.6.16).

10.6.24 How can conditional density (10.6.16) be estimated?

10.6.25 Verify (10.6.17).

10.6.26 Describe the simulation used in Figure 10.13.

10.6.27 Using Figure 10.13, explain how the E-estimator performs.

10.6.28 Using Figure 10.13, develop a numerical study to explore the effect of the availability likelihood function on estimation. Write a report.

10.6.29 Suggest a numerical study, using Figure 10.13, to explore the effect of parameter σ on quality of estimation. Write a report.

10.6.30 Conduct a numerical study, using Figure 10.13, to explore the effect of sample size on quality of estimation. Write a report.

10.6.31* Explain the regression estimator (10.6.18). Evaluate its MISE.

10.6.32* Consider the case of unknown distribution of ξ. Propose a consistent regression estimator. Hint: Use an extra sample.

10.6.33* It is known that for a standard regression with MAR responses a complete-case approach is consistent. Is it also the case for the MEP regression with missing response? Hint: Consider the two missing mechanisms discussed in Section 10.6.

10.6.34* The equality $f^{\xi|A}(z|1) = f^\xi(z)$ is used to establish (10.6.17). Prove this equality and then explain its meaning.

10.7.1 What is the empirical cumulative distribution function? Is it a bona fide cumulative distribution function?

10.7.2 Describe properties of the empirical cumulative distribution function.

10.7.3 Draw $\mathbb{V}(\hat{F}^X(x))$ as a function in x. Hint: First draw a cumulative distribution function and then draw the corresponding variance. Pay attention to values at boundary points.

10.7.4 Explain (10.7.6).

10.7.5 Find the mean and variance of Fourier estimator (10.7.8).

10.7.6 Verify (10.7.11).

10.7.7 Do we need to assume that the density is periodic for (10.7.13) to be valid?

10.7.8 Verify (10.7.14).

10.7.9 Verify (10.7.15).

10.7.10* Explain how to construct an estimator of the derivative of a density. Then calculate the MISE.

10.7.11 Verify (10.7.18). Do you need any assumptions?

10.7.12 Check validity of (10.7.20).

10.7.13 What are the assumptions of (10.7.20)?

10.7.14 Prove (10.7.22).

10.7.15 Density estimation is ill-posed with respect to the cumulative distribution function estimation, and density's derivative estimation is ill-posed with respect to density estimation. In your opinion, which ill-posedness is more severe?

10.7.16* Consider the problem of estimation of the derivative of a regression function. Propose a consistent E-estimator.

10.7.17* Evaluate the MISE of an E-estimator for the derivative of regression function. Hint: Make an assumption that simplifies the problem.

10.7.18* Consider the problem of estimation of derivative of the density when observations are available with measurement errors (the setting of Section 10.1). Propose a consistent E-estimator.

10.7.19* Consider the setting of Section 10.2 and propose a consistent E-estimator of the density's derivative.

10.7.20* Suppose that observations are truncated. Suggest a consistent estimator of the density's derivative.

10.7.21* Consider estimation of a kth derivative of the density. Formulate assumptions, propose E-estimator and evaluate its MISE.

10.9   Notes

There is a number of good mathematical books devoted to deterministic (not stochastic) ill-posed problems, for instance see the classical Tikhonov (1998) or more recent Kabanikhin (2011). Statistical theory of nonparametric deconvolution is considered in Efromovich and Ganzburg (1999), Efromovich and Koltchinskii (2001), and Meister (2009).

10.1 The problem of measurement errors in density estimation is considered in Section 3.5 of Efromovich (1999a) as well as in Efromovich (1994c, 1997a). Statistical analysis of directional data is as old as the analysis of linear data. As an example, Gauss developed the theory of measurement errors to analyze directional measurements in astronomy, see books by Mardia (1972) and Fisher (1993). Different methods for density deconvolution are discussed in Wand and Jones (1995, Section 6.2.4). For more recent results and settings, see Lepskii and Willer (2017), Pensky (2017), and Yi (2017).

10.2 Sequential estimation is a natural setting for the considered problem, see a discussion in Efromovich (2004d, 2015, 2017).

10.3 There is a number of natural extensions of the considered problem. For instance, hazard rate estimation for censored and measured with error data is discussed in Comte, Mabon, and Samson (2017).

10.4 Interval censoring is discussed in the books by Chen, Sun and Peace (2012), Sun and Zhao (2013), Groeneboom and Jongbloed (2014), and Klein et al. (2014).

10.5 Section 4.11 in Efromovich (1999a) considers the problem of MEP regression for the case of the Uniform design density, and it is assumed that the design density is known. The classical book by Carroll et al. (2006) is devoted to measurement errors in nonlinear models, and also see the more recent book by Grace (2017). For a nice and concise discussion of the interpolation problem see Demidovich and Maron (1981). Quantile regression with measurement errors is discussed in Chester (2017) where further references may be found.

10.6 Let us briefly explain another possible scenario when the missing mechanism is defined by the measurement error ξ. This is a more complicated case and here our intent is to comment on a possible solution. For the considered missing mechanism we have

$$\mathbb{P}(A = 1|U = u, Y = y, \xi = z) = \mathbb{P}(A = 1|\xi = z) =: w(z) > 0. \tag{10.9.1}$$

To understand how we may estimate the regression $m(x)$ under this missing scenario, let us look again at statistic (10.6.10). Write for its expectation,

$$\mathbb{E}\{\hat{\kappa}_j\} = \mathbb{E}\{AY\varphi_j(U)\} = \mathbb{E}\{\mathbb{E}\{AY\varphi_j(U)|Y, U, \xi\}\}$$

$$= E\{Y\varphi_j(U)\mathbb{E}\{A|Y, U, \xi\}\} = \mathbb{E}\{w(\xi)Y\varphi_j(U))\} = \mathbb{E}\{w(\xi)m(X)\varphi_j(U)\}. \tag{10.9.2}$$

Using (10.5.8) and independence of X and ξ we continue the analysis,

$$\mathbb{E}\{\hat{\kappa}_j\} = \mathbb{E}\{w(\xi)m(X)\varphi_j(X)\cos(\pi j\xi)\} - 2^{1/2}\mathbb{E}\{w(\xi)m(X)\sin(\pi jX)\sin(\pi j\xi)\}$$

$$= \mathbb{E}\{m(X)\varphi_j(X)\}\mathbb{E}\{w(\xi)\cos(\pi j\xi)\} - 2^{1/2}\mathbb{E}\{m(X)\sin(\pi jX)\}\mathbb{E}\{w(\xi)\sin(\pi j\xi)\}. \tag{10.9.3}$$

Now let us additionally assume that the likelihood availability function $w(z)$ is symmetric about zero, and recall that it is assumed that the distribution of ξ is symmetric about zero. This implies $\mathbb{E}\{w(\xi)\sin(\pi j\xi)\} = 0$ and we can continue,

$$\mathbb{E}\{\hat{\kappa}_j\} = [\int_0^1 [f^X(x)m(x)]\varphi_j(x)dx][2^{-1/2}\mathbb{E}\{w(\xi)\varphi_j(\xi)\}]. \tag{10.9.4}$$

This result is pivotal for the proposed solution. First of all, let us show how we can estimate $\mathbb{E}\{w(\xi)\varphi_j(\xi)\}$. Write following (10.9.2)-(10.9.3),

$$2^{1/2}\mathbb{E}\{A\varphi_j(U)\} = \mathbb{E}\{w(\xi)\varphi_j(\xi)\varphi_j(X)\} = \mathbb{E}\{w(\xi)\varphi_j(\xi)\}\mathbb{E}\{\varphi_j(X)\}. \tag{10.9.5}$$

According to Section 10.1, an unbiased sample mean estimator for Fourier coefficients

$$\theta_j := \mathbb{E}\{\varphi_j(X)\} = \int_0^1 \varphi_j(x) f^X(x) dx \tag{10.9.6}$$

of the design density f^X is

$$\hat{\theta}_j := (n\phi^\xi(\pi j))^{-1} \sum_{l=1}^n \varphi_j(U_l). \tag{10.9.7}$$

As a result, using (10.9.5) and (10.9.7) we can estimate the parameter

$$\gamma_j := 2^{-1/2} \mathbb{E}\{w(\xi)\varphi_j(\xi)\} \tag{10.9.8}$$

by estimator

$$\hat{\gamma}_j := \frac{n^{-1} \sum_{l=1}^n A_l \varphi_j(U_l)}{\hat{\theta}_j} = \frac{\phi^\xi(\pi j) \sum_{l=1}^n A_l \varphi_j(U_l)}{\sum_{l=1}^n \varphi_j(U_l)}. \tag{10.9.9}$$

Returning to (10.9.4), set

$$g(x) := f^X(x) m(x), \tag{10.9.10}$$

denote its Fourier coefficients as $g_j := \int_0^1 g(x)\varphi_j(x) dx$, and introduce an estimator for these Fourier coefficients

$$\hat{g}_j := \frac{n^{-1} \sum_{l=1}^n A_l Y_l \varphi_j(U_l)}{\hat{\gamma}_j}. \tag{10.9.11}$$

This Fourier estimator yields the E-estimator $\hat{g}(x)$. Finally, using the density E-estimator $\hat{f}^X(x)$ of Section 10.1, based on the observed sample $U_1, \ldots, U_n$, we get a regression estimator

$$\hat{m}(x) := \frac{\hat{g}(x)}{\max(\hat{f}^X(x), c/\ln(n))}, \quad x \in [0, 1]. \tag{10.9.12}$$

Another possible approach, left as an exercise, is to directly estimate Fourier coefficients of $m(x)$.

10.7 Statistical theory of estimation of derivatives is discussed in Efromovich (1998c, 1999a, 2010c). Using multiwavelets is an attractive alternative to trigonometric bases, see a discussion in Efromovich et al. (2004) and Efromovich (2009b).

References

Aalen, O, Borgan, O., and Gjessing, H. (2008). *Survival and Event History Analysis: A Process Point of View (Statistics for Biology and Health)*. New York: Springer.

Adak, S. (1998). Time-dependent spectral analysis of nonstationary time series. *Journal of the American Statistical Association* **93** 1488–1501.

Addison, P. (2017). *The Illustrated Wavelet Transform Handbook*. 2nd ed. Boca Raton: CRC Press.

Allen, J. (2017). A Bayesian hierarchical selection model for academic growth with missing data. *Applied Measurement in Education* **30** 147–162.

Allison, P. (2002). *Missing Data*. Thousand Oaks: Sage Publications.

Allison, P. (2010). *Survival Analysis Using SAS*. Cary: SAS Institute.

Allison, P. (2014). *Event History and Survival Analysis*. 2nd ed. Thousand Oaks: SAGE Publications.

Anderson, T. (1971). *The Statistical Analysis of Time Series*. New York: Wiley.

Andersen P.K., Borgan, O., Gill, R.D., and Keiding, N. (1993). *Statistical Models Based on Counting Processes*. New York: Springer.

Antoniadis, A., Gregoire, G., and Nason, G. (1999). Density and hazard rate estimation for right-censored data by using wavelets methods. *Journal of the Royal Statistical Society: Series B (Statistical Methodology)* **61** 63–84.

Aoshima, M. and Yata, K. (2011). Two-stage procedures for high-dimensional data. *Sequential Analysis* **30** 356–399.

Arnab, R. (2017). *Survey Sampling Theory and Applications*. London: Academic Press.

Austin, P. (2017). A Tutorial on multilevel survival analysis: methods, models and applications. *International Statistical Review* **85** 185–203.

Baby, P. and Stoica, P. (2010). Spectral analysis of non uniformly sampled data - a review. *Digital Signal Processing* **20** 359–378.

Bagkavos, D. and Patil, P. (2012). Variable bandwidths for nonparametric hazard rate estimation. *Communications in Statistics - Theory and Methods* **38** 1055–1078.

Baisch, S. and Bokelmann, G. (1999). Spectral analysis with incomplete time series: an example from seismology. *Computers and Geoscience* **25** 739–750.

Baraldi, A. and Enders, C. (2010). An introduction to modern missing data analyses. *Journal of School Psychology* **48** 5–37.

Bary, N.K. (1964). *A Treatise on Trigonometric Series*. Oxford: Pergamon Press.

Bellman, R.E. (1961). *Adaptive Control Processes*. Princeton: Princeton University Press.

Beran, J. (1994). *Statistics for Long-Memory Processes*. New York: Chapman & Hall.

Berglund, P. and Heeringa S. (2014). *Multiple Imputation of Missing Data Using SAS*. Cary: SAS Institute.

Berk, R. (2016). *Statistical Learning from a Regression Perspective*. New York: Springer.

Bickel, P.J. and Ritov, Y. (1991). Large sample theory of estimation in biased sampling regression models. *The Annals of Statistics* **19** 797–816.

Bickel, P.J. and Doksum, K.A. (2007). *Mathematical Statistics*. 2nd ed. London: Prentice Hall.

Birgé, L. and Massart, P. (1997). From model selection to adaptive estimation. *Festschrift for Lucien Le Cam.* (Pollard, D., ed.) New York: Springer, 55–87.

Bloomfield, P. (1970). Spectral analysis with randomly missing observations. *Journal of Royal Statistical Society, Ser. B* **32** 369–380.

Bloomfield, P. (2004). *Fourier Analysis of Time Series.* New York: Wiley.

Bodner T. (2006). Missing data: Prevalence and reporting practices. *Psychological Reports* **99** 675–680.

Borrajo, M., González-Manteiga, W., and Martínez-Miranda, M. (2017). Bandwidth selection for kernel density estimation with length-biased data. *Journal of Nonparametric Statistics* **29** 636–668.

Bott, A. and Kohler, M. (2017). Nonparametric estimation of a conditional density. *Annals of the Institute of Statistical Mathematics* **69** 189–214.

Bouza-Herrera, C. (2013). *Handling Missing Data in Ranked Set Sampling.* New York: Springer.

Box, G., Jenkins, G., Reinsel, G., and Ljung, G. (2016). *Time Series Analysis: Forecasting and Control.* 5th ed. Hoboken: Wiley.

Bremhorsta, V. and Lamberta, F. (2016). Flexible estimation in cure survival models using Bayesian p-splines. *Computational Statistics and Data Analysis* **93** 270–284.

Brockwell, P.J. and Davis, R.A. (1991). *Time Series: Theory and Methods.* 2nd ed. New York: Springer.

Brown, L.D. and Low, M.L. (1996). Asymptotic equivalence of nonparametric regression and white noise. *The Annals of Statistics* **24** 2384–2398.

Brunel, E. and Comte, F. (2008). Adaptive estimation of hazard rate with censored data. *Communications in Statistics - Theory and Methods* **37** 1284–1305.

Brunel, E., Comte, F., and Guilloux, A. (2009). Nonparametric density estimation in presence of bias and censoring. *Test* **18** 166–194.

Butcher, H. and Gillard, J. (2016). Simple nuclear norm based algorithms for imputing missing data and forecasting in time series. *Statistics and Its Interface* **10** (1) 19–25.

Butzer, P.L. and Nessel, R.J. (1971). *Fourier Analysis and Approximations.* New York: Academic Press.

Cai, T. and Low, M. (2006). Adaptive confidence balls. *The Annals of Statistics* **34** 202–228.

Cai, T. and Guo, Z. (2017). Confidence intervals for high-dimensional linear regression: Minimax rates and adaptivity. *The Annals of Statistics* **45** 615–646.

Cao, R., Janssen, P., and Veraverbeke, N. (2005). Relative hazard rate estimation for right censored and left truncated data. *Test* **14** 257–280.

Cao, W., Tsiatis, A., and Davidian, M. (2009). Improving efficiency and robustness of the doubly robust estimator for a population mean with incomplete data. *Biometrika* **96** 723–734.

Carpenter, J. and Kenward, M. (2013). *Multiple imputation and its application.* Chichester: Wiley.

Carroll, R. and Ruppert, D. (1988). *Transformation and Weighting in Regression.* Boca Raton: Chapman & Hall.

Carroll, R., Ruppert, D., Stefanski, L., and Crainceanu, C. (2006). *Measurement Error in Nonlinear Models: A Modern Perspective.* 2nd ed. Boca Raton: Champan & Hall.

Casella, G. and Berger, R. (2002). *Statistical Inference.* 2nd ed. New York: Duxbury.

Chan, K. (2013). Survival analysis without survival data: connecting length-biased and case-control data. *Biometrika* **100** 764–770.

Chaubey, Y., Chesneau, C., and Navarro, F. (2017). Linear wavelet estimation of the derivatives of a regression function based on biased data. *Communications in Statistics - Theory and Methods* **46** 9541–9556.

Cheema, J. (2014). A Review of missing data handling methods in education research. *Review of Educational Research* **84** 487–508.

Chen, D., Sun, J., and Peace, K. (2012). *Interval-Censored Time-to-Event Data: Methods and Applications*. Boca Raton: Chapman & Hall.

Chen, X. and Cai, J. (2017). Reweighted estimators for additive hazard model with censoring indicators missing at random. *Lifetime Data Analysis*, https://doi.org/10.1007/s10985-017-9398-z.

Chen, Y., Genovese, C., Tibshirani, R., and Wasserman, L. (2016). Nonparametric modal regression. *The Annals of Statistics* **44** 489–514.

Chen, Z.,Wang, Q., Wu, D., and Fan, P. (2016). Two-dimensional evolutionary spectrum approach to nonstationary fading channel modeling. *IEEE Transactions on Vehicular Technology* **65** 1083-1097.

Chentsov, N.N. (1962). Estimation of unknown distribution density from observations. *Soviet Math. Dokl.* **3** 1559–1562.

Chentsov, N.N. (1980). *Statistical Decision Rules and Optimum Inference*. New York: Springer-Verlag.

Chester, A. (2017). Understanding the effect of measurement error on quantile regressions. *Journal of Econometrics* **200** 223–237.

Choi, S. and Portnoy, S. (2016). Quantile autoregression for censored data. *Journal of Time Series Analysis* **37** 603–623.

Cohen, A.C. (1991). *Truncated and Censored Samples: Theory and Applications*. New York: Marcel Dekker.

Collett, D. (2014). *Modeling Survival Data in Medical Research*. 3rd ed. Boca Raton: Chapman & Hall.

Comte, F., and Rebafka, T. (2016). Nonparametric weighted estimators for biased data. *Journal of Statistical Planning and Inference* **174** 104–128.

Comte, F., Mabon, G., and Samson, A. (2017). Spline regression for hazard rate estimation when data are censored and measured with error. *Statistica Neerlandica* **71** 115–140.

Cortese, G., Holmboe, S., and Scheike, T. (2017). Regression models for the restricted residual mean life for right-censored and left-truncated data. *Statistics in Medicine* **36** 1803–1822.

Cox, D.R. and Oakes, D. (1984). *Analysis of Survival Data*. London: Chapman & Hall.

Crowder, M. (2012). *Multivariate Survival Analysis and Competing Risks*. Boca Raton: Chapman & Hall.

Dabrowska, D. (1989). Uniform consistency of the kernel conditional Kaplan-Meier estimate. *The Annals of Statistics* **17** 1157–1167.

Daepp, M., Hamilton, M., West, G., and Bettencourt, L. (2015). The mortality of companies. *Journal of the Royal Society Interface* **12** DOI: 10.1098/rsif.2015.0120.

Dahlhaus, R. (1997). Fitting time series Models to nonstationary processes. *Annals of Statistics* **25**, 1–37.

Dai, H., Restaino, M., and Wang, H. (2016). A class of nonparametric bivariate survival function estimators for randomly censored and truncated data. *Journal of Nonparametric Statistics* **28** 736–751.

Daniels, M. and Hogan, J. (2008). *Missing Data in Longitudinal Studies: Strategies for Bayesian Modeling and Sensitivity Analysis*. New York: Chapman & Hall.

Davey, A. and Salva, J. (2009). *Statistical Power Analysis with Missing Data*. London: Psychology Press.

Davidian, M., Tsiatis, A., and Leon, S. (2005). Semiparametric estimation of treatment effect in a pretest-posttest study with missing data. *Statistical Science: A Review Journal of the Institute of Mathematical Statistics* **20** 261–301.

Dean, A., Voss, D., and Draguljic, D. (2017). *Design and Analysis of Experiments*. 2nd ed. New York: Springer.

De Gooijer, J. (2017). *Elements of Nonlinear Time Series Analysis and Forecasting.* New York: Springer.

Dedecker, J., Doukhan P., Lang, G., Leon, J., Louhichi, S., and Prieur, C. (2007). *Weak Dependence: With Examples and Applications.* Springer: New York.

Del Moral, P. and Penev, S. (2014). *Stochastic Processes: From Applications to Theory.* Boca Raton: CRC Press.

Delecroix, M., Lopez, O., and Patilea, V. (2008). Nonlinear Censored Regression Using Synthetic Data. *Scandinavian Journal of Statistics* **35** 248–265.

Demidovich, B. and Maron, I. (1981). *Computational Mathematics.* Moscow: Mir.

De Una-Álvarez, J. (2004). Nonparametric estimation under length-biased and type I censoring: a moment based approach. *Annals of the Institute of Statistical Mathematics* **56** 667–681.

De Una-Álvarez, J. and Veraverbeke, N. (2017) Copula-graphic estimation with left-truncated and right-censored data. *Statistics* **51** 387–403.

DeVore, R.A. and Lorentz, G.G. (1993). *Constructive Approximation.* New York: Springer-Verlag.

Devroye, L. and Györfi, L. (1985). *Nonparametric Density Estimation: The L_1 View.* New York: Wiley.

Devroye, L. (1987). *A Course in Density Estimation.* Boston: Birkhäuser.

Diggle, P.J. (1990). *Time Series: A Biostatistical Introduction.* Oxford: Oxford University Press.

Dobrow, R. (2016). *Introduction to Stochastic Processes with R.* Hoboken: Wiley.

Donoho, D. and Johnstone, I. (1995). Adapting to unknown smoothness via wavelet shrinkage. *Journal of the American Statistical Association* **90** 1200–1224.

Doukhan, P. (1994). *Mixing: Properties and Examples.* New York: Springer-Verlag.

Dryden, I.L. and Mardia, K.V. (1998). *Statistical Shape Analysis.* New York: Wiley.

Dunsmuir, W. and Robinson, P. (1981a). Estimation of time series models in the presence of missing data. *Journal of the American Statistical Association* **76** 560–568.

Dunsmuir, W. and Robinson, P. (1981b). Asymptotic theory for time series containing missing and amplitude modulated observations, *Sankhya: The Indian Journal of Statistics, ser. A* **43** 260–281.

Durrett, R. (2016). *Essentials of Stochastic Processes.* New York: Springer.

Dym, H. and McKean, H.P. (1972). *Fourier Series and Integrals.* London: Academic Press.

Dzhaparidze, K. (1985). *Estimation of Parameters and Verification of Hypotheses in Spectral Analysis.* New York: Springer.

Efromovich, S. (1980a). Information contained in a sequence of observations. *Problems of Information Transmission* **15** 178–189.

Efromovich, S. (1980b). On sequential estimation under conditions of local asymptotic normality. *Theory of Probability and its Applications* **25** 27–40.

Efromovich, S. (1984). Estimation of a spectral density of a Gaussian time series in the presence of additive noise. *Problems of Information Transmission* **20** 183–195.

Efromovich, S. (1985). Nonparametric estimation of a density with unknown smoothness. *Theory of Probability and its Applications* **30** 557–568.

Efromovich, S. (1986). Adaptive algorithm of nonparametric regression. *Proc. of Second IFAC symposium on Stochastic Control.* Vilnuis: Science, 112–114.

Efromovich, S. (1989). On sequential nonparametric estimation of a density. *Theory of Probability and its Applications* **34** 228–239.

Efromovich, S. (1992). On orthogonal series estimators for random design nonparametric regression. *Computing Science and Statistics* **24** 375–379.

Efromovich, S. (1994a). On adaptive estimation of nonlinear functionals. *Statistics and Probability Letters* **19** 57–63.

Efromovich, S. (1994b). On nonparametric curve estimation: multivariate case, sharp-optimality, adaptation, efficiency. *CORE Discussion Papers* **9418** 1–35.

Efromovich, S. (1994c). Nonparametric curve estimation from indirect observations. *Computing Science and Statistics* **26** 196–200.

Efromovich, S. (1995a). Thresholding as an adaptive method (discussion). *Journal of Royal Statistical Society ser. B* **57** 343.

Efromovich, S. (1995b). On sequential nonparametric estimation with guaranteed precision. *The Annals of Statistics* **23** 1376–1392.

Efromovich, S. (1996a). On nonparametric regression for iid observations in general setting. *The Annals of Statistics* **24** 1126–1144.

Efromovich, S. (1996b). Adaptive orthogonal series density estimation for small samples. *Computational Statistics and Data Analysis* **22** 599–617.

Efromovich, S. (1997a). Density estimation for the case of supersmooth measurement error. *Journal of the American Statistical Association* **92** 526–535.

Efromovich, S. (1997b). Robust and efficient recovery of a signal passed through a filter and then contaminated by non-Gaussian noise. *IEEE Transactions on Information Theory* **43** 1184–1191.

Efromovich, S. (1997c). Quasi-linear wavelet estimation involving time series. *Computing Science and Statistics* **29** 127–131.

Efromovich, S. (1998a). On global and pointwise adaptive estimation. *Bernoulli* **4** 273–278.

Efromovich, S. (1998b). Data-driven efficient estimation of the spectral density. *Journal of the American Statistical Association* **93** 762–770.

Efromovich, S. (1998c). Simultaneous sharp estimation of functions and their derivatives. *The Annals of Statistics* **26** 273–278.

Efromovich, S. (1999a). *Nonparametric Curve Estimation: Methods, Theory, and Applications.* New York: Springer.

Efromovich, S. (1999b). Quasi-linear wavelet estimation. *The Journal of the American Statistical Association* **94** 189–204.

Efromovich, S. (1999c). How to overcome the curse of long-memory errors. *IEEE Transactions on Information Theory* **45** 1735–1741.

Efromovich, S. (1999d). On rate and sharp optimal estimation. *Probability Theory and Related Fields* **113** 415–419.

Efromovich, S. (2000a). Can adaptive estimators for Fourier series be of interest to wavelets? *Bernoulli* **6** 699–708.

Efromovich, S. (2000b). On sharp adaptive estimation of multivariate curves. *Mathematical Methods of Statistics* **9** 117–139.

Efromovich, S. (2000c). Sharp linear and block shrinkage wavelet estimation. *Statistics and Probability Letters* **49** 323–329.

Efromovich, S. (2001a). Density estimation under random censorship and order restrictions: from asymptotic to small samples. *The Journal of the American Statistical Association* **96** 667–685.

Efromovich, S. (2001b). Second order efficient estimating a smooth distribution function and its applications. *Methodology and Computing in Applied Probability* **3** 179–198.

Efromovich, S. (2001c). Multiwavelets and signal denoising. *Sankhya ser. A* **63** 367–393.

Efromovich, S. (2002). Discussion on random rates in anisotropic regression. *The Annals of Statistics* **30** 370–374.

Efromovich, S. (2003a). On the limit in the equivalence between heteroscedastic regression and filtering model. *Statistics and Probability Letters* **63** 239–242.

Efromovich, S. (2004a). Density estimation for biased data. *The Annals of Statistics* **32** 1137–1161.

Efromovich, S. (2004b). Financial applications of sequential nonparametric curve estimation. In

Applied Sequential Methodologies, eds. N.Mukhopadhyay, S.Datta, and S.Chattopadhyay. 171–192.

Efromovich, S. (2004c). Distribution estimation for biased data. *Journal of Statistical Planning and Inference* **124** 1–43.

Efromovich, S. (2004d). On sequential data-driven density estimation. *Sequential Analysis Journal* **23** 603–624.

Efromovich, S. (2004e). Analysis of blockwise shrinkage wavelet estimates via lower bounds for no-signal setting. *Annals of the Institute of Statistical Mathematics* **56** 205–223.

Efromovich, S. (2004f). Oracle inequalities for Efromovich–Pinsker blockwise estimates. *Methodology and Computing in Applied Probability* **6** 303–322.

Efromovich, S. (2004g). Discussion on "Likelihood ratio identities and their applications to sequential analysis" by Tze L. Lai. *Sequential Analysis Journal* **23** 517–520.

Efromovich, S. (2004h). Adaptive estimation of error density in heteroscedastic nonparametric regression. In: Proceedings of the 2nd International workshop in Applied Probability IWAP 2004, Univ. of Piraeus, Greece, 132–135.

Efromovich, S. (2005a). Univariate nonparametric regression in the presence of auxiliary covariates. *Journal of the American Statistical Association* **100** 1185–1201.

Efromovich, S. (2005b). Estimation of the density of regression errors. *The Annals of Statistics* **33** 2194–2227.

Efromovich, S. (2007a). A lower-bound oracle inequality for a blockwise-shrinkage estimate. *Journal of Statistical Planning and Inference* **137** 176–183.

Efromovich, S. (2007b). Universal lower bounds for blockwise-shrinkage wavelet estimation of a spike. *Journal of Applied Functional Analysis* **2** 317–338.

Efromovich, S. (2007c). Adaptive estimation of error density in nonparametric regression with small sample size. *Journal of Statistical Inference and Planning* **137** 363–378.

Efromovich, S. (2007d). Sequential design and estimation in heteroscedastic nonparametric regression. Invited paper with discussion. *Sequential Analysis* **26** 3–25.

Efromovich, S. (2007e). Response on sequential design and estimation in heteroscedastic nonparametric regression. *Sequential Analysis* **26** 57–62.

Efromovich, S. (2007f). Optimal nonparametric estimation of the density of regression errors with finite support. *Annals of the Institute of Statistical Mathematics* **59** 617–654.

Efromovich, S. (2007g). Conditional density estimation. *The Annals of Statistics* **35** 2504–2535.

Efromovich, S. (2007h). Comments on nonparametric inference with generalized likelihood ratio tests. *Test* **16** 465–467.

Efromovich, S. (2007i). Applications in finance, engineering and health sciences: Plenary Lecture. *Abstracts of IWSM-2007*, Auburn University, 20–21.

Efromovich, S. (2008a). Optimal sequential design in a controlled nonparametric regression. *Scandinavian Journal of Statistics* **35** 266–285.

Efromovich, S. (2008b). Adaptive estimation of and oracle inequalities for probability densities and characteristic functions. *The Annals of Statistics* **36** 1127–1155.

Efromovich, S. (2008c). Nonparametric regression estimation with assigned risk. *Statistics and Probability Letters* **78** 1748–1756.

Efromovich, S. (2009a). Lower bound for estimation of Sobolev densities of order less $1/2$. *Journal of Statistical Planning and Inference* **139** 2261–2268.

Efromovich, S. (2009b). Multiwavelets: theory and bioinformatic applications. *Communications in Statistics – Theory and Methods* **38** 2829–2842

Efromovich, S. (2009c). Optimal sequential surveillance for finance, public health, and other areas: discussion. *Sequential Analysis* **28** 342–346.

Efromovich, S. (2010a). Sharp minimax lower bound for nonparametric estimation of Sobolev densities of order $1/2$. *Statistics and Probability Letters* **80** 77–81.

Efromovich, S. (2010b). Oracle inequality for conditional density estimation and an actuarial example. *Annals of the Institute of Mathematical Statistics* **62** 249–275.

Efromovich, S. (2010c). Orthogonal series density estimation. *WIREs Computational Statistics* **2** 467–476.

Efromovich, S. (2010d). Dimension reduction and oracle optimality in conditional density estimation. *Journal of the American Statistical Association* **105** 761–774.

Efromovich, S. (2011a). Nonparametric regression with predictors missing at random. *Journal of the American Statistical Association* **106** 306–319.

Efromovich, S. (2011b). Nonparametric regression with responses missing at random. *Journal of Statistical Planning and Inference* **141** 3744–3752.

Efromovich, S. (2011c). Nonparametric estimation of the anisotropic probability density of mixed variables. *Journal of Multivariate Analysis* **102** 468–481.

Efromovich, S. (2012a). Nonparametric regression with missing data: theory and applications. *Actuarial Research Clearing House* **1** 1–15.

Efromovich, S. (2012b). Sequential analysis of nonparametric heteroscedastic regression with missing responses. *Sequential Analysis* **31** 351–367.

Efromovich, S. (2013a). Nonparametric regression with the scale depending on auxiliary variable. *The Annals of Statistics* **41** 1542–1568.

Efromovich, S. (2013b). Notes and proofs for nonparametric regression with the scale depending on auxiliary variable. *The Annals of Statistics* **41**, 1–29.

Efromovich, S. (2013c). Adaptive nonparametric density estimation with missing observations. *Journal of Statistical Planning and Inference* **143** 637–650.

Efromovich, S. (2014a). On shrinking minimax convergence in nonparametric statistics. *Journal of Nonparametric Statistics* **26** 555–573.

Efromovich, S. (2014b). Efficient nonparametric estimation of the spectral density in the presence of missing observations. *Journal of Time Series Analysis* **35** 407–427.

Efromovich, S. (2014c). Nonparametric regression with missing data. *Computational Statistics* **6** 265–275.

Efromovich, S. (2014d). Nonparametric estimation of the spectral density of amplitude-modulated time series with missing observations, *Statistics and Probability Letters* **93** 7–13.

Efromovich, S. (2014e). Nonparametric curve estimation with incomplete data, *Actuarial Research Clearing House* **15** 31–47.

Efromovich, S. (2015). Two-stage nonparametric sequential estimation of the directional density with complete and missing observations. *Sequential Analysis* **34** 425–440.

Efromovich, S. (2016a). Minimax theory of nonparametric hazard rate estimation: efficiency and adaptation. *Annals of the Institute of Statistical Mathematics* **68** 25–75.

Efromovich, S. (2016b). Estimation of the spectral density with assigned risk. *Scandinavian Journal of Statistics* **43** 70–82.

Efromovich, S. (2016c). What an actuary should know about nonparametric regression with missing data. *Variance* **10** 145–165.

Efromovich, S. (2017). Missing, modified and large-p-small-n data in nonparametric curve estimation. *Calcutta Statistical Association Bulletin* **69** 1–34.

Efromovich, S. and Baron, M. (2010). Discussion on "quickest detection problems: fifty years later" by Albert N. Shiryaev. *Sequential Analysis* **29** 398–403.

Efromovich, S. and Chu, J. (2018a). Efficient nonparametric hazard rate estimation with left truncated and right censored data. *Annals of the Institute of Statistical Mathematics*, in press. https://doi.org/10.1007/s10463-017-0617-x

Efromovich, S. and Chu, J. (2018b). Small LTRC samples and lower bounds in hazard rate estimation. *Annals of the Institute of Statistical Mathematics*, in press. https://doi.org/10.1007/s10463-017-0617-x

Efromovich, S. and Ganzburg, M. (1999). Best Fourier approximation and application in efficient blurred signal reconstruction. *Computational Analysis and Applications* **1** 43–62.

Efromovich, S., Grainger, D., Bodenmiller, D., and Spiro, S. (2008). Genome-wide identification of binding sites for the nitric oxide sensitive transcriptional regulator NsrR. *Methods in Enzymology* **437** 211–233.

Efromovich, S. and Koltchinskii, V. (2001). On inverse problems with unknown operators. *IEEE Transactions on Information Theory* **47** 2876–2894.

Efromovich, S., Lakey, J., Pereyra, M.C., and Tymes, N. (2004). Data-driven and optimal denoting of a signal and recovery of its derivative using multiwavelets. *IEEE Transactions on Signal Processing* **52** 628–635.

Efromovich, S. and Low, M. (1994). Adaptive estimates of linear functionals. *Probability Theory and Related Fields* **98** 261–275.

Efromovich, S. and Low, M. (1996a). On Bickel and Ritov's conjecture about adaptive estimation of some quadratic functionals. *The Annals of Statistics* **24** 682–686.

Efromovich, S. and Low, M. (1996b). Adaptive estimation of a quadratic functional. *The Annals of Statistics* **24** 1106–1125.

Efromovich, S. and Pinsker M.S. (1981). Estimation of a square integrable spectral density for a time series. *Problems of Information Transmission* **17** 50–68.

Efromovich, S. and Pinsker M.S. (1982). Estimation of a square integrable probability density of a random variable. *Problems of Information Transmission* **18** 19–38.

Efromovich, S. and Pinsker M.S. (1984). An adaptive algorithm of nonparametric filtering. *Automation and Remote Control* **11** 58–65.

Efromovich, S. and Pinsker M.S. (1986). Adaptive algorithm of minimax nonparametric estimating spectral density. *Problems of Information Transmission* **22** 62–76.

Efromovich, S. and Pinsker, M.S. (1989). Detecting a signal with an assigned risk. *Automation and Remote Control* **10** 1383–1390.

Efromovich, S. and Pinsker, M. (1996). Sharp-optimal and adaptive estimation for heteroscedastic nonparametric regression. *Statistica Sinica* **6** 925–945.

Efromovich, S. and Salter-Kubatko, L. (2008). Coalescent time distributions in trees of arbitrary size. *Statistical Applications in Genetics and Molecular Biology* **7** 1–21.

Efromovich, S. and Samarov, A. (1996). Asymptotic equivalence of nonparametric regression and white noise model has its limits. *Statistics and Probability Letters* **28** 143–145.

Efromovich, S. and Samarov, A. (2000). Adaptive estimation of the integral of squared regression derivatives. *Scandinavian Journal of Statistics* **27** 335–352.

Efromovich, S. and Smirnova, E. (2014a). Wavelet estimation: minimax theory and applications. *Sri Lankan Journal of Applied Statistics* **15** 17–31.

Efromovich, S. and Smirnova, E. (2014b). Statistical analysis of large cross-covariance and cross-correlation matrices produced by fMRI Images. *Journal of Biometrics and Biostatistics* **5** 1–8.

Efromovich, S. and Thomas, E. (1996). Application of nonparametric binary regression to evaluate the sensitivity of explosives. *Technometrics* **38** 50–58.

Efromovich, S. and Valdez-Jasso, Z.A. (2010). Aggregated wavelet estimation and its application to ultra-fast fMRI. *Journal of Nonparametric Statistics* **22** 841–857.

Efromovich, S. and Wu, J. (2017). Dynamic nonparametric analysis of nonstationary portfolio returns and its application to VaR and forecasting. *Actuarial Research Clearing House* 1–25.

Efron, B. and Tibshirani, R. (1996). Using specially designed exponential families for density estimators. *The Annals of Statistics* **24** 2431–2461.

El Ghouch, A. and Van Keilegom, I. (2008). Nonparametric regression with dependent censored data. *Scandinavian Journal of Statistics* **35** 228–247.

El Ghouch, A. and Van Keilegom, I. (2009). Local linear quantile regression with dependent censored data. *Statistica Sinica* **19** 1621–1640.

Enders, C. (2006). A primer on the use of modern missing-data methods in psychomatic medicine research. *Psyhosomatic Medicine* **68** 427–436.

Enders, C. (2010). *Applied Missing Data Analysis.* New York: The Guilford Press.

Eubank, R.L. (1988). *Spline Smoothing and Nonparametric Regression.* New York: Marcel and Dekker.

Everitt, B. and Hothorn, T. (2011). *An Introduction to Applied Multivariate Analysis with R.* New York: Springer.

Fan, J. and Gijbels, I. (1996). *Local Polynomial Modeling and its Applications - Theory and Methodologies.* New York: Chapman & Hall.

Fan, J. and Yao, Q. (2003). *Nonlinear Time Series: Nonparametric and Parametric Methods.* New York: Springer.

Fan, J. and Yao, Q. (2015). *The Elements of Financial Economics.* Beijing: Science Press.

Faraway, J. (2016). *Extending the Linear Model with R: Generalized Linear, Mixed Effects and Nonparametric Regression Models.* 2nd ed. New York: Chapman & Hall.

Fisher, N.I. (1993). *Statistical Analysis of Circular Data.* Cambridge: Cambridge University Press.

Fleming, T.R. and Harrington, D.P. (2011). *Counting Processes and Survival Analysis.* New York: Wiley.

Frumento, P. and Bottai, M. (2017). An estimating equation for censored and truncated quantile regression. *Computational Statistics and Data Analysis* **113** 53–63.

Genovese, C. and Wasserman, L. (2008). Adaptive confidence bands. *The Annals of Statistics* **36** 875–905.

Ghosal, S. and van der Vaart, A. (2017). *Fundamentals of Nonparametric Bayesian Inference.* Cambridge: Cambridge University Press.

Gill, R., Vardi, Y., and Wellner, J. (1988). Large sample theory of empirical distributions in biased sampling models. *The Annals of Statistics* **16** 1069–1112.

Gill, R. (2006). *Lectures on Survival Analysis.* New York: Springer.

Giné, E. and Nickl, R. (2010). Confidence bands in density estimation. *The Annals of Statistics* **38** 1122–1170.

Glad, I., Hjort, N., and Ushakov, N. (2003). Correction of density estimators that are not densities. *The Scandinavian Journal of Statistics* **30** 415–427.

Goldstein, H., Carpenter, J., and Browne, W. J. (2014). Fitting multilevel multivariate models with missing data in responses and covariates that may include interactions and non-linear terms. *Journal of the Royal Statistical Society, Ser. A* **177** 553–564.

Gou, J. and Zhang, F. (2017). Experience Simpson's paradox in the classroom. *American Statistician* **71** 61–66.

Grace, Y. (2017). *Statistical Analysis with Measurement Error or Misclassification: Strategy, Method and Application.* New York: Springer.

Graham, J. (2012). *Missing Data: Analysis and Design.* New York: Springer.

Green, P. and Silverman, B. (1994). *Nonparametric Regression and Generalized Linear Models: a Roughness Penalty Approach.* London: Chapman & Hall.

Greiner, A., Semmler, W., and Gong, G. (2005). *The Forces of Economic Growth: A Time Series Perspective.* Princeton: Princeton University Press.

Groeneboom, P. and Jongbloed, G. (2014). *Nonparametric Estimation under Shape Constraints: Estimators, Algorithms and Asymptotics.* Cambridge: Cambridge University Press.

Groves R., Dillman D., Eltinge J., and Little R. (2002). *Survey Nonresponse.* New York: Wiley.

Guo, S. (2010). *Survival Analysis.* Oxford: Oxford University Press.

Györfi, L., Kohler, M., Krzyzak, A., and Walk, H. (2002). *A Distribution-Free Theory of Nonparametric Regression.* New York: Springer.

Györfi, L., Härdle, W., Sarda, P., and Vieu, P. (2013). *Nonparametric Curve Estimation from Time Series*. New York: Springer.

Hagar,Y. and Dukic, V. (2015). Comparison of hazard rate estimation in R. arXiv: 1509.03253v1

Hall, P. and Hart, J.D. (1990). Nonparametric regression with long-range dependence. *Stochastic Processes and their Applications* **36** 339–351.

Han, P. and Wang, L. (2013). Estimation with missing data: beyond double robustness. *Biometrika* **100** 417–430.

Härdle, W. (1990). *Applied Nonparametric Regression*. Cambridge: Cambridge University Press.

Härdle, W., Kerkyacharian, G., Picard, D., and Tsybakov, A. (1998). *Wavelets, Approximation and Statistical Applications*. New York: Springer.

Harrell, F. (2015). *Regression Modeling Strategies: With Applications to Linear Models, Logistic and Ordinal Regression, and Survival Analysis*. 2nd ed. London: Springer.

Hart, J.D. (1997). *Nonparametric Smoothing and Lack-Of-Fit Tests*. New York: Springer.

Hastie, T.J. and Tibshirani, R. (1990). *Generalized Additive Models*. London: Chapman & Hall.

Helsel, D. (2011). *Statistics for Censored Environmental Data Using Minitab and R*. 2nd. ed. New York: Wiley.

Hoffmann, M. and Nickl, R. (2011). On adaptive inference and confidence bands. *The Annals of Statistics* **39** 2383–2409.

Hollander, M., Wolfe, D., and Chicken, E. (2013). *Nonparametric Statistical Methods*. New York: Wiley.

Honaker, J. and King, G. (2010). What to do about missing values in time-series cross-section data. *American Journal of Political Science* **54** 561–581.

Horowitz, J. and Lee, S. (2017). Nonparametric estimation and inference under shape restrictions. *Journal of Econometrics* **201** 108–126.

Hosmer D., Lemeshow, S., and May, S. (2008). *Applied Survival Analysis: Regression Modeling of Time-to-Event Data*. 2nd ed. New York: Wiley.

Ibragimov, I.A. and Khasminskii, R.Z. (1981). *Statistical Estimation: Asymptotic Theory*. New York: Springer.

Ibragimov, I.A. and Linnik, Yu. V. (1971). *Independent and Stationary Sequences of Random Variables*. Groningen: Walters-Noordhoff.

Ingster, Yu. and Suslina, I. (2003). *Nonparametric Goodness-of-Fit Testing Under Gaussian Models*. New York: Springer

Ivanoff, S., Picard, F., and Rivoirard, V. (2016). Adaptive Lasso and group-Lasso for functional Poisson regression. *Journal of Machine Learning Research* **17** 1–46.

Izbicki, R. and Lee, A. (2016). Nonparametric conditional density estimation in a high-dimensional regression setting. *Journal of Computational and Graphical Statistics* **25** 1297–1316.

Izenman, A. (2008). *Modern Multivariate Statistical Techniques: Regression, Classification, and Manifold Learning*. New York: Springer.

Jankowski, H. and Wellner, J. (2009). Nonparametric estimation of a convex bathtub-shaped hazard function. *Bernoulli* **15** 1010–1035.

Jiang, J. and Hui, Y. (2004). Spectral density estimation with amplitude modulation and outlier detection. *Annals of the Institute of Statistical Mathematics* **56** 611–630.

Jiang, P., Liu, F., Wang, J., and Song, Y. (2016). Cuckoo search-designated fractal interpolation functions with winner combination for estimating missing values in time series. *Applied Mathematical Modeling* **40** 9692–9718.

Johnstone, I. (2017). *Gaussian Estimation: Sequence and Wavelet Models*. Manuscript, Stanford: University of Stanford.

Kabanikhin, S. (2011). *Inverse and Ill-Posed Problems: Theory and Applications*. Berlin: De Gruyter.

Kalbfleisch, J. and Prentice, R. (2002). *The Statistical Analysis of Failure Time Data*. 2nd ed. New York: Springer.

Kitagawa, G. and Akaike, H. (1978). A procedure for the modeling of nonstationary time series. *Annals of the Institute of Statistical Mathematics* **30** 351–363.

Klein, J.P. and Moeschberger, M.L. (2003). *Survival Analysis: Techniques for Censored and Truncated Data*. New York: Springer.

Klein, J.P., van Houwelingen, H., Ibrahim, J., and Scheike, T. (2014). *Handbook of Survival Analysis*. Boca Raton: Chapman & Hall.

Kleinbaum, D. and Klein, M. (2012). *Survival Analysis*. 3rd ed. New York: Springer.

Klugman, S., Panjer, H., and Willmot, G. (2012). *Loss Models: From Data to Decisions*. 4th ed. New York: Wiley.

Kokoszka, P. and Reimherr, M. (2017). *Introduction to Functional Data Analysis*. New York: Chapman & Hall.

Kolmogorov, A.N. and Fomin, S.V. (1957). *Elements of the Theory of Functions and Functional Analysis*. Rochester: Graylock Press.

Kosorok, M. (2008). *Introduction to Empirical Processes and Semiparametric Inference*. New York: Springer.

Kou, J. and Liu, Y. (2017). Nonparametric regression estimations over L_p risk based on biased data. *Communications in Statistics - Theory and Methods* **46** 2375-2395.

Krylov, A.N. (1955). *Lectures in Approximate Computation*. Moscow: Science (in Russian).

Kutner, M., Nachtsheim, C., Neter, J., and Li, W. (2005). *Applied Linear Statistical Models*. 5th ed. Boston: McGraw-Hill.

Lang, K. and Little, T. (2016). Principled missing data treatments. *Prevention Science* 1–11, https://doi.org/10.1007/s11121-016-0644-5.

Lawless, J., Kalbfleisch, J., and Wild, C. (1999). Semiparametric methods for response selective and missing data problems in regression. *Journal of the Royal Statistical Society B* **61** 413–438.

Lee, E. and Wang, J. (2013). *Statistical Methods for Survival Analysis*. 4th ed. New York: Wiley.

Lee, M. (2004). Strong consistency for AR model with missing data, *J. Korean Mathematical Society* **41** 1071–1086.

Lehmann, E.L. and Casella, G. (1998). *Theory of Point Estimation*. New York: Springer.

Lepskii, O. and Willer, T. (2017). Lower bounds in the convolution structure density model. *Bernoulli* **23** 884–926.

Levit, B. and Samarov, A. (1978). Estimation of spectral functions. *Problems of Information Transmission* **14**, 61–66.

Li, J. and Ma, S. (2013). *Survival Analysis in Medicine and Genetics*. Boca Raton: Chapman & Hall.

Li, Q. and Racine, J. (2007). *Nonparametric Econometrics: Theory and Practice*. Princeton: Princeton University Press.

Liang H. and de Una-Álvarez J. (2011). Wavelet estimation of conditional density with truncated, censored and dependent data. *Journal of Multivariate Analysis* **102** 448–467

Little, R. and Rubin, D. (2002). *Statistical Analysis with Missing Data*. New York: Wiley.

Little, R. et al. (2016). The treatment of missing data in a large cardiovascular clinical outcomes study. *Clinical Trials* **13** 344–351.

Little, T.D., Lang, K.M., Wu, W. and Rhemtulla, M. (2016). Missing data. *Developmental Psychopathology: Theory and Method*. (D. Cicchetti, ed.) New York: Wiley, 760–796.

Liu, X. (2012). *Survival Analysis: Models and Applications*. New York: Wiley

Longford, N. (2005). *Missing Data and Small-Area Estimation*. New York: Springer.

Lorentz, G., Golitschek, M., and Makovoz, Y. (1996). *Constructive Approximation. Advanced Problems*. New York: Springer-Verlag.

Lu, X. and Min, L. (2014). Hazard rate function in dynamic environment. *Reliability Engineering and System Safety* **130** 50–60.

Luo, X. and Tsai, W. (2009). Nonparametric estimation for right-censored length-biased data: a pseudo-partial likelihood approach. *Biometrika* **96** 873–886.

Mallat, S. (1998). *A Wavelet Tour of Signal Precessing.* Boston: Academic Press.

Mardia, K.V. (1972). *Statistics of Directional Data.* London: Academic Press.

Martinussen, T. and Scheike, T. (2006). *Dynamic Regression Models for Survival Data.* New York: Springer.

Matloff, N. (2017). *Statistical Regression and Classification: From Linear Models to Machine Learning.* New York: Chapman & Hall.

Matsuda, Y. and Yajima, Y. (2009). Fourier analysis of irregularly spaced data on R^d. *Journal of Royal Statistical Society, Ser. B* **71** 191–217.

McKnight, P., McKnight, K., Sidani, S., and Figueredo, A. (2007). *Missing Data: A Gentle Introduction.* London: The Guilford Press.

Meister, A. (2009). *Deconvolution Problems in Nonparametric Statistics.* New York: Springer.

Miller, R.G. (1981). *Survival Analysis.* New York: Wiley.

Mills, M. (2011). *Introducing Survival and Event History Analysis.* Thousand Oaks: Sage.

Molenberghs, G., Fitzmaurice G., Kenward, M., Tsiatis A., and Verbeke G. (Eds.) (2014). *Handbook of Missing Data Methodology.* Boca Raton: Chapman & Hall.

Molenberghs, G. and Kenward, M. (2007). *Missing Data in Clinical Trials.* Hoboken: Wiley.

Montgomery, D., Jennings, C., and Kulahci, M. (2016). *Introduction to Time Series Analysis and Forecasting.* 2nd ed. Hoboken: Wiley.

Moore, D. (2016). *Applied Survival Analysis Using R.* New York: Springer.

Moore, D., McGabe, G., and Craig, B. (2009). *Introduction to the Practice of Statistics.* 6th ed. New York: W.H.Freeman and Co.

Mukherjee, R. and Sen S. (2018). Optimal adaptive inference in random design binary regression. *Bernoulli* **24** 699–739.

Mukhopadhyay, N. and Solanky, T. (1994). *Multistage Selection and Ranking Procedures.* New York: Marcel Dekker.

Müller, H. (1988). *Nonparametric Regression Analysis of Longitudinal Data.* Berlin: Springer-Verlag.

Müller, H. and Wang, J. (2007). Density and hazard rate estimation. *Encyclopedia of Statistics in Quality and Reliability.* (Ruggeri, F., Kenett, R. and Faltin, F., eds.) Chichester: Wiley, 517–522.

Müller, U. (2009). Estimating linear functionals in nonlinear regression with responses missing at random. *The Annals of Statistics* **37** 2245–2277.

Müller, U. and Schick, A. (2017). Efficiency transfer for regression models with responses missing at random. *Bernoulli* **23** 2693–2719.

Müller, U. and Van Keilegom, I. (2012). Efficient parameter estimation in regression with missing responses. *Electronic Journal of Statistics* **6** 1200–1219.

Nakagawa, S. (2015). Missing data: mechanisms, methods, and messages. *Ecological Statistics: Contemporary Theory and Application.* (Fox, G., Negrete-Yankelevich, S. and Sosa, V., eds.) Oxford: Oxford University Press, 81–105.

Nason, G. (2008). *Wavelet Methods in Statistics with R.* London: Springer.

Nemirovskii, A.S. (1999). *Topics in Non-Parametric Statistics.* New York: Springer.

Newgard, C. and Lewis, R. (2015). Missing data: how to best account for what is not known. *The Journal of American Medical Association* 2015 **314** 940–941.

Nickolas, P. (2017). *Wavelets: A Student Guide.* Cambridge: Cambridge University Press.

Nikolskii, S.M. (1975). *Approximation of Functions of Several Variables and Embedding Theorems*. New York: Springer-Verlag.

Ning, J. Qin, J., and Shen, Y. (2010). Nonparametric tests for right-censored data with biased sampling. *Journal of Royal Statistical Society, B* **72** 609–630.

Nussbaum, M. (1996). Asymptotic equivalence of density estimation and Gaussian white noise. *The Annals of Statistics* **24** 2399–2430.

O'Kelly, M. and Ratitch, B. (2014). *Clinical Trials with Missing Data: A Guide to Practitioners*. New York: Wiley.

Parzen, E. (1963). On spectral analysis with missing observations and amplitude modulation. *Sankhya, ser. A* **25** 383–392.

Patil, P. (1997). Nonparametric hazard rate estimation by orthogonal wavelet method. *Journal of Statistical Planning and Inference* **60** 153–168.

Patil, P. and Bagkavos, D. (2012). Histogram for hazard rate estimation. *Sankhya B* **74** 286–301.

Pavliotis, G. (2014). *Stochastic Processes and Applications*. New York: Springer.

Pensky, M. (2017). Minimax theory of estimation of linear functionals of the deconvolution density with or without sparsity. *The Annals of Statistics* **45** 1516–1541.

Petrov, V. (1975). *Sums of Independent Random Variables*. Springer, New York.

Pinsker, M.S. (1980). Optimal filtering a square integrable signal in Gaussian white noise. *Problems of Information Transmission* **16** 52–68.

Prakasa Rao, B.L.S. (1983). *Nonparametric Functional Estimation*. New York: Academic Press.

Priestly, M. (1965). Evolutionary spectra and non-stationary processes. *Journal of the Royal Statistical Society* **27** 204–237.

Pukelsheim, F. (1993). *Optimal Design of Experiments*. New York: Wiley.

Qian, J. and Betensky, R. (2014). Assumptions regarding right censoring in the presence of left truncation. *Statistics and Probability Letters* **87** 12–17.

Qin, J. (2017). *Biased Sampling, Over-Identified Parameter Problems and Beyond*. New York: Springer.

Rabhi, Y. and Asgharian, M. (2017). Inference under biased sampling and right censoring for a change point in the hazard function. *Bernoulli* **23** 2720–2745.

Raghunathan, T. (2016). *Missing Data Analysis in Practice*. Boca Raton: Chapman & Hall

Rio, E. (2017). *Asymptotic Theory of Weakly Dependent Random Processes*. New York: Springer.

Robinson, P. (2008). Correlation testing in time series, spatial and cross-sectional data. *Journal of Econometrics* **147** 5–16.

Rosen, O., Wood, S., and Stoffer, D. (2012). AdaptSPEC: adaptive spectral estimation for nonstationary time series. *The Journal of American Statistical Association* **107** 1575–1589.

Ross, S. (2014). *Introduction to Probability Models*. 11th ed. New York: Elsevier.

Ross, S. (2015). *A First Course in Probability*. 9th ed. Upper Saddle River: Prentice Hall.

Royston, P. and Lambert, P. (2011). *Flexible Parametric Survival Analysis Using Stata: Beyond the Cox Model*. College Station: Stata Press.

Rubin, D.B. (1976). Inference and missing data. *Biometrika* **63** 581–590.

Rubin, D.B. (1987). *Multiple Imputation for Nonresponse in Surveys*. New York: Wiley.

Sakhanenko, L. (2015). Asymptotics of suprema of weighted Gaussian fields with applications to kernel density estimators. *Theory of Probabability and its Applications* **59** 415–451.

Sakhanenko, L. (2017) In search of an optimal kernel for a bias correction method for density estimators. *Statistics and Probability Letters* **122** 42–50.

Samorodnitsky, G. (2016). *Stochastic Processes and Long Range Dependence*. Cham: Springer.

Sandsten, M. (2016). *Time-Frequency Analysis of Time-Varying Signals in Non-Stationary Processes*. Lund: Lund University Press.

Scheinok, P. (1965). Spectral analysis with randomly missed observations: binomial case. *Annals of Mathematical Statistics* **36** 971–977.

Scott, D.W. (2015). *Multivariate Density Estimation: Theory, Practice, and Visualization.* 2nd ed. New York: Wiley.

Shaw, B. (2017). *Uncertainty Analysis of Experimental Data with R.* NewYork: Chapman & Hall.

Shen, Y., Ning, J., and Qin, J. (2017). Nonparametric and semiparametric regression estimation for length-biased survival data. *Lifetime Data Analysis* **23** 3–24.

Shi, J., Chen, X., and Zhou, Y. (2015). The strong representation for the nonparametric estimation of length-biased and right-censored data. *Statistics and Probability Letters* **104** 49–57.

Shumway, R. and Stoffer, D. (2017). *Time Series Analysis and Its Applications with R Examples.* 4th ed. New York: Springer.

Silverman, B.W. (1986). *Density Estimation for Statistics and Data Analysis.* London: Chapman & Hall.

Simonoff, J.S. (1996). *Smoothing Methods in Statistics.* New York: Springer.

Srivastava, A. and Klassen, E. (2016). *Functional and Shape Data Analysis.* New York: Springer.

Stein, C. (1945). Two-sample test for a linear hypothesis whose power is independent of the variance. *Annals of Mathematical Statistics* **16** 243–258.

Stoica, P. and Moses, R. (2005). *Spectral Analysis of Signals.* New York: Prentice Hall.

Su, Y. and Wang, J. (2012). Modeling left-truncated and right censored survival data with longitudinal covariates. *The Annals of Statistics* **40** 1465–1488.

Sullivan, T., Lee, K., Ryan, P., and Salter, A. (2017). Treatment of missing data in follow-up studies of randomized controlled trials: a systematic review of the literature. *Clinical Trials* **14** 387–395.

Sun, J. and Zhao, X. (2013). *The Statistical Analysis of Interval-Censored Failure Time Data.* New York: Springer.

Takezawa, K. (2005). *Introduction to Nonparametric Regression.* Hoboken: Wiley.

Talamakrouni, M., Van Keilegom, I., and El Ghouch, A. (2016). Parametrically guided nonparametric density and hazard estimation with censored data. *Computational Statistics and Data Analysis* **93** 308–323.

Tan, M., Tian, G., and Ng, K. (2009). *Bayesian Missing Data Problems: EM, Data Augmentation and Noniterative Computation.* Boca Raton: Chapman & Hall.

Tanaka, K. (2017). *Time Series Analysis: Nonstationary and Noninvertible Distribution Theory.* New York: Wiley.

Tarczynski, A. and Allay, B. (2004). Spectral analysis of randomly sampled signals: Suppression of aliasing and sampler jitter. *IEEE Transactions on Signal Processing* **52** 3324-3334.

Tarter, M.E. and Lock, M.D. (1993). *Model-Free Curve Estimation.* London: Chapman & Hall.

Temlyakov, V.N. (1993). *Approximation of Periodic Functions.* New York: Nova Science Publishers.

Thompson, J.R. and Tapia, R.A. (1990). *Nonparametric Function Estimation, Modeling, and Simulation.* Philadelphia: SIAM.

Thompson, S. and Seber, A. (1996). *Adaptive Sampling.* New York: Wiley.

Tikhonov, A.N. (1998). *Nonlinear Ill-Posed Problems.* New York: Springer.

Tsai, W. (2009). Pseudo-partial likelihood for proportional hazards models with biased-sampling data. *Biometrika* **96** 601-615.

Tsay, R.S. (2005). *Analysis of Financial Time Series.* 2nd ed. New York: Wiley.

Tsiatis, A. (2006). *Semiparametric Theory and Missing Data.* New York: Springer.

Tsybakov, A. (2009). *Introduction to Nonparametric Estimation.* New York: Springer.

Tutz, G. and Schmid, M. (2016). *Modeling Discrete Time-to-Event Data.* Cham: Springer.

Tymes, N., Pereyra, M.C., and Efromovich, S. (2000). The Application of Multiwavelets to Recovery of Signals. *Computing Science and Statistics* **33** 234–241.

Uzunogullari, U. and Wang, J. (1992). A comparison of hazard rate estimators for left truncated and right censored data. *Biometrika* **79** 297–310.

van Buuren, S. (2012) *Flexible Imputation of Missing Data.* Boca Raton: Chapman & Hall.

van Houwelingen, H. and Putter, H. (2011). *Dynamic Prediction in Clinical Survival Analysis.* Boca Raton: Chapman & Hall.

Vidakovic, B. (1999). *Statistical Modeling by Wavelets.* New York: Wiley.

Vorotniskaya, T. (2008). Estimates of covariance function and spectral density of stationary stochastic process with Poisson gaps in observations. *Vestnik* **31** 3–11.

Wahba, G. (1990). *Spline Models for Observational Data.* Philadelphia: SIAM.

Wald, A. (1947). *Statistical Analysis.* New York: Wiley.

Wald, A. (1950). *Statistical Decision Functions.* New York: Wiley.

Walter, G.G. (1994). *Wavelets and Other Orthogonal Systems with Applications.* London: CRC Press.

Wand, M.P. and Jones, M.C. (1995). *Kernel Smoothing.* London: Chapman & Hall.

Wang, C. and Chan, K. (2018). Quasi-likelihood estimation of a censored autoregressive model with exogenous variables. *Journal of the American Statistical Association*, in press. dx.doi.org/10.1080/01621459.2017.1307115

Wang, J.-L. (2005). Smoothing hazard rate. *Encyclopedia of Biostatistics*, 2nd ed. (Armitage, P. and Colton, T., eds.) Chichester: Wiley, **7** 4486-4497.

Wang, M. (1996). Hazards regression analysis for length-biased data. *Biometrika* **83** 343–354.

Wang, Y. (1995). Jump and sharp cusp detection by wavelets. *Biometrica* **82** 385–397.

Wang, Y., Zhou, Z., Zhou, X., and Zhou, Y. (2017). Nonparametric and semiparametric estimation of quantile residual lifetime for length-biased and right-censored data. *The Canadian Journal of Statistics* **45** 220–250.

Wasserman, L. (2006). *All of Nonparametric Statistics.* New York: Springer.

Watson, G. and Leadbetter, M. (1964). Hazard rate analysis. *Biometrika* **51** 175–184.

Wienke, A. (2011). *Frailty Models in Survival Analysis.* Boca Raton: Chapman & Hall.

Wilks, S. (1962). *Mathematical Statistics.* New York: John Wiley.

Wood, S. (2017). *Generalized Additive Models: An Introduction with R.* Boca Raton: Chapman & Hall.

Woodroofe, M. (1985). Estimating a distribution function with truncated data. *The Annals of Statistics* **13** 163–177.

Wu, S. and Wells, M. (2003). Nonparametric estimation of hazard functions by wavelet methods. *Nonparametric Statistics* **15** 187–203.

Wu, W. and Zaffaroni, P. (2018). Asymptotic theory for spectral density estimates of general multivariate time series. *Econometric Theory*, in press. https://doi.org/10.1017/S0266466617000068

Yang, D. (2017). *Handbook of Regression Methods.* New York: Chapman & Hall.

Yang, Y. (2001). Nonparametric regression with dependent errors. *Bernoulli* **7** 633–655.

Yi, G. (2017). *Statistical Analysis with Measurement Error or Misclassification: Strategy, Method and Application.* New York: Springer.

Yoo, W. and Ghosal, S. (2016). Supremum norm posterior contraction and credible sets for nonparametric multivariate regression. *The Annals of Statistics* **44** 1069–1102.

Young W., Weckman G., and Holland, W. (2011). A survey of methodologies for the treatment of missing values within datasets: Limitations and benefits. *Theoretical Issues in Ergonomics Science* **12** 15–43.

Zhang, F. and Zhou, Y. (2013). Analyzing left-truncated and right-censored data under Cox model with long-term survivors. *Acta Mathematicae Applicatae Sinica* **29** 241–252.

Zhou, M. (2015). *Empirical Likelihood Method in Survival Analysis*. Boca Raton: Chapman & Hall.

Zhou, X., Zhou, C., and Ding, X. (2014). *Applied Missing Data Analysis in the Health Sciences*. New York: Wiley.

Zhu, T. and Politis, D. (2017). Kernel estimates of nonparametric functional autoregression models and their bootstrap approximation. *Electronic Journal of Statistics* **11** 2876–2906.

Zou, Y. and Liang, H. (2017). Wavelet estimation of density for censored data with censoring indicator missing at random. *A Journal of Theoretical and Applied Statistics* **51** 1214–1237.

Zucchini, W., MacDonald, I., and Langrock, R. (2016). *Hidden Markov Models for Time Series: An Introduction Using R*. Boca Raton: CRC Press.

Zurbenko, I. (1991). Spectral analysis of non-stationary time series. *International Statistical Review* **59** 163–173.

Author Index

Aalen, O., 29, 240
Adak, S., 370
Addison, P., 242
Akaike, H., 370
Allay, B., 320
Allen, J., 280
Allison, P., 28, 29, 142, 240, 280
Andersen, P.K., 241
Anderson, T., 320, 369
Antoniadis, A., 241
Aoshima, M., 370
Arnab, R., 370
Asgharian, M., 241
Austin, P., 242

Baby, P., 320
Bagkavos, D., 240
Baisch, S., 320
Baraldi, A., 142
Baron, M., 370
Bary, N.K., 65
Bellman, R.E., 66
Beran, J., 321, 369
Berger, R., 14
Berglund, P., 28, 143
Berk, R., 66
Betensky, R., 241
Bettencourt, L., 240
Bickel, P.J., 65, 97, 240
Birgé, L., 65
Bloomfield, P., 320, 369
Bodenmiller, D., 65
Bodner, T., 142
Bokelmann, G., 320
Borgan, O., 29, 240, 241
Borrajo, M., 97
Bott, A., 144
Bottai, M., 242
Bouza-Herrera, C., 28
Box, G., 320, 369
Bremhorsta, V., 241
Brockwell, P.J., 320, 369
Brown, L., 370
Browne, W., 144

Brunel, E., 97, 241
Butcher, H., 320
Butzer, P.L., 65

Cai, J., 280
Cai, T., 66
Cao, R., 143, 241
Carpenter, J., 142, 144
Carroll, R., 66, 421
Casella, G., 14, 98
Chan, K., 97, 242, 321
Chaubey, Y., 97
Cheema, J., 142
Chen, D., 29, 240, 421
Chen, X., 241, 280
Chen, Y., 98
Chen, Z., 370
Chentsov, N.N., 65
Chesneau, C., 97
Chester, A., 421
Chicken, E., 65
Choi, S., 321
Chu, J., 66, 241
Cohen, A.C., 241
Collett, D., 29, 240
Comte, F., 97, 241, 421
Cortese, G., 242
Cox, D.R., 29, 240, 241
Craig, B., 370
Crainceanu, C., 421
Crowder, M., 29, 240

Dabrowska, D., 241
Daepp, M., 240
Dahlhaus, R., 370
Dai, H., 241
Daniels, M., 142
Davey, A., 142
Davidian, M., 143
Davis, R.A., 320, 369
De Gooijer, J., 369
De Una-Álvarez, J., 241
Dean, A., 370
Dedecker, J., 320, 369

Del Moral, P., 369
Delecroix, M., 242
Demidovich, B., 421
DeVore, R.A., 65
Devroye, L., 65
Diggle, P.J., 320, 369
Dillman, D., 142
Ding, X., 28, 142
Dobrow, R., 369
Doksum, K.A., 65, 240
Donoho, D., 65
Doukhan, P., 320
Draguljic, D., 370
Dryden, I.L., 369
Dukic, V., 241
Dunsmuir, W., 320
Durrett, R., 369
Dym, H., 65
Dzhaparidze, K., 320

Efromovich, S., 29, 65, 66, 97, 143, 179,
        240, 241, 280, 320, 321, 369, 370,
        421, 422
Efron, B., 65
El Ghouch, A., 241
Eltinge, J., 142
Enders, C., 28, 142, 179
Eubank, R.L., 66
Everitt, B., 66

Fan, J., 144, 320, 321, 369
Fan, P., 370
Faraway, J., 66
Figueredo, A., 143
Fisher, N.I., 421
Fitzmaurice, G., 142, 179
Fleming, T.R., 240
Flumento, P. , 242
Fomin, S.V., 65

Ganzburg, M., 179, 421
Genovese, C., 66, 98
Ghosal, S., 66, 144, 240
Gijbels, I., 144, 321
Gill, R., 29, 97, 240, 241
Gillard, J., 320
Giné, E., 66
Gjessing, H., 29, 240
Glad, I., 65
Goldstein, H., 144
Golitschek, M., 65, 66
Gong, G., 320

González-Manteiga, W., 97
Gou, J., 370
Grace, Y., 421
Graham, J., 28, 142, 143
Grainger, D., 65
Green, P., 66
Gregoire, G., 241
Greiner, A., 320
Groeneboom, P., 29, 66, 241, 421
Groves, R., 142
Guilloux, A., 97
Guo, S., 29, 240
Guo, Z., 66
Györfi, L., 65, 66, 320

Härdle, W., 65
Hagar, Y., 241
Hall, P., 369
Hamilton, M., 240
Harrell, F., 144, 240, 280
Harrington, D.P., 240
Hart, J.D., 65, 369
Hastie, T.J., 66, 144
Heeringa, S., 28, 143
Helsel, D., 321
Hjort, N., 65
Hoffmann, M., 66
Hogan, J., 142
Holland, W., 142
Hollander, W., 65
Holmboe, S., 242
Honaker, J., 142
Horowitz, J., 66
Hosmer, D., 29, 240
Hothorn, T., 66
Hui, Y., 320

Ibragimov, I.A., 65, 369
Ibrahim, J., 240, 280
Ingster, Yu., 66
Ivanoff, S., 144
Izbicki, R., 144
Izenman, A., 66, 144

Jankowski, H., 240
Janssen, P., 241
Jenkins, G., 320, 369
Jennings, C., 320
Jiang, J., 320
Jiang, P., 320
Johnstone, I., 28, 65, 370
Jones, M.C., 66, 97, 421

Jongbloed, G., 29, 66, 241, 421

Kabanikhin, S., 29, 421
Kalbfleisch, J., 29, 97, 240
Keiding, N., 241
Kenward, M., 28, 142, 179
Kerkyacharian, G., 65
Khasminskii, R.Z., 65, 369
King, G., 142
Kitagawa, G., 370
Klassen, E., 241
Klein, J., 29, 240, 280, 421
Klein, M., 29, 240
Kleinbaum, D., 29, 240
Klugman, S., 240
Kohler, M., 66, 144
Kokoszka, P., 280
Kolmogorov, A.N., 65
Koltchinskii, V., 421
Kosorok, M., 240
Kou, J., 97
Krylov, A.N., 65
Krzyzak, A., 66
Kulahci, M., 320
Kutner, M., 370

Lambert, P., 29, 240
Lamberta, F., 241
Lang, G., 320
Lang, K., 142
Langrock, R., 320
Lawless, J., 97
Leatbetter, M., 240
Lee, A., 144
Lee, E., 29, 240
Lee, M., 320
Lee, P., 142
Lee, S., 66
Lehmann, E.L., 98
Lemeshow, S., 29, 240
Leon, J., 320
Leon, S., 143
Lepskii, O., 421
Levit, B., 320
Lewis, R., 142
Li, J., 29, 240
Li, Q,, 66
Li, Q., 321
Li, W., 370
Liang, H., 241, 280
Linnik, Yu.V., 369
Little, R., 28, 142, 179, 280

Little, T.D., 280
Liu, X., 29, 240
Liu, Y., 97
Ljung, G., 320, 369
Lock, M.D., 65
Longford, N., 28
Lopez, O., 242
Lorentz, G.G., 65, 66
Louhichi, S., 320
Low, M., 66, 179, 370
Lu, X., 240
Luo, X., 97

Müller, H., 66, 240
Müller, U., 143
Ma, S., 29, 240
Mabon, G., 421
MacDonald, I., 320
Makovoz, Y., 65, 66
Mallat, S., 65, 242, 369
Mardia, K.V., 369, 421
Maron, I., 421
Martínez-Miranda, M., 97
Martinussen, T., 29, 240
Massart, P., 65
Matloff, N., 66
Matsuda, Y., 370
May, S., 29, 240
McGabe, G., 370
McKean, H., 65
McKnight, K., 143
McKnight, P., 143
Meister, A., 29, 421
Miller, R.G., 29
Mills, M., 29, 240
Min, L., 240
Moeschberger, M.L., 29, 240
Molenberghs, G., 28, 142, 179
Montgomery, D., 320
Moore, D., 29, 240, 370
Moses, R., 370
Mukherjee, R., 66
Mukhopadhyay, N., 370

Nachtsheim, C., 370
Nakagawa, S., 142
Nason, G., 65, 241, 242
Navarro, F., 97
Nemirovskii, A., 66
Nessel, R.J., 65
Neter, J., 370
Newgard, C., 142

Ng, K., 28
Nickl, R., 66
Nickolas, P., 242
Nikolskii, S.M., 65, 66
Ning, J., 97
Nussbaum, M., 369

O'Kelly, M., 28
Oakes, D., 29, 240, 241

Pamjer, H., 240
Parzen, E., 320
Patil, P., 240
Patilea, V., 242
Pavliotis, G., 369
Peace, K., 29, 240, 421
Penev, S., 369
Pensky, M. , 421
Pereyra, M.C, 242
Petrov, V., 29
Picard, D., 65, 144
Pinsker, M.S., 65, 66, 179, 320, 369
Politis, D., 321
Portnoy, S., 321
Prakasa Rao, B.L.S., 98, 240, 370
Prentice, R., 29, 240
Priestly, M. , 370
Prieur, C., 320
Pukelsheim, F., 370
Putter, H., 29, 240

Qian, J., 241
Qin, J., 97

Rabhi, Y., 241
Racine, J., 66, 321
Raghunathan, T., 28, 142, 144, 179
Ratitch, B., 28
Rebafka, T., 97
Reimherr, M., 280
Reinsel, G., 320, 369
Restaino, M., 241
Rhemtulla, M., 142
Rio, E., 369
Ritov, Y., 97
Rivoirard, V., 144
Robinson, P., 320, 321
Rosen, O., 370
Ross, S., 14, 320
Royston, P., 29, 240
Rubin, D.B., 28, 142, 179
Ruppert, D., 66, 421

Ryan, P., 142

Sakhanenko, L., 65
Salter, A., 142
Salter-Kubatko, L., 65
Salva, J., 142
Samarov, A., 179, 320, 370
Samorodnitsky, G., 369
Samson, A., 421
Sansten, M., 370
Scheike, T., 29, 240, 242, 280
Scheinok, P., 320
Schick, A., 144
Schmid, M., 240
Scott, D.W., 65, 66
Seber, A., 370
Semmler, W., 320
Sen, S., 66
Shaw, B., 180
Shen, Y., 97
Shi, J., 241
Shumway, R. , 320
Sidani, S., 143
Silverman, B., 65, 66, 240
Simonoff, J., 66, 98
Smirnova, E., 65, 242, 370
Solanky, T., 370
Spiro, S., 65
Srivastava, A., 241
Stefanski, L., 421
Stein, C., 370
Stoffer, D., 320, 370
Stoica, P., 320, 370
Su, Y., 242
Sullivan, T., 142
Sun, J., 29, 240, 280, 321, 421
Suslina, I., 66

Takezawa, K., 66
Talamakrouni, M., 241
Tan, M., 28
Tanaka, K., 369
Tapia, R.A., 65
Tarczynski, A., 320
Tarter, M.E., 65
Temlyakov, V.N., 65, 66
Thomas, E., 66
Thompson, J.R., 65
Thompson, S., 370
Tian, G., 28
Tibshirani, R., 65, 66, 98, 144
Tikhonov, A.N., 29, 421

Tsai, W., 97
Tsay, R.S., 321, 369
Tsiatis, A., 28, 142, 143, 179
Tsybakov, A., 28, 65, 369
Tutz, G., 240
Tymes, N., 242

Ushakov, N., 65
Uzunogullari, U., 241

Valdez-Jasso, Z.A., 65, 370
van Buuren, S., 28, 143, 280
van der Vaart, A., 144, 240
van Houwelingen, H., 29, 240
Van Keilogom, I., 143, 241
Vardi, Y., 97
Veraverbeke, N., 241
Verbeke, G., 142, 179
Vidakovic, B., 65, 242
Vorotniskaya, T., 320, 370
Voss, D., 370

Wahba, G., 66
Wald, A., 370
Walk, H., 66
Walter, G.G., 65
Wand, M.P., 66, 97, 421
Wang, C., 242, 321
Wang, H., 241
Wang, J., 29, 240–242
Wang, M., 242
Wang, Q., 370
Wang, Y., 97, 241
Wasserman, L., 28, 65, 66, 98
Watson, G., 240
Weckman, G., 142
Wellner, J., 97, 240
Wells, M., 240
West, G., 240
Wienke, A., 240
Wild, C., 97
Wilks, S., 370
Willer, T., 421
Willmot, G., 240
Wolfe, D., 65
Wood, S., 144, 370
Woodroofe, M., 241
Wu, D., 370
Wu, J., 370
Wu, S., 240
Wu, W., 142, 320

Yajima, Y., 370

Yang, D., 66
Yang, Y., 369
Yao, Q., 320, 321, 369
Yata, K., 370
Yi, G., 421
Yoo, W., 66
Young, W., 142

Zaffaroni, P., 320
Zhang, F., 242, 370
Zhao, X., 280, 321, 421
Zhou, C., 28, 142
Zhou, M., 29, 240
Zhou, X., 28, 142, 241
Zhou, Y., 241, 242
Zhou, Z., 241
Zhu, T., 321
Zou, Y., 280
Zucchini, W., 320
Zurbenko, I., 370

Subject Index

:=, 14
=:, 14
A, 20, 100
$CJ0$, 42
$CJ1$, 42
CTH, 42
$F^X(x)$, 14
$G^X(x)$, 14
$H^X(x)$, 183
$I(\cdot)$, 14
N, 102
$X[A == 1]$, 114
$X_{(k)}$, 17
Δ, 188
$\mathbb{E}\{\cdot|\cdot\}$, 16
$\mathbb{E}\{\cdot\}$, 14
$\mathbb{P}(\cdot)$, 14
$\mathbb{V}(\cdot)$, 14
α_X, 15
β_X, 15
$\hat{\theta}$, 17
ISB, 41
ISE, 36
MISE, 19
MSE, 17
$\phi^X(t)$, 373
$\psi_j(x)$, 184
θ_j, 19
$\varphi_j(x)$, 19
c, 49
d, 69
$f^X(x)$, 15
$g^X(\lambda)$, 286
$h^X(x)$, 183
$o_s(1)$, 42
$w(\cdot)$, 20
$\mathcal{A}(Q, q, \beta, r)$, 284
$\mathcal{A}_{r,Q}$, 37
$\mathcal{S}_{\alpha,Q}$, 37

Adaptation, 40
Additive regression, 133
Amplitude-modulated missing, 297, 343
Analytic class, 37

AR, 283
ARMA, 283
Autocovariance, 282
    Nonstationary, 351
Auxiliary variable, 156
Availability, 20, 100, 101
Availability likelihood, 20, 100

Basis
    Cosine, 19, 33
    Cosine on interval, 184
    Orthonormal, 32
    Tensor-product, 53
Batch-Bernoulli missing, 295, 334
Bernoulli missing, 291, 334
Bernoulli regression, 51
    Unavailable failures, 87
Bernoulli variable, 51
Biased data, 67
Biased predictors and responses, 74
Biased responses, 72
Biasing function, 68, 72, 75
Bivariate density, 54
Bivariate regression, 129
Bona fide projection, 39
Boundary effect, 34, 37
Brownian motion, 328

Case
    Complete, 100
    Incomplete, 100
Categorical data, 80
Censored predictors, 262
Censored responses, 258
Censored time series, 301
Censoring, 188
    Left, 192
    Right, 6, 188
Change-point problem, 83
Characteristic function, 373
    Empirical, 373
Coefficient of difficulty, 46, 69, 106, 112,
    157, 185, 196, 200, 205, 210, 221,
    246, 290

Complete-case approach, 109
Conditional density, 16, 117, 217, 224
    Estimation for LT data, 217
    Estimation of, 117, 358
Conditional expectation, 16
Conditional survival function, 216, 224
    Estimation for LT data, 216
Confidence band
    Pointwise, 58
    Simultaneous, 59
Confidence interval, 56
Consistent estimation, 47
Convolution, 372
Corner (test) function, 31
    Bimodal, 32
    Custom-made, 34
    Normal, 32
    Strata, 32
    Uniform, 31
CSC problem, 390
Cumulative distribution function, 14
    Joint, 16
Current status censoring, 388
Cutoff, 33

Deconvolution, 372, 375
    Censored data, 386
    Missing data, 379
Deductible, 5
Density estimation, 2, 38, 246
    Auxiliary variable, 156
    Complete data, 38
    Dependent observations, 304
    Extra sample, 151
    Known availability, 146
    LTRC data, 221
    MCAR data, 101
    Measurement error, 372
    MNAR indicator of censoring, 251
    RC data, 204
Derivative
    Estimation of, 409
Design
    Fixed, 48
    Random, 48
    Sequential, 359
Design density, 48
Directional variable, 378
Distribution
    Bernoulli, 11, 14
    Binomial, 15
    Exponential, 184

    Laplace, 15
    Normal, 15
    Poisson, 10, 15
    Uniform, 15
    Weibull, 184

E-estimation, 39
    Multivariate, 53
E-estimator, 39
E-sample, 145
Ellipsoid
    Analytic, 61
    Periodic function, 61
    Sobolev, 37
Empirical autocovariance function, 286
Empirical cumulative distribution function, 410
Empirical cutoff, 39
Estimand, 17
Estimation of distribution
    LT, 208
    LTRC, 219
    MAR indicator of censoring, 243
    MNAR indicator of censoring, 249
    RC, 203
Estimator, 17
    Consistent, 18
    Unbiased, 18
Example
    Actuarial, 5, 9, 10, 188, 193
    Age and wage, 10
    Biased data, 2
    Clinical study, 193
    Epidemiological, 389
    Housing starts, 8
    Intoxicated drivers, 2
    Lifetime of bulb, 188
    Mortgage loan, 188
    Startup, 193
Expectation, 14

Failure rate, 183
Filtering signal, 330
Force of mortality, 183
Fourier coefficient, 19

H-sample, 20, 100, 145
Hazard rate, 183
    Estimation of, 186, 188, 192, 197, 246
    Properies, 183
    Ratio-estimator, 186
Histogram, 2

Ill-posed problem, 371, 399, 403
Indicator, 14
Indicator of censoring
    MAR, 243
    MNAR, 249
Inequality
    Bernstein, 19
    Cauchy, 21
    Cauchy–Schwarz, 19, 21
    Chebyshev, 18
    Generalized Chebyshev, 18
    Hoeffding, 18
    Jensen, 19
    Markov, 18
    Minkowski, 21
Integrated square error (ISE), 36
Integrated squared bias (ISB), 54
Integration by parts, 36
Interpolation, 399
Inverse problem, 372

Kaplan–Meier estimator, 203

Left truncation, 193
Limit on payment, 5
Linear regression, 7
Long memory, 306
LT, 208
LTRC, 197
    Formulas, 198
    Generator, 197
LTRC predictors, 269

M-sample, 20, 100, 145
MA, 283
MAR, 20
MAR censored responses, 253
MAR responses, 262, 264, 269
Markov chain, 292
Markov–Bernoulli missing, 292
MCAR, 20, 101
Mean integrated squared error (MISE),
    19, 38
Mean squared error (MSE), 17
Measurement errors, 372, 399
Memory
    Long, 283
    Short, 283
MEP regression, 400
    Missing responses, 404
Missing
    Amplitude-modulated, 297

At random (MAR), 20
Batch-Bernoulli, 295, 333, 334
Completely at random (MCAR), 20,
    101
Destructive, 145
Markov–Bernoulli, 291, 334
Nondestructive, 99
Not at random (MNAR), 20
Poisson amplitude-modulation, 299
Mixing coefficient, 285
Mixing theory, 284
Mixture, 83
Mixtures regression, 83
MNAR, 20

Nelson–Aalen estimator, 241
Nonnegative projection, 41
Nonparametric autoregression, 310
Nonparametric estimation, 38
Nonparametric regression, 47
    Additive, 133
    Bernoulli, 51
        Unavailable failures, 87
    Biased predictors and responses, 74
    Biased responses, 72
    Bivariate, 129
    Censored predictors, 262
    Censored responses, 258
    Dependent responses, 323
    Direct data, 7, 47
    Fixed-design, 48
    Heteroscedastic, 48
    Homoscedastic, 48
    LTRC predictors, 269
    MAR censored responses, 253
    MAR predictors, 112, 258
    MAR responses, 107, 262, 264, 269
    MEP, 399
    MEP with missing responses, 404
    Missing cases, 172
    MNAR predictors, 169
    MNAR responses, 160
    Random-design, 48
    RC predictors, 229
    RC responses, 225
    Truncated predictors, 264
Nonstationary amplitude-modulation, 348
Nonstationary autocovariance, 351
Nonstationary spectral density, 351
Nonstationary time series, 333
    Missing data, 333
Nuisance function, 13, 85

Estimation of, 86

Optimal design density, 112
Optimal rate of convergence, 54
Ordered statistics, 17

Parseval's identity, 20
Periodogram, 286
Poisson amplitude-modulation, 299
Poisson regression, 124
Principle of equivalence, 332
Probability density, 15
     Mixed, 17
Projection estimator, 40

R package, 21
Random variable
     Continuous, 15
     Discrete, 14
Rate of convergence
     MISE, 46
     Optimal, 46, 47
RC, 188
Regression function, 48
Repeated observations, 377

Sample
     Direct, 2, 17
     Extra, 151
     Hidden (H-sample), 108
Sample mean estimator, 18
     Plug-in, 18, 39
Sample variance estimator
     Plug-in, 39
Scale function, 127, 333
     Estimation of, 127, 339
Scattergram, 49
Seasonal component, 333
     Estimation of, 338
     Period, 286
Sequential design, 359
Shape, 297
Short memory, 306
Simpson's paradox, 355
Sobolev class, 37
Software, 21
Spatial data, 281
Spectral density, 286
     ARMA, 287
     Estimation of, 286
     Nonstationary, 351
     Shape, 297

Standard deviation, 15
Stochastic process, 327
Support, 15
     Unknown, 91
Survival analysis, 7, 181
Survival function, 14
     Estimation of, 185, 260

Thresholding, 41
Time domain approach, 284
Time series, 281
     m-dependent, 285
     Amplitude-modulation, 343, 404
     ARMA, 283
     Causal, 284
     Censored, 301
     Decomposition, 333
     Long-memory, 283
     Nonstationary, 333
     Nonstationary amplitude-modulation, 348
     Scale function, 333
     Seasonal component, 333
     Second-order stationary, 282
     Short-memory, 283
     Strictly stationary, 282
     Trend, 333
     Zero-mean, 282
Trend, 333
     Estimation of, 334
Truncated predictors, 264
Truncated series, 33
Truncation
     Left, 5, 192

Unavailable failures, 87

Variance, 14

Weak dependence, 306
White noise, 328
     Frequency limited, 328
Wiener process, 328
Wrapped density, 378